D1545622

PREPARATIVE CHEMISTRY USING SUPPORTED REAGENTS

PREPARATIVE CHEMISTRY USING SUPPORTED REAGENTS

Edited by

PIERRE LASZLO

Institut de Chimie
Université de Liège
Liège, Belgium
and
Ecole Polytechnique
Palaiseau, France

ACADEMIC PRESS, INC.
Harcourt Brace Jovanovich, Publishers
San Diego   New York   Berkeley   Boston
London   Sydney   Tokyo   Toronto

ACADEMIC PRESS, INC.
1250 Sixth Avenue, San Diego, California 92101

United Kingdom Edition published by
ACADEMIC PRESS INC. (LONDON) LTD.
24–28 Oval Road, London NW1 7DX

Library of Congress Cataloging in Publication Data

Preparative chemistry using supported reagents.

    Includes index.
    1. Chemistry, Organic—Synthesis.   2. Chemical tests
and reagents.   I. Laszlo, Pierre.   II. Title: Supported
reagents.
QD261.P66   1987   547'.2   87-1442
ISBN 0—12—437105—1 (alk. paper)

PRINTED IN THE UNITED STATES OF AMERICA

87 88 89 90   9 8 7 6 5 4 3 2 1

CONTENTS

**3   Metal Oxides and Their Physico-Chemical
      Properties in Catalysis and Synthesis**

KENNETH J. KLABUNDE, M. FAZLUL HOG,
FALEH MOUSA, AND HIROMI MATSUHASHI

4   The Photochemistry of Adsorbed Molecules

P. DE MAYO AND L. J. JOHNSTON

**5   Electrochemistry at Modified Electrode
      Surfaces**

ALLEN J. BARD AND WALTER E. RUDZINSKI

**6   Practical Considerations: How to Set Up
      a Supported Reagent**

ANDRÉ CORNÉLIS

**PART II   PHYSICO-CHEMICAL STUDIES OF THE
                STRUCTURE OF THE SOLID SUPPORTS**

PART III   POLYMER-SUPPORTED REAGENTS

12   Polymer-Supported Reagents, Polymer-Supported Catalysts, and Polymer-Supported Coupling Reactions

WARREN T. FORD AND ERICH C. BLOSSEY

13   Polymer-Supported Oxidations

GEORGES GELBARD

14   Polymer-Supported Reductions

JOSEPH LIETO AND HMAÏD MARRAKCHI

PART IV   GRAPHITE INTERCALATED

15   Intercalation Compounds of Graphite and Their Reactions

RALPH SETTON

PART V   ALUMINA-SUPPORTED REAGENTS

16   Alumina and Alumina-Supported Reagents

GARY H. POSNER

17   Anionic Activation, Reactions in Dry Media

ANDRÉ FOUCAUD, GEORGES BRAM, AND ANDRÉ LOUPY

PART VI   SILICA-SUPPORTED AND SILICA-GEL-SUPPORTED REAGENTS

18   Reductions, Oxidations, and Anionic Activations: Catalytic Reductions and Oxidations

ENZO SANTANIELLO

19   Silica-Supported Reagents: Polymerizations

PETER W. LEDNOR

20   Silica-Supported Reagents: Reactions in Dry Media

GEORGES BRAM AND ANDRÉ LOUPY

PART VII   USE OF ZEOLITES AS SUPPORTS

21   Use of Molecular Sieves as Supports: Novel Aluminophosphate-Based Molecular Sieves

BRENT M. LOK AND STEVE T. WILSON

22   Shape-Selective Catalysis

SIGMUND M. CSICSERY AND PIERRE LASZLO

PART VIII   CLAY-ACTIVATED ORGANIC REACTIONS

23   Clay-Activated Isomerization Reactions

F. J. A. KELLENDONK, J. J. L. HEINERMAN, AND R. A. VAN SANTEN

24   Oxidations and Catalytic Oxidations

A. MCKILLOP AND D. W. CLISSOLD

25   Clay-Activated Catalytic Hydrogenations: Catalyst Synthesis and Factors Influencing Selectivity

THOMAS J. PINNAVAIA

26   Anionic Activation

ANDRÉ FOUCAUD

27 Cationic Reactions

J. M. Adams

PREFACE

I know a wonderful little inn on a hillside, in the eastern part of France. It sports on its facade a feature out of a painting by Magritte: glass doors with deer antlers as handles. Reflections in these are of white clouds, plus the dark blue of the mountains, wavelet after wavelet, merging into the light blue of the sky.

In like manner, the subject matter of this book is practical, eerie, harmonious. It tells of a new dimension in the methodology of organic reactions. Adsorption of reactants and of reagents on solid supports upgrades the processes into efficiency, selectivity, and trivially easy work up.

The reasons for such routinely achieved successes are still somewhat mysterious—indeed this volume could quite profitably be used as a source book in surface science. This is also why quite a few chapters have been included on the physical methods for study and characterizing surfaces and their adsorbates, and on chemical reactivity at interfaces.

The eeriness has also to do with the harmony: that chapters on apparently distant topics such as polymer-supported reagents, shape-selectivity within zeolithes (spelled, less correctly but more according to tradition, zeolites in the rest of the book), and graphite intercalates could nevertheless fall into place like stained glass of various colors in a church window.

Medieval windows and, say, the *Wall Street Journal* have one thing in common—they make concise statements about complex matters. Each chapter here tells a different story, but one stripped to the essentials.

The whole is a rather comprehensive description (within the limited format allowed by the publishers) of a budding new field of chemistry, as well as a compendium of useful new techniques. The entire range, from the essential theoretical concepts to the "how to" technical aspects, is here.

One further word about this book is organization: I have cast the net wide, because this is a new field, the limits of which, at present, are beyond the horizon.

My fellow authors could have been drawn from Rembrandt's "Anatomy Lesson": skilled experts, intent on doing a highly professional job. Many are world-famous scientists, very much esteemed and very much in the public eye. They have all accepted their assignments with grace and with gusto. That they have all treated my editorial emendations and cuts with benign tolerance is further proof that they are gentlemen and scholars.

Some of this book was prepared while I benefited from visiting professorships, which gave me the calm, the library browsing, and the outdoors backpacking—all mandatory ingredients for this individual. I am grateful to the University of Hamburg (Professor Armin de Meijere), to the Institute of Molecular Science, in Okazaki, Japan (Professor Iwamura Hiizu), and to Cornell University (Professor Roald Hoffmann) for these generous invitations.

Finally, I have heeded the call of clean and delicate graphics. While the in-house staff at Academic Press have been up to their usual high standards, I am indebted for the cover design to Valerie Annette Seelig, of WFW Productions. I take this opportunity to thank warmly my co-worker of many years, Madame Lucienne Souka, for typing the index on very short notice.

Part I

Supported Reagents:
General Principles

1 HOMOGENEOUS AND HETEROGENEOUS REACTION CONDITIONS*

Pierre Laszlo

Institut de Chimie Organique
Université de Liège
Liège, Belgium
and
Ecole Polytechnique
Palaiseau, France

I. THE ALCHEMICAL HERITAGE

The distinction goes back to the alchemists and, beyond them, is traceable to Chinese categories of thought, which have thus fertilized Western science. Chemical transformations can be made to occur according to either the wet way or the dry way. The wet way comes close to what continues to be known in today's laboratories as wet chemistry, i.e., solution chemistry. The dry way was chosen by the alchemists for processes such as tincture of metals. It was effected typically by heating a mix of various substances as solid powders.

In modern times, and in most preparations of fine chemicals, whether in academic or in industrial laboratories, the wet way is dominant. It has become the highway chemists travel along. The dry way, if not totally abandoned, has become a byway, at least for such preparations. Yet, as

* Research on preparative chemistry using supported reagents in my laboratory was made possible by a grant from Programmation de la Politique Scientifique, Brussels (Action Concertée 82/87-34).

3

Copyright © 1987 by Academic Press, Inc.
All rights of reproduction in any form reserved.

this book will document, it is one of the multiple origins for preparative chemistry using supported reagents.

Our alchemist ancestors bequeathed to us another handling procedure for chemicals which, besides being basic to our science, is also another distant relative of modern supported preparative chemistry. Filtering is perhaps the most elementary act in separation science, hence its practical importance for chemistry. Filtering is also pregnant with the whole of chromatography.

There is yet another lesson from history. Think of it for a minute, and you might share my initial bewilderment (which led to my entering this field). We still run chemical reactions in much the same way as in the seventeenth and eighteenth centuries. In our age of sophisticated electronic instrumentation, sensors, microprocessors, and robots starting to see use in chemical synthesis, reaction flasks stand on our benches like relics from a distant past.

The reason is that they embody premodern concepts (dating back to the last quarter of the eighteenth century) about the virtue of dissolution, the attendant freeing from one another of the particles of matter. This is one of the most primitive and one of the most general means of activating chemicals: by dissolving a solid, we remove the attractive intermolecular forces present in the crystalline lattice. Since many dissolutions are often exothermic, these forces are not fully compensated by the resulting solvation forces. Such an idea was, one might say, both the dominant paradigm and the trademark of the chemists of yesteryear.

Very often textbooks of the time stressed this concept. I shall take as my example a book from 1803, embodying the chemical concepts taught at the young Ecole Polytechnique in Paris (founded in October 1794) by leading French chemists such as Berthollet, Fourcroy, Chaptal, and Guyton de Morveau. The author, Ségur, starts his book with the laws of chemical attraction. The first law is that attraction, leading to chemical combination, can only be present between the ultimate particles of matter. The second law states that this combinational attraction must destroy the aggregation forces between molecules. The third law states accordingly that for two chemicals to be submitted to combinational attraction at least one must be a fluid. Ségur writes that the "phenomenon of dissolution is due to this law." He makes this astute and perceptive comment (in my free translation): "the expressions of 'solvent' and 'solute' are improper. These terms seem to imply that one of the bodies is passive, while the other one, the *fluid*, is the only one to be active, the only one whose influence determines the change of state. The truth of the matter however is that the solid tends as much to solidify the liquid, as the liquid tends to liquefy the solid" (Ségur, 1803).

II. THE EXAMPLE OF INDUSTRIAL CHEMISTRY

Like most scientific discoveries, catalysis was a multiple discovery. One of the scientists involved was Johann Wolfgang Döbereiner, and the series of experiments he did in 1823 led him to discover heterogeneous catalysis (Collins, 1976). Ever since, it has remained in the forefront of chemical technology. Nowadays, most industrial chemical units being built are based on a heterogeneous catalytic process. A paradigm for industrial chemistry is the flow inside a reactor of a reaction mixture over a catalytic bed.

Examples (Kemball, 1984) include partial oxidation of naphthalene and *o*-xylene into phthalic anhydride, with supported V_2O_5 catalysts, oxidation of *p*-xylene into terephthalic acid, the formation of acrylonitrile from propene, dioxygen, and ammonia with mixed oxides such as tin–antimony or bismuth phosphomolybdate as catalysts (Keulks *et al.*, 1978; Gates *et al.*, 1979). Oxidations of ammonia into nitric acid (over 10% Rh–90% Pt catalysts), of sulfur dioxide into sulfuric acid (catalyzed by V_2O_5 promoted with K_2SO_4) also occur with heterogeneous catalysis (Dixon and Longfields, 1960). Steam reforming of hydrocarbons is done with nickel-supported catalysts (Bridger and Chinchen, 1970). Ziegler–Natta polymerization of dienes and trienes is performed with titanium–triethyl-aluminum catalysis (Berger *et al.*, 1969). Metallic iron stabilized on an alumina support and in the presence of potassium ions catalyzes ammonia synthesis (Ertl, 1980), for which another catalyst is ruthenium metal supported on carbon and promoted by an alkaline metal (Ozaki, 1981).

The catalytic cracking of hydrocarbons has gained increased selectivity with the onset of zeolite catalysis in 1964 (Csicsery, 1984, 1985). The conversion of carbon monoxide and hydrogen into methanol uses copper-containing catalysts (Denny and Whan, 1978). Conversely, the Mobil methanol-to-gasoline process relies on catalysis by HZSM-5 zeolites (Csicsery, 1984, 1985). Use of zeolites is further documented in Chapters 21 and 22, which are devoted to these remarkable materials and to their molecular sieve action, both static (separation) and dynamic (reactivity).

III. THE LESSONS FROM CHROMATOGRAPHY

At least initially, chemists intent upon trying supported reactions chose supports familiar to them from chromatography (column and thin-layer) such as kieselguhr, alumina, magnesia, silica, silica gel, clays, charcoal, sucrose, starch, cellulose, or paper. (The main supports of present use to chemistry are alumina, silica, and silica gel.)

Indeed, a significant motive for further investigating reactivity on solid supports was the accumulated body of knowledge about secondary reactions caused by the adsorbent itself (Lederer and Lederer, 1957). For instance, in the steroid series, alumina promotes the β-elimination of acetic acid, or the Darzens reaction of halohydrins. Silica gel rearranges a methyl-bearing epoxide into a methyl ketone; it also gives rise to an interesting ring-opening of a furan (Novotny and Kotva, 1974).

IV. THE LEAD FROM BIOCHEMISTRY

Too often, chemists tend to be a little condescending towards biochemistry. Yet, biochemistry, if a sister science to organic chemistry, has much for chemists to admire and emulate.

Biochemistry provides us with the examples of enzymes, the microscopic reactors for the reactions of metabolism. Enzymes teach us the principles of a supramolecular chemistry in which the mutual positioning of the reactants, when they have encountered one another after diffusion on the enzyme surface, prepares the high yields and selectivities which make us pale. I am certain that there are many more biochemical processes begging to be learned and copied, as so many analogic computers.

Another lesson from biochemistry is found in its various techniques. Many are the prototypes for what could be done in preparative organic and inorganic chemistry. Biochemistry has been leading the way with various imaginative uses of solid supports. The Merrifield automated peptide synthesis on a resin (justly awarded the 1984 Nobel Prize for chemistry), supported enzymes that enable the transformation of porcine insulin into human insulin (Mosbach, 1971), and affinity chromatography are but three routine biochemical techniques to be envied and copied by chemists.

V. SHIFTING HABITS

Prejudice marries the objective and the subjective. Oftentimes, this alloy resists easy analysis into logical thought and mere expression of opinion. Chemists prefer homogeneous reaction conditions for good reasons and for not-so-good reasons.

Homogeneous reaction conditions are deemed advantageous because of reproducibility, which is much easier to achieve; because of the greater ease of maintaining a proper and constant reaction temperature, when heat dissipation can be ensured as a regular flow; and because of the facilitated elucidation of the reaction mechanism.

Another factor in the preference for homogeneous over heterogeneous reaction conditions stems from the importance of visual changes as a reaction takes place. A change of color, the appearance of turbidity, deposition of a precipitate, formation of crystals, or appearance of an oily deposit on the walls of the flask are all signs marking various steps along a complex reaction pathway. They are the episodes of the story later told in the experimental part of a paper.

All such visual observations are rendered impossible, or they are made considerably less noticeable, if the reaction is run in a heterogeneous system, e.g., as a suspension of solid particles in a liquid. There may also be an instinctive preference for the limpidity of a liquid, an ingrained reluctance on our part when offered or confronted with a turbid liquid, perhaps because the purity of drinking water was and continues to be vital.

One could line up a plethora of other good reasons for the preference of chemists for homogeneous reaction conditions: equality of the chemical potentials throughout the reaction volume and the lack of rationale for much of the (truly fascinating) chemistry taking place on surfaces of inorganic solids. Also, as a not-so-good reason, colloidalists of the first half of the century may have given a bad name to heterogeneous chemistry.

This book aims to reverse the dominant paradigm. It is a tract for the bold exploration of heterogeneous conditions and for the reaping of the benefits thus accruing. Innovation can be just to dare and do something different. In this sense, innovation shares with heresy its rebellious spirit, i.e., the flaunting of conceptual enclosures.

Innovation can be just the lifting of an inhibition. An inhibition is an internalized prohibition. This book will list in systematic manner the numerous advantages of solid-supported procedures.

Innovation can soar with transfer from another sphere. To a large extent, the idea to try inorganic supports for chemical reactions came from chromatography, and it was given renewed impetus by various biochemical techniques.

Finally innovation can be the imaginative response to a perceived need. Supported reagents answer the need for selective, easily worked-up and efficient reactions. Their widespread applicability will be obvious from the varied areas of chemistry to be covered in the following chapters.

VI. EASIER WORK UP

Supported reagents as a rule make life considerably easier for the preparative chemist. Setting and running a reaction are time effective, but the working out of a reaction mixture is not. Very often the separation and purification, not to mention the identification of the reaction products, are the slow steps. Attachment of the reagents to an insoluble support, such as a polymer or an inorganic solid, allows one to speed them up. The reasons are threefold and discussed in the following.

First a simple filtration removes the reagents from the reaction mixture. This elementary process switches off the reaction and it removes any labile reaction products whose contact with the reagents may lead to unwanted side products. In other words, impregnation of the reagents onto a solid effects a compartmentalization; while starting material and reaction products diffuse freely in the solution, in three-dimensional reaction space, the reagents are kept separate, on a solid surface. Thus, it is trivially easy to remove them from the battleground, once the battle has been waged and won.

Second, removal of the supported reagents from the reaction mixture allows for easier and speedier regeneration and recovery of the reagents or catalysts. This can be not only time effective (because of the time gained in setting up, later on, another reaction of the same type), but also cost effective since the same reagent or catalyst can be reused a number of times.

Third, and perhaps most important, is the frequent observation of greater reaction selectivity in the presence of supported reagents. Cutting down on the number of reaction products and accordingly increasing the proportion of the major product obviously makes for easier separation and work up, at the end of a reaction.

VII. FASTER REACTIONS UNDER MILDER CONDITIONS

Running reactions on solid surfaces very often spurs them from a crawl to a dash. A number of factors can be responsible for such accelerations. In the following only three factors are discussed; Chapter 2 by Peter Pfeifer provides a much more thorough and detailed analysis.

The first factor is local accumulation of reactant molecules. These molecules find their diffusive way to clefts, crannies, and pockets within the finely divided solid. In this manner, stagnant pools of chemicals form within the cavities of the polymeric or, more frequently, inorganic support. Accordingly, the probability of a reactive encounter is greatly increased because the local concentration of one or both reaction partners has been boosted. It has only been recently that this effect has begun to be described and analyzed properly, in the context of Benoit Mandelbrot's fractal theory. The interested reader will find the results from this powerful formalism in Chapter 2. Intuitively, they correspond to the notion of pockets small enough to capture reactant molecules and greatly restrict their translational diffusion, and yet large enough so that the trapped molecules are able to escape in order to react or after the reaction has taken place.

The second factor is also an heightened concentration (or activity) due to the proximity of a solid surface, whenever the solid, whether inorganic or polymeric, is a polyelectrolyte. Hence, one should be prepared for such effects not only with synthetic polyelectrolytes (e.g., polyacrylic acid), but also with natural polyelectrolytes, ranging from DNA to alumina, silica, silicates, and aluminosilicates. If the solid surface is a polyanion, positively charged counterions will congregate next to the surface. Conversely, condensation of anions in the vicinity of a polycationic surface will take place. These effects can be large—typically, when the average concentration of an electrolyte is millimolar, the order of magnitude of the local concentration next to the charged surface can be molar or more. There are many reactions that proceed faster due to either a greater ionic strength or an increased surface acidity or surface basicity. In this book, Chapters 16–20 on alumina- and silica-supported reagents and Chapters 23–27, on clay-supported chemistry amply document this concept. Obviously, redox reactions, in particular, those performed under electrochemical conditions, also have much to gain from the use of solid polyelectrolytes. Chapter 5 provides a state-of-the-art tantalizing glimpse into the prospects of such applications of synthetic (Nafion) as well as naturally occurring (montmorillonite clays) polyelectrolytes coated on electrodes.

A third factor is the restriction in the dimensionality of reaction space. In a bimolecular reaction, when diffusing reaction partners are constrained to the surface, instead of roaming freely within a three-dimensional reaction volume, encounter rates greatly increase. This key observation was first made by Adam and Delbrück, in the context of an explanation of enzymatic catalysis. The enzyme forces its substrates to evolve on its surface, thereby greatly increasing the frequency of collisions. Adam and Delbrück (see Chapter 2) gave an analysis in terms of the Smoluchowsky–Debye diffusional equations, further detailed by implicit consideration of the dimension of diffusional space. The basis on which they built their analysis was the results obtained by the mathematicians, Polya and Montroll, in the solution of the so-called drunkard problem—a random walker trying to retrace his steps back to his original point. The problem is an obvious analogue to diffusion by Brownian motion of a chemical particle till it hits a target. As a problem in pure mathematics, the "drunkard problem" has a simply stated solution, although it is very difficult to work out. In three-dimensional space, the drunkard has a probability of only 0.34 of finding the way back to his or her abode. However, in both one- and two-dimensional space, the probability jumps back to unity. Translated into chemical terms, this means that, everything else being equal, a vastly increased pre-exponential factor can stem from running a reaction on a surface (preferably a smooth surface) rather than in a volume.

Even in the presence of only one of these accelerating factors, chemical reactions can undergo very significant reductions in the activation energy when they are forced to occur on the surface. This allows the experimentalist to lower the reaction temperature and/or pressure and to use much milder conditions. A typical example is the Diels–Alder reaction between unactivated hydrocarbon dienes and dienophiles. To take place it demands heating at 180°–200°C for a day or two under homogeneous conditions. With catalysis from a modified clay, it can become quantitative in a few minutes at room temperature!

VIII. IMPROVED SELECTIVITY

The price to pay for the generally observed increase in reaction rates, one might guess, has to be proliferation of products. This is what a naive application of the reactivity–selectivity principle would lead one to expect. How then can we explain that reaction conditions involving supported reagents can be conducive to increases *both* in reactivity and in selectivity?

This is an experimental fact, in very many cases. An enlightened scientist knowing the value of peripheral, serendipitous collection of information will find dozens of such examples by browsing through this book.

The author wishes to point out the paradox here, he does not claim to understand in detail how it comes about. An answer, equally naive as the reference to the reactivity–selectivity principle, consists of pointing out the reduced reaction temperatures as one of the explanations for the improved product selectivity. Another answer is shape selectivity. It occurs in reactions placed under geometrical constraints, such as those that stem from the use of a solid support. Chapters 21 and 22 summarize the impressive selectivities (i.e., reactant-type, transition-state-type, and product-type) that can be achieved with zeolites. Zeolites are not the only class of supports that can function effectively in this manner. Pillared clays and normal clays (See Chapters 23 and 24) can also endow guest reactants with highly structured microreactors that will favor the selective formation of a single product. More general yet, and even though there is much to be learned about the factors involved, chemistry on surfaces tends to be clean, efficient, and selective. Is it due to the detailed packing of reactant and solvent molecules next to a surface? Does it come from slight variations in the interaction energy between adsorbates and the adsorbing surface? Are they privileged modes of chemical activation of a substrate sitting on a surface, due to coupling of vibrations with the solid lattice? Again these are questions to which the answers have yet to be found. Framing the questions however can already be a step forward.

IX. SAFER REAGENTS

Alfred Nobel's name has to be quoted in the opening of this paragraph. He was the great pioneer who domesticated wild nitroglycerine by affixing it onto kieselguhr (diatomaceous earth) and, thus, invented dynamite. Likewise, a number of dangerous or toxic chemicals can be approached much more safely after having been deposited on insoluble solid. Heavy metal ions will remain trapped in the electrostatic atmosphere of a polyion, such as the anionic sheets in phyllosilicates. In this manner, leaching of such toxic chemicals can be greatly reduced; thus leading to the concept of radioactive waste storage within clays. The laboratory chemist can also make use of this concept when he needs to use chemicals that are a potential threat to the environment.

I shall quote here from our own research.

> Preparation of "clayfen" (clay-supported ferric nitrate) draws upon the outstanding oxidizing properties of covalent metallic nitrates, in the anhydrous state. Once ferric nitrate, dissolved in acetone, has shed its hydration, it is powerfully activated by such dehydration. The acetonic solvate of $Fe(NO_3)_3$ (presumably a diacetonic solvate) has one-third of the explosive potential of TNT, according to thermochemical calculations. It is such a powerful oxidant that we have seen it oxidize Nujol! It would be irresponsible to handle such a wild beast. It has first to be tamed. A montmorillonite clay was used for this purpose. The stabilization thus achieved, at some price in reactivity obviously, is such that freshmen at the University of Liège have been routinely doing oxidations of alcohols using "clayfen" for the last half-a-dozen years, without a single incident.

X. SURFACE CHEMISTRY AS A NEW FRONTIER IN CHEMISTRY

The author started this chapter, with some reflexions on prejudice: chemists, for good or bad reasons, are strongly biased toward homogeneous reaction conditions. The author has continued with expressions of pride, as a practitioner of heterogeneous reaction conditions, and enthusiasm— at the danger of sounding like a patent medicine salesman! Hence, the author wishes to close this introductory chapter with a more moderate and more neutral pitch. The final point is that there is ample new chemical knowledge to be gleaned from heterogeneous reaction conditions.

So many things become different when reactions are run on solid surfaces! The regioselectivity can change. The stereoselectivity can be modified. Catalytic activity can be altered too. For instance, the sequence of catalytic power of Lewis acids (such as $AlCl_3$, $TiCl_4$, etc.) in Friedel–Crafts alkylations is totally changed when the reaction, instead of being run under homogeneous conditions, is performed in the presence of a clay intercalated with such Lewis acids (Laszlo and Mathy, 1987).

REFERENCES

Berger, M. N., Boocock, G., and Hayward, R. N. (1969). *Adv. Catal.* **19**, 211.

Bridger, G. W., and Chinchen, G. C. (1970). "Catalyst Handbook," p. 64. Wolfe Sci. Books, London.

Collins, P. (1976). *Ambix* **23**, 96.

Csicsery, S. M. (1984). *Zeolites* **4**, 202.

Csicsery, S. M. (1985). *Chem. Br.* p. 473.

Denny, P. J., and Whan, D. A. (1978). *Catalysis (London)* **2**, 46.

Dixon, J. K., and Longfield, J. E. (1960). In "Chemistry of Catalytic Processes" (P. H. Emmett, ed.), Vol. 7, p. 281. McGraw-Hill, New York.

Ertl, G. (1980). *Catal. Rev.—Sci. Eng* **21**, 201.

Gates, B. C., Katzer, J. R. and Schuit, G. C. A. (1979). "Chemistry of Catalytic Processes." McGraw-Hill, New York.

Kemball, C. (1984). *Chem. Soc. Rev.* **13**, 375.

Keulks, G. W., Krenzhe, L. D., and Notermann, T. M. (1978). *Adv. Catal.* **27**, 183.

Laszlo, P, and Mathy, A. (1987). *Helv. Chim. Acta*, in press.

Lederer, E., and Lederer, M. (1957). "Chromatography," 2nd Ed., pp. 61–67. Elsevier, Amsterdam.

Mosbach, K. (1971). *Sci. Am.* **225**(3), 26.

Novotny, L., and Kotva, K. (1974). *Collect. Czech. Chem. Commun.* **39**, 2949.

Ozaki, A. (1981). *Acc. Chem. Res.* **14**, 6.

Ségur, O (1803). "Lettres élémentaires sur la chimie," Vol. I, pp. 12–14. Migneret, Paris.

2   CHARACTERIZATION OF SURFACE IRREGULARITY

Peter Pfeifer*

Department of Physics and Astronomy
University of Missouri
Columbia, Missouri 65211

I. INTRODUCTION

Traditional characterizations of surface irregularity for chemical purposes can be found in the texts of Thomas and Thomas (1967), Aris (1975), Parfitt and Sing (1976), Lecloux (1981), Adamson (1982), Gregg and Sing (1982), Thomas (1982), Mikhail and Robens (1983), and Delannay (1984). This chapter describes the unique approach in terms of fractal dimension from its first surface-chemical exploration (Avnir and Pfeifer, 1983) to the present state.

Fractal analysis is a resolution analysis. It tracks the recurrence of topographical features of the surface at different length scales. Many materials display surface irregularity over several length scales. Such many-scale topography controls chemical reactions in very specific ways. It determines the distribution of active sites, static and dynamic accessibility of sites, surface diffusion, flow through pores, and even the structure of surface–molecule complexes.

The general reference on fractals is the text by Mandelbrot (1982). For fractal properties of specific systems (polymers, gels, colloidal aggregates,

* The author is indebted to D. Avnir and D. Farin for joint work seminal to this research and for critical reading of the manuscript. Discussions with A. Blumen, J. Klafter, and P. Laszlo were very helpful. This work was supported in part by the Research Council of the University of Missouri.

13

and numerous other disordered structures) see Shlesinger *et al.* (1984), Family and Landau (1984), Stanley and Ostrowsky (1986), and Pietronero and Tosatti (1986). For a review focused on surface problems, see Pfeifer (1985).

II. NOTION OF FRACTAL DIMENSION

A. Fundamentals

The concept of fractal dimension is based on interrogation of the surface at different length scales. The most primitive, but indeed definitive, interrogation is to divide the embedding space into cells (boxes) of side length r and to count how many are intersected by the surface (Fig. 1). Let this number of nonempty cells, i.e., the number of pixels needed to represent the surface to within resolution r, be $N(r)$. The function $N(r)$ contains everything that we need: (1) It gives the object size (diameter) L as the r value at which $N(r)$ levels off to unity. (2) It gives the space-filling ability of the object as the power by which $N(r)$ increases for decreasing r, $r \ll L$:

$$N(r) = \text{const} \begin{cases} (r/L)^{-1} & \text{for a straight line (Fig. 1a)} & \text{(1a)} \\ (r/L)^{-\ln 4/\ln 3} & \text{for the curve in Fig. 1b} & \text{(1b)} \\ (r/L)^{-2} & \text{for a square (Fig. 1c)} & \text{(1c)} \\ (r/L)^{-3} & \text{for a cube} & \text{(1d)} \end{cases}$$

[The outcome (1b) is explained in the following]. (3) It gives the macroscopic shape of the object as the prefactor in Eqs. (1). The point is [see Eq. (1b)] that the exponent in the power law for $N(r)$ for small r need not be an integer. If

$$N(r) = \text{const} \, (r/L)^{-D} \tag{2}$$

with fractional D, the object has by definition the fractal dimension D. An irregular curve with $1 < D < 2$ has a space-filling ability between that of a straight line and a square [Eqs. (1a–c), Fig. 1]. The higher D, the more irregular is the curve. Similarly, a surface with $2 < D < 3$ interpolates between a plane and a volume. Section III.A will implement Eq. (2) in terms of monolayer coverage.

Instead of the box-counting relation, Eq. (2), some authors prefer the equivalent definition of D via the site distribution (mass–radius relation). It states that, for a fixed box size, $r \ll L$, the number $M(R)$ of nonempty boxes ("sites"), within distance R from some chosen origin on the object,

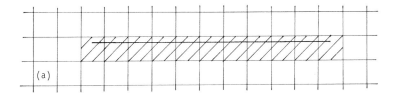

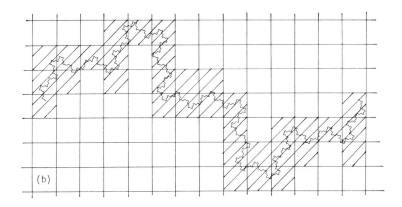

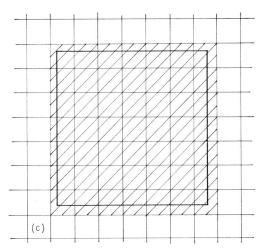

Fig. 1. Counting nonempty cells (shaded) for (a) a straight line, (b) an irregular curve, and (c) a square.

grows with R, $r \ll R \ll L$, as

$$M(R) = \text{const} \, (R/r)^D \qquad (3)$$

For the straight line ($D = 1$) and the square ($D = 2$), Eq. (3) is again trivially true. In practice, the sites are identified as adsorption sites if the system is a surface (Pfeifer and Avnir, 1983), as monomers if it is a polymer (Mandelbrot, 1982; Family, 1984; Stanley, 1984), or as primary particles if it is a colloidal aggregate (Forrest and Witten, 1979; Weitz and Huang, 1984). For a translation of Eq. (3) into density correlation functions, see Witten and Sander (1981).

The value $D = \ln 4/\ln 3$ ($=1.26$) for the curve in Fig. 1b results as follows. By construction (Fig. 2), the curve consists of four copies of itself reduced by a factor $\frac{1}{3}$. So, if the whole curve fits into one box of size L, it

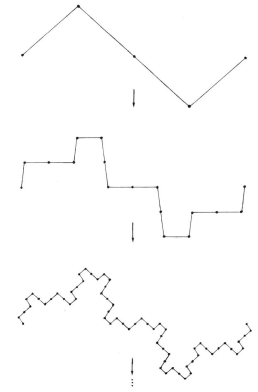

Fig. 2. Iterative construction of the curve in Fig. 1b. In each step, every straight-line segment of the previous stage is replaced by four segments of $\frac{1}{3}$ of its length. For many other constructions of this type, both nonrandom (as here) and random, see Mandelbrot (1982).

will fit into four boxes of size $L/3$: $N(L/3) \approx 4N(L)$. What holds for the whole curve, holds for any of its reduced copies. There is a whole cascade of copies within copies, and thus $N(L/3^k) \approx 4^k N(L)$, $k = 1, 2, 3, \ldots$. Comparing this with Eq. (2), we obtain $D = \ln 4/\ln 3$. This self similarity is the key property of fractals. In real systems, self similarity is always restricted to a finite range of length scales, $r_{min} < r < r_{max}$. These limits, r_{min} and r_{max} for Eq. (2), are termed inner and outer cutoff or crossover scale. In Fig. 1, r_{max} is of the order of the object size L. However, for surface fractality in the molecular domain ($r_{min} \approx$ atomic lengths) we often have $r_{max} \ll L$ and the surface is smooth on scales betweeen r_{max} and L. In that case, the surface has dimension $D > 2$ at scales below r_{max}, and dimension $D = 2$ above r_{max}. To accept an experimental D value over a range $r_{min} < r < r_{max}$ as a well-defined dimension, we have the condition that the ratio r_{max}/r_{min} should be greater than $2^{1/D}$ (Pfeifer, 1984). If this is not the case, i.e., if $-d \ln N(r)/d \ln r$ exhibits no pronounced plateau, still very useful information is available [see Cederbaum *et al.* (1985), Tsurumi and Takayasu (1986), and Eq. (16)].

B. Classification of Fractal Interfaces

Table I lists the basic fractal classifiers. Entries 1–5 are the parameters introduced so far. Many modes of interaction of a fractal with its environment are not governed solely by the interface (boundary). Hence, sets other than the interface and dimensions other than D come into play. Some are subsets of the interface (entries 8–10); others are not (entries 6 and 7). Here we are concerned with D_{mass}, D_{pore}, and D_{edge}. They divide D-dimensional interfaces into four categories which are important for Section III.

1. Surface Fractals ($D_{mass} = 3$, $D_{pore} = 3$)

Let $M_x(R)$ be the number of x-type sites within radius R from a fixed origin of the same type. Each site type gives rise to a separate exponent D_x for Eq. (3). A surface fractal (Fig. 3a) then is characterized by

$$M_{surf}(R) \propto R^D$$

$$M_{mass}(R) \propto R^3 \qquad (4)$$

$$M_{pore}(R) \propto R^3$$

So the surface sites form but a small fraction of the whole solid, and only the surface is fractal. Material examples ordered according to increasing D include fume silica at length scales smaller than 10 nm, graphite (Avnir

TABLE I

Parameters that Distinguish Different Fractal Materials (Solids with Fractal Interface)

Parameter	Meaning	Remarks	More details in
D	Fractal dimension of interface	$D_{top} \leq D < 3$	Section II.A
D_{top}	Topological dimension of interface (2 for surface, 1 for curve)	Models for ramified materials may be curves	Mandelbrot (1982, p. 136)
r_{min}, r_{max}	Cutoffs	Domain of Eq. (2)	Section II.A
Constant in Eq. (2)	Macroscopic shape factor of interface	If $r_{max} \sim L$	Section II.A
Constant in Eq. (3)	(Lacunarity of interface)$^{-1}$	If $r > r_{min}$	Mandelbrot (1982, p. 314)
D_{mass}	Fractal dimension of solid	3 for surface and pore fractals; D for mass fractals	Section II.B
D_{pore}	Fractal dimension of pore space	3 for surface and mass fractals; D for pore fractals	Section II.B
D_{edge}	Fractal dimension of edges/ corners; distinguishes interfaces with $D = D_{top}$	$D = \max\{D_{edge}, D_{top}\}$	Section II.B
$\bar{D}$	Spectral dimension ($\sim$ connectivity) of interface	$D_{top} \leq \bar{D} \leq D$	Section IV.B
D_{exp}	Fractal dimension of exposed interface (with respect to diffusion from outside)	$D_{exp} \leq D$	Section V.B

et al., 1983), the lysozyme molecule (Pfeifer *et al.*, 1985), crushed quartz (Avnir *et al.*, 1984, 1985), metal fracture surfaces (Mandelbrot *et al.*, 1984), carbon black (Avnir *et al.*, 1983), lignite (Bale and Schmidt, 1984), Ca/Mg carbonates (Avnir *et al.*, 1984, 1985), kaolinite (Pfeifer, 1984; Avnir *et al.*, 1985), and porous silica (Avnir and Pfeifer, 1983; Farin *et al.*, 1985a; Rojanski *et al.*, 1987).

2. Mass Fractals ($D_{mass} = D$, $D_{pore} = 3$)

Here (Fig. 3b), surface and mass sites are comparable and thus scale identically, while the pore space is still three-dimensional:

$$M_{surf}(R) \propto R^D$$
$$M_{mass}(R) \propto R^D \tag{5}$$
$$M_{pore}(R) \propto R^3$$

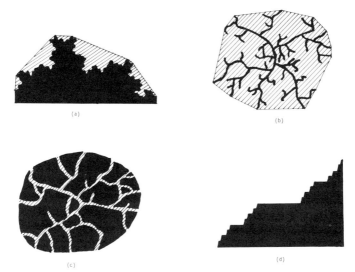

Fig. 3. (a) Surface fractals, (b) mass fractals, (c) pore fractals, and (d) subfractal surfaces. Black areas are mass sites, shaded areas are pore sites (see Pfeifer and Salli, 1987, for the general definition of the pore space). The boundary between the two makes the surface sites. The surface (d) with Cantor's staircase (Mandelbrot, 1982) as profile has $D = 2$ and $D_{edge} = 1 + \ln 2/\ln 3$.

Material examples include fume silica aggregates at length scales greater than 10 nm (Sinha *et al.*, 1984), gold–colloid aggregates (Weitz and Huang, 1984), polymers (Schaefer and Keefer, 1984), electrolytic copper deposits (Brady and Ball, 1984), and vapor-deposited metals (Forrest and Witten, 1979; Kapitulnik and Deutscher, 1984; Elam *et al.*, 1985).

3. Pore Fractals ($D_{mass} = 3$, $D_{pore} = D$)

These (Fig. 3c) are mass fractals in which mass and pores have been interchanged:

$$M_{surf}(R) \propto R^D$$
$$M_{mass}(R) \propto R^3 \qquad (6)$$
$$M_{pore}(R) \propto R^D$$

They are solids criss-crossed by channels of uniform width which form a fractal network. Material examples include: Sandstones (Katz and Thompson, 1985), Vycor glass (Dozier *et al.*, 1986). Zeolites and expandable clays, on length scales above the respective pore width, have $D = 3$ and thus are equally pore, mass, and surface fractals [cf. Eqs. (4)–(6)]. A material with D close to three is identified as a pore, mass, or surface fractal depending on whether $r_{\min} \sim \delta \ll r_{\max}$, $r_{\min} \ll \delta \ll r_{\max}$, or $r_{\max} \ll \delta$, where δ is the average wall thickness.

4. Subfractal Surfaces ($D = 2$, $D_{\text{edge}} < 2$)

A surface with $D > 2$ has irregularities everywhere, i.e., every point is an edge/corner point (Fig. 1b). So, for $D > 2$, we always have $D_{\text{edge}} = D$. For subfractal surfaces, in contrast, irregularities are not surface-filling (Fig. 3d):

$$M_{\text{surf}}(R) \propto R^2,$$

$$M_{\text{edge}}(R) \propto R^{D_{\text{edge}}}, \qquad D_{\text{edge}} < 2 \tag{7}$$

For example, a cube as $D_{\text{edge}} = 1$. Platinum crystallites on SiO_2 have $D_{\text{edge}} = 1.7$ (Pfeifer, 1985). Pillared clays (Van Damme and Fripiat, 1985) in which D_{edge} describes the base points of pillars [model: Cantor set in the plane (Mandelbrot, 1982)] may be another example.

III. EXPERIMENTAL METHODS OF FRACTAL SURFACE ANALYSIS

This section surveys experimental manifestations of surface fractality and ensuing methods for measuring D. Most of them do not apply equally to surface/mass/pore fractals. This yields information beyond D. For example, a successful observation of Eq. (8) implies that the system is a surface or mass fractal, i.e., that $D_{\text{pore}} = 3$. Experimental examples are but a selection from many more in the references to Section II.B.

A. Adsorption (Monolayer Capacity)

If in Eq. (2) boxes intersecting the surface are replaced by molecules forming a monolayer on the surface, then the number of molecules n on an adsorbent volume V_0 drops with increasing molecular cross-sectional area σ and adsorbent particle diameter L ($\ll V_0^{1/3}$) as

$$n = \text{const } V_0 \, \sigma^{-D/2} L^{D-3} \tag{8}$$

(Pfeifer and Avnir, 1983; Pfeifer and Salli, 1987). This holds for surface and mass fractals. It assumes that the shape of the molecules and of the adsorbent particles remain constant as σ or L is varied, making all shape and packing factors coalesce into the unitless prefactor. Equation (8) may be used in two ways:

(a) If only σ is varied, size, shape, and density of adsorbent particles are irrelevant, and D is obtained from n (per g adsorbent) $\propto \sigma^{-D/2}$ (Fig. 4). For methods to get (on-surface conformation) σ, see Pfeifer and Avnir

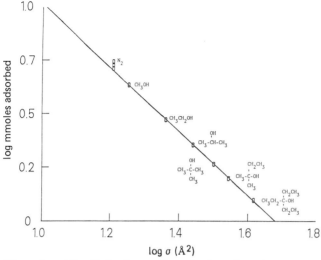

Fig. 4. Illustration of Eq. (8) for the paradigm of a three-dimensional surface, silica gel. Shown are the most recent measurements (Rojanski *et al.*, 1987) giving $D = 2.97 \pm 0.02$ from n_{alcohols} versus σ, and $D = 3.00 \pm 0.01$ from n_{nitrogen} versus L. In accordance with Eq. (8) and $D = 3$, the six values of n_{nitrogen} are independent of L and therefore fall essentially onto a single point in the figure. The so-observed range of self similarity is $\sigma = 16$–114 Å^2.

(1983) and Avnir *et al.* (1983, 1984). Conversely, unknown σ values can be determined from known D and n values (Farin *et al.*, 1985a; Meyer *et al.*, 1986). For chain polymers (Pfeifer, 1986), the conformations of coils in solution and on the surface admit of fractal dimensions D_{c}^{sol} and D_{c}^{ads}, leading to

$$n \propto \sigma^{-(D/2)D_{c}^{\text{sol}}/D_{c}^{\text{ads}}} \tag{9}$$

where σ is the cross section in solution. Unfolding ($D_{c}^{\text{ads}} = D$) occurs for $D \sim 2$, while for $D \sim 3$ the conformation is maintained ($D_{c}^{\text{ads}} = D_{c}^{\text{sol}}$).

(b) If σ is kept constant and L is varied, Eq. (8) reduces to

$$n \text{ (per ml adsorbent)} \propto L^{D-3} \tag{10a}$$

$$n \text{ (per g adsorbent)} \propto L^{D-D_{\text{mass}}} \tag{10b}$$

(which applies also to pore fractals if $\sigma < r_{\text{min}}^2$). For an experimental example, see Fig. 4 again. For translation of L ranges in which Eqs. (10) hold, into σ ranges in which $n \propto \sigma^{-D/2}$ holds, see Pfeifer and Avnir (1983).

B. Adsorption (Multilayers)

The determination of n in Section III.A requires as many adsorption isotherms (low-coverage) as data points for n are desired. Instead, D

can also be obtained from a single isotherm if high-coverage data is included (Cole *et al.*, 1986). The fractal extension of the classical BET model predicts that the total number of gas molecules adsorbed (multi-layers), n_{tot}, grows with relative pressure x ($0 \leqslant x \leqslant 1$) according to

$$n_{tot} = n_{sites} \frac{cx}{1 - x} \frac{g_D(x)}{g_D(0)} \tag{11}$$

$$g_D(x) = \int_1^\infty m^{1-D} \frac{1 - (m + 1)x^m + mx^{m+1}}{1 + (c - 1)x - cx^{m+1}} dm \tag{12}$$

where n_{sites} is the number of surface sites, c the adsorption strength, and $r_{max} \gg$ adsorbate diameter is assumed. For $D = 2$, Eqs. (11) and (12) yield the classical BET isotherm. For $x \to 0$, the same is true for any D. At higher x, the fractal isotherm stays below the classical one, due to the reduction of effective surface as ever larger pores are filled (Section III.E).

C. Reaction Rates

The rate v of a surface reaction is proportional to the number of available active sites, which in turn is n in Eq. (10). So if v is taken as the number of molecules converted per unit time and volume of the solid, its dependence on the particle size L is

$$v \propto L^{D_{act}-3} \tag{13}$$

Table II explains D_{act} and gives examples. Other ways how fractal analysis can identify pathways and optimal rates are described in Sections IV and V.

D. Photochemical Energy Transfer

Electronic energy transfer between adsorbed donor and acceptor molecules is sensitive to D because it depends on the number of acceptors

TABLE II

The Dimension of Active Sites D_{act} for Eq. (13)

D_{act}	Active sites	Example	Reference
D	All surface	Dissolution of quartz	Avnir *et al.* (1985)
D_{edge}	Edges/corners	Hydrogenation on Pt	Pfeifer *et al.* (1984); Pfeifer (1985)
D_{exp}	Exposed sites	Dissolution of carbonates	Silverberg *et al.* (1986)

within distance R from a donor [Eq. (3)]. For dipole–dipole interaction the fluorescence intensity (donor) at time t, $I(t)$, is given by (Klafter and Blumen, 1984)

$$I(t) = \exp[(-t/\tau_0) - \text{const } \chi_2 \, (t/\tau_0)^{D/6}] \tag{14}$$

where τ_0 is the radiative life time of the donor, and χ_1, χ_2 are the fraction of sites occupied by donors and acceptors, respectively ($\chi_1 \ll \chi_2 \ll 1$). This assumes identical across-pore and across-wall interactions, i.e., that the solid and the pores have the same dielectric constant (phase matching). It applies to surface, mass, and pore fractals. Vycor glass (Even et al., 1984; Yang et al., 1985) and the silica gel in Fig. 4 were studied by this method.

E. Porosimetry

If pores are understood as "off-shore" space accessible only to "yard-sticks" (molecules, liquid menisci, etc.) sufficiently small, then any fractal is necessarily porous (the lower D, the shallower are the pores). Self similarity entails pores of all sizes, increasing in number with decreasing size. The pore-size distribution $V_{\text{pore}}(r)$, defined as the volume of pores of radius less than r, reflects this for surface and mass fractals through (Pfeifer and Avnir, 1983; Pfeifer and Salli, 1987)

$$\frac{d}{dr} V_{\text{pore}}(r) \propto r^{2-D} \tag{15}$$

Figure 5 shows a case in which $dV_{\text{pore}}(r)/dr$ was measured by adsorption/desorption hysteresis.

$V_{\text{pore}}(r)$ is the empty volume between the surface and a monolayer of molecules of radius r. Let the volume of the monolayer be $V_{\text{mono}}(r)$. These two complementary volumes can be shown to obey

$$\frac{d}{dr} V_{\text{pore}}(r) = \frac{1}{2}\left(\frac{1}{r} - \frac{d}{dr}\right) V_{\text{mono}}(r) \tag{16}$$

which is valid for any surface. For fractals, $V_{\text{mono}}(r) \propto r^{3-D}$ (Section III.A), so that Eq. (15) is a special case of Eq. (16). It shows that a smooth surface is nonporous by the vanishing of the prefactor in Eq. (15) for $D = 2$.

F. Small-Angle Scattering

Like porosimetry, small-angle scattering (x-rays or neutrons) is particu-larly suited to detect self-similarity extending over a broad range of length

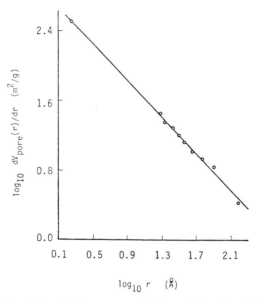

Fig. 5. Illustration of Eq. (15) for porous alumina, yielding $D = 3.04 \pm 0.02$ over the range $r = 2$–195 Å. [From Pfeifer (1986). Copyright 1987 North-Holland Publ. Co., Amsterdam.]

scales ($\sim 10^0 - 10^4$ Å). Measurement of the scattered intensity I at decreasing scattering angle θ, or length q of the scattered wave vector, where $q = 4\pi\lambda^{-1}\sin(\theta/2)$ and λ is the wavelength, is equivalent to probing the solid by increasing wavelengths, i.e., at increasing length scales. Viewing the scattering as Bragg-type reflection at the boundary between solid and pores, it is clear that $I(q)$ depends on the distributions of surface, mass, and pore sites together. This is why different scattering laws hold for surface fractals (Bale and Schmidt, 1984; Kjems and Schofield, 1985) and mass/pore fractals (Sinha *et al.*, 1984; Schaefer and Keefer, 1984; Schaefer *et al.*, 1984). They are all contained in the single law (Pfeifer, 1987),

$$I(q) \propto \sin\left(\frac{D-1}{2}\pi\right) q^{D - 2D_{mass} - 2D_{pore} + 6} \tag{17}$$

which is valid for $r_{max}^{-1} \ll q \ll r_{min}^{-1}$. For surface fractals, the exponent reduces to $D - 6$, for mass/pore fractals it reduces to $-D$. It takes values less than -3 and greater than -3, respectively, so the method automatically distinguishes surface and mass/pore fractals. Experimental examples are in the above-mentioned references. For $D = 3$, the prefactor in Eq. (17) vanishes and other terms become leading (Rojanski *et al.*, 1987). For self-affine interfaces, see Wong (1985) and Pfeifer (1987).

G. Miscellany

When micrographs are available, the following image-analytical methods may be used:

(1) counting of features (Katz and Thompson, 1985; Krohn and Thompson, 1986);

(2) length measurements of surface profiles with different yardsticks (Kaye, 1981; Farin et al., 1985b; Pfeifer et al., 1985);

(3) Fourier analysis of profiles (Mandelbrot et al., 1984; Pfeifer, 1984); and

(4) slit-island technique (Mandelbrot et al., 1984).

Electrochemical methods to study fractal electrode properties are described in de Gennes (1982), Brady and Ball (1984), Liu (1985), Kaplan and Gray (1985), and Nyikos and Pajkossy (1986). Also the electrical skin effect may be utilized (Tsallis, 1986). Partial filling of surface fractals by a wetting fluid is analyzed in de Gennes (1985).

IV. RESTRICTED DIFFUSION ON A SURFACE AND ENCOUNTER OF REAGENTS: THE EFFECTS OF DIMENSIONALITY

A. Diffusion in d-Dimensional Euclidean Space

To see how the fractal dimension affects diffusion of reagents on the surface, it is useful to first recall the effect of dimensionality when the host space is ordinary Euclidean space of dimension $d = 1, 2, 3$. We adopt a lattice viewpoint again and treat diffusion as a random walk on the lattice (see, e.g., Montroll and West, 1979). From the classical relation

$$\xi^2 \propto t \qquad (d = 1, 2, 3) \qquad (18)$$

for the mean diffusion distance ξ (root mean square displacement) as a function of time t (number of steps of the random walk), it follows that most points of diffusion space are visited infinitely often for $d = 1$, essentially once for $d = 2$, and never for $d = 3$:

$$\frac{\text{no. of sites (not necessarily distinct) visited}}{\text{no. of sites within radius } \xi \text{ from origin}} \propto \frac{t}{(\sqrt{t})^d} \xrightarrow{t \to \infty} \begin{cases} \infty & \text{for } d = 1 \\ 1 & \text{for } d = 2 \\ 0 & \text{for } d = 3 \end{cases}$$

$$(19)$$

Equivalent statements are (a) diffusion is recurrent for $d = 1$, just barely recurrent for $d = 2$, and nonrecurrent for $d = 3$; (b) exploration of the diffusion space is compact for $d = 1, 2$, and noncompact for $d = 3$; and

(c) the mean number of distinct visited sites S obeys (omitting logarithmic corrections for $d = 2$)

$$S \propto \begin{cases} t^{1/2} & \text{for } d = 1 \\ t & \text{for } d = 2, 3 \end{cases} \qquad (20)$$

A celebrated consequence is (Adam and Delbrück, 1968) that a diffusing particle finds a given (fixed) target more rapidly in a diffusion space of dimension $d = 1, 2$ than for $d = 3$: The mean diffusion time τ to reach the target in a diffusion space of diameter L is

$$\tau \propto \begin{cases} L^2 & \text{for } d = 1, 2 & (21a) \\ L^3 & \text{for } d = 3 & (21b) \end{cases}$$

It suggests that diffusion restricted to $d = 1, 2$ dimensions should lead to an increased encounter of reactants and therefore to shorter reaction times (see, e.g., Hardt, 1979). Section IV.C will show that this is only conditionally true.

B. Diffusion on Fractals

Consider a random walk on the singly connected curve in Fig. 1b and a walk on a network (multiply connected curve) where both curves have the same D. On the network, ξ will grow faster with t because the walker can take crosslinks (shortcuts) nonexistent on the singly connected curve to move a distance ξ from the origin. Thus, D alone cannot describe diffusion on fractals. The additional required characteristic is the spectral, or fracton, dimension $\tilde{D}$, defined via the probability $p \propto r^{-\tilde{D}/2}$ that a random walk returns to the origin at time t (Rammal, 1984). The more connected or ramified the structure, the higher is $\tilde{D}$. It obeys $D_{\text{top}} \leqslant \tilde{D} \leqslant D$, where the low end ($\tilde{D} = D_{\text{top}}$) is realized for nonramified structures, and the high end ($\tilde{D} = D$) for structures like Menger's sponge (Mandelbrot, 1982). For d-dimensional Euclidean space, $D_{\text{top}} = \tilde{D} = D = d$.

So, diffusion should be fast for high $\tilde{D}$ and low D [compare also $\xi(t)$ for Figs. 1a,b]. This is exactly the fractal generalizations of Eqs. (18), (20), and (21)

$$\xi^2 \propto t^{\tilde{D}/D} \qquad [\tilde{D} = D: \text{normal (Euclidean-like) diffusion},$$
$$\tilde{D} < D: \text{anomalous (slower) diffusion}] \qquad (22)$$

$$S \propto t^{\min\{1, \tilde{D}/2\}} \qquad [\tilde{D} \leqslant 2: \text{diffusion is recurrent},$$
$$\tilde{D} > 2: \text{diffusion is nonrecurrent}] \qquad (23)$$

$$\tau \propto L^{D \max\{1, 2/\tilde{D}\}} \qquad (24)$$

(Rammal and Toulouse, 1983; Rammal, 1984; Pfeifer, 1985) imply. Here of particular interest is Eq. (24). It shows that the diffusion time τ is minimized by $\tilde{D} = D \leqslant 2$, in which case $\tau \propto L^2$. So, the classical optimum [Eq. (21a)] cannot be improved.

Equations (22)–(24) are microscopic diffusion laws. For example, Eq. (22) holds for diffusion distances ξ between r_{min} and r_{max}. Suitable for measurement of $\tilde{D}$ are the following methods. (a) Quasi-elastic neutron scattering (incoherent dynamic structure factor) yields the ratio $\tilde{D}/D$ (Pfeifer *et al.*, 1987). (b) If the solid is smooth on lengths above r_{max}, then on time scales such that $\xi \gg r_{max}$, diffusion obeys the classical law [Eq. (18)] with a diffusion coefficient Δ_{macro} [= prefactor in Eq. (18)] given by

$$\Delta_{macro} = \left(\frac{r_{min}}{r_{max}}\right)^{2(D/\tilde{D}-1)} \Delta_{micro} \tag{25}$$

(Dozier *et al.*, 1986), where Δ_{micro} is the coefficient on an otherwise identical, but smooth (down to r_{min}) surface. That is to say that macroscopically, $\tilde{D} < D$ shows up in an anomalously small diffusion coefficient. (c) Chemical kinetic signatures of $\tilde{D}$ are the subject of Section IV.C. (d) If, as is true for the low-D surface fractals in Section II.B, there are no through-pores and branchings, then trivially $\tilde{D} = 2$.

C. Langmuir–Hinshelwood Model on Fractals

This section summarizes theoretical studies of the diffusion-controlled reaction $A_{ads} + B_{ads} \rightarrow$ products on fractals (for experiments see Anacker *et al.*, 1984 and Kopelman *et al.*, 1986). It is assumed that initial conditions are homogeneous, with A and B randomly distributed on the fractal, and that whenever A and B walk onto the same site, they react instantaneously and leave the system. Concentrations $[A]_t$ and $[B]_t$ (t = time) are expressed as the fraction of occupied sites. Since $[A]_0 - [A]_t = [B]_0 - [B]_t$, the quantity

$$\Phi(t) = [A]_t/[A]_0 \tag{26}$$

gives the full description of the kinetics. While the decay of $\Phi(t)$ may vary from stretched exponential to power law (Table III), it depends in each case only on $[B]_0$ and the spectral dimension $\tilde{D}$. For fixed $[B]_0$, all decay rates increase with $\tilde{D}$. This is even true when α in Table III no longer varies with $\tilde{D}$ ($\tilde{D} \geqslant 2$): Simulations of $A_{ads} + A_{ads} \rightarrow$ products on Euclidean lattices (Zumofen *et al.*, 1985) give values for $\Phi(t)$, $t = 100$ steps, that amount to a reaction almost twice as fast in three dimensions than in two dimensions.

That reactions are fast in high dimensions is not in conflict with Eqs. (21) and (24) if we note the different comparison frames: Eq. (24) compares systems having the same diameter L; Table III compares systems

TABLE III

Decay Laws for $\Phi(t)$, Eq. (26), under Various Conditions[a]

Situation	$\Phi(t)$ for small t[b]	$\Phi(t)$ for medium and large t[b]	Remarks
A = immobile, B = mobile, $[A]_0 \ll [B]_0$	$\exp(-C[B]_0 t^\alpha)$	as for small t	Target problem (Adam–Delbrück). pseudounimolecular
A = mobile, B = immobile, $[A]_0 \ll [B]_0$	$\exp(-C[B]_0 t^\alpha)$	$\exp(-C[B]_0^{2/(\tilde{D}+2)} t^{\tilde{D}/(\tilde{D}+2)})$	Trapping problem, pseudounimolecular[c]
A, B = mobile, $[A]_0 = [B]_0$	$\dfrac{C}{[B]_0} t^{-\alpha}$	$\dfrac{C}{\sqrt{[B]_0}} t^{-\tilde{D}/4}$	Bimolecular[c]
A = B	$\dfrac{C}{[B]_0} t^{-\alpha}$	as for small t	Bimolecular

[a] The results (see, e.g., Zumofen *et al.*, 1985; Anacker and Kopelman, 1984) are confirmed by simulations.

[b] $\alpha \equiv \min\{1, \tilde{D}/2\}$. C stands generically for a constant which may be different in different entries.

[c] The slow-down of the decay for large t results from the rare but decisive event when an A molecule moves in a large region free of B molecules (fluctuations determine the large-t behavior).

that have the same number of adsorption sites N_{sites}. The two frames are related by

$$[B]_0 = N_B/N_{\text{sites}} \propto N_B/L^D \qquad (27)$$

[cf. Eq. (2)], where N_B is the initial number of B molecules. So, if in Table III we express $[B]_0$ in terms of L, the same conclusions as from Eq. (24) are obtained, and vice versa. These conclusions are amplified in Table IV.

TABLE IV

Optimization with Respect to D and $\tilde{D}$ of Reaction Time for $A_{\text{ads}} + B_{\text{ads}} \rightarrow$ Products and Fixed N_B

	Fixed L	Fixed N_{sites}
If optimization is subject to		
Then reaction time is minimal if[a]	$\tilde{D} = D \leq 2$ ($\tilde{D} = D = 1$)	$\tilde{D} \geq 2$ ($\tilde{D} = D = 3$)
Reaction time T_d in d-dimensional Euclidean space	$T_1 \lesssim T_2 \ll T_3$	$T_1 \gg T_2 \gtrsim T_3$
Reference frame is pertinent to	Biological systems	Preparative chemistry

[a] Parenthetical values result when prefactors are included.

V. DIFFUSION OF REAGENTS IN PORE SPACE

The diffusion of reactants into pores and to the surface controls the rate in the Eley–Rideal reaction model, the discharge of ions at an electrode, paramagnetic relaxation, etc. For pore fractals, it is equivalent to diffusion on the fractal (Sections IV.B and IV.C). Therefore, only surface and mass fractals are at issue here (three-dimensional diffusion space with fractal boundary).

A. Rate of Arrival at the Surface

Consider diffusion of a reactant A from an initially homogeneous solution or gas phase to the solid; and suppose that A reacts to a product P on first contact with a wall of the solid. Then, for short times (de Gennes, 1982; Pfeifer *et al.*, 1984; Nyikos and Pajkossy, 1986) and for long times, the concentration $[P]_t$ increases as

$$[P]_t \propto t^{(3-D)/2} \qquad \text{for} \quad t < t_c \qquad (28a)$$

$$[P]_t \propto t^{1/2} \qquad \text{for} \quad t \gg t_c \qquad (28b)$$

respectively, where $t_c \sim r_{max}^2/\text{diffusivity}$ of A in the fluid. In Eq. (28a), successively larger pores up to the outer cutoff r_{max} are depleted of A. In Eq. (28b), regions outside pore space are depleted. On those length scales, the surface is two-dimensional and Eq. (28b) is obtained by letting $D = 2$ in Eq. (28a). An analysis of the prefactor in Eq. (28a) shows that the reaction is accelerated by high D if the number of pore sites is kept fixed, and by low D if the number of surface sites is kept fixed.

B. Positions of Arrival at the Surface

Equations (28) do not imply that all parts of the surface are reached equally by the diffusing A molecules. For high D, much of the surface is screened and thus is dynamically inaccessible to molecules starting from the outside of a grain. They are captured at exposed sites only. This subset of the surface has a fractal dimension $D_{exp} \leq D$. Coniglio and Stanley (1984) have advanced a theory leading to

$$D_{exp} = (D + 1)/2 \qquad (29)$$

special cases of which have been confirmed by simulations (Meakin *et al.*, 1985). However, for a smooth surface, where all the surface is exposed and hence $D_{exp} = D = 2$, Eq. (29) predicts $D_{exp} = 1.5$. It is also at variance with Grassberger (1984) who suggests that $D_{exp} = 2$ for nonramified surfaces. Experimental studies (Table II) yield $D_{exp} = 2.0$ to within experimental accuracy, for D values varying between 2.7 and 3.0.

The concept of D_{exp} also bears on Section IV.C. Recent simulations on a flat surface, allowing for attractive interaction between the B molecules, have shown that the reaction proceeds at the boundary, presumable fractal, of islands of B molecules (Silverberg *et al.*, 1985). There, D_{exp} refers to the exposed sites of a B island, and the short-time kinetics should obey the planar analog of Eq. (28a), $[P]_t \propto t^{(2-D)/2}$, where D is the fractal dimension of the island boundary.

VI. CONCLUSIONS AND STRATEGIES FOR SUPPORTED REACTIONS

An overview of fractal surface characterization with focus on the most recent conceptual, experimental, and applicational developments has been given. Trusting that the reader will look up the references to extensive experimental work, the emphasis has been on principles rather than on a review of case studies. A thorough analysis has been given of the dependence of diffusion-controlled reaction rates on dimensionality. The often-invoked Adam–Delbrück result, that reactions are faster in low dimensions, has been shown to be rather misleading in the context of preparative chemistry. Instead, in diffusion spaces with fixed number of diffusion sites, it is high dimensionality throughout (spectral dimension $\bar{D}$ for the Langmuir–Hinshelwood model, fractal dimension D for the Eley–Rideal model) that favors fast reactions. The following conclusions for optimization of supported reactions emerge:

(a) If the solid offers active sites for which there is no equivalent in the unsupported comparison reaction (most of heterogeneous catalysis belongs here), then a high-D surface is optimal (i) because it carries the largest number of active sites per number of atoms in the solid [Eq. (10b)], and (ii) because a high D allows a high $\bar{D}$.

(b) If the surface of the solid allows higher concentrations $[A]_0$ and $[B]_0$ than are possible in solution or gas phase [e.g., due to surface adsorbate interactions comparable to thermal energies, which includes the possibility of a phase transition in the adsorbed layer (Pfeifer and Avnir, 1983; Pfeifer, 1985)], then similarly to (a) a surface with high D and $\bar{D}$ is optimal, and superior to reaction without support.

(c) If surface concentrations are small (e.g., because of slow formation of intermediates) so that reactant molecules must diffuse across length scales greater than r_{max} in order to meet, it is crucial to have a large diffusion coefficient $\triangle_{macro}$ [Eq. (25)], i.e., to have $D/\bar{D}$ close to 1.

(d) If transport to the surface is the rate-limiting step, the initial rate is controlled by D [Eq. (28)], and the steady-state rate by D_{exp} [Eq. (13)].

REFERENCES

Adam, G., and Delbrück, M. (1968). *In* "Structural Chemistry and Molecular Biology" (A. Rich and N. Davidson, eds.) pp. 198–215. Freeman, San Francisco, California.

Adamson, A. W. (1982). "Physical Chemistry of Surfaces," 4th Ed. Wiley, New York.

Anacker, L. W., and Kopelman, R. (1984). *J. Chem. Phys.* **81**, 6402.

Anacker, L. W., Klymko, P. W., and Kopelman, R. (1984). *J. Lumin.* **31/32**, 648.

Aris, R. (1975). "Mathematical Theory of Diffusion and Reaction in Permeable Catalysts," Vols. 1 and 2. Oxford Univ. Press (Clarendon), London and New York.

Avnir, D., and Pfeifer, P. (1983). *Nouv. J. Chim.* **7**, 71.

Avnir, D., Farin, D., and Pfeifer, P. (1983). *J. Chem. Phys.* **79**, 3566.

Avnir, D., Farin, D., and Pfeifer, P. (1984). *Nature (London)* **308**, 261.

Avnir, D., Farin, D., and Pfeifer, P. (1985). *J. Colloid Interface Sci.* **103**, 112.

Bale, H. D., and Schmidt, P. W. (1984). *Phys. Rev. Lett.* **53**, 596.

Brady, R. M., and Ball, R. C. (1984). *Nature (London)* **309**, 225.

Cederbaum, L. S., Haller, E., and Pfeifer, P. (1985). *Phys. Rev. A* **31**, 1869.

Cole, M. W., Holter, N. S., and Pfeifer, P. (1986). *Phys. Rev. B* **33**, 8806.

Coniglio, A., and Stanley, H. E. (1984). *Phys. Rev. Lett.* **52**, 1068.

de Gennes, P. G. (1982). *C. R. Hebd. Seances Acad. Sci., Ser. 2* **295**, 685.

de Gennes, P. G. (1985). *In* "Physics of Disordered Materials" (D. Adler, H. Fritzsche, and S. R. Ovshinsky, eds.), pp. 227–241. Plenum, New York.

Delannay, F., ed. (1984). "Characterization of Heterogeneous Catalysts." Dekker, Basel.

Dozier, W. D., Drake, J. M., and Klafter, J. (1986). *Phys. Rev. Lett.* **56**, 197.

Elam, W. T., Wolf, S. A., Sprague, J., Gubser, D. U., Van Vechten, D., Barz, G. L., and Meakin, P. (1985). *Phys. Rev. Lett.* **54**, 701.

Even, V., Rademann, K., Jortner, J., Manor, N., and Reisfeld, R. (1984). *Phys. Rev. Lett.* **52**, 2164.

Family, F. (1984). *J. Stat. Phys.* **36**, 881.

Family, F., and Landau, D. P., eds. (1984). "Kinetics of Aggregation and Gelation." Elsevier, Amsterdam.

Farin, D., Volpert, A., and Avnir, D. (1985a). *J. Am. Chem. Soc.* **107**, 3368.

Farin, D., Peleg, S., Yavin, D., and Avnir, D. (1985b). *Langmuir* **1**, 399.

Forrest, S. R., and Witten, T. A. (1979). *J. Phys. A* **12**, L109.

Grassberger, P. (1984). Unpublished work.

Gregg, S. J., and Sing, K. S. W. (1982). "Adsorption, Surface Area and Porosity." Academic Press, New York.

Hardt, S. L. (1979). *Biophys. Chem.* **10**, 239.

Kapitulnik, A., and Deutsher, G. (1984). *J. Stat. Phys.* **36**, 815.

Kaplan, T., and Gray, L. J. (1985). *Phys. Rev. B.* **32**, 7360.

Katz, A. J., and Thompson, A. H. (1985). *Phys. Rev. Lett.* **54**, 1325.

Kaye, B. H. (1981). "Direct Characterization of Fineparticles," Ch. 10.3. Wiley, New York.

Kjems, J. K., and Schofield, P. (1985). *In* "Scaling Phenomena in Disordered Systems" R. Pynn and A. Skjeltorp, eds.), NATO Adv. Study Inst. B, Vol. 133, pp. 141–149. Plenum, New York.

Klafter, J., and Blumen, A. (1984). *J. Chem. Phys.* **80**, 875.

Kopelman, R., Parus, S., and Prasad, J. (1986). *Phys. Rev. Lett.* **56**, 1742.

Krohn, C. E., and Thompson, A. H. (1986). *Phys. Rev. B* **33**, 6366.

Lecloux, A. J. (1981). *In* "Catalysis Science and Technology" (J. R. Anderson and M. Boudart, eds.), Vol. 2, pp. 171–230. Springer-Verlag, Berlin and New York.

Liu, S. H. (1985). *Phy. Rev. Lett.* **55**, 529.

Mandelbrot, B. B. (1982). "The Fractal Geometry of Nature." Freeman, San Francisco, California.

Mandelbrot, B. B., Passoja, D. E., and Paullay, A. J. (1984). *Nature (London)* **308**, 721.

Meakin, P., Stanley, H. E., Coniglio, A., and Witten, T. A. (1985). *Phys. Rev. A* **32**, 2364.

Meyer, A. Y., Farin, D., and Avnir, D. (1986). *J. Am. Chem. Soc.* **108**, 7897.

Mikhail, R. S., and Robens, E. (1983). "Microstructure and Thermal Analysis of Solid Surfaces." Wiley, New York.

Montroll, E. W., and West, B. J. (1979). *Stud. Stat. Mech.* **7**, 61.

Nyikos, L., and Pajkossy, T. (1986). *Electrochim. Acta* **31**, 1347.

Parfitt, G. D., and Sing, K. S. W., eds. (1976). "Characterization of Powder Surfaces." Academic Press, New York.

Pfeifer, P. (1984). *Appl. Surf. Sci.* **18**, 146.

Pfeifer, P. (1985). *Chimia* **39**, 120.

Pfeifer, P. (1986). *In* "Fractals in Physics" (L. Pietronero and E. Tosatti, eds.), pp. 47–54. Elsevier, Amsterdam.

Pfeifer, P. (1987). *In* "Multiple Scattering of Waves in Random Media and Random Rough Surfaces" (V. K. Varadan and V. V. Varadan, eds.), pp. 45–49. The Penn. State Univ., University Park, Pennsylvania.

Pfeifer, P., and Avnir, D. (1983). *J. Chem. Phys.* **79**, 3558; **80**, 4573.

Pfeifer, P., and Salli, A. (1987). Submitted.

Pfeifer, P., Avnir, D., and Farin, D. (1984). *J. Stat. Phys.* **36**, 699; **39**, 263.

Pfeifer, P., Welz, U., and Wippermann, H. (1985). *Chem. Phys. Lett.* **113**, 535.

Pfeifer, P., Stella, A. L., Toigo, F., and Cole, M. W. (1987). *Europhys. Lett.* **3**, 717.

Pietronero, L., and Tosatti, E., eds. (1986). "Fractals in Physics." Elsevier, Amsterdam.

Rammal, R. (1984). *J. Stat. Phys.* **36**, 547.

Rammal, R., and Toulouse, G. (1983). *J. Phys. Lett. (Orsay, Fr.)* **44**, L13.

Rojanski, D., Huppert, D., Bale, H. D., Dacai, X., Schmidt, P. W., Farin, D., Seri-Levy, A., and Avnir, D. (1987). In "Proc. 2nd Int. Conf. Unconventional Photoactive Materials" (H. Scher, ed.), in press. Plenum, New York.

Schaefer, D. W., and Keefer, K. D. (1984). *Phys. Rev. Lett.* **53**, 1383.

Schaefer, D. W., Martin, J. E., Wiltzius, O., and Cannell, D. S. (1984). *Phys. Rev. Lett.* **52**, 2371.

Shlesinger, M. F., Mandelbrot, B. B., and Rubin, R. J., eds. (1984). *J. Stat. Phys.* **36**, 519–921.

Silverberg. M., Ben-Shaul, A., and Rebentrost, F. (1985). *J. Chem. Phys.* **83**, 6501.

Silverberg, M., Farin, D., Ben-Shaul, A., and Avnir, D. (1986). *Ann. Isr. Phys. Soc.* **8**, 451.

Sinha, S. K., Freltoft, T., and Kjems, J. (1984). *In* "Kinetics of Aggregation and Gelation" (F. Family and D. P. Landau, eds.), pp. 87–90. Elsevier, Amsterdam.

Stanley, H. E. (1984). *J. Stat. Phys.* **36**, 843.

Stanley, H. E., and Ostrowsky, N., eds. (1986). "On Growth and Form." Nijhoff, Boston, Massachusetts.

Thomas, J. M., and Thomas, W. J. (1967). "Introduction to the Principles of Heterogeneous Catalysis." Academic Press, New York.

Thomas, T. R. (1982). "Rough Surfaces." Longman, London.

Tsallis, C. (1986). *In* "Fractals in Physics" (L. Pietronero and E. Tosatti, eds.), pp. 65–69. Elsevier, Amsterdam.

Tsurumi, S., and Takayasu, H. (1986). *Phys. Lett. A* **113**, 449.

Van Damme, H., and Fripiat, J. J. (1985). *J. Chem. Phys.* **82**, 2785.

Weitz, D. A., and Huang, J. S. (1984). *In* "Kinetics of Aggregation and Gelation" (F. Family and D. P. Landau, eds.), pp. 19–28. Elsevier, Amsterdam.

Witten, T. A., and Sander, L. M. (1981). *Phys. Rev. Lett.* **47**, 1400.

Wong. P. Z. (1985). *Phys. Rev. B* **32**, 7417.

Yang, C. L., Evesque, P., and El-Sayed, M. A. (1985). *J. Phys. Chem.* **89**, 3442.

Zumofen, G., Blumen, A., and Klafter, J. (1985). *J. Chem. Phys.* **82**, 3198.

3

METAL OXIDES AND THEIR PHYSICO-CHEMICAL PROPERTIES IN CATALYSIS AND SYNTHESIS*

Kenneth J. Klabunde, M. Fazlul Hoq, Faleh Mousa, and Hiromi Matsuhashi

Department of Chemistry
Kansas State University
Manhattan, Kansas 66506

The purpose of this chapter is to critically, but not exhaustively, review the uses of common metal oxides as reagents for carrying out a variety of catalytic and stoichiometric transformations of small molecules and organic compounds. In order to understand why metal oxides such as Al_2O_3 or MgO can possess or be given highly reactive sites, structural features and defects will be discussed. Activation of metal oxides by irradiative and thermal means is a very important aspect of this work, thus allowing metal oxides to be active in acid–base chemistry, telomerization and polymerization, electron transfer, C—C bond breaking, H–D exchange, and other important processes.

I. STRUCTURAL ASPECTS OF METAL OXIDES

A. Crystal Structures of Common Oxides

In order to represent the structural differences in metal oxides, several common oxides will be considered, e.g., Na_2O, MgO, SiO_2, and Al_2O_3.

*The support of the Army Research Office under Grant No. DAAG29-84-K-0051 is gratefully acknowledged.

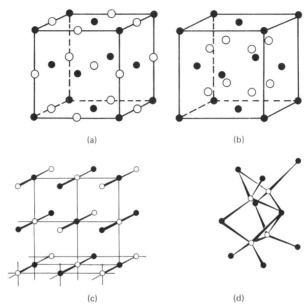

(a)   (b)

(c)   (d)

Fig. 1. Unit cell of (a) NaCl and (b) Na$_2$O. Coordination numbers of ions (c) NaCl and (d) Na$_2$O. [After West 1984. Reprinted by permission of John Wiley and Sons, Ltd.]

Antifluorite (Na$_2$O) has a cubic close packed (face centered cubic) arrangement of anions, the same as rock salt (NaCl). The difference between NaCl and Na$_2$O is in the placement of the cations. These structures are compared in Fig. 1. Note the octahedral (six-coordinated) coordination of all ions in NaCl while in Na$_2$O the cations are tetrahedrally coordinated, and the anions are eight-coordinated. A general rule is that for any structure of formula A$_x$X$_y$, the coordination numbers for A and X must be in the ratio of $y : x$ (West, 1984). Thus, in Na$_2$O the formula A$_2$X calls for coordination numbers of cation and anion in the ratio of $1 : 2$ (or $4 : 8$ in this particular case).

Many metal oxides of stoichiometry AX possess the rock salt structure with all ions octahedrally coordinated. The list includes MgO, CaO, SrO, BaO, TiO, MnO, FeO, CoO, NiO, and CdO (West, 1984). Interestingly, BeO possesses a Wurtzite (ZnS) structure, with the common hexagonal close packed arrangement of anions while the cations differ in position (Fig. 2). For BeO or ZnS the coordinations are made of networks of tetrahedra. As might be expected the difference in structures for BeO and MgO is due to the small size of the beryllium cation, which is too small to fit nicely within its anion array. Therefore, BeO adopts a different structure to partially alleviate this problem. This structural difference

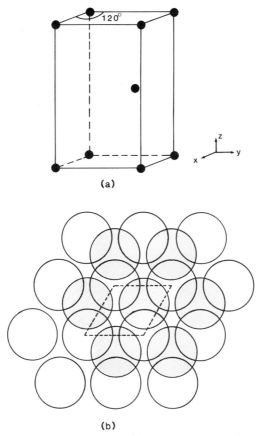

(a)

(b)

Fig. 2. (a) Hexagonal unit cell of BeO and (b) hexagonal close packed anion array. [After West (1984). Reprinted by permission of John Wiley and Sons, Ltd.]

between BeO and MgO can have a significant effect on chemical reactivity (Morris and Klabunde, 1983).

Let us deal now with the very important and complex oxide SiO_2. Since the Si=O bond is not energetically favorable, and since silicon is of intermediate electronegativity, SiO_2 should not be considered highly ionic, and bonding is a mixture of ionic and covalent. Thus, SiO_2 and other glass-forming oxides such as B_2O_3, GeO_2, and P_2O_5 can be regarded as three-dimensional polymeric structures. This often results in rather complex structures where SiO_4 tetrahedra are arranged in various ways. West (1984) has stated the attendant rules as follows: (1) almost all silicate structures are built up of tetrahedra; (2) the SiO_4 tetrahedra may bind together by sharing corners; (3) only two SiO_4 tetrahedra may share a common oxygen; and (4) SiO_4 tetrahedra never share edges or faces with each other. In these structures the coordination of silicon is four while that for oxygen is two.

Thus, SiO_2 offers.a complex polymorphism dependent on temperature and pressure. α-Quartz converts to β-quartz at 573°C, and at higher temperatures yet other structures have been observed.

A brief consideration of Al_2O_3 structure is also in order. Both α- and β-Al_2O_3 are common. β-Al_2O_3 is particularly interesting because the oxide layers are close packed stacked in three dimensions, but every fifth layer has three quarters of its oxygens missing, and so tunnels exist through which ions such as Li^+ or Na^+ can easily move. Thus, doping with Li_2O or Na_2O leads to formulations $(M_2O)_n \cdot Al_2O_3$ which are ionic conductors.

B. Defect Sites

1. Types of Defects

Ionic, high-melting oxides can be prepared in high-surface-area forms. The crystallites can show a wide variety of defect sites, such as coordination deficient anions or cations, missing anions or cations, or various combinations of these (Rees, 1954; Henderson and Wertz, 1968, 1977; Taylor, 1968; Sondor and Silby, 1972; Morris, 1981). In addition, surface "impurities," especially adsorbed H_2O, OH groups, H^+ and other species can have effects (Morris, 1981). Listed below are some of the important defects known.

(1) Steps and kinks on surfaces. This refers to crystal dislocations at various positions (Lempfuhl and Uchida, 1979). The consequence of these dislocations is the presence of coordinately unsaturated oxide anions or metal cations, leading to higher Lewis base or acid character (Tench and Nelson, 1967; Malinski *et al.*, 1974; Garrone and Stone, 1984). Such sites are usually covered with adsorbed H_2O, CO_2, or other species which usually need to be cleaned off to enhance the defect site reactivity.

(2) Another class of structural defects are called Schottky and Frenkel defects after the scientists that described them (Schottky, 1935a,b). A Frenkel defect is an ion vacancy with the ion located in an interstitial site (Hughes and Henderson, 1972; Coluccia *et al.*, 1978). The Schottky defect is a pair of vacancies (anion and cation) and the missing ions are located in another part of the crystallite (Fig. 3). The consequence of such defects is to generate highly basic or highly acidic sites. Thus, a missing Mg^{2+} in solid MgO would generate a very electron-rich basic site due to the several O^{2-} species in close proximity. The location of the ion vacancy on the surface of

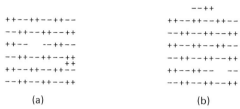

```
                                      -- ++
  + + - - + + - - + + - -      + + - - + + - - + + - -
  - - + + - - + + - - + +      - - + + - - + + - - + +
  + + - -     - - + + - -      + + - - + + - - + + - -
  - - + + - - + + - - + +      - - + + - - + + - - + +
                       + +
  + + - - + + - - + + - -      + + - - + + - -     - -
  - - + + - - + + - - + +      - - + + - - + + - - + +

        (a)                          (b)
```

Fig. 3. Schematic of (a) Frenkel and (b) Schottky defects in alkaline earth oxides. [After Morris and Klabunde (1983). Copyright 1983 American Chemical Society.]

a crystallite could lead to "mildly defective" (e.g., moderately basic) or "highly defective" (e.g., strongly basic) sites (Fig. 4) (Klabunde *et al.*, 1978). The concentrations of Frenkel and Schottky defects depends on (a) thermodynamics where free energy is increased by the occurrence of point defects, but entropy increases, (b) impurities, and (c) methods of preparation (Boudart *et al.*, 1972; Spitsyn *et al.*, 1977; Shaporev and Bulat, 1980).

(3) Paramagnetic defects are readily formed from Schottky and Frenkel (diamagnetic) defects. These are generally referred to as F or V centers, and the nomenclature for different types of F and V centers has been summarized (Taylor, 1968; Morris 1981) and will not be discussed here. For our purposes simply note that anion vacancies (high Lewis acidity) can capture electrons to form e^- centers whereas cation vacancies (high Lewis basicity) can give up electrons to form + centers (Fig. 5) (Nelson and Tench, 1964; Lunsford and Jayne, 1965; Nelson *et al.*, 1967; Chen and Sibley, 1969; Kappers *et al.*, 1970). Electron-excess centers (F-centers) are capable of undergoing oxidation and reduction reactions with adsorbed reagents. There are different F centers possible depending on the type and geometry of the surrounding groups (Wertz *et al.*, 1959, 1960). V centers, or electron difficient centers (positive holes), are generally cation vacancies

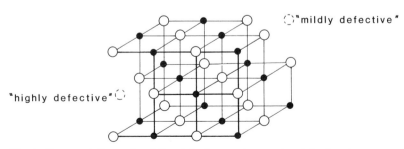

Fig. 4. Representation of mildly defective and strongly defective ion vacancies (Frenkel/Schottky defects). [After Klabunde, *et al.* (1978). Copyright 1978 American Chemical Society.]

(a)   (b)

Fig. 5. Schematic of anion and cation vacancies and (a) their reduced and (b) their oxidized forms. [After Morris (1981).]

with a hole located on an adjacent oxygen moiety (Culvahouse *et al.*, 1965; Abraham *et al.*, 1979). This hole can migrate from one oxygen to another, and V centers are easily eliminated thermally or photochemically (electron–hole recombination). Again, a variety of V centers are possible due to different positioning of neighboring cation and anion vacancies or by adsorption of H_2O or other impurities (Henderson and Tomlinson, 1969; Abraham *et al.*, 1975).

To summarize, defect sites are structural such as cleavage planes, kinks and steps, and cation or anion vacancies (Frenkel and Schottky). They are also electronic in nature such as electron excess or electron deficiency centers. And a myriad of variations on these defect types are possible depending on the environment of the defect.

2. How Defects Can Be Produced

How can certain defects be encouraged or discouraged?

The method of preparation of metal oxides has a dramatic effect on the crystallinity and surface area of the final material (Tanabe, 1970; Morris and Klabunde, 1983). Although such preparations are fairly straightforward chemically speaking (i.e., hydroxide or carbonate decomposition), the actual formation and precipitation of the oxide is a complex process, and there is little chemical understanding of this step. A great deal of mechanistic chemical work is warranted.

One series of experiments will suffice as an example. Morris and Klabunde (1983) prepared MgO by (a) Mg combustion leading to well-formed cubic and hexagonal plates with little surface reactivity, (b) thermal decomposition of $Mg(NO_3)_2$ giving beautiful crystalline hexagonal and octagonal plates of low reactivity and low surface area, (c) thermal decomposition of $MgCO_3$ yielding a high-surface-area porous powder of high reactivity, and (d) thermal decomposition of $Mg(OH)_2$ yielding a very reactive, high-surface porous powder (almost amorphous). Figure 6 shows electron micrographs of two of these samples.

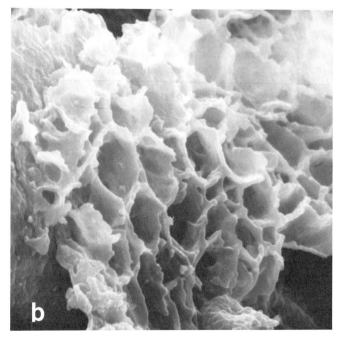

Fig. 6. Electron micrographs of magnesium oxide powders prepared by (a) decomposition of $Mg(NO_3)_2$ and (b) decomposition of $Mg(OH)_2$. (After Morris, 1981).

Cation and anion vacancies can be encouraged by doping. For example, if MgO is doped with Al_2O_3, more cation vacancies would be expected since stoichiometrically more O^{2-} species are being added than M^{2+}/M^{3+} species (West, 1984) If a rock salt structure is still adopted by the $MgO-Al_2O_3$ sample, some extra cation vacancies would exist. Another approach would be to dope with Li_2O which would encourage anion vacancies by similar reasoning. Likewise, electron-excess centers can be increased by doping with metals such as Li vapor or Na vapor. By similar reasoning, electron-deficient centers may be encouraged by treatment of the oxide with highly electronegative radicals, perhaps oxygen atoms or fluorine atoms. Note that all of these doping procedures do not change the overall electronic neutrality of the starting oxide.

Discouraging the formation of defects can generally be done by maintaining high purities, choosing the correct oxide preparation method, and annealing at high temperatures followed by slow cooling. Interestingly, annealing followed by rapid cooling occasionally allows the formation of new defects, which may be short-lived (Spitsyn *et al.*, 1977; Shaporev and Bulat, 1980).

3. Modified Metal Oxides

One example will be given here, that of MgO doped with Li_2CO_3 or Li_2O. Driscoll *et al.* (1985) and Ito *et al.* (1985) believe that the active site for H abstraction from CH_4 (in a CH_4/O_2 environment) is a cation vacancy and electron deficient site. By addition of Li_2O higher activities were achieved, likely due to the natural tendency to form similar Li^+O^- sites:

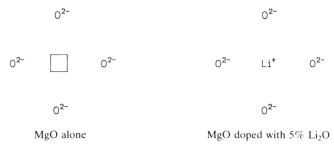

MgO alone   MgO doped with 5% Li_2O

Good selectivity for conversion of CH_4 to C_2H_4 and C_2H_6 were achieved over Li_2CO_3/MgO systems.

C. Activation Processes

1. Irradiation

Two very important processes can be used to activate metal oxides. The first is irradiation with either high-energy particles (e.g., neutrons) or photons (e.g., x-rays) (Wertz *et al.*, 1960; Walters and Estle, 1961; Lunsford and Jayne, 1966a; Nelson *et al.*, 1967; Tench and Nelson, 1968). The energies are high enough that electrons can be dislodged from any O^{2-} species which can then be trapped at anion vacancy sites. In this way both electron-excess and electron-deficient centers are generated. Upon such treatment the oxide changes color. Thus, MgO upon treatment with neutrons or x-rays turns from white to red–blue. Interestingly even UV photons can cause color changes in MgO.

The energy necessary to ionize an O^{2-} species varies greatly with its coordinative unsaturation (Duley, 1984). Thus, only coordinatively deficient O^{2-} could be ionized by UV irradiation, suggesting that defect sites are involved. So we conclude that structurally defective O^{2-} sites are ionized leaving electron-deficient sites and that the electron is then trapped at other defects such as anion vacancies:

$$
\begin{array}{ccccc}
O^{2-} & Mg^{2+} & O^{2-} & & \\
O^{2-} & Mg^{2+} & O^{2-} & Mg^{2+} & \\
Mg^{2+} & & Mg^{2+} & O^{2-} & \xrightarrow{h\nu} \\
O^{2-} & Mg^{2+} & O^{2-} & & \\
 & & O^{2-} & & \\
\end{array}
\qquad
\begin{array}{cccc}
O^{2-} & Mg^{2+} & O^- & \\
O^{2-} & Mg^{2+} & O^{2-} & Mg^{2+} \\
Mg^{2+} & - & Mg^{2+} & O^{2-} \\
O^{2-} & Mg^{2+} & O^{2-} & \\
 & O^{2-} & & \\
\end{array}
$$

2. Thermal Means

The second activation method for metal oxides involves thermal treatment sometimes at temperatures as high as 1200°C. Usually vacuum/thermal activation is best although rapid gas flow over the sample is sometimes effective. Rapid removal of adsorbed H_2O, CO_2, and OH groups is important, especially H_2O since even at low pressures superhot steam can cause dramatic structural changes of the surfaces of metal oxide microcrystals (Dell and Weller, 1959; Anderson and Morgan, 1964).

The removal of adsorbed species by vacuum/thermal activation is a cleaning process. In doing so structural defects (kinks, steps, and cation and anion vacancies) become available for other chemical interactions. In addition, chemical processes such as H_2 elimination from surface OH groups can lead to additional reactive sites such as electron deficient centers: Simultaneous oxidation of transition-metal impurities may occur (Wertz *et al.*, 1959; Sibley and Chen, 1967; Sibley *et al.*, 1969; Chen *et al.*, 1976).

$$
\begin{array}{cccc}
{}^-OH & M^{2+} & {}^-OH & \\
O^{2-} & M^{2+} & O^{2-} & M^{2+}
\end{array}
\quad
\begin{array}{c}
\xrightarrow[{-H_2}]{heat}
\end{array}
\quad
\begin{array}{cccc}
O^- & M^{2+} & O^- & \\
O^{2-} & M^{2+} & O^{2-} & M^{2+}
\end{array}
$$

The temperature of activation is an important parameter. The activity of the resultant oxide for certain catalytic processes and for generation of maximum solid acidity/basicity can be governed by the activation temperature (Tanabe, 1970; Tanabe *et al.*, 1975; Onishi and Tanabe, 1974; Hattori *et al.*, 1976a,b; Iizuka *et al.*, 1976). The optimum activation temperature depends on the desorption of adsorbed H_2O, CO_2, etc. from those sites that are active for the desired chemical process. For example, activation of CaO at 700–900°C generates maximum Lewis basicity and optimum activity for 2-butene isomerization. Activation at lower temperature does not allow the cleaning of these active sites, and activation at higher temperatures frees other active sites that cause undesirable chemical process with 2-butene. The optimum activation temperature also depends on avoidance of annealing of active sites, which can often occur at too high an activation temperature.

The wide variety of methods for generating active sites on metal oxides is encouraging for the chemist. Now we will consider adsorption processes on activated oxides.

II. CHEMISORPTION PROCESSES ON METAL OXIDES

A. Acid–Base

Metal oxides can behave as Lewis or Brønsted acids or bases. Basicity of oxide anions can vary depending on the coordinative nature of the oxide species in question. Likewise, Lewis acid sites are available at metal cations, and protons are available from surface OH groups. The strength of acid–base sites varies from one oxide to another, and the number of acid–base sites can vary greatly with preparative method.

Tanabe (1970) has described titration procedures for determining the number of sites and the approximate acid or base strength of these sites. These titration techniques employ color indicators of known pH or pK_a and nonaqueous solvents where the oxide is titrated in slurry form.

Solid acids and solid bases can be prepared with super strength. This sometimes dramatically high basicity is partly due to structural defects that are particularly basic because of the high ionic character of the $M^{2+}O^{2-}$ moiety. (A great deal of charge resides on O^{2-}, and if M^{2+} cation vacancies exist very high basicity results.) Even greater basicity, or "super basicity" can be generated by doping with metals such as Na^0 (Ushikubo *et al.*, 1984; West, 1984). In this way free electrons are added into a system that already possesses very electronically rich sites.

Extreme acidity is also possible, for example when Fe_2O_3 is doped with $Fe_2(SO_4)_3$ so that SO_4 groups perturb already acidic sites (perhaps anion vacancies). Acidity functions approaching that of concentrated sulfuric acid are possible. Likewise, ZrO_2 and TiO_2 treated with H_2SO_4 can be made very acidic and indeed can be considered as solid super acids (Hino et al., 1979; Hino and Arata, 1979a,b, 1980, 1985; Tanabe et al., 1981; Jin et al., 1984).

There seems to be little doubt that sites of extreme acidity or basicity are due to unusual coordinative unsaturation or defects. However, even an ordered metal oxide surface possesses acidity and basicity. For example, an oxidized Mg(001) surface caused dissociative chemisorption (H^+X^-) of C_2H_2, CH_3COOH, CH_3OH, C_2H_5OH, i-C_3H_7OH, and H_2O (Martinez and Barteau, 1985). High-surface-area polycrystalline MgO is also capable of such heterolytic chemisorption processes:

$$
\begin{array}{c}
\text{O—Mg—O—Mg} \\
\text{| | | |} \\
\text{Mg—O—Mg—O}
\end{array}
\xrightarrow{\text{ROH}}
\begin{array}{c}
\text{H OR H OR} \\
\text{| | | |} \\
\text{O—Mg—O—Mg} \\
\text{| | | |} \\
\text{Mg—O—Mg—O}
\end{array}
$$

Thus, we can understand why metal oxides possess many acid–base sites, the weaker ones being due to the surface cations or anions of highest coordination and the stronger ones due to defects of varying environments.

Garrone and Stone (1984) have given further proof that a wide variety of XH molecules (hydrogen, water, alcohols, ammonia, alkenes, acetylene, and aromatics) heterolytically dissociate on activated MgO. Interestingly they find that the acidities of XH over MgO correlate better with solution-phase pK_a values than with gas-phase determinations, which indicates that the surface of MgO can "solvate" and thereby stabilize a variety of anions. Evidence was also presented indicating that some of the adsorbed anions, specifically $\bar{N}H_2$, H^-, and $CH_2{=}CHC\bar{H}_2$, react with added CO. Garrone and Stone have proposed that coordinatively unsaturated surface sites are responsible for these acid–base reactions.

The Lewis acid CO_2 has been studied extensively on NiO (Eischens and Pliskin, 1957), TiO_2 (Yates, 1961), ZnO (Taylor and Amberg, 1961), BeO (Stuart and Whateley, 1965), MgO (Gregg and Ramsey, 1970), and CaO (Fukada and Tanabe, 1973). Both unidentate and bidentate complexes have been observed, and the base strength of the oxide in question has an effect:

unidentate   bidentate

B. Electron Transfer

There are two kinds of sites capable of transferring electrons from the metal oxide surface to an adsorbed molecule. The first is an electron-excess site that is paramagnetic. The second is a diamagnetic cation vacancy that relieves its high electron density by transferring one of its paired electrons to a chemisorbed molecule. Both of these sites can be referred to as "reducing sites." Let us deal with electron-excess-reducing sites first (recall that these can be formed by irradiation of the oxide or by doping the oxide with alkali metals).

Several metal oxides have been studied (Barry and Stone, 1960; Gravelle *et al.*, 1971; Meriaudeau and Vedrine, 1976; Che *et al.*, 1976). Dioxygen addition has been of particular interest (Che and Tench, 1973; Lunsford, 1973). Some typical reactions that occur at room temperature or below are shown (Lunsford and Jayne, 1965, 1966a,b; Tench and Lawson, 1970; Naccache, 1971; Meriaudeau *et al.*, 1975, 1977; Aika and Lunsford, 1977):

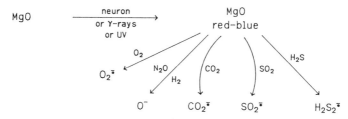

Even more interesting are diamagnetic cation vacancies as reducing sites. These are more selective; for example CO_2 and O_2 are not reduced, but more electron demanding adsorbed molecules such as nitrobenzene, *p*-benzoquinone, benzophenone, acetophenone, napthalene, phenanthrene, anthracene, and 2,2-bipyridine are reduced to their respective adsorbed anion radicals (Morris and Klabunde, 1983). A reduction potential greater than -2.50 V (with respect to the standard calomel electrode) is necessary for electron transfer to occur. In most cases near monolayer anion radical formation occurred (Klabunde, *et al.*, 1978).

C. Telomerization of Small Molecules

With certain small molecules, coupling and telomerization can be induced. For example, pyridine forms the 4,4'-bipyridine anion radical with loss of hydrogen (Iizuka and Tanabe, 1975; Che *et al.*, 1978). Carbon monoxide forms a variety of interesting cyclic species $C_2O_2^{\cdot}$, $C_4O_4^{2-}$, $C_5O_5^{2-}$, $C_6O_6^{2-}$ and $C_6O_6^{3\cdot}$ (Morris and Klabunde, 1983). It appears that cyclic dianions form first and then are slowly converted to trianion radicals, especially $C_6O_6^{2-} \rightarrow C_6O_6^{3-}$ is prevalent. These induced coupling reactions are of considerable interest, but so far attempts to devise catalytic cycles for production of CO coupled products have not been successful, although Wang *et al.*, (1982, 1983) have shown that adsorbed CO will interact with H_2 to yield adsorbed formaldehyde.

D. Carbonyl and Other Organic Molecules

Several workers have examined the adsorption of acetaldehyde on porous glass, silica (Sidorov, 1954; Gryazev and Solyanova, 1965; Young and Sheppard, 1967; Robinson and Ross, 1968), silica–alumina (Bonino and Fabbri, 1964), magnesium oxide (Bykhovskii and Krylov, 1972), and tin oxide (Thornton and Harrison, 1975). In the condensation of acetaldehyde over alumina above 200°C essentially degradation occurs, producing a range of gases, while below 200°C only condensation takes place;

$$2CH_3CHO \longrightarrow CH_3CH(OH)CH_2CHO \longrightarrow CH_3CH{=}CHCHO + H_2O$$

The rate of conversion of acetaldehyde and the yield of crotonaldehyde were functions of catalyst acidity, and aldol was not isolated as an intermediate. Infrared spectroscopic studies of the adsorbed layer indicate that the adsorption of acetaldehyde on silica proceeds through the carbonyl group of the aldehyde and surface hydroxy groups.

Robinson and Ross (1968) obtained isotherms of the adsorption of acetaldehyde on silica gel from 21 to 51°C. They also carried out a study of the effect of heat treatment and alumina impregnation on the nature of the products obtained from the gel at 190°C. The degree of conversion into crotonaldehyde on silica gel at 190°C was shown to depend on the pretreatment, up to 1000°C, and on the aluminum content of the gel. It is likely that the condensation reactions which gave aldol and then crotonaldehyde took place in pores after the adsorption of a complete monolayer.

Bykhovskii and Krylov (1972) studied the adsorption of acetaldehyde on MgO under dynamic conditions, and crotonaldehyde was formed. They showed the amount of acetaldehyde adsorbed on the catalysts remained constant in the temperature interval 98–250°C and until almost two monolayers were adsorbed. All the acetaldehyde was adsorbed irreversibly and was not removed from the catalyst when the temperature was increased.

Evidence for an adsorbed alkoxide species on silica gel was discussed by Young and Sheppard (1967) and bands in the infrared for acetaldehyde adsorbed on γ-alumina have been assigned to a carbonylate group and an unstable alkoxide form of chemisorbed acetaldehyde (Sokolskii *et al.*,

1976). Thornton and Harrison (1975) reported IR bands attributed to acetate ions at 1530–1520, 1430, and 1347 cm^{-1} for acetaldehyde adsorbed on tin oxide. Several other authors have assigned bands to products of aldol condensation reactions. The carbonyl stretching vibration of saturated aldehyde gives a strong band at approximately 1720 cm^{-1} in both IR and Raman spectra. Bands near 1682 and 1641 cm^{-1} in spectra of acetaldehyde adsorbed on silica and MgO (Low *et al.*, 1973) were attributed to adsorbed crotonaldehyde resulting from the self-condensation of acetaldehyde. Multiple condensation reactions leading to the formation of conjugated double bands have been inferred from the presence of infrared bands between 1640 and 1560 cm^{-1} in IR spectra of acetaldehyde adsorbed MgO. Klassen and Hill (1981) studied adsorbed species from the reaction of acetaldehyde on NaOH-treated silica gel using Raman spectroscopy. Catalytic activity for acetaldehyde condensation reactions increased with coverage of Na$^+$ ions. At room temperature and an acetaldehyde vapor pressure of 300 torr, the Raman spectrum of acetaldehyde adsorbed on NaOH-treated silica gel was characteristic of adsorbed acetaldehyde and of 6-hydroxy-2,4-dimethyl-1,3-dioxane (bands at 816,535 and 489 cm^{-1}), the latter compound being a trimer of acetaldehyde. At lower pressure the above dioxane decomposed and chemisorbed acetaldol was the dominant surface species. At elevated temperature Raman bands of crotonaldehyde, which were produced by the dehydration of absorbed species, appeared in spectra of the acetaldehyde reaction system. They showed that the reaction goes through four distinct stages:

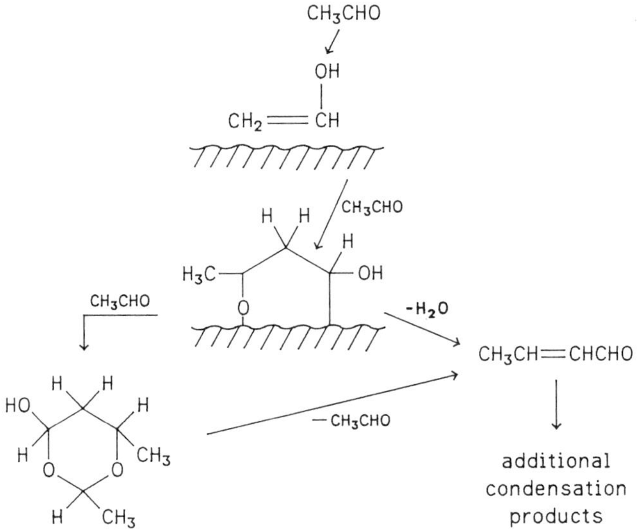

Adsorption of acetone on alumina, rutile, magnesia, beryllia, and nickel and calcium oxides has been studied. Three common modes of reaction have been distinguished. Surface acetate is the most common chemisorption product (Hair and Chapman, 1965; Kiselev and Uvarov, 1967; Deo et al., 1971; Knozinger, 1976; Miyata et al., 1974a,b; Griffiths and Rochester, 1978), although coordinated acetone is also observed (Kadushin et al., 1967; McManus et al., 1969). Both coordinated acetone and surface enolate appear to be intermediates in the formation of both mesityl oxide and surface acetate. No evidence has been found for the dissociative chemisorption of acetone on silica or germania (Low et al., 1969; Griffiths and Rochester, 1977), although hydrogen bonding to surface hydroxyl groups occurs. Other ketones studied are halogenated acetones and symmetrical ketones such as diisopropyl ketone (Tretyakov and Filimonovy, 1970; Griffiths and Rochester, 1977). Surface bound carbonyl compounds have often been proposed as intermediates in the oxidation of hydrocarbons over tin(IV) oxide-containing catalysts. Thornton and Harrison (1975) examined the chemisorption of acetone on tin oxide, where surface acetate was also observed. Harrison and Mavaders (1984) reported the chemisorption of a number of unsymmetrical ketones, RCOMe [$R = C_3H_7$, C_2H_5, $(CH_3)_2CH$, $(CH_3)_3C$, Ph] on tin oxide. In every case, the final product was the surface carboxylate, RCO_2^-. A mechanism involving initial coordination of the ketone to a surface tin site, followed by nucleophilic attack of a neighboring hydroxyl group at the carbonyl carbon appears as likely.

III. IMPORTANT CATALYTIC PROCESSES ON METAL OXIDES

A. Bond Breaking Processes

1. Protium–Deuterium Exchange

Boudart et al. (1972) found that paramagnetic defect sites on MgO catalyze the H_2–D_2 exchange at a temperature as low as $-196°C$ with an activation barrier of about 2 kcal/mole. Larson and Hall (1965) found that thermally activated Al_2O_3 samples caused the $CH_4 + D_2 \rightarrow CH_3D + HD$ reaction to take place at room temperature, and proposed that OH groups on the Al_2O_3 surface were involved in the active sites. Likewise, Flockhart et al. (1971) have studied the propene–D_2 system and have correlated activity with oxidizing sites on Al_2O_3. However, with alkanes the acidity of the hydrocarbon appears to be most important (Robertson et al., 1975), and activated MgO, CaO, SrO, BaO, and La_2O_3 are basic enough to effect CH_4–D_2 exchange (Utiyama et al., 1978).

A common observation in many of these studies is that oxide activities can vary with preparation and/or pretreatment (outgassing) temperature. Also, XH acidity plays an important role. Based on these ideas Hoq and Klabunde (1986) investigated certain MgO samples as catalysts for selective RH–D_2 exchange. Some hydrocarbons exchange at temperatures below $-78°C$; for example, the methyl group of toluene and the methylene group of ethylbenzene can be completely deuterated selectively. Simply by raising the temperature the phenyl group and then the methyl group of ethylbenzene can be deuterated selectively. At still higher temperatures alkanes, even neopentane ($pK_a = 49$) can be deuterated. Using activated MgO as a deuteration catalyst appears to be superior to conventional methods due to ease of separation of products (heterogeneous) and easily controlled high selectivities.

2. Elimination Reactions

a. Dehydrogenation. Generally, metal oxide catalysts show activity for dehydrogenation of paraffins and alkenes. A relationship exists between activity for the dehydrogenation of cyclohexane and the d electronic configuration of metals (Whittingham, 1977). Highest activity was associated with d^3, d^2, and d^1 configurations of the metal ions, for example, Cr_2O_3 ($3d^3$), V_2O_3 ($3d^2$), and Ti_2O_3 ($3d^1$), while low activity was associated with the more stable d^0, d^5, and d^{10} configurations, for example MgO, CaO, Al_2O_3 (d^0), MnO ($3d^5$), and ZnO ($3d^{10}$) (Richardson and Rossington, 1969).

b. Dehydration and Dehydrogenation of Alcohols. Acidic solid catalysts are capable of dehydrating alcohols. Primary and secondary alcohols are dehydrated by a concerted mechanism in which both acid and base sites on Al_2O_3 are involved, thus allowing a trans elimination (Pines and Manassen, 1966). Generally, dehydrogenation of alcohols into ketones or aldehydes is the most important reaction over base catalysts, while dehydration is of secondary importance; the apparent activation energies for dehydration of 2-propanol on alkaline earth oxides are always greater than those for dehydrogenation and vary from 38 kcal/mole for BaO to 54 kcal/mole for CaO and SrO (McCaffreg *et al.*, 1972).

c. Deamination and Dehydrogenation of Amines. Satoh *et al.* (1983) reported on the deamination and dehydrogenation of butylamine at 400°C on various metal oxide catalysts. *n*-Butylamine was dehydrogenated to give nitrile. In the case of 2-butylamine, where nitrile could not be formed, deamination proceeded yielding 1-butene selectively on ZnO, La_2O_3, ThO_2, and MgO.

d. Dehydrohalogenation. 2,3-Dibromobutane, 2,3-dichlorobutane, and 2-bromobutane were dehydrohalogenated over MgO, CaO, SrO, and Al_2O_3 at 100–250°C and yielded 2-bromo-2-butene, 2-chloro-2-butene, and butene, respectively (Misono and Yoneda, 1974). The elimination reaction of hydrogen halide proceeded selectively in a trans-concerted fashion over fresh oxide surfaces, and mechanistically resembles an E2 reaction. The trans elimination selectivity was in the order $SrO \geqq CaO > MgO > Al_2O_3$ which is also the order of basicity of these materials. After the catalyst surfaces became contaminated with hydrogen halide the trans/cis ratio decreased.

C. C—C Bond Breaking

Perhaps the most important application of metal oxides, especially Al_2O_3, is for catalytic reforming of hydrocarbons (Germain, 1969):

(a) Styrenes are obtained from dehydrogenation of short aromatic side-chains with $ZnO–Al_2O_3–CaO$ and $Cr_2O_3–Al_2O_3$ (for example, ethylbenzene to styrene).

(b) Alkanes with carbon chains of at least six carbons have been converted to aromatic hydrocarbons with $Cr_2O_3–Al_2O_3$, $MoO_3–AlO_3$, and $V_2O_5–Al_2O_3$:

$$n\text{-}C_7H_{16} \xrightarrow[\text{cat.}]{450–550°C} C_6H_5\text{—}CH_3 + H_2$$

Platinized alumina ($Pt–Al_2O_3$) allows such conversions to occur at even lower temperatures, i.e., 300–400°C.

Alumina treated with acid forming or acidic materials, such as CCl_4, $AlCl_3$, or SiO_2 yields a modified Al_2O_3 with great catalytic activity for a variety of important industrial processes:

(a) Alkylation of isobutane by 2-butene to give C_8 high-octane blending materials has been performed by chlorided alumina (Al_2O_3 was chlorided with CCl_4 and/or $AlCl_3$).

(b) Cracking of isooctane on crystalline and amorphous aluminosilicate (CaHY) afforded $(CH_3)_3CH$ and 1-butene. Amorphous alumina silicate gave more propene and CH_4 than isobutane (Yushcenco *et al.*, 1979).

In spite of little success in finding useful C—C coupling or C—C cleavage reactions catalyzed by solid bases, a variety of interesting and useful rearrangement reactions have been found to occur (carbanions are important intermediates); these include (a) isomerization of butenes over activated MgO and (b) isomerization of cyclic unsaturated hydrocarbons, such as 1-*p*-menthene (Pines, 1981).

The most active base catalysts appear to be alkali metals supported on highly surface-activated alumina, they are effective even at room temperature (Pines, 1981). For example, a very active catalyst composed of 5% Na on alumina was prepared by impregnating alumina with an ammoniacal solution of sodium. When passed over this material, 1-butene was readily isomerized at 20°C to 2-butene with a nonequilibrium ratio of cis to trans isomers equal to 5 : 1. Similarly, 1-heptene was isomerized to a mixture of cis-2-, cis-3-, trans-2-, and trans-3-heptene.

B. Bond Making Processes

1. Isomerizations

Alkenes are readily isomerized over activated MgO, CaO, and BaO (Schachter and Pines, 1968; Baird and Lunsford, 1972; Tani *et al.*, 1973; Shimazu *et al.*, 1977). Interestingly, 1-butene was readily isomerized into 2-butene over BaO (pretreated at 400°C) but butene–D_2 exchange did not occur until higher pretreatment temperatures were employed (Hattori *et al.*, 1976a,b). Carbanions are intermediates in both processes, but different active sites are apparently involved.

Activated MgO and CaO are effective for isomerization of 3-carene to 2-carene (and additional products) (Shimazu *et al.*, 1977):

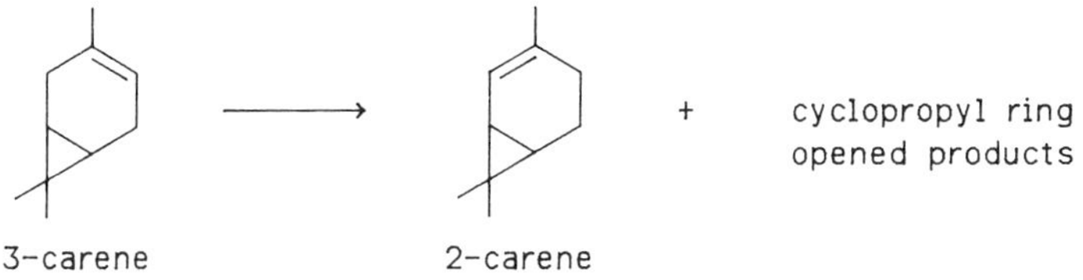

3-carene   2-carene   +   cyclopropyl ring opened products

Double bond isomerizations for conversion of allylamine to its enamine are also effective over MgO and CaO; anionic intermediates were again proposed (Hattori *et al.*, 1980; Hattori and Tanabe, 1981).

2. Hydrogenations

Thermally activated MgO, CaO, and BaO also serve as alkene and diene hydrogenation catalysts when high activation temperatures are employed (Hattori *et al.*, 1975; Tanaka *et al.*, 1976). A novel example was reported by Hattori *et al.* (1976a,b). Activation of MgO at 1100°C yielded a catalyst capable of hydrogenation of 1,3-butadiene to cis-2-butene at 0°C. They were able to 1,4-dideuterate 1,3-butadiene to 1,4-dideutero-2-butene without H/D scrambling at other positions. This is quite different from metallic-based catalysts, and indicates that heterolytic D_2 dissociation $(D_2 \to D^+_{ads} + D^-_{ads})$ is an important step, as opposed to homolytic dissociation on metals. Interestingly, this catalyst did not show activity for H_2–D_2 equilibration, even though MgO activated at lower temperatures does show such activity. It would appear that 1100°C activation generates sites that dissociate D_2 but do not allow facile migration of D^+/D^- on the surface. These differences are not understood at the present time.

3. Polymerizations

Iizuka et al. (1971) and Hattori et al. (1972–1973) have also shown that styrene can be polymerized at 30°C over activated alkaline earth oxides. A 500°C activation temperature was optimum and anion-radical intermediates were proposed.

4. Esterifications

Benzaldehyde interacts with activated oxides to yield the "Tishchenko" product, benzylbenzoate;

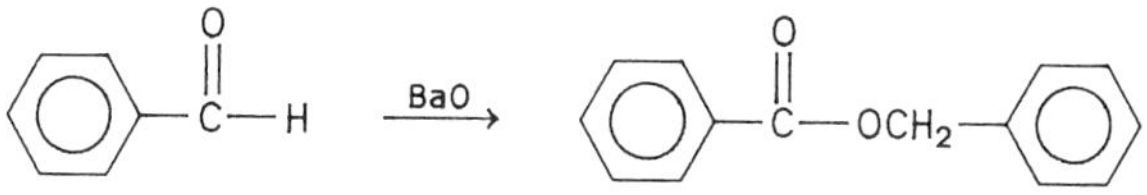

the oxides effectiveness for this reaction varied as BaO $\gg$ SrO $\gg$ CaO $>$ MgO, and a mechanism was proposed (Tanabe and Saito, 1974):

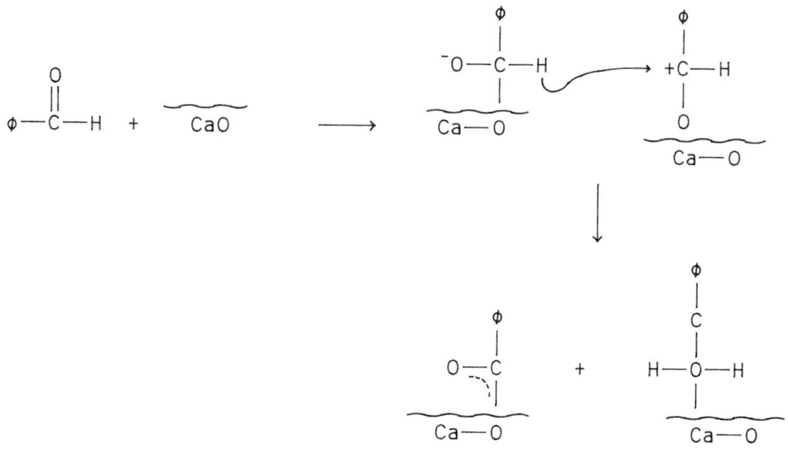

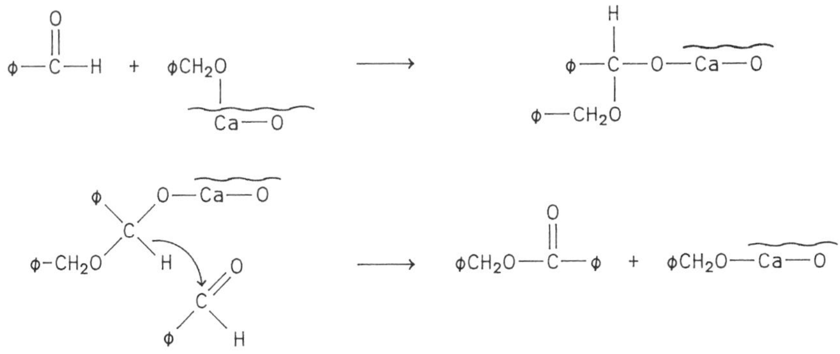

5. Additions

Kakuno *et al.* (1982) have reported the first heterogeneous diene–amine addition reactions over MgO, CaO, SrO, La_2O_3, and ThO_2. Of these, CaO was best. The amine anion is believed to be a key intermediate:

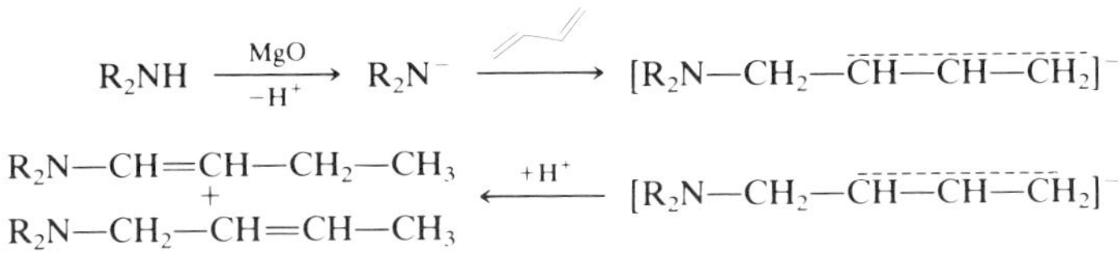

IV. METAL OXIDES AS SYNTHETIC REAGENTS

We have seen in the preceeding section how activated metal oxides are effective and unusual catalysts. Can they be used as synthetic stoichiometric reagents in organic chemistry? These studies are still in their initial stages. Proper activation is again a key step.

A. Carbonyl Compounds

Thermally activated MgO interacts strongly with aldehydes at room temperature. When no α-hydrogens are present, as in benzaldehyde, an esterification reaction takes place, as discussed in Section III. However, acetaldehyde reacts smoothly to give the aldol coupling product (Mousa and Klabunde, 1987):

$$CH_3-\overset{O}{\overset{\|}{C}}-H \xrightarrow{MgO} CH_3-\underset{H}{\underset{|}{\overset{OH}{\overset{|}{C}}}}-CH_2-\overset{O}{\overset{\|}{C}}-H$$

A series of carbonyl compounds is under study, and preliminary results are illustrated below; they show that the coupling reaction has little selectivity:

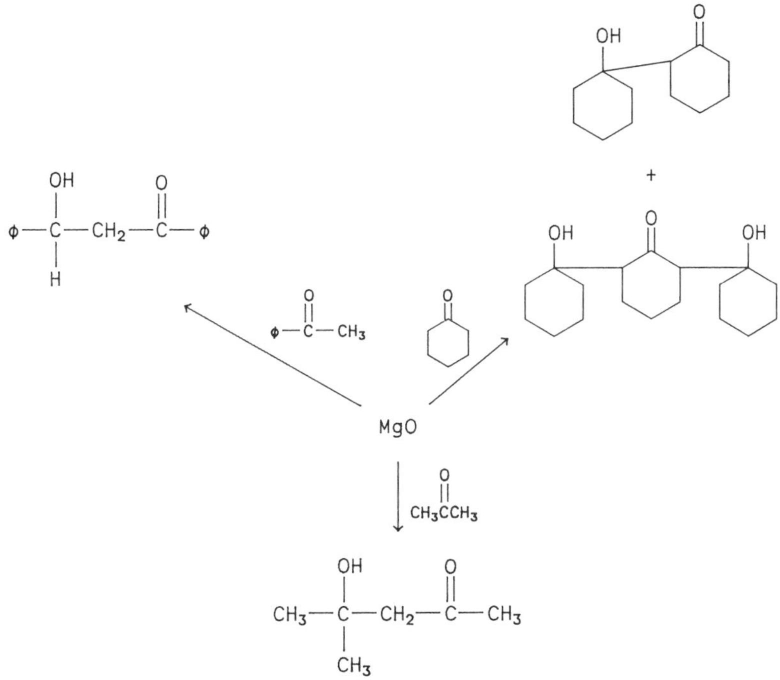

The most likely mechanism is carbanion formation at the α-carbon and attack of this anion on the carbonyl group of additional reagent molecules. Currently it is unclear what determines whether mono-alkylation or dialkylation results. Most of these reactions are catalytic and can be carried out in boiling hexane. (The use of solvent aids the removal of product from the surface exposing the active surface sites again.)

B. Phosphorus Compounds

Templeton and Weinberg (1985) have studied the decomposition of phosphonate esters $(RO)_2(R)P{=}O$ over ordered Al_2O_3 surfaces and were able to identify two important reactions: (1) nucleophilic substitution at phosphorus and (2) nucleophilic substitution at the alkyl carbon of an alkoxy group. These substitutions are due to surface oxide species.

Lin and Klabunde (1985) have recently studied a series of phosphorus compounds with activated oxides. Organophosphates $(RO)_3P{=}O$, organophosphites $(RO)_3P$, and organophosphines R_3P were found to adsorb strongly on MgO and CaO. High surface concentrations, generally about one molecule to 10 surface MgO moieties were found at room temperature. Under mild conditions the phosphorus compounds were labilized and decomposed to ethers, alcohols, and alkenes while the phosphorus residue remained with the MgO. Metathesis, like C—C bond forming processes, also took place. The amounts of phosphorus compounds decomposed by thermally activated MgO were large, essentially stoichiometric. For example, one mole of MgO decomposed 0.5 mole $(CH_3CH_2O)_3P$. In general organophosphorus compounds are highly labilized on the active MgO surface. Defect sites (Coordinatively unsaturated O^{2-} species or cation vacancies) are believed to be responsible for these reactions. Adsorption/decomposition mechanisms were determined by IR coupled with product analyses. An example for $(CH_3CH_2O)_3P{=}O$ is shown:

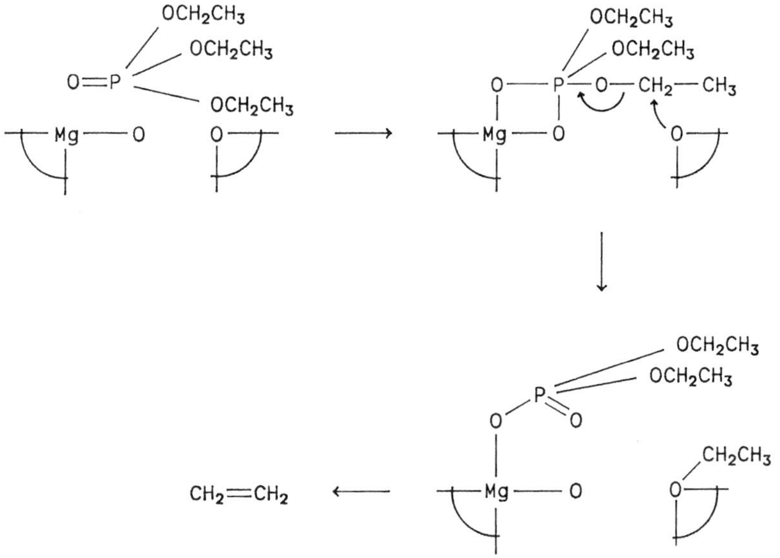

V. CONCLUSIONS

Extreme chemical activities are possible when metal oxides are prepared and activated properly. Defect sites of various types are important. Yet many unanswered questions remain: (1) What are the solid state mechanisms for metal oxide formation from nitrates, hydroxides and other precursors? (2) Why are activation energies so low for some reactions over metal oxide surfaces? (3) What active sites are involved in these processes and can they be selectively prepared? (4) Is industrial-scale thermal activation (800–1200°C) of metal oxides possible?

Considering these questions and the fact that many metal oxides are relatively abundant and inexpensive, their further development as catalysts and chemical reagents is an important endeavor.

REFERENCES

Abraham, M. M., Chen, Y., Boatner, L. A., and Reynolds, R. W. (1975). *Solid State Commun.* **16**, 209.

Abraham, M. M., Chen, Y., Peters, W. C., Rubio, J., and Unruh, W. P. (1979). *J. Chem. Phys.* **71**, 3658.

Aika, K. I., and Lunsford, J. H. (1977). *J. Phys. Chem.* **81**, 1393.

Anderson, P. J., and Morgan, P. L. (1964). *Trans. Faraday Soc.* **60**, 930.

Baird, M. J., and Lunsford, J. H. (1972). *J. Catal.* **26**, 440.

Barry, T. I., and Stone, F. S. (1960). *Proc. R. Soc., London, Ser. A* **255**, 124.

Bonino, G. B., and Fabbri, G. (1964). *Atti Accad. Naz. Lincei, Cl. Sci. Fis., Mat. Nat., Rend.* **35**, 218.

Boudart, M., Delbouille, A., Derouane, E. G., Indovina, V., and Walters, A. B. (1972). *J. Am. Chem. Soc.* **94**, 6622.

Bykhovski, M.-y., and Krylov, O. V. (1972). *Kinet. Catal (Engl. Transl.)* **13**, 686.

Che, M., and Tench, A. J. (1973). *Chem. Phys. Lett.* **18**, 199.

Che, M., Tench, A. J., Coluccia, S., and Zecchina, A. (1976). *J.C.S. Faraday I* **72**, 1553.

Che, M., Coluccia, S., and Zecchina, A. (1978). *J.C.S. Faraday I* **74**, 1324.

Chen, Y., and Sibley, W. A. (1969). *Philos. Mag.* **20**, 217.

Chen, Y., Abraham, M. M., Templeton, L. C., and Sondor, E. (1976). *Solid State Commun.* **18**, 61.

Coluccia, S., Deane, A. M., and Tench, A. J. (1978). *J.C.S. Faraday I* **74**, 2913.

Culvahouse, J. W., Holyroyd, L. V., and Kolopus, J. L. (1965). *Phys. Rev. A* **140**, 1181.

Dell, R. M., and Weller, S. W. (1959). *Trans. Faraday Soc.* **55**, 2203.

Deo, A. V., Chuang, T. T., and Dalla Lana, J. G. (1971). *J. Phys. Chem.* **75**, 234.

Driscoll, D. J., Martir, W., Wang, J. X., and Lunsford, J. H. (1985). *J. Am. Chem. Soc.* **107**, 58.

Duley, W. W. (1984). *J.C.S. Faraday I* **80**, 1173, and references therein.

Eischens, R. P., and Pliskin, W. A. (1957). *Adv. Catal.* **9**, 662.

Flockhart, B. D., Uppal, S. S., and Pink, R. C. (1971). *Trans. Faraday Soc.* **67**, 513.

Fukada, Y., and Tanabe, K. (1973). *Bull. Chem. Soc. Jpn.* **46**, 1616.

Garrone, J. E., and Stone, F. S. (1984). *Proc. Int. Congr. Catal., 8th* **3**, 441.

Germain, J. E. (1969). "Catalytic Conversion of Hydrocarbons," pp. 94–95. Academic Press, New York.

Gravelle, P. C., Juillet, F., Meriaudeau, P., and Teichner, S. J. (1971). *Discuss. Faraday Soc.* **52**, 140.

Gregg, S. J., and Ramsey, J. D. (1970). *J. Chem. Soc. A*, p. 2784.

Griffiths, D. M., and Rochester, C. H. (1977). *J.C.S. Faraday I* **73**, 1913.

Griffiths, D. M., and Rochester, C. H. (1978). *J.C.S. Faraday I* **74**, 403.

Gryazev, N. N., and Solyanova, L. N. (1965). *Dokl. Akad. Nauk SSSR* **161**, 380.

Hair, M. L., and Chapman, I. D. (1965). *J. Phys. Chem.* **69**, 3949.

Harrison, P. G., and Mavaders, B. M. (1984). *J.C.S. Faraday I* **80**, 1329.

Hattori, H., and Tanabe, K. (1981). *Heterocycles* **16**, 1863.

Hattori, H., Yoshii, N., and Tanabe, K. (1972–1973). *Catal. Proc. Int. Congr., 5th* Pap. 10, p. 232.

Hattori, H., Tanaka, Y., and Tanabe, K. (1975). *Chem. Lett.* p. 659.

Hattori, H., Maruyama, K., and Tanabe, K. (1976a). *J. Catal.* **44**, 50.

Hattori, H., Tanaka, Y., and Tanabe, K. (1976b). *J. Am. Chem. Soc.* **98**, 4652.

Hattori, A., Hattori, H., and Tanabe, K. (1980). *J. Catal.* **65**, 245.

Henderson, B., and Tomlinson, A. C. (1969). *J. Phys. Chem. Solids* **30**, 1801.

Henderson, B., and Wertz, J. E. (1968). *Adv. Phys.* **70**, 749.

Henderson, B., and Wertz, J. E. (1977). "Defects in Alkaline Earth Oxides," Halsted, New York.

Hino, M., and Arata, K. (1979a). *Chem. Lett.* p. 477.

Hino, M., and Arata, K. (1979b). *J.C.S. Chem. Commun.* 1148.

Hino, M., and Arata, K. (1980). *J.C.S. Chem. Commun.* 851.

Hino, M., and Arata, K. (1985). *J.C.S. Chem. Commun.* 112.

Hino, M., Kobayashi, S., and Arata, K. (1979). *J. Am. Chem. Soc.* **101**, 6439.

Hoq, M. F., and Klabunde, K. J. (1986). *J. Am. Chem. Soc.* **108**, 2114.

Hughes, A. E., and Henderson, B. (1972). In "Defects in Crystalline Solids" (J. H. Crawford and L. M. Slifkin, eds.), p. 1. Plenum, New York.

Iizuka, T., and Tanabe, K. (1975). *Bull. Chem. Soc. Jpn.* **48**, 2527.

Iizuka, T., Hattori, H., Ohno, Y., Sohma, J., and Tanabe, K. (1971). *J. Catal.* **22**, 130.

Iizuka, T., Endo, Y., Hattori, H., and Tanabe, K. (1976). *Chem. Lett.* p. 803.

Ito, T., Wang, J. X., Lin, C. H., and Lunsford, J. H. (1985). *J. Am. Chem. Soc.* **107**, 5062.

Jin, T., Machida, M., Yamaguchi, T., and Tanabe, K. (1984). *Inorg. Chem* **23**, 4396.

Kadushin, A. A., Rufov, Y. N., and Roginskii, S. Z. (1967). *Kinet. Catal. (Engl. Transl.)* **8**, 1356.

Kakuno, Y., Hattori, H., and Tanabe, K. (1982). *Chem. Lett.* p. 2015.

Kappers, L. A., Kroes, R. L., and Hensley, E. B. (1970). *Phys. Rev. B* **1**, 4151.

Kayo, A., Yamaguchi, T., and Tanabe, K. (1983). *J. Catal.* **83**, 99.

Kiselev, A. V., and Uvarov, A. V. (1967). *Surf. Sci.* **6**, 399.

Klabunde, K. J., Kaba, R. A., and Morris, R. M. (1978). *Inorg. Chem.* **17**, 2684.

Klassen, A. W., and Hill, C. G., Jr. (1981). *J. Catal.* **69**, 299.

Knozinger, H. (1976). *Adv. Catal.* **25**, 184.

Larson, J. G., and Hall, W. K. (1965). *J. Phys. Chem.* **69**, 3080.

Lempfuhl, G., and Uchida, Y. (1979). *Ultramicroscopy* **4**, 275.

Lin, S. T., and Klabunde, K. J. (1985). *Langmuir* **1**, 600.

Low, M. S.D., Madison, N., and Ramanurthy, P. (1969). *Surf. Sci.* **13**, 238.

Low, M. J. D., Jacobs, H., and Takezawa, H. (1973). *Water, Air, Soil Pollut.* **2**, 61.

Lunsford, J. H. (1973). *Catal. Rev.* **8**, 135.

Lunsford, J.H., and Jayne, J.P. (1965). *J. Phys. Chem.* **69**, 2182.

Lunsford, J. H., and Jayne, J. P. (1966a). *J. Chem. Phys.* **44**, 1487.

Lunsford, J. H., and Jayne, J. P. (1966b). *J. Phys. Chem.* **70**, 3464.

McCaffreg, E. F., Micka, T. A., and Ross, R. A. (1972). *J. Phys. Chem.* **76**, 3372.

McManus, J. C. Harano, Y., and Low, M. J. D. (1969). *Can. J. Chem.* **47**, 2545.

Malinski, E. N., Belotserkovskaya, N. G., and Dobychin, D. P. (1974). *Zh. Prikl. Khim.* **47**, 2401.

Martinez, R., and Barteau, M. A. (1985). *Langmuir* **1**, 684.

Meriaudeau, P., and Vedrine, J. C. (1976). *J.C.S. Faraday II* **72**, 472.

Meriaudeau, P., Vedrine, J. C., Taarit, Y. B., and Naccache, C. (1975). *J.C.S. Faraday II* **71**, 736.

Meriaudeau, P., Taarit, Y. B., Vedrine, J. C., and Naccache, C. (1977). *J.C.S. Faraday II* **73**, 76.

Misono, M., and Yoneda, Y. (1974). *J. Catal.* **33**, 474.

Miyata, H., Toda, Y., and Kubokawa, Y. (1974a). *J. Catal.* **32**, 155.
Miyata, H., Wakamiya, M., and Kubokawa, Y. (1974b). *J. Catal.* **34**, 117.
Morris, R. M. (1981). Ph. D. Thesis, Univ. of North Dakota, Grand Forks.
Morris, R. M., and Klabunde, K. J. (1983). *Inorg. Chem.* **22**, 682.
Mousa, F., and Klabunde, K. J. (1987). *Synthesis* (submitted).
Naccache, C. (1971). *Chem. Phys. Lett.* **11**, 323.
Nelson, R. L., and Tench., A. J. (1964). *J. Chem. Phys.* **40**, 2736.
Nelson, R. L., Tench, A. J., and Harmsworth, B. J. (1967). *J.C.S. Faraday I* **63**, 1427.
Ohnishi, R., and Tanabe, K. (1974). *Chem. Lett.* p. 207.
Pines, H. (1981). "The Chemistry of Catalytic Hydrocarbons," pp. 126–128. Academic Press, New York.
Pines, H., and Manassen, J. (1966). *Adv. Catal.* **16**, 49.
Rees, A. L. G. (1954). "Chemistry of the Defect Solid State," Methuen Landon and Wiley, New York.
Richardson, P. C., and Rossington, D. R. (1969). *J. Catal.* **14**, 175.
Robertson, P. J., Scurnell, M. S., and Kimball, C. (1975). *J.C.S. Faraday I* **71**, 903.
Robinson, E., and Ross, R. A. (1968). *J. Chem. Soc. A* p. 2137.
Satoh, A., Hattori, H., and Tanabe, K. (1983). *Chem. Lett.* p. 497.
Schachter, Y., and Pines, H. (1968). *J. Catal.* **11**, 147.
Schottky, W. (1935a). *Z. Phys. Chem. B* **29**, 335
Schottky, W. (1935b). *Naturwissenschaften* **23**, 656.
Shaporev, V. P., and Bulat, A. E. (1980). *Izv. Akad. Nauk SSSR, Neorg. Mater.* **16**, 1430.
Shimazu, K., Hattori, H., and Tanabe, K. (1977). *J. Catal.* **48**, 302.
Sibley, W. A., and Chen, Y. (1967). *Phys. Rev.* **160**, 712.
Sibley, W. A., Klopus, J. L., and Mallard, W. C. (1969). *Phys. Status Solidi* **31**, 223.
Sidorov, A. N. (1954). *Dokl. Akad. Nauk SSSR* **95**, 1235.
Sokolskii, D. V., Vozdvizhenskii, V. F., Kuang-Shev, A.-S., and Kobetz, A. V. (1976). *React. Kinet. Catal. Lett.* **5** 163.
Sondor, E., and Sibley, W. A. (1972). "Point Defects in Solids" (J. H. Crawford and L. M. Slifken, eds., 201. Plenum, New York.
Spitsyn, V. I., Barsova, L. I., and Zyazula, I. I. (1977). *Izv. Akad. Nauk SSSR, Ser. Chim.* **11**, 2422.
Stuart, W.I., and Whateley, T. L. (1965). *Trans. Faraday Soc.* **61**, 2763.
Tanabe, K. (1970). "Solid Acids and Bases." Academic Press, New York.
Tanabe, K., and Saito, K. (1974). *J. Catal.* **35**, 247.
Tanabe, K, Shilmazu, K., and Hattori, H. (1975). *Chem. Lett.*, p. 507.
Tanabe, K., Kayo, A., and Yamaguchi, T. (1981). *J.C.S. Chem. Commun.* p. 602.
Tanaka, H., Hattori, H., and Tanabe, K. (1976). *Chem. Lett.* p. 37.
Tani, N., Misono, M., and Yoneda, Y. (1973). *Chem. Lett.* p. 591.
Taylor, E. H. (1968). *Adv. Catal. Relat. Subj.* **18**, 111.
Taylor, J. H., and Ambeng, C. H. (1961). *Can. J. Chem.* **39**, 537.
Templeton, M. K., and Weinberg, W. H. (1985). *J. Am. Chem. Soc.* **107**, 774.
Tench, A. J., and Lawson, T. (1970). *Chem. Phys. Lett.* **7**, 459.
Tench, A. J., and Nelson, R. L. (1967). *Trans. Faraday Soc.* **63**, 2254.
Tench, A. J., and Nelson, R. L. (1968). *J. Colloid Interface Sci.* **26**, 364.
Thornton, E. W., and Harrison, P. G. (1975). *J.C.S. Faraday* **71**, 2468.
Tretyakov, N. E., and Filimonovy, V. N. (1970). *Kinet. Catal.* (*Engl. Transl.*) **11**, 815.
Ushikubo, T., Hattori, H., and Tanabe, K. (1984). *Chem. Lett.* p. 649.
Utiyama, M., Hattori, H., and Tanabe, K. (1978). *J. Catal.* **53**, 237.
Walters, G. K., and Estle, T. L. (1961). *J. Appl. Phys.* **32**, 1854.

Wang, G. W., Hattori, H., Itoh, H., and Tanabe, K. (1982). *J.C.S. Chem. Commun.* p. 1256.

Wang, G. W., Itoh, H., Hattori, H., and Tanabe, K. (1983). *J.C.S. Faraday I* **79**, 1371.

Wertz, J. E., Auzins, P., Griffiths, J. H .E., and Orton, J. W. (1959). *Discuss. Faraday Soc.* **28**, 136.

Wertz, J. E., Orton, J. W., and Auzins, P. (1960). *Discuss. Faraday Soc.* **31**, 140.

West, A. T. (1984). "Solid State Chemistry and Its Applications," pp. 231, 232, 235, 243, 258, 264, 362, 375, 382, 468. Wiley, New York.

Whittingham, M. S. (1977). *J. Catal* **50**, 549.

Yates, D. J. C. (1961). *J. Phys. Chem.* **65**, 746.

Young, R. P., and Sheppard, N. (1967). *J. Catal.* **7**, 223.

Yuschcenco, V. V., Topchieva, K. V., Imanov, F. M., Zulfugarov, Z. K. (1979). *Neftekhimiya* **19** (1), 55.

4

THE PHOTOCHEMISTRY OF ADSORBED MOLECULES

P. de Mayo

Photochemistry Unit
Department of Chemistry
The University of Western
 Ontario
London, Ontario
Canada N6A 5B7

L. J. Johnston

Division of Chemistry
National Research Council of Canada
Ottawa, Ontario
Canada K1A OR6

I. INTRODUCTION

Recent years have seen a striking increase in the study of the photochemistry and photophysics of organic molecules adsorbed on metal oxide surfaces (see Chapter 3) such as silica gel and alumina. Many of these investigations have been aimed at examining the nature of the surface itself and discovering what effects the surface has on the behavior of the adsorbed molecules. The second section of this chapter attempts to summarize the information that photophysical studies have provided about the nature of the surface–adsorbate interaction, the mobility of adsorbed species and the modification, by adsorption, of the electronic properties of molecules and their excited states. An understanding of these factors is essential for the interpretation of the photochemistry of adsorbed species and for the eventual prediction of useful surface reactions.

The following section outlines the known photochemical reactions of adsorbed molecules. Emphasis is placed on those reactions that, because of changes in the adsorbate's mobility or conformation, follow substantially different reaction pathways or give significantly different products from those observed in solution. The chapter deals primarily with adsorbed organic molecules on silica and alumina surfaces. These supports are, for the most part, unchanged upon irradiation. This chapter does not attempt

PREPARATIVE CHEMISTRY
USING SUPPORTED REAGENTS

to provide a comprehensive review of adsorbed species on other surfaces such as metals or semiconductors, in which the adsorbent participates in the overall chemistry, e.g., by electron transfer. The occurrence (or not) of electron transfer may be a matter of degree since it has been reported to occur with silica gel dried at 700°C (Yeremenko *et al.*, 1985, and references cited therein) and with alumina (see, e.g., Flockhart *et al.*, 1969, 1980; see also other papers in this series). The photochemistry and photophysics of adsorbed molecules has been reviewed briefly (de Mayo, 1982).

II. PHOTOPHYSICS

A number of the early studies examined the shifts in UV absorption spectra for silica gel–solvent slurries of a variety of ketones (Leermakers and Thomas, 1965; Weis *et al.*, 1968; Herz *et al.*, 1977), aromatic hydrocarbons (Robin and Trueblood, 1957; Leermakers *et al.*, 1966), amines, phenols and alkyl pyridinium iodides (Leermakers *et al.*, 1966), and retinals (Irving and Leermakers, 1968; Zawadzki and Ellis, 1983a). In general, n, π^* states showed bathochromic shifts while π, π^* states were hypsochromically shifted; these results are those expected for the effect of a polar environment on the adsorbed molecules. In one case the shifts were large enough to invert the energetic levels of the n, π^* and π, π^* states and, thus, modify the observed photochemistry (Griffiths and Hart, 1968). The preceding studies were complicated by the fact that the organic substances were distributed between the solvent and the surface, usually to an unknown degree. However, molecules such as thioindigo (Fassler *et al.*, 1984), *p*-nitroanisole (Flowers *et al.*, 1984), a spiropyran (de Mayo *et al.*, 1984a), and dibenzotropone (Grauer *et al.*, 1983) adsorbed on dry silica gel showed similar behavior.

Many steady-state and time-resolved emission measurements for adsorbed molecules have been reported. Pyrene has been a particularly useful probe since it has a long fluorescence lifetime, forms emissive excimers, and has a fluorescence spectrum sensitive to the polarity of the environment. Pyrene adsorbed on silica gel exhibited monomer fluorescence which did not show single exponential decay kinetics, was shorter-lived than in solution, and was indicative of a polar environment (Hara *et al.*, 1980, Bauer *et al.*, 1982a,b, 1984; Francis *et al.*, 1983). Multi-exponential excimer fluorescence was also observed and was shown to arise, in part, from ground-state complexes (Hara *et al.*, 1980; Bauer *et al.*, 1982a; Francis *et al.*, 1983; Avnir *et al.*, 1985); the complexes are a consequence of surface interaction since pyrene is not associated in solution. These results provided evidence for an inhomogeneous distribution of adsorbed molecules. Pyrenyl–silane molecules chemically bound to silica also gave monomer and excimer fluorescence which was interpreted in terms of an inhomogeneous distribution where clusters of pyrene predominated (Lochmuller *et al.*, 1983, 1984). Modification of the silica gel surface by the addition of polar coadsorbates or by various types of pretreatment produced a more homogeneous surface on which pyrene fluorescence decays approached solution values and dynamic excimer formation was observed (Bauer *et al.*, 1982a, 1984). A decanol-covered silica gel, for instance, gave pyrene monomer and excimer emissions typical of a liquid environment (de Mayo *et al.*, 1985), as did the chemically modified silica surfaces RP-2 and RP-18 (Stahlberg and Almgren, 1985). Similarly, dynamic excimer formation was shown by 1,3-di-l-pyrenylpropane on silica gel with coadsorbed l-octanol and on reversed-phase octadecylsilica (Avnir *et al.*, 1985).

Quenching of pyrene monomer fluorescence by halonaphthalenes has been shown to occur by both static and dynamic mechanisms (Bauer *et al.*, 1982b; de Mayo *et al.*, 1984b, 1985); the latter requires that diffusion of the adsorbed molecules occur. Analysis of the fluorescence quenching at various temperatures on "dry" and decanol-covered silica gel gave activation energies for diffusion of ~4 and ~2 kcal/mole, respectively (de Mayo *et al.*, 1985).

Pyrene adsorbed on surfaces other than silica showed similar results. For example, on alumina emission from pyrene monomers, excimers, and ground-state complexes occurred, and additive effects and quenching have been observed (Oelkrug *et al.*, 1974; Beck and Thomas, 1983). Three emitting species were present for pyrene in zeolites and were affected by the type of zeolite, the sample preparation, and the coverage (Suib and Kostapapas, 1984). In contrast, pyrenealdehyde exhibited radical-cation as well as excimer fluorescence on several zeolites (Baretz and Turro, 1984). On porous glass adsorbed pyrene showed both monomer and excimer emission, but only the former was observed on calcium fluoride (Bauer *et al.*, 1982a). Pyrene adsorbed on titanium oxide in aqueous suspension showed fluorescence which differed from that in solution, and which varied both spectrally and kinetically with added coadsorbates (Chandrasekaran and Thomas, 1984). The lack of pyrene excimer emission in colloidal montmorillonite clay indicated that diffusion of molecules was quite limited in this system (Della Guardia and Thomas, 1984).

The photophysics of a variety of other adsorbed organic molecules have been investigated. For example, enhanced room-temperature phosphorescence (RTP) (for reviews of RTP on paper, see Ward *et al.*, 1981; Parker *et al.*, 1980; Ford and Hurtubise, 1980) has been observed for ionic organic species on silica and alumina (Schulman and Walling, 1972), acriflavine on silica gel (Rosenberg and Humphries, 1967), aromatic hydrocarbons on alumina (Oelkrug *et al.*, 1975, 1979; Honnen *et al.*, 1983), and β-arylpropiophenones and other ketones in zeolites (Casal and Scaiano, 1984, 1985; Scaiano *et al.*, 1985). The latter results indicated restrictions on molecular rotational mobility and pointed to multiple inclusion sites for ketones on zeolites. Similarly, the monomer and intramolecular charge transfer fluorescence from 9,9'-bianthryl adsorbed on porous glass suggested multiple adsorption sites and a lack of rotational reorientation on the time scale of the fluorescence lifetime (Nakashima and Phillips, 1983). In contrast, fluorescence from 1,1'-binaphthyl (Bauer *et al.*, 1982b) and from 1-(*N*,*N*,-dimethylamino)-4-benzonitrile (Levy *et al.*, 1985) adsorbed on silica gel indicated that rotational movement for these molecules was relatively unrestricted. Clearly, the requirements for restriction on rotation remain undefined.

The emission from aromatic hydrocarbons adsorbed on alumina, gallium oxide, and titanium oxide has been extensively examined (Oelkrug *et al.*, 1974, 1975, 1979; Kessler *et al.*, 1981; Honnen *et al.*, 1983). The observed fluorescence was attributed to charge transfer transitions of complexes between the adsorbate and active surface sites and showed the nonexponential decays characteristic (see, however, James *et al.*, 1985) of multiple adsorption sites Kessler *et al.*, 1981; Honnen *et al.*, 1983). The surface adsorbate interaction was strongest for alumina surfaces (Oelkrug and Radjaipour, 1980). Fluorescence from 1-naphthol adsorbed on several surfaces (Oelkrug *et al.*, 1977) and oxygen quenching of aromatic hydrocarbon emission on porous glass and silica gel (Ishida *et al.*, 1970) also indicated a range of surface–adsorbate charge-transfer interactions.

Several reports on the photophysics of adsorbed inorganic species have appeared. Uranyl ions adsorbed on colloidal silica had a modified transient absorption from that in solution and an unusually long excited-state lifetime (Wheeler and Thomas, 1984). Both static and dynamic quenching processes occur in the quenching of ruthenium tris(bipyridine) cations adsorbed on porous Vycor (Basu *et al.*, 1983). Quenching of the emission of these cations in colloidal clay systems occurs by a dynamic process that indicates facile motion of the adsorbed species (Della Guardia and Thomas, 1983). Similarly, energy transfer between uranyl and europium-(III) ions in zeolites occurs by both electron exchange and long-range interactions with efficiencies which vary with the zeolite (Suib and Carrado, 1985). Fluorescence quenching of various cationic probes in a silica gel matrix has also been examined (Thomas and Wheeler, 1985).

The development of diffuse reflectance laser flash photolysis has permitted the direct detection of transients generated from adsorbed molecules. For example, triplet–triplet absorption spectra for several aromatic hydrocarbons adsorbed on alumina showed large shifts that were attributed to transitions to charge transfer states of surface–adsorbate systems (Kessler and Wilkinson, 1981). Pyrene adsorbed on porous glass also showed a triplet absorption spectrum shifted substantially from that in solution (Piciulo and Sutherland, 1979). Enhanced triplet lifetimes and nonexponential decays were observed in both these systems as well as for triplet ketones in zeolites (Wilkinson *et al.*, 1986). The latter also provided evidence for restrictions to motion. Energy transfer from benzophenone triplet to naphthalene by both "static" and dynamic processes on a silica gel surface has been measured by monitoring the decay and growth of the two triplet states (Turro *et al.*, 1985a). The overall quenching data resembled that for the quenching of acenaphthylene triplets by ferrocene (Bauer *et al.*, 1982b) (a study which did not recognize this "static" process).

Many of these results have indicated that adsorbed molecules are inhomogeneously distributed on the surface, and attempts have been made to associate the number of exponentials with the number of local environments. These inhomogeneous distributions, however, would be expected to result in a much wider range of lifetimes for an excited species, and such complications have hitherto been neglected. A communication by Ware and co-workers (James *et al.*, 1985) has demonstrated that the single or double (or higher) exponential fluorescence decays often reported for adsorbed molecules may equally well fit a broad distribution of rate constants and, in fact, may fit several such distributions. This will require rethinking of much of the data pertaining to emission from mixtures of conformers or environmental situations. It remains to be seen if the original distribution can be recovered from the experimental data or whether data can be collected which can be so analyzed.

III. PHOTOCHEMISTRY

A. Isomerizations

A number of photoisomerizations of adsorbed molecules are summarized in Table I. The cis–trans isomerization of stilbene in a silica gel slurry gave a photostationary state different from that in solution (Weis *et al.*, 1968) whereas on alumina a largely reversible photochromism was observed (Hecht and Jensen, 1978; Hecht and Crackel, 1981). The latter was suggested to involve an intermediate with substantial, but not well-defined, bonding to an active surface site. A possibly related surface interaction had been earlier invoked to explain geometric isomerization of

TABLE I

Photoisomerizations of Adsorbed Molecules

Surface	Substrate (isomerization)	References
Silica gel/cyclohexane	Stilbene (cis–trans)	Weis et al. (1968)
Alumina	Stilbene (photochromic)	Hecht and Jensen (1978); Hecht and Crackel (1981)
Porous Vycor glass	2-Butene (cis–trans)[a]	Otsuka and Morikawa (1974, 1975); Otsuka et al. (1978)
Porous Vycor glass	2-Butene (cis–trans)	Morikawa et al. (1977)
Silica gel/benzene	Piperylene (cis–trans)[a]	Weis et al. (1966)
Silica gel/cyclohexane	Retinals (cis–trans)	Zawadzki and Ellis (1983a)
Alumina	Thioindigo (cis–trans)	Breuer and Jacob (1980)
Alumina	Phenylethylenes (cyclization)	Moesta (1975)
Silica gel/cyclohexane	Spiropyrans (photochromic)	Evans et al. (1967); Weis et al. (1968)
Silica gel	Spiropyrans (photochromic)	Balny and Douzou (1967); Balny et al. (1967)
Silica	Norbornadiene (cyclization)[a]	Hautala et al. (1981)
Silica gel	1-(9-Anthryl)-4,4-diphenyl-2,3-diazabutadiene (E–Z)	Fassler and Guenther (1977, 1978); Fasler et al. (1980)
Alumina	Chlorinated polycyclic dienes (valence)	Moesta et al. (1971)
Aluminosilica/methylene chloride	α,β-Unsaturated ketones (cis–trans)	Childs et al. (1984)

[a] Sensitized.

2-butene on porous Vycor (Morikawa *et al.*, 1977). Changes in the photostationary states for the cis–trans isomerization of piperylene (Weis *et al.*, 1966) and retinal isomers (Zawadzki and Ellis, 1983b) were rationalized on the basis of changes in excited-state energy levels for the adsorbed molecules. Although trans–cis photoisomerization of adsorbed thioindigo occurred readily, the reverse reaction was not detected, perhaps because the cis isomer was stabilized by strong hydrogen bonding on the surface (Breuer and Jacob, 1980). The photochromic behavior of spiropyrans was also modified by surface interactions (Weis *et al.*, 1968; Balny *et al.*, 1967).

B. Photodimerizations and Photocycloadditions

Table II lists the photodimerizations and photocycloadditions that have reported for adsorbed molecules. Product ratios that were different from those obtained in solution resulted from the dimerization of a styrylisoxazole (Donati *et al.*, 1979) and the cycloaddition of alkenes to steroidal enones (Farwaha *et al.*, 1983, 1985). For the latter, adsorption of the enone on the less hindered face directs the reaction to the more hindered side of the molecule; in one case, a complete reversal of the solution stereochemistry was observed. Adsorption may also restrict conformational inversion in the biradical. The sensitized dimerization of acenaphthylene as well as the quenching of the reaction have demonstrated that translational motion of both monomer and dimer occur on the time scale of the triplet lifetime (de Mayo *et al.*, 1981; Bauer *et al.*, 1982b). 9-Cyanophenanthrene, on the other hand, has a sufficiently long singlet lifetime that molecules can diffuse together and dimerize; the shorter-lived acenaphthylene singlet afforded dimer only by nearest-neighbor interactions (Bauer *et al.*, 1982b).

TABLE II

Photodimerizations and Photocycloadditions of Adsorbed Molecules

Surface	Substrate(s)	References
Silica gel	Acenaphthylene	Bauer *et al.* (1982b); de Mayo *et al.* (1981)
Silica gel	9-Cyanophenanthrene	Bauer *et al.* (1982b)
Silica gel; silica gel/cyclohexane	3-Methyl-4-nitro-5-styrylisoxazole	Donati *et al.* (1979)
Silica gel; alumina	Alkenes + steroidal enones	Farwaha *et al.* (1983, 1985)
Silica gel	2-Methyl-1,4-naphthoquinone	Werbin and Strom (1968)
Silica gel	Dibenzotropone	Grauer *et al.* (1983)
Silica gel	Tricyclic dienone	Lazare *et al.* (1981)

C. Radical Reactions

Photolysis of a variety of ketones, esters, and azo compounds on silica surfaces generates radical pairs whose behavior is, in some cases, modified by interaction with the surface. These are listed in Table III, along with a variety of other radical reactions on surfaces.

Enhanced geminate recombination of short-lived radical pairs occurred on silica gel, as illustrated by low reaction efficiencies (Leermakers et al., 1965, 1966), substantial isotope enrichment in recovered irradiated dibenzyl ketone (Johnston and Wong, 1984; Turro et al., 1985b), and the formation of rearranged starting materials (Frederick et al., 1984). The rotational motion of cyanopropyl (Johnston et al., 1984) and cumyl (Leffler and Zupancic, 1980) radicals was not restricted, although the neophyl radical rearrangement was suppressed on the surface (Leffler and Barbas, 1981). Cage effects for benzyl radicals were larger for shorter lived singlet pairs than for the longer lived triplet pairs (Frederick et al., 1984) and depended on the silica pore size, the strength of an external magnetic field, and the temperature (Baretz and Turro, 1983; Turro et al., 1984, 1985b). Although the singlet photo-Fries radical pairs apparently showed no nongeminate combination, the products did require that minimal translational motion of the radicals occur (Avnir et al., 1978; Abdel-Malik and de Mayo, 1984). Major deviations from results obtained in solution have been observed for the photolysis of benzoin ethers on silica gel (de Mayo et al., 1982) and have been attributed to restricted motion, presumably by hydrogen bonding, for the adsorbed species.

Alkyl ketones adsorbed on porous Vycor (Kubokawa and Anpo, 1974; Anpo and Kubokawa, 1974, 1975, 1976; Anpo et al., 1975a,b, 1977) and phenyl alkyl ketones in zeolites (Turro and Wan, 1984) have shown substantial changes in the relative amounts of Type I/Type II processes because of surface interactions. Photolysis of dibenzyl ketones in zeolites has been reported to yield very high cage effects relative to results in solution, to give large amounts of rearranged ketone products, and to be unusually sensitive to additive effects (Turro et al., 1985c,d; Turro and Wan, 1985).

D. Other Reactions

A number of photochemical reactions which do not fit into the three preceding categories have also been reported and are tabulated in Table IV. In several of the oxidations and reductions the surface has been observed to have a catalytic effect (Kortum and Braun, 1960) or to be directly involved in the reaction (Gerasimov and Filimonov, 1983;

TABLE III

Radical Reactions of Adsorbed Molecules

Surface	Substrate	References
Silica gel/cyclohexane	Tetramethyl-1,3-cyclobutanedione	Leermakers et al. (1965, 1966)
Silica gel	Dibenzyl ketones (Type I)	Baretz and Turro (1983); Turro et al. (1984, 1985b); Frederick et al. (1984); Johnston and Wong (1984); Epling and Florio (1981)
Silica gel	Benzyl phenyl acetates	Frederick et al. (1984)
Silica gel/benzene	Azobisisobutyronitrile	Leermakers et al. (1965, 1966); Johnston et al. (1984)
Silica gel	Azobisisobutyronitrile	Johnston et al. (1984)
Silica gel	Azocumene	Leffler and Zupancic (1980)
Silica gel	Diacyl peroxides	Leffler and Barbas (1981)
Silica gel; silica gel/pentane	Aromatic esters (photo-Fries)	Avnir et al. (1978)
Silica gel	Aromatic amides (photo-Fries)	Abdel-Malik and de Mayo (1984)
Silica gel	Benzoin ethers (Type I/Type II)	de Mayo et al. (1982)
Porous Vycor	Alkyl ketones (Type I/Type II)	Kubokawa and Anpo (1974); Anpo and Kubokawa (1974, 1975); Anpo et al. (1977)
Zeolites	Phenyl alkyl ketones (Type I/Type II)	Turro and Wan (1984)
Zeolites	Dibenzyl ketones (Type I)	Turro et al. (1985c); Turro and Wan (1985)
Silica gel	Dibenzyl sulfones	Frederick et al (1984)
Silica gel	2,3,3-trimethyl-1-penten-4-one	Kiefer and Carlson (1967)
Silica gel; alumina	N,N-Dialkyl-α-oxoamides	Aoyama et al. (1983)
Silica gel	Arenediazonium salts	Bazold et al. (1982)
Silica gel	Methyl iodide	Baixes et al. (1985)
Silica	Chloromethanes	Ausloos et al (1977)

TABLE IV

Reactions of Adsorbed Molecules

Surface	Substrate (reaction)	References
Alumina; silica gel	Anthracene (oxidation)	Kortum and Braun (1960)
Silica gel/vanadium(V)	Hexane, 2-propanol (oxidation)	Gerasimov and Filimonov (1983)
Silica gel; alumina; florisil	Substituted phenylethylenes (oxidative cleavage)	Aronovitch and Mazur (1985)
Colloidal silica	Propyl viologensulfonate (reduction)	Calvin et al. (1981); Laane et al. (1981); Willner et al. (1981)
Colloidal silica/water	Methylene blue (reduction)	Litsov et al. (1981)
Porous Vycor	Metal oxides (reduction)	Anpo et al. (1982)
Silica gel	Methane (oxidation)	Surin et al. (1977)
Alumina	Stearic acid (chlorination)	Eden and Shaked (1975)
Alumina/ferric chloride	Aromatic hydrocarbons (chlorination)	Hasebe et al. (1981)
Porous Vycor glass	Isopropoxide decomposition	Yun et al. (1979)
Alumina/benzene	Heterocyclic N-oxides	Hata (1975)
Silica	Ruthenium carbonyl complexes	Liu and Wrighton (1982)
Silica	Iron pentacarbonyl	Jackson and Trusheim (1982); Trusheim and Jackson (1983)
Porous Vycor	Metal hexacarbonyl (loss carbon monoxide)	Simon et al. (1983)
Silica gel	Arene tricarbonyl chromium(0) complexes (loss carbon monoxide)	Zawadzki and Ellis (1983b)

Litsov *et al.*, 1981). Also of interest are several reports concerning the substantial stabilization against back electron transfer in the reduction of viologen derivatives adsorbed on charged colloidal silica particles (Calvin *et al.*, 1981; Laane *et al.*, 1981; Willner *et al.*, 1981). Both modified reactivities (Zawadzki and Ellis, 1983b) and new products (Jackson and Trusheim, 1982; Trusheim and Jackson, 1983) have been observed in the photochemistry of adsorbed metal complexes.

IV. CONCLUSIONS

Studies of the photophysics of adsorbed molecules have demonstrated that surfaces such as silica and alumina are polar in nature. Adsorbed molecules are usually inhomogeneously distributed on the surface, and, as a result, a distribution of excited state lifetimes is often observed. Rotational motion of the adsorbates is usually unrestricted, but the amount of translational movement varies with the lifetime of the excited species. Addition of coadsorbates to the surface produces a more uniform surface on which a more solution-like behavior is observed.

Many of the photochemical reactions of adsorbed molecules that have been reported have shown deviations in the reaction products or pathways. These are generally caused by restrictions on the mobility (translational, but not usually rotational) of the adsorbed species or by the stabilization on the surface of a particular molecular conformation. Several examples of the observation of new reaction products and the reversal of normal solution stereochemistry have been observed, and there is obviously the potential for many more useful modifications of photochemical reactivity for adsorbed species. However, although useful results have been obtained, prediction of chemical behavior (except perhaps in the crudest sense) remains exceedingly poor.

At present, except in a few limited cases, the prospects for "predictable" preparative work are not overly bright. As understanding grows, probably from physical studies, prediction will become possible; the conditions for performing the operations are already available.

REFERENCES

Abdel-Malik, M. M., and de Mayo, P. (1984). *Can. J. Chem.* **62**, 1275–1278.
Anpo, M., and Kubokawa, Y. (1974). *J. Phys. Chem.* **78**, 2446–2449.
Anpo, M., and Kubokawa, Y. (1975). *Bull. Chem. Soc. Jpn.* **48**, 3085–3087.
Anpo, M., and Kubokawa, Y. (1976). *Bull. Chem. Soc. Jpn.* **49**, 2623–2624.
Anpo, M., Hirohashi, S., and Kubokawa, Y. (1975a). *Bull. Chem. Soc. Jpn.* **48**, 985–987.

Anpo, M., Wada, T., and Kubokawa, Y. (1975b). *Bull. Chem. Soc. Jpn.* **48**, 2663–2666.
Anpo, M., Wada, T., and Kubokawa, Y. (1977). *Bull. Chem. Soc. Jpn.* **50**, 31–35.
Anpo, M., Tanahashi, I., and Kubokawa, Y. (1982). *J. Phys. Chem.* **86**, 1–3.
Aoyama, H., Miyazaki, K., Sakamoto, M., and Omote, Y (1983). *Chem. Lett.* 1583–1586.
Aronovitch, C., and Mazur, Y. (1985). *J. Org. Chem.* **50**, 149–150.
Ausloos, P., Rebbert, R. E., and Glasgow, L. (1977). *J. Res. Natl. Bur. Stand. (U.S.)* **82**, 1–8.
Avnir, D., de Mayo, P., and Ono, I. (1978). *J.C.S. Chem. Commun.* 1109–1110.
Avnir, D., Busse, R., Ottolenghi, M., Wellner, E, and Zachariasse, K.A. (1985). *J. Phys. Chem.* **89**, 3521–3526.
Balny, C., and Douzou, P. (1967). *C. R. Hebd. Seances Acad. Sci., Ser. C* **264**, 477–479.
Balny, C., Djaparidze, K., and Douzou, P. (1967). *C. R. Hebd. Seances Acad. Sci., Ser. C* **265**, 1148–1150.
Baretz, B. H., and Turro, N. J. (1983). *J. Am. Chem. Soc.* **105**, 1309–1316.
Baretz, B. H., and Turro, N. J. (1984). *J. Photochem.* **24**, 201–205.
Barnes, J. D., Oduwole, D. A., Trivedi, M. A., and Wiseall, **B.** (1985). *Appl. Surf. Sci.* **20**, 249–256.
Basu, A., Gafney, H. D., Perettle, D. J., and Clark, J. B. (1983). *J. Phys. Chem.* **87**, 4532–4538.
Bauer, R. K., de Mayo, P., Ware, W. R., and Wu, K. C. (1982a). *J. Phys. Chem.* **86**, 3781–3789.
Bauer, R. K., Borenstein, R., de Mayo, P., Okada, K., Rafalska, M., Ware, W. R., and Wu, K. C. (1982b). *J. Am. Chem. Soc.* **104**, 4635–4644.
Bauer, R. K., de Mayo, P., Natarajan, L. V., and Ware, W. R. (1984). *Can J. Chem.* **62**, 1279–1286.
Bazold, D., Fassler, D., and Kunert, R. (1982). *J. Prakt. Chem.* **324**, 209–216.
Beck, G., and Thomas, J. K. (1983). *Chem. Phys. Lett.* **94**, 553–557.
Breuer, H. D., and Jacob, H. (1980). *Chem. Phys. Lett.* **73**, 172–174. .
Calvin, M., Willner, I., Laane, C., and Otvos, J. W. (1981). *J. Photochem.* **17**, 195–206.
Casal, H. L., and Scaiano, J. C. (1984). *Can. J. Chem.* **62**, 628–629.
Casal, H. L., and Scaiano, J. C., (1985). *Can. J. Chem.* **63**, 1308–1314.
Chandrasekaran, K., and Thomas, J. K. (1984). *J. Colloid Interface Sci.* **100**, 116–120.
Childs, R. F., Duffey, B., and Mika-Gibala, A. (1984). *J. Org. Chem.* **49**, 4352–4358.
Della Guardia, R. A., and Thomas, J. K. (1983). *J. Phys. Chem.* **87**, 990–998.
Della Guardia, R. A., and Thomas, J. K. (1984). *J. Phys. Chem.* **88**, 964–970.
de Mayo, P. (1982). *Pure Appl. Chem.* **54**, 1623–1632.
de Mayo, P., Okada, K., Rafalska, M., Weedon, A. C., and Wong, G. S. K. (1981), *J.C.S. Chem. Commun.* 820–821.
de Mayo, P., Nakamura, A., Tsang, P. W. K., and Wong, S. K. (1982). *J. Am. Chem. Soc.* **104**, 6824–6825.
de Mayo, P., Safarzadeh-Amiri, A., and Wong, S. K. (1984a). *Can. J. Chem.* **62**, 1001–1002.
de Mayo, P., Natarajan, L. V., and Ware, W. R. (1984b). *Chem. Phys. Lett.* **107**, 187–192.
de Mayo, P., Natarajan, L. V., and Ware, W. R. (1985). *J. Phys. Chem.* **89**, 3526–3530.
Donati, D., Fiorenza, M., and Sarti-Fantoni, P. (1979). *J. Heterocycl. Chem.* **16**, 253–256.
Eden, C., and Shaked, Z. (1975). *Isr. J. Chem.* **13**, 1–4.
Epling, G. A., and Florio, E. (1981). *J. Am. Chem. Soc.* **103**, 1237–1238.
Evans, T. R., Toth, A. F., and Leermakers, P. A. (1967). *J. Am Chem. Soc.* **89**, 5060–5061.
Farwaha, R., de Mayo, P., and Toong, Y. C. (1983). *J.C.S. Chem. Commun.* 739–740.
Farwaha, R., de Mayo, P., Schauble, J. H., and Toong, Y. C. (1985). *J. Org. Chem.* **50**, 245–250.

Fassler, D., and Guenther, W. (1977). Z. Chem. 17, 429–430.
Fassler, D., and Guenther, W. (1978). Z. Chem. 18, 69–70.
Fassler, D., Gade, R., and Guenther, W. (1980). J. Photochem. 13, 49–54.
Fassler, D., Maren, R., and Dietmar, B. (1984). Z. Chem. 24, 411–412.
Flockhart, B. D., Leith, I. R., and Pink, R. C. (1969). Trans. Faraday Soc. 65, 542–551.
Flockhart, B. D., Sesay, I. M., and Pink, R. C. (1980). J.C.S. Chem. Commun. 735–737.
Flowers, G. C., Lindley, S., and Leffler, J. E. (1984). Tetrahedron Lett. 25, 4997–4998.
Ford, C. D., and Hurtubise, R. J. (1980). Anal. Chem. 52, 656–662.
Francis, C., Lin, J., and Singer, L. A. (1983). Chem. Phys. Lett. 94, 162–167.
Frederick, B., Johnston, L. J., de Mayo, P., and Wong, S. K. (1984). Can. J. Chem. 62, 403–410.
Gerasimov, S. F., and Filimonov, V. N. (1983). React. Kinet. Catal. Lett. 22, 371–374.
Grauer, Z., Daniel, H., and Avnir, A. (1983). J. Colloid Interface Sci. 96, 411–414.
Griffiths, J., and Hart, H. (1968). J. Am. Chem. Soc. 90, 5296–5298.
Hara, K., de Mayo, P., Ware, W. R., Weedon, A. C., Wong, G. S. K., and Wu, K. C. (1980). Chem. Phys. Lett. 69, 105–108.
Hasebe, M., Lazare, C., de Mayo, P., and Weedon, A. C. (1981). Tetrahedron Lett. 22, 5149–5152.
Hata, N. (1975). Chem. Lett. 401–404.
Hautala, R. R., King, R. B., Sweet, E. M., Little, J. L., and Shields, A. W. (1981). J. Organomet. Chem. 216, 281–286.
Hecht, H. G., and Crackel, R. L. (1981). J. Photochem. 15, 263–264.
Hecht, H. G., and Jensen, J. L. (1978). J. Photochem. 9, 33–42.
Herz, W., Iyer, V. S., Nair, M. G., and Saltiel, J. (1977). J. Am. Chem. Soc. 99, 2704–2713.
Honnen, W., Krabichler, G., Uhl. S., and Oelkrug, D. (1983). J. Phys. Chem. 87, 4872–4877.
Irving, C. S., and Leermakers, P. A., (1968). Photochem. Photobiol. 7, 665–670.
Ishida, H., Takahashi, H., and Tsubomura, H. (1970). Bull. Chem. Soc. Jpn. 43, 3130–3136.
Jackson, R. L., and Trusheim, M. R. (1982). J. Am. Chem. Soc. 104, 6590–6596.
James, D. R., Liu, Y.-S., de Mayo, P., and Ware, W. R. (1985). Chem. Phys. Lett. 120, 460–465.
Johnston, L. J., and Wong, S. K., (1984). Can. J. Chem. 62, 1999–2005.
Johnston, L. J., de Mayo, P., and Wong, S. K. (1984). J. Org. Chem. 49, 20–26.
Kessler, R. W., and Wilkinson, F. (1981). J.C.S. Faraday Trans. I 77, 309–320.
Kessler, R. W., Uhl, S., Honnen, W., and Oelkrug, D. (1981). J. Lumin. 24/25, 551–554.
Kiefer, E. F., and Carlson, D. A. (1967). Tetrahedron Lett. 17, 1617–1622.
Kortum, G., and Braun, W. (1960) Justus Liebigs Ann. Chem. 632, 104–115.
Kubokawa, Y., and Anpo, M. (1974). J. Phys. Chem. 78, 2442–2446.
Laane, C., Willner, I., Otvos, J. W., and Calvin, M. (1981). Proc. Natl. Acad. Sci. U.S.A. 78, 5928–5932.
Lazare, S., de Mayo, P., and Ware, W. R. (1981). Photochem. Photobiol. 34, 187–190.
Leermakers, P. A., and Thomas, H. T. (1965). J. Am. Chem. Soc. 87, 1620–1622.
Leermakers, P. A., Weis, L .D., and Thomas, H. T. (1965). J. Am. Chem. Soc. 87, 4403–4404.
Leermakers, P. A., Thomas, H. T., Weis, L. D., and James, F. C. (1966). J. Am. Chem. Soc. 88, 5075–5083.
Leffler, J. E., and Barbas, J. T. (1981). J. Am. Chem. Soc. 103, 7768–7773.
Leffler, J. E., and Zupancic, J. J. (1980). J. Am. Chem. Soc. 102, 259–267.
Levy, A., Avnir, D., and Ottolenghi, M. (1985). Chem. Phys. Lett. 121, 233–238.
Litsov, N. I. Nikolaevskaya, V. I., and Kachan, A. A. (1981). High Energy Chem. (Engl. Transl.) 15, 178–182.

Liu, D. K., and Wrighton, M. S. (1982). *J. Am. Chem. Soc.* **104**, 898–901.

Lochmuller, C. H., Colborn, A. S., Hunnicutt, M. L., and Harris, J. M. (1983). *Anal. Chem.* **55**, 1344–1348.

Lochmuller, C. H., Colborn, A. S., Hunnicutt, M. L., and Harris, J. M. (1984). *J. Am. Chem. Soc.* **106**, 4077–4082.

Moesta, H. (1975). *Faraday Discuss. Chem. Soc.* **58**, 244–252.

Moesta, H., Zarif-Sarban, M., and Schmitz, B. (1971). *Tetrahedron Lett.* **42**, 3857–3860.

Morikawa, A., Hattori, M., Yagi, K., and Otsuka, K. (1977). *Z. Phys. Chem. (Wiesbaden)* **104**, 309–320.

Nakashima, N., and Phillips, D. (1983). *Chem. Phys. Lett.* **97**, 337–341.

Oelkrug, D., and Radjaipour, M. (1980). *Z. Phys. Chem. (Wiesbaden)* **123**, 163–172.

Oelkrug, D., Radjaipour, M., and Erbse, H. (1974). *Z. Phys. Chem. (Wiesbaden)* **88**, 23–36.

Oelkrug, D., Erbse, H., and Plauschinat, M. (1975). *Z. Phys. Chem. (Wiesbaden)* **96**, 283–296.

Oelkrug, D., Schrem, G., and Andrä, I. (1977). *Z. Phys. Chem. (Wiesbaden)* **106**, 197–210.

Oelkrug, D., Plauschinat, M., and Kessler, R. W. (1979). *J. Lumin.* **18/19**, 434–438.

Otsuka, K., and Morikawa, A. (1974). *Bull. Chem. Soc. Jpn.* **47**, 2335–2336.

Otsuka, K., and Morikawa, A. (1975). *Bull. Chem. Soc. Jpn.* **48**, 3025–3027.

Otsuka, K., Fukaya, M., and Morikawa, A. (1978). *Bull. Chem. Soc. Jpn.* **51**, 367–371.

Parker, R. T., Freelander, R. S., and Dunlop, R. B. (1980). *Anal. Chim. Acta* **119**, 189–205; **120**, 1–17.

Piciulo, P. L., and Sutherland, J. W. (1979). *J. Am. Chem. Soc.* **101**, 3123–3125.

Robin, M., and Trueblood, K. N. (1957). *J. Am. Chem. Soc.* **79**, 5138–5142.

Rosenberg, J. L., and Humphries, F. S. (1967). *J. Phys. Chem.* **71**, 330–338.

Scaiano, J. C., Casal, H. L., and Netto-Ferreira, J. C. (1985). *ACS Symp. Ser.* **278**, 211–222.

Schulman, E. M., and Walling, C. (1972). *Science* **178**, 53–54.

Simon, R., Gafney, H. O., and Morse, D. L. (1983). *Inorg. Chem.* **22**, 573–574.

Stahlberg, J., and Almgren, M. (1985). *Anal. Chem.* **57**, 817–821.

Suib, S. L., and Carrado, K. A. (1985). *Inorg. Chem.* **24**, 200–202.

Suib, S. L., and Kostapapas, A. (1984). *J. Am. Chem. Soc.* **106**, 7705–7710.

Surin, S. A., Shuklov, A. D., Shelimov, B. N., and Kazanskii, V. B. (1977). *High Energy Chem. (Engl. Transl.)* **11**, 116–120.

Thomas, J. K., and Wheeler, J. (1985). *J. Photochem.* **28**, 285–295.

Trusheim, M. R., and Jackson, R. L. (1983). *J. Phys. Chem.* **87**, 1910–1916.

Turro, N. J., and Wan, P. (1984). *Tetrahedron Lett.* **25**, 3655–3658.

Turro, N. J., and Wan, P. (1985). *J. Am. Chem. Soc.* **107**, 678–682.

Turro, N. J., Cheng, C.-C., and Mahler, W. (1984). *J. Am. Chem. Soc.* **106**, 5022–5023.

Turro, N. J., Zimmt, M.B., Gould, I. R., and Mahler, W. (1985a). *J. Am. Chem. Soc.* **107**, 5826–5827.

Turro, N. J., Cheng, C.-C., Wan, P., Chung, C.-J., and Mahler, W. (1985b). *J. Phys. Chem.* **89**, 1567–1568.

Turro, N. J., Lei, X., Cheng, C.-C., Corbin, D. R., and Abrams, L. (1985c). *J. Am. Chem. Soc.* **107**, 5824–5826.

Turro, N. J., Cheng, C.-C., Lei, X., and Flanigen, E. M. (1985d). *J. Am. Chem. Soc.* **107**, 3739–3741.

Ward, J. L., Walden, G. L., and Winefordner, J. D. (1981). *Talanta* **28**, 201–206.

Weis, L. D., Bowen, B. W., and Leermakers, P. A. (1966). *J. Am. Chem. Soc.* **88**, 3176–3177.

Weis, L. D., Evans, T. R., and Leermakers, P. A. (1968). *J. Am. Chem. Soc.* **90**, 6109–6118.

Werbin, H., and Strom, E. T. (1968). *J. Am. Chem. Soc.* **90**, 7296–7301.

Wheeler, J., and Thomas, J. K. (1984). *J. Phys. Chem.* **88**, 750–754.

Wilkinson, F., Willsher, C. J., Casal, H. L., Johnston, L. J., and Scaiano, J. C. (1986). *Can. J. Chem.* **64**, 539–544.

Willner, I., Otvos, J. W., and Calvin, M. (1981). *J. Am. Chem. Soc.* **103**, 3203–3205.

Yeremenko, A. M., Blagoeshtchenskii, V. V., Smirnov, N. P., Kholmogorov, V. Y., and Tchuiko, A. A. (1985). *Theor. Exp. Chem. (Engl. Transl.)* **21**, 118–123.

Yun, C., Anpo, M., and Kubokawa, Y. (1979). *Chem. Lett.* 631–634.

Zawadzki, M. E., and Ellis, A. B. (1983a). *J. Org. Chem.* **48**, 3156–3161.

Zawadzki, M. E., and Ellis, A. B. (1983b). *Organometallics* **3**, 192–197.

5 ELECTROCHEMISTRY AT MODIFIED ELECTRODE SURFACES

Allen J. Bard

Department of Chemistry
University of Texas
Austin, Texas 78712

Walter E. Rudzinski

Department of Chemistry
Southwest Texas State University
San Marcos, Texas 78666

1. INTRODUCTION

Synthesis on both the laboratory and industrial scales can now be carried out electrochemically. Moreover, electrochemical methods are used to probe reaction mechanisms and to study solid–solution interfaces. While most electrochemistry is carried out on unmodified electrodes of metal, carbon, or semiconductor, there has been a greater emphasis on modifying the electrode surface in a purposeful manner, e.g., by coating with a polymer layer or specifically adsorbing a strongly adhering component. Electrode surface modification is often motivated by a desire to catalyze electrode reactions or make them more selective. Other applications not dealt with here include analysis and preconcentration, display devices, passivation and corrosion protection of the underlying substrate, photosensitization, and electrochemiluminescence. The literature on the preparation, characterization, and application is extensive, and no attempt at exhaustive coverage of these topics will be attempted in this chapter. Other reviews of chemically modified electrodes have appeared (Murray, 1980, 1983; Snell and Keenen, 1979; Albery and Hillman, 1981; Faulkner, 1984; Bard, 1983; Fujihira, 1986).

PREPARATIVE CHEMISTRY
USING SUPPORTED REAGENTS

A. Monolayers

A monolayer of a species on an electrode surface amounts to roughly 1×10^{-10} mol/cm^2 or 6×10^{13} molecules/cm^2. Such layers sometimes form by strong (irreversible) specific adsorption (or chemisorption) from solution (Anson, 1975; Lane and Hubbard, 1973a,b). They can also be prepared by pretreatment of the substrate followed by covalent bond formation between the substrate and the molecules of interest.

B. Polymer Layers

Thicker layers of polymers can be coated on the electrode surface. Several different types of polymers have been used; these include electroactive, polyelectrolyte, and electronically conductive. Electroactive polymers contain groups that can undergo oxidation or reduction reactions by charge transfer from the electrode [e.g., poly(vinylferrocene), Merz and Bard, 1978]. Polyelectrolytes such as protonated poly(vinylpyridine) (Oyama and Anson, 1979) and Nafion (Rubinstein and Bard, 1980) bear cationic or anionic centers and bind ions from the contacting solution. Layers of electronically conductive polymers, for example polypyrrole and polyaniline, can also be formed on electrodes. More complicated polymer layers, e.g., those that involve both polyelectrolyte and electronically conductive zones (Henning et al., 1981; Fan and Bard, 1986) or bilayers of two different polymers (Denisevich et al., 1981) have also been investigated.

C. Inorganic Films

Layers of this type involve clays (Ghosh and Bard, 1983), zeolites (Murray et al., 1984), and alumina (Zak and Kuwana, 1983). Other types of inorganic layers can be electrochemically deposited; these include films of Prussian Blue (Neff, 1978; Itaya et al., 1982) or nickel iron cyanide species (Bocarsly and Sinha, 1982).

II. METHODS OF PREPARATION

A. Adsorption

Many species are strongly and sometimes irreversibly adsorbed on an electrode surface. The extent of adsorption depends on the nature of the electrode material, the solvent, and the electrode potential. For example, hydrophobic organic species are often strongly adsorbed on metal or carbon surfaces from aqueous solutions. Similarly, compounds containing sulfur often adsorb on mercury. Preparation of a monolayer of such compounds on an electrode surface simply involves cleaning the substrate material and immersing it in a solution containing this species. Thus, essentially monolayer amounts of dicobalt porphyrin dimers on pyrolytic graphite (Collman *et al.*, 1979) can be obtained by immersion of the electrode in a dilute (ca. 0.5 mM) solution in dichloromethane. Sometimes such layers tend to desorb slowly with use (Collman *et al.*, 1980), but this can often be retarded by addition of the compound to the electrolyte solution. Protein and enzyme layers have also been formed in a similar manner (see, e.g., Stankovich and Bard, 1978; Santhanam *et al.*, 1977; Lee *et al.*, 1984).

B. Covalent Attachment

To enhance the stability of immobilized layers, species have been covalently bonded to the electrode. For example, Moses *et al.* (1975) utilized silylation chemistry to immobilize amine, pyridyl, and ethylene-diamine ligands to an SnO_2 electrode surface:

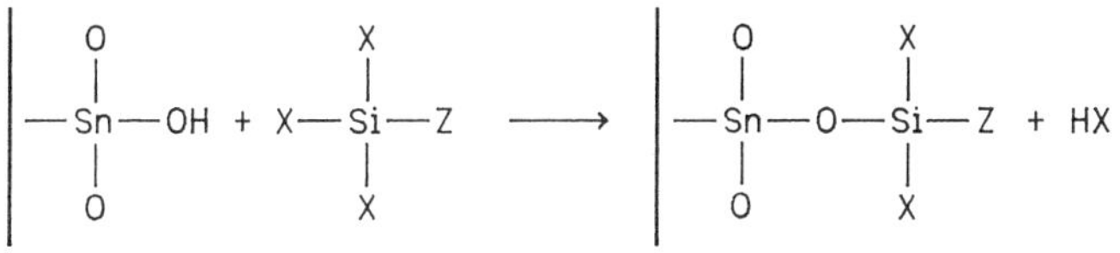

$$I$$

where X = Cl, OCH_3, and Z is a reactive functionality. When this technique is applied to platinum or carbon electrodes, prior oxidation of the surface is usually required. Thus, to bond tetra(aminophenyl)porphyrin (TAPP) to a carbon surface (Jester *et al.*, 1980), the surface is first oxidized, treated with $SOCl_2$, and finally bonded to $(NH_4)_2$TAPP:

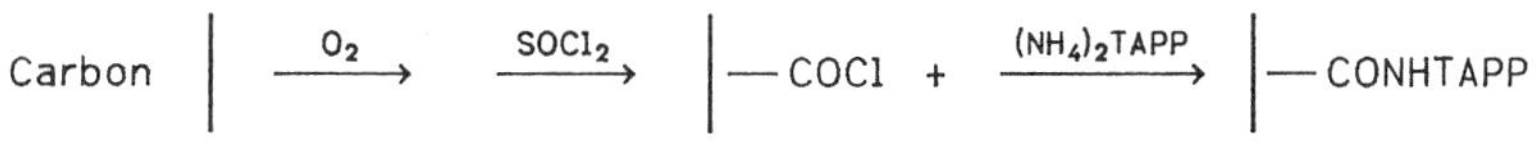

$$II$$

A similar method was employed by Watkins *et al.* (1975) to attach phenylalanine to a graphite surface. Cyanuric chloride has also been employed as a coupling agent for carbon or metal oxide surfaces (Yacynych and Kuwana, 1978).

C. Dip and Spin Coating

Multilayers can be cast by dipping the electrode in a polymer solution. More reproducible films, however, are formed by spin coating, wherein a known quantity of polymer is deposited on a rapidly rotating and often heated electrode (Tachikawa and Faulkner, 1976). Examples illustrating these approaches include preparation of films of poly(p-nitrostyrene) (Van De Mark and Miller, 1978), Nafion (Rubinstein and Bard, 1980), poly-(vinylferrocene) (Merz and Bard, 1978), and poly(vinylpyridine) (Oyama and Anson, 1979). Note that polymer layers can also be covalently attached to the electrode surface. For example, Itaya and Bard (1978) attached hydroxymethyl ferrocene to a poly(methacrylchloride) layer that was linked to the SnO_2 electrode via a silanization procedure (**III**).

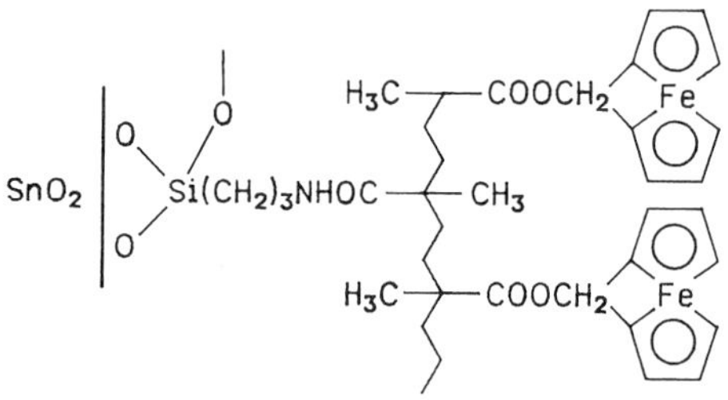

III

D. Electrodeposition

Multilayers can also be prepared by electrodepositing an insoluble film. For example, Merz and Bard (1978) produced a layer of poly-(vinylferrocene) (PVF) on a Pt electrode by oxidation of PVF in dichloromethane/tetra-n-butylammonium perchlorate to form an insoluble layer of $PVF^+ClO_4^-$. The electrode was then transferred to acetonitrile where neither the oxidized form nor PVF itself is soluble, thus producing an immobilized layer on the electrode surface.

E. Polymerization of Monomer

Films can also be produced by inducing polymerization of the monomer on the electrode surface, e.g., by electrochemical oxidation or reduction or via plasma or glow discharge polymerization. For example, polymerization of films containing metal-2,2'-bipyridine (bpy) moieties takes place by the reduction of the precursor complex (Abruna et al., 1981; Denisevich et al., 1982). Thus, upon cycling a Pt electrode between 0 and -1.8 V versus a saturated NaCl calomel electrode (SSCE) in a solution of

$[Ru(bpy)_2(vpy)_2]^{2+}$ (where vpy is vinylpyridine) in acetonitrile results in the growth of cyclic voltammetric waves attributable to surface immobilized ruthenium complex. By carrying out this polymerization reaction successively in solutions containing two different monomers, bilayer structures on the electrode surface can be produced (Denisevich et al., 1981). Polymer layers can also form spontaneously on electrodes immersed in solutions of monomers. For example, an oxidized Pt electrode when immersed in an isooctane solution of (trichlorosilyl)ferrocene produces a polymeric multilayer (Wrighton et al., 1978). Films of electrochemically conducting polymers, such as polypyrrole and polyaniline, can also be prepared by anodic deposition from solutions of pyrrole and aniline, respectively (Diaz et al., 1981; Bull et al., 1982). Plasma or glow-discharge polymerization is carried out with the monomer in the vapor phase (Nowak et al., 1978; Doblhofer et al., 1978; Dautartas and Evans, 1980).

III. METHODS OF CHARACTERIZATION OF MODIFIED ELECTRODES

A. Electrochemical

Techniques like cyclic voltammetry (CV) and chronocoulometry (Bard and Faulkner, 1980) are widely used to obtain information about films on electrode surfaces. The sensitivity and ease of measurement of small currents makes studies of even fractions of monolayers possible; a monolayer of 10^{-10} mol/cm^2 is equivalent, for a one-electron transfer reaction, to about 10 μC/cm^2. The locations of the CV waves provide information about the redox potential of the species on the electrode surface, whereas the shapes of the waves are related to molecular interactions within the films and the kinetics of electron and mass transfer processes (Peerce and Bard, 1980a,b; Laviron, 1981). Electrochemical measurements can also be employed to determine apparent diffusion coefficients for the transport of charge through the films and the rate of heterogeneous electron transfer to electroactive species at the film/electrode interface (see, e.g., Leddy and Bard, 1985; Oyama et al., 1983). Of greater interest in preparative chemistry at modified electrodes is the rate of charge transfer at the film–solution interface and the ability of the modified electrodes to catalyze reactions. Measurements with the rotating disk electrode (RDE) and rotating ring-disk electrode (RRDE) have been especially useful. A rather extensive treatment of the effect of rotation rate and solution concentration on the current of a modified RDE that allows determination of reaction rate constants in some cases (see, e.g., Andrieux et al., 1982; Anson et al., 1983b; Leddy et al., 1985) has appeared (see Section IV. A. 1).

At the RRDE, the disk electrode is modified and the concentric ring electrode is used to analyze products electrogenerated at the disk. For example, in the electroreduction of oxygen, the voltammetric response of the ring electrode can be employed to determine whether the reduction at the disk proceeds via a two-electron process to peroxide or a four-electron process to water (Collman et al., 1980).

B. Spectroscopic Techniques (See Chapters 7–9)

The full range of surface spectroscopic techniques have been applied to modified electrodes (Karweik et al., 1982). Scanning electron microscopy, x-ray photoelectron (XPS) and Auger electron spectroscopy have been most widely applied. While these have proven very useful in showing the structure of the dry surfaces and in providing atomic speciation, they require removal of the electrode from its solution environment and transfer to a high vacuum (i.e., they are ex-situ methods). Obviously this can lead to large changes in surface structure and possible contamination. Among in-situ methods, visible and infrared reflectance and Raman spectroscopic methods have been used. For example, Jester et al. (1980) employed reflectance measurements to show changes of the surface bound CoTPP species with potential, and surface enhanced Raman spectroscopy (SERS) has been employed with adsorbed biological species on silver surfaces (Cotton et al., 1980). Luminescence measurements have also been useful in probing processes in polymer films (see, e.g., Majda and Faulkner, 1982).

C. Electron Spin Resonance Methods

Electron spin resonance (ESR) measurements are useful in obtaining information about the environment and mobility of surface-confined paramagnetic species, as well as some insight into the mechanism of charge transport within films. Thus, Izelt et al. (1983) studied tetracyanoquino-dimethane (TCNQ)-modified Pt electrodes and Albery et al. (1984) investigated polymer layers of poly(nitrostyrene) and poly(vinylan-thracene). The spectra of the radical ions in both of these cases were singlets because of restricted rotation of the species. On the other hand, Gaudiello et al. (1985) showed that the methyl viologen cation radical held in Nafion shows hyperfine splitting, providing evidence for rapid tumbling in the polymer layer. In this latter study, evidence was found for slower electron transfer in the polymer film compared to the same species in solution.

D. Thickness Measurements

The thickness of the layer attached to an electrode is important in determining the mass transport and charge transport rates through a film. However, accurate measurement of the thickness (nanometer to micrometer), especially *in situ*, is difficult. For dry films, relative intensities of XPS lines can be used to estimate thickness (Untereker *et al.*, 1977; Itaya and Bard, 1978; Hawn and Armstrong, 1978; Fischer *et al.*, 1979), especially when angular resolved XPS can be used (Abruna *et al.*, 1981; Umana *et al.*, 1981). Surface profilometry of wet and dry films can also be employed (Rubinstein and Bard, 1981). Ellipsometric measurements are probably the most powerful for *in-situ* measurements, when they can be employed (Babai and Gottesfeld, 1980; Carlin *et al.*, 1985). Accurate determination by ellipsometry is possible when the films are uniform, homogeneous, and nonabsorbing.

IV. APPLICATIONS

A. Catalysis and Synthesis

1. Mediation

Modified electrocatalysis involves the mediated transfer of electrons between the electrode surface and some substrate that would otherwise undergo a slow electron transfer at a naked electrode (Murray, 1983). Mediation of electron transfer by an immobilized redox couple can be represented by the following scheme:

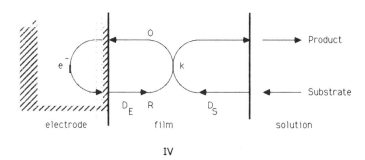

IV

The kinetics of mediated reactions at modified electrodes depend on various charge transport, mass transport, and electron exchange processes. A kinetic model describing the catalysis of electrochemical reactions at redox polymer electrodes has been developed (Andrieux *et al.*, 1982; Buttry and Anson, 1984; Leddy *et al.*, 1985; Sharp *et al.*, 1985, and references therein). Electrocatalysis depends upon the interplay of several processes:

(a) Transport of the substrate from the bulk of the solution to the electrode film.

(b) Transport of substrate across the interface.

(c) Transport of the substrate within the film; the rate depends on the nature of the film (its microstructure, charge, porosity, etc.) and is measured by D_s, the diffusion coefficient of the substrate within the film.

(d) Transport of charge through the film; the rate depends upon the apparent diffusion coefficient D_E, and both D_S and D_E can be limited by the diffusivity of counterions within the film.

(e) The rate of reaction k between the substrate and the catalyst within the film.

Much of the research thrust on modified electrode catalysis has been directed towards enhancing the rate of substrate and charge transport.

The theoretical treatments previously described have provided guidelines for maximizing the effectiveness of modified electrodes, e.g., by optimization of film thickness. Polymer films have the advantage that the surface coverage (Γ°) can be increased from that of a monolayer (ca. 10^{-10} mol/cm^2) up to 20,000 monolayers (ca. 5×10^{-6} mol/cm^2). The increased coverage enhances the rate of reaction when the second-order rate constant k is low. In this case, increasing the amount of redox catalysts maximizes the number of sites for catalyst–substrate reaction. Kuo and Murray (1982) demonstrated that the electron self-exchange rate for ferrocyanide and ferricyanide and the rate of oxidation of ascorbic acid by ferricyanide were both larger in an alkylaminesiloxane polymer film on a Pt electrode than they were in a homogeneous solution. On the other hand, if the catalytic rate constant is large compared to the mass and charge transfer rates, the reaction zone is then confined to a thin layer at a film interface. Many electrocatalytic sites are not accessible and the catalytic efficiency may actually be decreased because of the onset of kinetic control by the rate of charge transport within the film (Kuo and Murray, 1982). Under these conditions the redox polymer films behave as a single monolayer located either at the film–solution interface (Ikeda *et al.*, 1982; Bettelheim *et al.*, 1980; Chao *et al.*, 1983) or at the electrode–film interface (Fuki *et al.*, 1982).

Anson *et al.* (1983a) have optimized the reaction between $Co(tpy)_2^{2+}$ (tpy = 2,2′,2″-terpyridine) and $Mo(CN)_8^{3-}$ or $W(CN)_8^{3-}$ by using a protonated poly-L-lysine copolymer on a graphite disk electrode. The octacyano complexes are bound within the polycationic film, and the $Co(tpy)_2^{2+}$ oxidation is enhanced by the presence of the mediating $Mo(CN)_8^{3-}$ and $W(CN)_8^{3+}$ complexes. A number of other studies have also been concerned with the factors affecting mediated reactions at modified electrodes (see, e.g., Buttry and Anson, 1983; Krishnan *et al.*, 1984, and references therein).

With redox polymer films based on polyelectrolytes, the reaction between substrate and the catalyst or mediating species often takes place throughout the polymer (Sharp *et al.*, 1985). This should lead to enhanced catalytic rates for a catalyst-impregnated polymer-modified electrode as compared to those that can be achieved in a homogeneous solution. Some comparisons have been made (Sharp *et al.*, 1985), and the rate constants have differed significantly. Collisional restrictions were implicated in the reduced rate of electron self-exchange for $Co(bpy)_3^{2+}$ when incorporated within Nafion (Buttry and Anson, 1983).

Some of the work in developing improved catalytic systems has focused on increasing the porosity of the polymer film. Van Koppenhagen and Majda (1985) developed a copolymer based on acrylamide (AC) and vinyl pyridine (V-P) and then monitored the ferricyanide-mediated oxidation of ascorbic acid. There was no apparent hindrance of the ascorbic acid transport within the film. Sharp *et al.* (1985) showed that the ternary copolymer **V** exhibited unusually high diffusional rates in the oxidation of catechol and L-dopa as mediated by $IrCl_6^{2-}$.

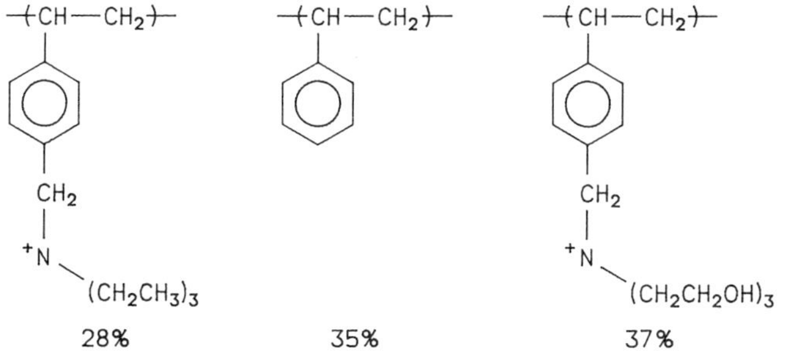

V

The rate of charge transport through the surface layer (characterized by the "charge hoping" diffusion coefficient D_E) often limits the rate of a mediated process. To improve the electronic conductivity, conductive zones (e.g., metals, carbon, and electronically conductive polymers) can be incorporated within the film (Henning *et al.*, 1981; Bookbinder *et al.*, 1980; Abruna and Bard, 1981; Burgmayer and Murray, 1982; Pickup *et al.*, 1983).

2. Oxygen Reduction

Interest in the electroreduction of O_2 at inexpensive electrode materials (e.g., for use in fuel cells) has motivated numerous investigations of possible catalysts; metal phthalocyanine and porphyrin complexes on carbon are used frequently. Bettelheim *et al.* (1980) attached Fe(III) tetra(*o*-aminophenyl)porphyrin (FeTAPP) and Fe(III) tetra(*N*-(2-hydroxylethyl)pyridyl)porphyrin (FeTEpyP) to glassy carbon through amidization or esterification of the methylacrylchloride polymer (see Fig. 1). This extended their previous work on the electrocatalysis of oxygen reduction (Bettelheim *et al.*, 1979) and applied the direct covalent bonding strategy, employed for bonding tetra(aminophenyl)porphyrin (TAPP) and cobalt tetra(aminophenyl)porphyrin (CoTAPP) (Lennox and Murray, 1978), to the analogous iron complex. The authors succeeded in reducing O_2 to H_2O using FeTAPP, attributing the successful reduction to the high surface coverage achieved for the porphyrin polymer. The overall $4e^-$ reduction proceeds via a two-step process involving the reduction of O_2 to H_2O_2, followed by a porphyrin catalyzed dismutation of H_2O_2, forming O_2 and H_2O (Yeager, 1976, 1981).

Collman *et al.* (1980) prepared a series of dimeric metalloporphyrin molecules that contained two face-to-face porphyrin rings constrained by two amide bridges of varying length. They demonstrated by RRDE

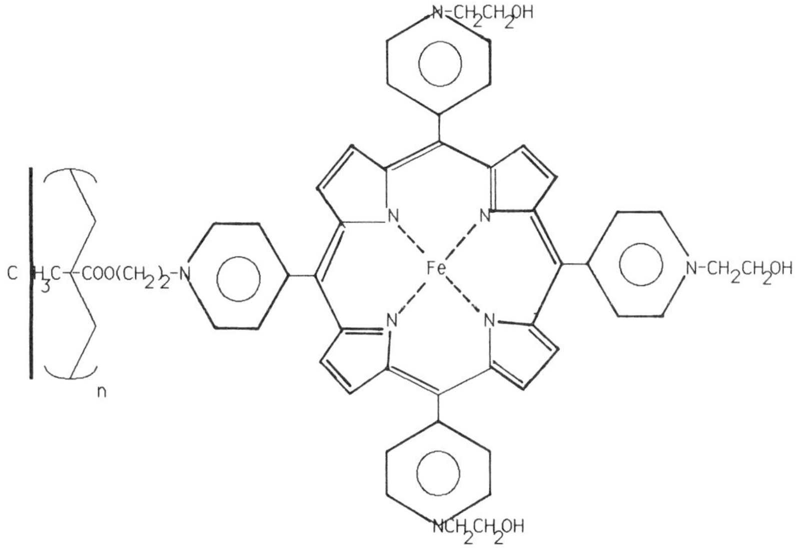

Fig. 1. Iron(III)tetra(N-(2-hydroxyethyl)pyridyl)porphyrin attached to a poly(methacryl-chloride)–glassy carbon electrode.

methods that the dicobalt cofacial porphyrin, linked by two four-atom bridges, catalyzed the reduction of O_2 to H_2O. Very little anodic current was observed at the ring, indicating minimal H_2O_2 production. Liu *et al.* (1983) demonstrated the utility of a dicobalt cofacial dimer with an active porphyrin catalyst in a cis configuration which facilitates solvent access, multiple proton transfer, and dioxygen bond cleavage. Further work on these types of catalysts has elucidated the details of the mechanism of O_2 reduction to H_2O (Anson *et al.*, 1985; LeMest *et al.*, 1985, and references therein).

Work with tetraphenylporphyrin (TPP) catalysts has focused on new methods of preparation. Buttry and Anson (1984) incorporated TPP into Nafion and then metallated the polymer. Cobalt TPP catalyzes the electroreduction of dioxygen so long as a redox mediator such as $Ru(NH_3)_6^{2+}$ is also present in the film. Bull *et al.* (1984) incorporated iron phthalocyanine into polypyrrole, while White and Murray (1985) prepared porphyrin films by electropolymerizing metallotetra(*o*-aminophenyl)porphyrin. Takeuchi and Murray (1985) incorporated metalloporphyrins into carbon paste electrodes and thus built on the work of Korfhage *et al.* (1984) who used cobalt phthalocyanine in a carbon paste as a detector in liquid chromatography. The immobilization of an extended ring structure in a carbon paste offers the advantages of simplicity, access to aqueous electrochemistry, and the ability to regenerate a clean electrode surface easily.

3. Hydrogen Evolution and CO_2 Reduction

Much of the work on H_2 evolution on modified electrode surfaces has been connected with solar energy utilization at semiconductor materials. A complete discussion of this area is beyond the scope of this chapter and only a few selected examples will be given. In general the H_2 evolution reaction is slow at a semiconductor surface and catalysts (most frequently, Pt) must be added to promote this reaction. Typically a viologen in a polymer film is reduced by photogenerated electrons (e.g., at irradiated *p*-Si or *p*-GaAs) and Pt incorporated in the polymer layer catalyzes proton reduction by the reduced viologen (Bookbinder *et al.*, 1979; Bookbinder and Wrighton, 1980; Abruna and Bard, 1981). Typical viologen polymers are obtained from **VI** and **VII**.

$$[(CH_3O)_3Si(CH_2)_3 \overset{\overset{\displaystyle H}{|}}{-^+N} \hspace{-0.3em}-\hspace{-0.3em}\underset{}{\bigcirc}\hspace{-0.3em}-\hspace{-0.3em}\underset{}{\bigcirc}\hspace{-0.3em}-\hspace{-0.3em}\underset{\underset{\displaystyle H}{|}}{N^+}(CH_2)_3Si(OCH_3)_3]Br^2$$

VI

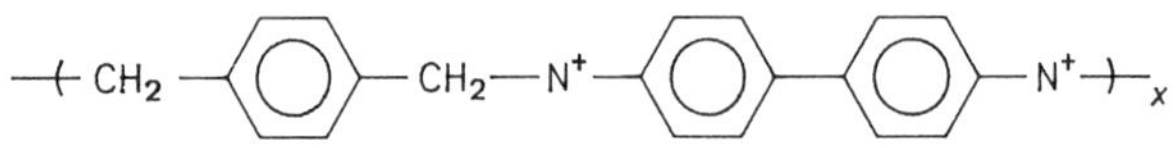

$$-(CH_2 - \langle \bigcirc \rangle - CH_2 - N^+ - \langle \bigcirc \rangle - \langle \bigcirc \rangle - N^+ -)_x$$

VII

Such electrode systems and related ones involving semiconductor particles where a number of different components (e.g., polymer, semiconductor, redox couple, and catalyst) are combined to carry out a given reaction have been named "integrated chemical systems" (see, e.g., Bard *et al.*, 1983; Krishnan *et al.*, 1983, and references therein).

Kao and Kuwana (1984) dispersed Pt microparticles in polyvinylacetic acid/glassy carbon electrodes; the particles catalyzed the evolution of hydrogen and the reduction of oxygen with the same overpotentials as those observed on a smooth Pt electrode. Daube *et al.* (1985) described polymer film–noble metal combinations as catalysts for the reduction of H_2O and aqueous CO_2. The cobalticenium-based reagent **VIII** immobolized on *p*-InP and impregnated with rhodium can be used to generate H_2.

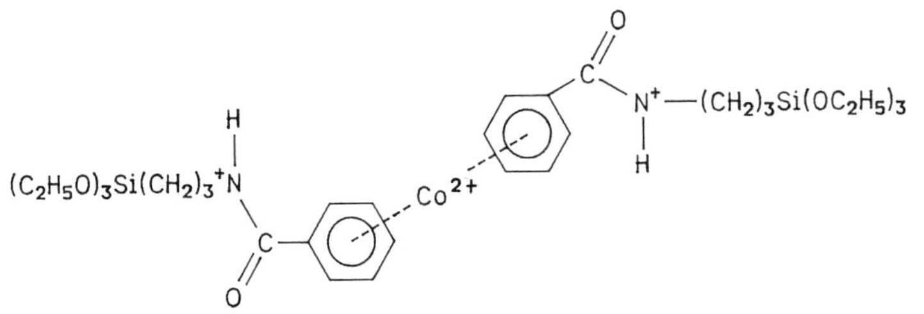

$$(C_2H_5O)_3Si(CH_2)_3\overset{H}{\overset{|}{N}} \quad \langle \bigcirc \rangle \cdots Co^{2+} \cdots \langle \bigcirc \rangle \overset{O}{\overset{\|}{C}} N^+ - (CH_2)_3Si(OC_2H_5)_3$$

VIII

Reduction of CO_2 at modified electrodes has also been of interest; see, e.g., the work of O'Toole *et al.* (1985) on electropolymerized Re(vbpy)(CO)$_3$Cl (vbpy = 4-vinyl-4'-methyl-2,2'-bipyrdine) and Meshitsuka *et al.* (1974) and Lewis and Lieber (1984) on adsorbed Co(II) phthalocyanines.

4. Synthesis

The major effort in synthesis at modified electrodes involves catalysis of desired reactions at lower overpotentials with enhanced selectivity. From a commercial standpoint, the most important modified electrode is probably the dimensionally stable anode (DSA) employed in the chloralkali process for chlorine evolution and based on titanium, with a coating of ruthenium oxide containing other transition-metal oxides (De Nora, 1970).

Organic synthetic reactions have also been described at modified electrodes. Watkins *et al.* (1975) immobilized S-(−)-phenylalanine methyl ester to the thionyl chloride-activated edge plane of pyrolytic graphite.

Reductive electrolysis of a solution of 4-acetylpyridine or ethylphenyl-glyoxylate, yielded an enantiomeric excess of the product from the corresponding prochiral substrate; the highest optical yield was 14.5% for the reduction of 4-acetylpyridine. Kerr *et al.* (1980) immobilized poly(*p*-nitrostyrene) (PNS) on platinum and investigated the catalyzed reduction of 1,2-dibromo-1,2-diphenylethane to *trans*-stilbene:

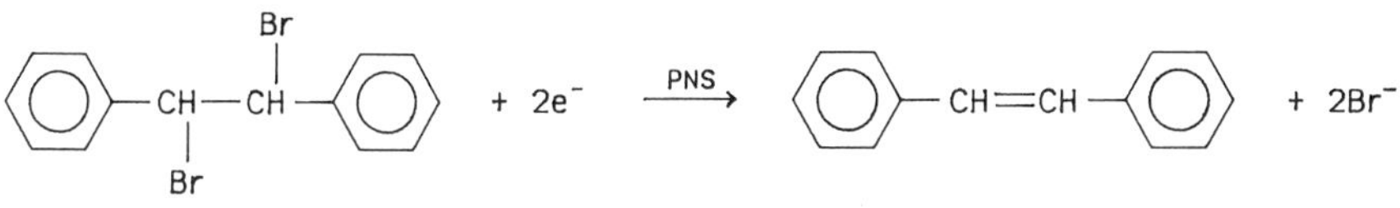

IX

High efficiencies for reduction of this dihalide could be observed. Rocklin and Murray (1981) similarly reduced 1,2-dibromo-1,2-diphenylethane, 1,2-dibromophenylethane, and 1,2-dibromopropane at the surface of electrodes to which copper(II) or cobalt(II) tetra(*p*-aminophenyl)porphyrin had been attached. Samuels and Meyer (1981) prepared an electrocatalyst by depositing a thin polymeric film of $(bpy)_2(H_2O)Ru(II)$ bound to PVP on a glassy carbon electrode. Upon oxidation of the ruthenium to Ru(IV), *p*-toluic acid, 2-propanol, and solubilized xylenes were all oxidized. The catalyst lifetime was limited to approximately 30 turnovers/site because the oxidized Ru(IV) attacked the polymer. Komori and Nonaka (1984) deposited layers of poly(amino acid) on electronically conducting polypyrrole electrodes. An enantiomeric excess of 73% was obtained for the assymetric oxidation of isopropylphenyl sulfide to the sulfoxide when using a poly-L-valine/polypyrrole/Pt electrode.

A different approach for achieving stereoselectivity in electroorganic synthesis was demonstrated by Osa and co-workers (Matsue *et al.*, 1979). Using cyclodextrins (cyclic 1,4-linked D-glucopyranose oligomers) as hosts, the authors demonstrated the stereoselective anodic chlorination of anisole. Electrogenerated HOCl attacks the inclusion complex formed by the anisole and cyclodextrin, preferentially forming the para isomer, since the ortho position is sterically blocked by the inner wall of the cyclodextrin cavity:

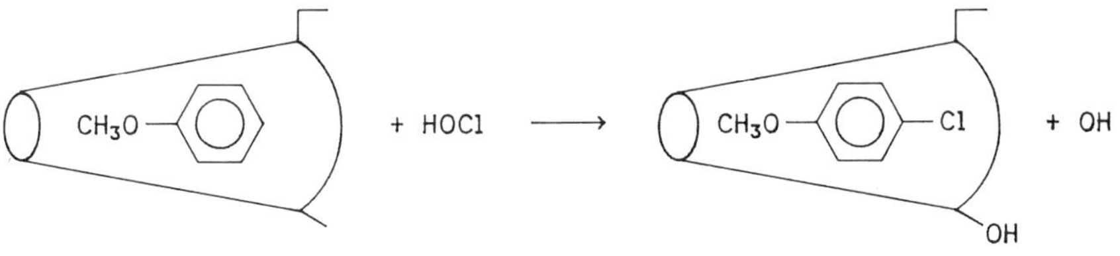

X

The ratio of para/ortho isomer was enhanced by a factor of approximately four.

Yamagishi and Aramata (1984) developed a novel asymetric synthesis scheme involving Δ-Ru(phen)$_3^{2+}$ (phen = 1,10-phenanthroline) incorporated into a montmorillonite clay on a SnO_2 glass electrode. They oxidized racemic Co(phen)$_3^{2+}$ producing an excess of ΛCo(phen)$_3^{3+}$. Labile polymetalloxides have also been incorporated within clay modified electrodes. After heating, the clay becomes pillared, introducing rigidity and size selectivity to the framework. Itaya and Bard (1985) showed that a pillared clay-modified electrode will retain its activity in aprotic solvents, whereas Rudzinski and Bard (1986) demonstrated that the electrodes can exhibit a size selectivity.

B. Enzyme Modified Electrodes

Redox enzymes have been investigated actively by electrochemical techniques. While some enzymes appear to undergo direct electron transfer reactions with electrodes (see, e.g., Yeh and Kuwana, 1977), in other cases the active site may be inaccessible to the electrode. Two approaches have been taken to promote electron transfer in these latter systems:

(a) An electron transfer mediator (i.e., a redox system) is interposed between the electrode and the enzyme, and the mediator shuttles charge between the electrode and the active site.

(b) A surface modifier is adsorbed on the electrode surface and promotes charge transfer from the electrode to the enzyme.

Electron transfer is promoted by dissolved mediators (e.g., the methyl viologen system, $MV^{2+/+}$ (Heineman et al., 1972, 1973). In such studies spectroelectrochemistry at an optically transparent electrode is often used to monitor the optical spectra of the enzymes during the course of a coulometric titration; the ratio of oxidized to reduced form is obtained and aids in the estimation of the formal potential of the enzyme. Lewis and Wrighton (1981) immobilized a viologen polymer on Au, Pt, and p-Si surfaces. The rate of reduction of horse heart ferricytochrome c was enhanced with this modification when compared to that of the underivatized electrode. In this case the potential of the surface mediator (-0.55 V versus SCE) was considerably more negative than that of the enzyme ($+0.02$ V versus SCE). Chao et al. (1983) immobilized ferrocene ($E^{\circ\prime} = 0.04$ V versus SCE) on a Pt electrode surface and showed that both oxidation and reduction of cytochrome c could be carried out without a large overvoltage.

Modification of an electrode by specific adsorption of small molecules can also promote electron transfer to a dissolved enzyme. Eddowes et al.

(1979) adsorbed 4,4′-bipyridine and 1,2-bis(4-pyridyl)ethylene on a gold electrode and reported enhanced electron transfer to cytochromes. The modifiers contain a pyridyl moiety that is thought to hydrogen bond to the positively charged lysine residues of the cytochrome c, so that the enzyme adheres to the surface in a preferred orientation for charge transfer to the electrode (Allen et al., 1984). In a further development, Hill et al. (1985) described the direct electrochemistry of redox proteins with negatively charged binding domains. The authors modified gold electrodes with 2-aminoethanethiol, which promoted the redox chemistry of both plastocyanin and multisubstituted carboxydinitrophenyl cytochrome c (a cytochrome modified to have a negatively charged binding domain).

Electrodes can be modified with many types of enzymes as well. For example, urease strongly adsorbed on a mercury electrode surface is capable of catalyzing the hydrolysis of urea (Santhanam et al., 1977). Electroreduction of disulfide bonds in the enzyme was shown to inhibit the catalysis, with some activity being regenerated upon reoxidation. Enzyme electrodes involving many different enzymes held in rather thick layers on electrode surfaces are widely used in analysis; discussion of these is beyond the scope of this chapter (see, e.g., Kobos, 1980).

C. Solid Polymer Electrolyte

A solid polymer electrolyte (SPE) contains porous metal electrode layers deposited on an ion-exchange membrane (usually Nafion). Such SPE technology has been used in fuel cells and water electrolyzers (Lu and Srinivasan, 1979). More recently SPE systems have been applied to synthesis applications. In these, the electrochemical cell (Fig. 2) consists of a polymer membrane (SPE) sandwiched between two solutions. The porous working electrode deposited on one side of the membrane faces the solution containing substrate. The counter electrode and supporting electrolyte are in the other solution. In the configuration of Fig. 2, the substrate is oxidized and electroneutrality is maintained by protons passing through the cation exchange membrane into the counter electrode chamber. The working electrode can consist of Pt particles impregnated within the membrane or a Pt-grid. The counter electrode can either be affixed to the membrane or simply dipped into the supporting electrolyte solution. The advantages of the SPE configuration are (1) the direct mounting of an electrode on a SPE membrane may improve the electrodes properties (Aramata and Ohnishi, 1984) and (2) the working electrode compartment need not contain supporting electrolyte. Solid polymer electrode technology has been used for the electrolysis of a variety of organic compounds (Ogumi et al., 1985; Sarrazin and Tallec, 1982).

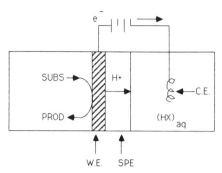

Fig. 2. Electrochemical cell reaction at a working electrode/SPE membrane (W.E., working electrode; C.E., counter electrode; SUBS, substrate; PROD, product; HX, supporting electrolyte where X = anion).

Katayama-Aramata and Ohnishi (1983) bonded Pt on Nafion and observed the electrooxidation of methanol in $HClO_4$. The reaction is of interest because of possible fuel cell applications. Anodically activated platinum initially shows high catalytic activity, but it cannot be maintained during long-term polarization. By impregnating Pt on an SPE, the authors prepared an electrocatalyst more durable than a bare Pt electrode. A surface mediator containing Pt^0 and Pt^{2+} was proposed as the couple promoting catalytic activity, and the system was durable because the SPE matrix stabilized Pt in a higher oxidation state (Aramata and Ohnishi, 1984).

Investigators have used the SPE electrolysis approach to synthesize organic compounds in solvent systems free of supporting electrolyte. Ogumi et al. (1981) electroreduced a series of olefins (cyclooctene, α-methylstyrene, and diethylmaleate) at Pt–SPE and Pt–Au–SPE electrodes. Raoult et al. (1984) affixed a platinum grid to a cation exchange membrane and oxidized solutions of olefins in methanol. The product yield depended strongly on olefin structure, but up to 80% of 2,5-dimethoxy-2,5-dihydrofuran was obtained. Finally, a halogen mediatory system was developed for use with a Pt SPE (Ogumi et al., 1985). Cyclohexanol was oxidized to cyclohexanone in the presence of iodine or iodide. The mediator accelerated the oxidation of the alcohol, via a unipositive iodine species that was continually regenerated at the Pt–SPE anode.

D. Controlled Release Systems

Controlled release systems refer to modified polymer assemblies that can release compounds into solution at a controlled rate or at a given time. Such systems can be useful in the controlled delivery of biomolecules (e.g., neurotransmitters or drugs) to a specific location. Lau and Miller (1983a) developed a controlled release system, consisting of a biomolecule attached to a polymer backbone on a glassy carbon disk. When an appropriate potential was applied, the biomolecule was released into the solution. The investigators prepared two different controlled release systems from polystyrenes attached to isonicotinamide derivatives of neurotransmitters, γ-aminobutyric, or glutamic acids (see **XI**).

R=COOH, H

XI

Upon application of a pulse at approximately -0.9 V versus SCE, small amounts of the neurotransmitters, glutamate, or γ-aminobutyrate, were released into an aqueous electrolyte solution. The cathodic pulse served to break the amide linkage. The rate of this reaction was generally slow, but the addition of MV^{2+} was found to accelerate the rate of production of the neurotransmitter.

Lau and Miller (1983b) also extended their previous work with a dopamine polymer (Miller *et al.*, 1982) and proposed this modified electrode as a primitive analog of a synapse. The electrode released dopamine into a pH 7 solution at -0.9 V. At 0.2 V the electrode promoted the electrocatalytic oxidation of solution-phase NADH.

In each of the previous examples of controlled-release systems, the rate of release of the biomolecule was limited by the conductivity of the polymer and the overall rate of reduction. To improve the efficiency, Zinger and Miller (1984) reported the timed release of ferrocyanide and glutamate from polypyrrole films. The results clearly demonstrated the utility of using conductive polymers that can switch from a cationic conductive form to a neutral insulating state. The total amount of glutamate released was 200 times greater than an earlier system and functioned well during the imposition of a series of repetitive pulses. Espenscheid and Martin (1985), using electroactive cation exchange polymers based on styrene, styrenesulfonate, and vinyl ferrocene, electromodulated the ion exchange capacity of the polymer and demonstrated that a potential of $+0.8$ V could inhibit the penetration of $Ru(NH_3)_6^{3+}$ from an external solution or expel $Ru(NH_3)_6^{3+}$ from a loaded film.

Finally, in related developments, electric field and electrochemical control of transport through membranes has been demonstrated (see, e.g., Grodzinsky and Weiss, 1985, and references therein). Polymer-modified electrodes have been used to investigate permeation through polymer films and to produce membranes of controlled ionic permeability, "ion gate electrodes" (see, e.g., Burgmayer and Murray, 1984; Ewing *et al.*, 1985; Krishnan *et al.*, 1984, and references therein).

REFERENCES

Abruna, H. D., and Bard, A. J. (1981). *J. Am. Chem. Soc.* **103**, 6898–6901.
Abruna, H. D., Denisevich, P., Umana M., Meyer, T. J., and Murray, R. W. (1981). *J. Am. Chem. Soc.* **103**, 1–5.
Albery, W. J., and Hillman, A. R. (1981). *Annu. Rep. C, R. Soc. Chem.* **10**, 377–436.
Albery, W. J., Compton, R. G., and Jones, C. C. (1984). *J. Am. Chem. Soc.* **106**, 469–473.
Allen, P. M., Hill, H. A. O., and Walton, N. J. (1984). *J. Electroanal. Chem.* **178**, 69.
Andrieux, C. P., Dumas-Bouchiat, J. M., and Saveant, J. M. (1982). *J. Electroanal. Chem.* **131**, 1–35.
Anson, F. C. (1975). *Acc. Chem. Res.* **8**, 400.
Anson, F. C., Ohsaka, T., and Saveant, J.-M. (1983a). *J. Am. Chem. Soc.* **105**, 4883–4890.
Anson, F. C., Saveant, J. M., and Shigehara, K. (1983b). *J. Phys. Chem.* **87**, 214.
Anson, F. C., Ni, C.-L., and Saveant, J.-M. (1985). *J. Am. Chem. Soc.* **107**, 3442–3450.
Aramata, A., and Ohnishi, R. (1984). *J. Electroanal. Chem.* **162**, 153–162.
Babai, M., and Gottesfeld, S. (1980). *Surf. Sci.* **96**, 461–475.
Bard, A. J. (1983). *J. Chem. Educ.* **60**, 302–304.
Bard, A. J., and Faulkner, L. R. (1980). "Electrochemical Methods: Fundamentals and Applications". Wiley, New York.
Bard, A. J., Fan, F.-R., Hope, G. A., and Keil, R. G. (1983). *ACS Symp. Ser. No.* 211, 93.
Bettelheim, A., Chan, R. J. H., and Kuwana, T. (1979). *J. Electroanal. Chem.* **99**, 391–397.
Bettelheim, A., Chan, R. J. H., and Kuwana, T. (1980). *J. Electroanal. Chem.* **110**, 93–102.
Bocarsly, A. B., and Sinha, S. (1982). *J. Electroanal, Chem.* **137**, 157.
Bookbinder, D. C., and Wrighton, M. S. (1980). *J. Am. Chem. Soc.* **102**, 5125–5127.
Bookbinder, D. C., Lewis, N. S., Bradley, M. G., Bocarsly, A. B., and Wrighton, M. S. (1979). *J. Am. Chem. Soc.* **101**, 7721–7723.
Bookbinder, D. C., Bruce, J. A., Dominey, R. N., Lewis, N. S., and Wrighton, M. S. (1980). *Proc. Natl. Acad. Sci. U.S.A.* **77**, 6280.
Bull, R. A., Fan, F.-R., and Bard, A. J. (1982). *J. Electrochem. Soc.* **129**, 1009–1015.
Bull, R., Fan, F.-R., and Bard, A. J. (1984). *J. Electrochem. Soc.* **131**, 687–689.
Burgmayer, P., and Murray, R. W. (1982). *J. Electroanal. Chem.* **135**, 335–342.
Burgmayer, P., and Murray, R. W. (1984). *J. Phys. Chem.* **88**, 2515–2521.
Buttry, D. A., and Anson, F. C. (1983). *J. Am. Chem. Soc.* **105**, 685–689.
Buttry, D. A., and Anson, F. C. (1984). *J. Am. Chem. Soc.* **106**, 59–64.
Carlin, C. M., Kepley, L. J., and Bard, A. J. (1985). *J. Electrochem. Soc.* **132**, 353–359.
Chao, S., Robbins, J. L., and Wrighton, M. S. (1983) *J. Am. Chem. Soc.* **105**, 181–188.
Collman, J. P., Marrocco, M., Denisevich, P., Koval, C., and Anson, F. C. (1979). *J. Electroanal. Chem.* **101**, 117–122.
Collman, J. P., Denisevich, P., Konai, Y., Marrocco, M., Koval, C., and Anson, F. C. (1980). *J. Am. Chem. Soc.* **102**, 6027–6036.
Cotton, T. M., Schultz, S. A., and Van Duyne, R. P. (1980). *J. Am. Chem. Soc.* **102**, 960.
Daube, K. A., Harrison, D. J., Mallouk, T. E., Ricco, A. J., Chao, S., and Wrighton, M. S. (1985). *J. Photochem.* **29**, 71–88.

Dautartas, and Evans, J. F. (1980). *J. Electroanal. Chem.* **109**, 301–315.

Denisevich, P., Willman, K. W., and Murray, R. W. (1981). *J. Am. Chem. Soc.* **103**, 4727–4737.

Denisevich, P., Abruna, H. D., Leidner, C. R., Meyer, T. J., and Murray, R. W. (1982). *J. Am. Chem. Soc.* **21**, 2153–2161.

De Nora, O. (1970). *Chem.-Ing.-Tech.* **42**, 222.

Diaz, A. F., Castillo, J. I., Logan, J. A., and Lee, W.-Y. (1981). *J. Electroanal. Chem.* **129**, 115–132.

Doblhofer, K., Nolte, D., and Ulstrup, J. (1978). *Ber. Bunsenges. Phys. Chem.* **82**, 403.

Eddowes, M. J., Hill, H. A. O., and Uosaki, K. (1979). *J. Am. Chem. Soc.* **101**, 7113–7114.

Espenscheid, M. W., and Martin, C. R. (1985). *J. Electroanal. Chem.* **188**, 73–84.

Ewing, A. G., Feldman, B. J., and Murray, R. W. (1985). *J. Phys. Chem.* **89**, 1263–1269.

Fan, F. R., and Bard, A. J. (1986). *J. Electrochem. Soc.* **133**, 293.

Faulkner, L. R. (1984). *Chem. Eng. News* **62**, 28–45.

Fischer, A. B., Wrighton, M. S., Umana, M., and Murray, R. W. (1979). *J. Am. Chem. Soc.* **101**, 3442–3446.

Fujihira, M. (1986). *In* "Topics in Organic Electrochemistry" (A. J. Fry and W. E. Britton, eds.), pp. 255–294. Plenum, New York.

Fuki, M., Kitani, A., Degrand, C., and Miller, L. L. (1982). *J. Am. Chem. Soc.* **104**, 28–33.

Gaudiello, J. A., Ghosh, P. K., and Bard, A. J. (1985). *J. Am. Chem. Soc.* **107**, 3027–3032.

Ghosh, P. K., and Bard, A. J. (1983). *J. Am. Chem. Soc.* **105**, 5691.

Grodzinsky, A. J., and Weiss, A. M. (1985). *Sep. Purif. Methods* **14**, 1–40.

Hawn, D. D., and Armstrong, N. R. (1978). *J. Phys. Chem.* **82**, 1288–1295.

Heineman, W. R., Kuwana, T., and Hartzell, C. R. (1972). *Biochem. Biophys. Res. Commun.* **49**, 1.

Heineman, W. R., Kuwana, T., and Hartzell, C. R. (1973). *Biochem. Biophys. Res. Commun.* **50**, 892–900.

Henning, T. P., White, H. S., and Bard, A. J. (1981). *J. Am. Chem. Soc.* **103**, 3937.

Hill, H. A. O., Page, D. J., Walton, N. J., and Whitford, D. (1985). *J. Electroanal. Chem.* **187**, 315–324.

Ikeda, T., Leidner, L. R., and Murray, R. W. (1982). *J. Electroanal, Chem.* **138**, 343–365.

Itaya, K., and Bard, A. J. (1978). *Anal. Chem.* **50**, 1487–1489.

Itaya, K., and Bard, A. J. (1985). *J. Phys. Chem.* **89**, 5565.

Itaya, K., Shibayama, K., Akahoshi, H., and Toshima, S. (1982). *J. Appl. Phys.* **53**, 1498.

Izelt, G., Day, R. W., Kinstle, J. F., and Chambers, J. Q. (1983). *J. Phys. Chem.* **87**, 4592–4598.

Jester, C. P., Rocklin, R. D., and Murray, R. W. (1980). *J. Electrochem. Soc.* **127**, 1979–1985.

Kao, W.-H. and Kuwana, T. (1984). *J. Am. Chem. Soc.* **106**, 473–476.

Karweik, D. H., Miller, C. W., Porter, M. D., and Kuwana, T. (1982). *ACS Symp. Ser. No.* **199**, 89–119.

Katayama-Aramata, A., and Ohnishi, R. (1983). *J. Am. Chem. Soc.* **105**, 658–659.

Kerr, J. B., Miller, L. L., and Van De Mark, M. R. (1980). *J. Am. Chem. Soc.* **102**, 3383–3389.

Kobos, R. K. (1980). *In* "Ion-Selective Electrodes in Analytical Chemistry" (H. Freiser, ed.), Vol. 2, pp. 1–84. Plenum, New York.

Komori, T., and Nonaka, T. (1984). *J. Am. Chem. Soc.* **106**, 2656–2659.

Korfhage, K. M., Ravichandran, K., and Baldwin, R. P. (1984). *Anal. Chem.* **56**, 1514.

Krishnan, M., White, J. R., Fox, M. A., and Bard, A. J. (1983). *J. Am. Chem. Soc.* **105**, 7002–7003.

Krishnan, M., Zhang, X., and Bard, A. J. (1984). *J. Am. Chem. Soc.* **106**, 7371.

Kuo, K.-N., and Murray, R. W. (1982). *J. Electroanal. Chem.* **131**, 37–60.

Lane, R. F., and Hubbard, A. T. (1973a). *J. Phys. Chem.* **77**, 1401–1410.

Lane, R. F., and Hubbard, A. T. (1973b). *J. Phys. Chem.* **77**, 1411.

Lau, A. N. K., and Miller, L. L. (1983a). *J. Am. Chem. Soc.* **105**, 5271–5277.

Lau, A. N. K., and Miller, L. L. (1983b). *J. Am. Chem. Soc.* **105**, 5278–5284.

Laviron, E. (1981). *J. Electroanal. Chem.* **122**, 37–44.

Leddy, J., and Bard, A. J. (1985). *J. Electroanal. Chem.* **189**, 203–219.

Leddy, J., Bard, A. J., Maloy, J. T., and Saveant, J. M. (1985). *J. Electroanal. Chem.* **187**, 205–227.

Lee, C. W., Gray, H. B., Anson, F. C., and Malmstrom, B. G. (1984). *J. Electroanal. Chem.* **172**, 289.

LeMest, Y., L'Her, M., Courtot-Coupez, J., Collman, J. P., Evitt, E. R., and Bencosme, C. S. (1985). *J. Electroanal. Chem.* **184**, 331–346.

Lennox, J. C., and Murray, R. W. (1978). *J. Am. Chem. Soc.* **100**, 3710.

Lewis, N. S., and Lieber, C. M. (1984). *J. Am. Chem. Soc.* **106**, 5033.

Lewis, N. S., and Wrighton, M. S. (1981). *Science* **211**, 944–947.

Liu, H. Y., Weaver, M. J., Wang, C. B., and Chang, C. K. (1983). *J. Electroanal. Chem.* **145**, 43.

Lu, P. W. T., and Srinivasan, S. (1979). *J. Appl. Electrochem.* **9**, 269–283.

Majda, M., and Faulkner, L. F. (1982). *J. Electroanal. Chem.* **137**, 149–156.

Matsue, T., Fujihara, M., and Osa, T. (1979). *J. Electrochem. Soc.* **126**, 500.

Merz, A., and Bard, A. J. (1978). *J. Am. Chem. Soc.* **100**, 3222–3223.

Meshitsuka, S., Ichikawa, M., and Tamaru, K. (1974). *J.C.S. Chem. Commun.* p. 158.

Miller, L. L., Lau, A. N. K., and Miller, E. K. (1982). *J. Am. Chem. Soc.* **104**, 5242.

Moses, P. R., Wier, L., and Murray, R. W. (1975). *Anal. Chem.* **47**, 1882.

Murray, R. W. (1980). *Acc. Chem. Res.* **13**, 135–141.

Murray, R. W. (1983). *Electroanal. Chem.* **13**, 191–363.

Murray, C. G., Nowak, R. J., and Rolison, D. R. (1984). *J. Electroanal. Chem.* **150**, 205.

Neff, V. D. (1978). *J. Electrochem. Soc.* **125**, 886.

Nowak, R., Shultz, F. A., Umana, M., Abruna, H. D., and Murray, R. W. (1978). *J. Electroanal. Chem.* **94**, 219.

Ogumi, Z., Nishio, K., and Yoshizawa, S. (1981). *Electrochim. Acta* **26**, 1779–1782.

Ogumi, Z., Ohashi, S., and Takehara, Z. (1985). *Electrochim. Acta* **30**, 121–124.

O'Toole, T. R., Margerum, L. D., Westmoreland, T. D., Vining, W. J., Murray, R. W., and Meyer, T. J. (1985), *J.C.S. Chem. Commun.* 1416–1417.

Oyama, N., and Anson, F. C. (1979). *J. Am. Chem. Soc.* **101**, 3450–3456.

Oyama, N., Ohsaka, T., Kaneko, M., Sato, K., and Matsuda, H. (1983). *J. Am. Chem. Soc.* **105**, 6003.

Peerce, P. J., and Bard, A. J. (1980a). *J. Electroanal. Chem.* **112**, 97–115.

Peerce, P. J., and Bard, A. J. (1980b). *J. Electroanal. Chem.* **114**, 89–111.

Pickup, P. G., Kuo, K. N., and Murray, R. W. (1983). *J. Electrochem. Soc.* **130**, 2205–2216.

Raoult, E., Sarrazin, J., and Tallec, A. (1984). *J. Appl. Electrochem.* **14**, 639–643.

Rocklin, R. D., and Murray, R. W. (1981). *J. Phys. Chem.* **85**, 2104–2112.

Rubinstein, I., and Bard, A. J. (1980). *J. Am. Chem. Soc.* **102**, 6641–6642.

Rubinstein, I., and Bard, A. J. (1981). *J. Am. Chem. Soc.* **103**, 5007–5013.

Rudzinski, W. E., and Bard, A. J. (1986). *J. Electroanal. Chem.* **199**, 323–340.

Samuels, G. J., and Meyer, T. J. (1981). *J. Am. Chem. Soc.* **103**, 307–312.

Santhanam, K. S. V., Jespersen, N., and Bard, A. J. (1977). *J. Am. Chem. Soc.* **99**, 274–275.

Sarrazin, J., and Tallec, A. (1982). *J. Electroanal. Chem.* **137**, 183.

Sharp, M., Montgomery, D. D., and Anson, F. C. (1985). *J. Electroanal. Chem.* **194**, 247–259.

Snell, K. D., and Keenan, A. G. (1979). *Chem. Soc. Rev.* **8**, 259–282.

Stankovich, M. T., and Bard, A. J. (1978). *J. Electroanal. Chem.* **86**, 188–199.

Tachikawa, H., and Faulkner, L. R. (1976). *Chem. Phys. Lett.* **39**, 436–441.

Takeuchi, E. S., and Murray, R. W. (1985). *J. Electroanal, Chem.* **188**, 49–57.

Umana, M., Denisevich, P., Rolison, D. R., Nakahama, S., and Murray, R. W. (1981). *Anal. Chem.* **53**, 1170–1175.

Untereker, D. T., Lennox, J. C., Wier, L. M., Moses, P. R., and Murray, R. W. (1977). *J. Electroanal. Chem.* **81**, 309–318.

Van De Mark, M. R., and Miller, L. L. (1978). *J. Am. Chem. Soc.* **100**, 3223–3225.

Van Koppenhagen, J. E., and Majda, M. (1985). *J. Electroanal. Chem.* **189**, 379–388.

Watkins, B. F., Behling, J. R., Kariv, E., and Miller, L. L. (1975). *J. Am. Chem. Soc.* **97**, 3549.

White, B. A., and Murray, R. W. (1985). *J. Electroanal. Chem.* **189**, 345–352.

Wrighton, M. S., Austin, R. G., Bocarsly, A. B., Bolts, J. M., Haas, O., Legg, K. D., Nadjo, L., and Palazzoto, M. C. (1978). *J. Electroanal. Chem.* **87**, 429–433.

Yacynych, A. M., and Kuwana, T. (1978). *Anal. Chem.* **50**, 640–645.

Yamagishi, A., and Aramata, A. (1984). *J.C.S. Chem. Commun.* p. 452.

Yeager, E. (1976). *NBS Spec. Publ. (U.S.)* No. 455, 203.

Yeager, E. (1981). *J. Electrochem. Soc.* **128**, 160C–171C.

Yeh, P., and Kuwana, T. (1977). *Chem. Lett.* 1145–1148.

Zak, J., and Kuwana, T. (1983). *J. Am. Chem. Soc.* **104**, 5514–5515

Zinger, B., and Miller, L. L. (1984). *J. Am. Chem. Soc.* **106**, 6861–6863.

6 PRACTICAL CONSIDERATIONS: HOW TO SET UP A SUPPORTED REAGENT

André Cornélis

Université de Liège
Institut de Chimie
Sart-Tilman par B-4000
Liège, Belgium

I. INTRODUCTION

Utmost practical simplicity is the most readily apparent feature of the new supported reagents and catalysts (McKillop and Young, 1979a,b). The reactions are carried at ambient temperature and in open flasks. A simple filtration eliminates the inorganic residues, leaving an uncontaminated organic phase. The solvents are easy to recover and recycle safely. Due to the nearly quantitative yields and high selectivities, mere evaporation often provides pure products. Most examples to be discussed here use alumina, silica gel, or clays as inorganic supports; this choice (somewhat arbitrary) is also representative of current practice.

II. SETTING UP A SUPPORTED REAGENT

The supports themselves can be active through their Brønsted acidity or basicity. Astute organic chemists have upturned a drawback of chromatographic alumina, which is prone to undesirable secondary reactions during separations, to design easy and efficient processes (Posner, 1978; Texier-Boullet and Foucaud, 1982). Similar use has been made of silica gel (Hojo and Masuda, 1975a,b). Residual water within the support is often

PREPARATIVE CHEMISTRY
USING SUPPORTED REAGENTS

essential, and prior dehydration of the support may be detrimental (Degl'Innocenti *et al.*, 1980).

A first degree of modification is indeed dehydration of the support by heat and/or vacuum. Dried overnight at 120°C, the basic calciosilicate xonotlite efficiently catalyzes Knoevenagel condensations (Chalais *et al.*, 1985b). Dried silica gel promotes the Beckmann rearrangement of oxime tosylates (Costa *et al.*, 1982). Dehydration may be essential to selectivity: while secondary acyclic tosylates are converted into 2 : 1 mixtures of olefins and alcohols by commercial activity I alumina, this ratio increases to 10 : 1 with a dehydrated support (Posner and Gurria, 1976). Texier-Boullet *et al.* (1985) report examples in which the reaction is switched from Knoevenagel to Wittig–Horner by increasing the humidity of alumina.

Deuterium–protium exchange is another minimal alteration effected by thermal dehydration and rehydration with D_2O. Exchange of the acidic hydrogens of alumina (Mislow *et al.*, 1964; Peri, 1965) must be performed in repeated cycles; heating, even at elevated temperature, does not remove all hydrogen atoms. Exchange efficiency has been studied, both experimentally and on a simple model, and a device has been proposed to perform it (Gaetano *et al.*, 1985). When using deuterated alumina to introduce deuterium in organic substrates, chromatography is advisable, except with very slow exchange rates.

Ion exchange, a bigger perturbation, is performed on certain clays (Foster, 1951). Cation hydration, ionic size, charge, and polarizability govern convertibility (Swartzen-Allen and Matijevic, 1974). After exchange by a salt solution, the material recovered by centrifugation is suspended in deionized water and is reisolated by centrifugation; these washing cycles are repeated until disappearance of the exchanging salt anion from discarded water (Tennakoon *et al.*, 1974; Adams *et al.*, 1983). The clay is then dried at a moderate temperature (40–100°C) and finely ground. In the exchange of quaternary alkylammonium cations, a significant amount intercalates as ion pairs together with their original counterion, and this requires very thorough washing (Lagaly, 1984). Pinnavaia *et al.* (1979) and Raythatha and Pinnavaia (1983) exchange sodium in hectorite by two different methods: (1) rhodium cations are first inserted, prior to their phosphine ligands; and (2) preformed cationic complexes of rhodium triphenylphosphine are directly introduced in the intersticial space of this mineral.

Various techniques exist for the deliberate introduction of a reagent into or onto an inorganic solid. The first reported supported reagent (Fétizon and Golfier, 1968) was made by precipitation of silver carbonate on celite. A much more frequent procedure involves evaporation of the solvent (aqueous or organic) from a slurry of the support in a solution of the reagent. Typical examples are the "deposition" (whatever its exact nature may be) of potassium *t*-butoxide on xonotlite from a *t*-butanol solution (Chalais *et al.*, 1985b) or deposition of copper(II) nitrate on K10 clay ("claycop") from an acetonic solution of the hydrated salt (Balogh *et al.*, 1984). Use of water as solvent demands thorough drying, by prolonged heating *in vacuo* or by carrying away the residual water through distillation of an auxiliary solvent such as ethanol; alumina–KF has been prepared using both ways (Yamawaki and Ando, 1979; Villemin, 1985). Adequate solvent elimination and drying is often a necessity; K10 clay supported ferric nitrate ("clayfen") is extremely sensitive to such factors (Cornélis *et al.*, 1984; Cornélis and Laszlo, 1985). Dry dispersion, using mechanical grinding or the simple mixing of the solid reagent with the support is also feasible (Fadel and Salaün, 1985a). With low-melting-point hydrated salts, such as ferric chloride hexahydrate, both deposition and drying may be performed by azeotropic distillation of water from a suspension of all the components in a suitable liquid (Chalais *et al.*, 1985a).

In some instances the substrate, instead of the reagent, is first deposited on the support. For dry ozonations, the organic reactant is preadsorbed on silica gel, then saturated with ozone gas at low temperature ($-78°C$), and finally allowed to warm slowly to room temperature (Cohen *et al.*, 1975). Deposition of the substrate on the inorganic support can be done by direct mixing, by impregnation using a volatile solvent, or from the gas phase. De Meijere *et al.* designed a convenient device using impregnation by evaporation or sublimation; they also describe an apparatus for isothermal ozonation on solid supports (Zarth and de Meijere, 1985).

III. REPRESENTATIVE EXPERIMENTAL PROCEDURES

A. Anhydrous Ferric Chloride Dispersed on Silica Gel: Ring Enlargement of Cycloalkanols

Fadel and Salaün prepare this reagent by simple mechanical stirring of the two solids.

Preparation of Anhydrous Ferric Chloride on Silica Gel

> In a 250 mL flask, chromatographic grade silica gel (50 g) (70–230 mesh) and anhydrous ferric chloride (4 g) were vigorously stirred, without solvent, at room temperature for 24 hours. A pale yellowish green powder was obtained. [Reprinted with permission from Fadel and Salaün (1985a), Copyright 1985, Pergamon Journals Ltd.]

This reagent induces the specific ring enlargement of tertiary cyclobutanols and cyclopentanols into cyclopentene and cyclohexene derivatives.

Dehydration and Ring Enlargement of 1-t-Butylcyclopentanol

To 2.4 g of anhydrous $FeCl_3$–SiO_2 in a 10 mL flask was added 0.3 g of 1-*t*-butylcyclopentanol and the mixture was stirred at room temperature for 3 hours. The product was then distilled directly under vacuum (15 mm) into a trap cooled by liquid nitrogen to yield 230 mg (92.7%) of 1,6,6-trimethyl-1-cyclohexene. [Reprinted with permission from Fadel and Salaün (1985b), Copyright 1985, Pergamon Journals Ltd.]

B. Alumina–Potassium Fluoride: Anionic Condensation by Percolation

This solid base offers two attractive practical advantages: it promotes anionic reactions by simple percolation, and it can be easily regenerated. The Knoevenagel condensation of phenylacetonitrile with benzaldehyde (Villemin, 1985) is most illustrative. It combines speed, efficiency, convenience, cheapness, and stereospecificity (yield 95%; 100% *Z*).

Preparation of the Reagent

Potassium fluoride (20 g) was dissolved in water (200 mL) and mixed with neutral chromatographic alumina (Woelm-N, 2087; 30 g). The mixture was evaporated with a rotary evaporator under vacuum at 100°C. Ethanol (4×50 mL) was added for carrying away water. The dry white solid was placed in an oven at 100°C for 12 h and stored without change, in a closed flask. [From Villemin (1985).]

Knoevenagel Condensation

Phenylacetonitrile (10 mmol) and benzaldehyde (10 mmol) in methylene chloride (2 cm³) were percolated through a column of alumina–potassium fluoride (height 35 mm, $\phi = 12$ mm, 5 g; 49,2 mmol of KF). The product was eluted with methylene chloride after 300 s. [From Villemin (1985).]

Regeneration of the Solid Base

Alumina–potassium fluoride was recovered after washing with methanol, activated at 200°C during 16 h, and then became usable for further anionic reactions. [From Villemin (1985).]

C. Trimethyl Orthoformate–Methanol Impregnated on Montmorillonite K10: Conversion of Carbonyl Compounds to Acetals

Taylor and Chiang (1977) prepare methyl ketals on montmorillonite K10 impregnated with trimethyl orthoformate, or with a mixture of trimethyl orthoformate and methanol suspended in an inert solvent such as carbon tetrachloride. Most reactions are completed within a few minutes. The recovered acidic clay can be recycled.

Conversion of Carbonyl Compounds to Acetals: General Procedure

The K10/trimethyl orthoformate reagent is readily prepared by stirring K10 montmorillonite clay (20 g) with trimethyl orthoformate (30 mL) or with a mixture of trimethyl orthoformate (15 mL) and methanol (15 mL), followed by filtration. The resultant wet filter cake is used directly without further treatment. The carbonyl substrate (10 g) is stirred with a suspension of the K10/trimethyl orthoformate reagent (27–30 g) in an inert solvent (60 mL) such as carbon tetrachloride or hexane until T.L.C. analysis shows complete conversion. The product is isolated by filtration, washing of the filtrate with sodium hydrogen carbonate solution followed by water, drying, evaporation of the solvent, and distillation or crystallization. The clay which is removed by filtration can be reactivated by stirring it again with trimethyl orthoformate or trimethyl orthoformate/methanol. [From Taylor and Chiang (1977).]

D. Thallium Trinitrate Deposited on Montmorillonite K10: Oxidative Rearrangement of Olefins and Enols via Oxythallation

The deposition of thallium trinitrate on K10 clay affords a powerful and selective reagent (Taylor *et al.*, 1976). This procedure also minimizes the risks of scattering of thallium residues.

Oxidative Rearrangement of Olefins and Enols: General Procedure

Trimethyl orthoformate (125 mL) and methanol (100 mL) are placed in a 500 mL round-bottomed flask equipped with a magnetic stirring bar. The mixture is stirred, and thallium(III) nitrate trihydrate (49 g, 0.11mol) is added in one portion. Stirring is continued until all of the thallium salt has dissolved (approximately 2 min) and then montmorillonite K10 (110 g) is introduced in one portion through a powder funnel. The reaction is slightly exothermic. The powder funnel is rinsed with methanol (25mL) and the mixture is stirred for a further 5 min. The flask is then transfered to a rotary evaporator and the excess solvents are removed by distillation under reduced pressure (bath temperature 60–70°C and 15–25 torr). This is complete in 1.5 h and 170 g of the reagent system is obtained as a dry, free-flowing, pale yellow powder. The supported reagent (1.1 eq.) is added to a solution of the olefin or enolizable ketone (1 eq.) in an inert solvent (dichloromethane, carbon tetrachloride, toluene, dioxane, or heptane) and the reaction mixture is stirred at room temperature until a starch iodide test for thallium (III) is negative. Products are then isolated by removal of the spent reagent system by filtration, washing of the filtrate with aqueous sodium hydrogen carbonate solution, then water, drying, evaporation of the solvent and recrystallization or distillation of the crude product. [From McKillop and Young (1976b).]

E. Adsorbed Substrates Reacting with Gaseous Reagents

1. Ozone Oxidation of Primary Amines Deposited on Silica Gel

Using heterogeneous conditions, Keinan and Mazur (1977) obtain better yields of nitro compounds than in homogeneous conditions. Moreover, their method minimizes the amount of by-products.

Dry Ozonation of Primary Amines: General Procedure

The amines were adsorbed on the silica gel by thorough mixing with dry silica gel (Merck, Kieselgel 60, 70–230 mesh, dried for 24 h at 450 °C). The silica gel (ca. 30 g) containing the adsorbed material (0.1–0.2% w/w) was cooled to −78°C and a stream of 3% ozone (in oxygen) passed through it. The silica gel was allowed to warm to room temperature, and the material was eluted with an organic solvent. [Excerpted with permission from Keinan and Mazur (1977). Copyright 1977 American Chemical Society.]

2. Acetylation by Ketene of Alcohols Supported on Solid Adsorbents

Silica gel, celite, alumina, magnesium and titanium oxides, or chacoal, are some solids on which gaseous ketene readily acetylates alcohols at room temperature.

Acetylation of 1-Hexanol

1-Hexanol (1 g) was placed in a flask containing 10 g of silica gel powder and allowed to stand (1 h) until the lumps disappeared. Then the silica gel powder was trans-fered to a glass fritted filter of 6.5 cm diameter equipped with a rubber stopper which had a small outlet column containing a small amount of alumina powder. The alumina powder was used to reduce the evaporation losses of the alcohol. The acetylation was conducted by introduction of ketene in a stream of nitrogen from the bottom of the glass filter. The end of the reaction was detected by glc analyses of the sample or by change in color of the alumina powder from white to yellow. Ketene (2 mL) kept in a dry ice trap is required to complete the reaction. The silica gel and the alumina powder were washed with a solvent. After concentration of the eluate under reduced pressure, the product was purified by column chromatography to remove the polymerization products of ketene. [From Chihara *et al.* (1982).]

F. Substrate and Reagent Deposited on Separate Supports: Nitration of Polynuclear Aromatics by Cerium (IV) Ammonium Nitrate

Polynuclear arenes and hydroxynaphthalenes deposited on silica gel are selectively mononitrated by cerium(IV) ammonium nitrate deposited on another batch of silica gel, thus avoiding the polynitration and the overoxidation often observed in solution.

Nitration of Anthracene

Anthracene (1.5 g, 5 mmol) and cerium(IV) ammonium nitrate (3 g, 5,5 mmol) are separately dissolved in acetonitrile (4 mL) and these solutions are mixed to a slurry each with silica gel (2 g and 5 g, respectively). The slurries are dried in a hot air oven at 60–65°C for 1 hour. The dried mass is mixed well and this mixture is added to a prepacked column of silica gel (80 g). The column is eluted with petroleum ether/benzene (9/1). Concentration of the eluate affords 9-nitroanthracene as yellow needles; yield 0.72 g (55%). [From Chawla and Mittal (1985).]

G. Stabilization of a Reagent: Clay-Supported Ferric Nitrate (Clayfen)

The preparation of this reagent requires the strict observance of its set-up procedure. It deactivates upon aging, and most applications require freshly prepared clayfen.

Whereas these procedures worked out very safely in the hands of the author, nitrates are potentially dangerous compounds, and appropriate caution is to be applied in each step. In particular, avoid confinement conditions, and proceed to significant scaling-up only after appropriate safety tests.

Preparation of Clayfen

Iron(III) nitrate nonahydrate (22.5 g) is added to acetone (375 mL) in a 1 L pear-shaped evaporating flask. The mixture is stirred vigorously for ca. 5 min, until complete dissolution of the crystals of hydrated iron(III) nitrate. The first-formed homogeneous rusty brown solution turns, after a short time, into a muddy, light brown suspension. K10 clay (30 g) is added in small amounts and stirring is continued for another 5 min. The solvent is then eliminated from the resulting suspension, under reduced pressure (rotary evaporator) on a water bath at 50°C. After 30 minutes, the dry solid crust adhering to the walls of the flask is flaked off and crushed with a spatula, and further rotary evaporator drying, in the same conditions as above, is resumed for another 30 min. This procedure yields about 50 g of K10 clay supported iron(III) nitrate ("clayfen") as a yellow, floury powder. [From Cornélis and Laszlo (1985).]

WARNINGS (Cornélis and Laszlo, 1985)

1. Prolongated heating, or use of a constant temperature bath over 50°C, may yield an unstable reagent, which decomposes in a vigorous exothermic reaction, with emission of a large amount of nitrogen dioxide fumes. This decomposition generally takes place within 15 min after the end of an incorrect preparation of clayfen.

2. Preparation of clayfen includes flaking off of the dry solid crust formed in the evaporator. The physical state of the reagent at this step is crucial. If this removal is performed while the solid is still a little muddy, it will aggregate in spheres which remain wet inside, and the resulting reagent will display little or no activity, or will decompose as described earlier. This is best avoided by strict observance of the experimental conditions described (quantities, flask volume, and shape).

3. Clayfen spontaneously and slowly loses nitrous fumes, and should never been stored in a closed flask. [From Cornélis and Laszlo (1985).]

This versatile reagent reacts as a nitrosonium ion source, or as a mild nitrating agent, for example towards phenols.

General Procedure for Phenols Nitration

To a suspension of 5 g of freshly prepared "clayfen" in 50–100 mL of solvent (according to the solubility of the reaction products) are added 10 mmol of the phenolic compound. The mixture is stirred vigorously at room temperature, and, at the end of the reaction (monitored by TLC or by GLC) typically from 2 hours to 3 days) the inorganic residue is separated by filtration on sintered glass. The solid is thoroughly washed with 25 mL of solvent. The solvent is eliminated on a rotary evaporator, and the crude products are purified or separated by column chromatography on silica, using a cyclohexane/ethyl acetate mixture (75 : 25 v/v) as eluent [From Cornélis and Laszlo (1985).]

IV. GLOSSARY OF TERMS, IN THE FORM OF PRACTICAL HINTS

A. Centrifugation

Centrifugation is convenient for separating the liquid from the solid phase when filtration is prevented by the rheology of the mixture.

B. Choice of the Support

This is still largely pragmatic; a preliminary selection is made by first considering the chemical requirements of the reaction (e.g., acid or base) and the chemical compatibility with reactants and products, then the rheological properties of the supported reagent. The final choice is helped by experimental screening. A better understanding of the role of the support, lacking at present, is desirable.

C. Clogging and Coking

Obstructions of the reagent microstructure are seldom observed, as a result of the mild conditions. Problems have been reported with zinc permanganate deposited on silica gel (Wolfe and Ingold, 1983).

D. Column, Film, or Slurry

Column procedures offer the advantage of a continuous process, but percolation may be extremely slow with low granulometry supports (clays for instance), for which the slurry method is usually preferred. Films anchored on the walls of a flow system offer the advantages of a continuous process without the problem of clogging. Due to the relatively minute amounts reagent, they can be used only in catalytic processes or in stoichiometric transformation where regeneration of the film is possible. The mechanical resistance of the coating is another limit to their use. Most results using films still remain unpublished; this technology is most promising for the near future.

E. Creeping

Creeping easily happens with clays. Pinnavaia *et al.* (1979) use special flasks to minimize this phenomenon.

F. Filtration

Vacuum filtration on porosity-2 large-size sintered-glass filters is convenient in most cases (the thickness of the cake should not exceed 1 cm). These filters must be washed immediately after use, otherwise they will become clogged and hard to clean.

G. Glassware Care

To avoid permanent furring, the flasks must be washed as soon as possible. Alternating water and an hydrophilic organic solvent (acetone, alcohol) is often very efficient.

H. Product Recovery

Due to the high specific area of the support, substantial amounts of the products may remain adsorbed after filtration. Washing the cake with suitable solvents often allows complete recovery. This is best performed by resuspending the solid in the solvent. Due to channel formation, washing the cake directly on the filter is often inefficient.

I. Regeneration

Due to the low cost of supported reagents, little attention has been paid to their regeneration. Villemin (1985) regenerates alumina–KF used in Knoevenagel condensations.

J. Rheology

Good rheological properties of the reaction medium (in other words, the mechanical properties, the flow properties, etc. of the muds that form) are essential to easy setup and workup, and to maximum efficiency. The use of solvents giving light, dispersed, and easy to stir suspensions avoids having reagents stick to the walls of the reactors and to the stirring bar. It is often cumbersome to dissolve or to disrupt a compact solid made of the coagulated products of the reaction. Thixotropy is also a source of major problems.

K. Supported Reagent

This general name covers a broad range of reagents and catalysts using at least one solid phase under heterogeneous conditions. This phase may be a pure, well-characterized compound or a mixture; it may act as a stoichiometric agent or as a catalyst. This chapter is restricted to reagents on inorganic solid supports. Polymer-supported reagents require the use of a different set of techniques.

L. Stability

No rule is available yet to predict the stability of a supported reagent. Two kindred systems such as clayfen and claycop (clay-supported ferric or cupric nitrates) behave very differently; whereas the half-life of the former is only a few hours, the latter sustains storage for several months without any special care (Cornélis and Laszlo, 1985). This feature, of great practical import, should be investigated carefully.

M. Stirring (Magnetic)

Abrasion by the slurries scrapes off teflon stirring bars, eventually exposing their metallic core. Muds may be hard to set in motion; best results are obtained with dumbbell-shaped stirring bars.

N. Test Reaction

A test reaction should be inexpensive, easy to set up, rapid, and easily interpreted (ideally, its completion should be detected visually). An example is the thiophenol test of reactivity for nitrosating reagents (Cornélis and Laszlo, 1985).

O. Thixotropy

With hydrophilic swelling clays, it is advisable to avoid aqueous media, in which thixotropic gels form.

P. Vacillating Reaction

This refers to a synergistic and multiple coupling of two reagents in a "wolf-and-lamb" system. Kim and Regen (1983) applied this methodology to perdeuteration of the functional carbon of alcohols.

Q. Wolf-and-Lamb System

Compatibility brought about by immobilization on support(s) of two reagents otherwise reactive towards one another (Patchornik, 1982; Bergbreiter and Chandran, 1985). (See Chapter 18, Section II.H.)

V. SONICATION: A TOOL FOR THE NEAR FUTURE?

The success of ultrasonic irradiation in heterogeneous organic synthesis involving metals (Luche, 1982) makes it tempting for other reactions with other inorganic solids. Unfortunately, little is known about the mechanisms; some of the likely factors include energy transfer to the reacting partners by cavitation (Margulis, 1976), stripping of interfering species from the interface, disruption of the solid lattice which increases exposed surfaces and provides reactive sites, and improvement of mass exchanges between bulk solution and reaction areas. Whatever the exact cause, sonic waves exert a positive effect on some "dry media" and "supported reagents" processes, increasing their yields or their rates (Ando *et al.*, 1983), or even switching them to other pathways (Ando *et al.*, 1984). One has to be very careful in judging the resulting improvement (Muzart, 1985); moreover vagueness in description of procedures too often forbids comparison between results from different laboratories and prevents easy duplication. Common ultrasonic cleaners can serve as makeshifts for the more efficient sonication reactors described in the literature (Suslick and Schubert, 1983; Petrier *et al.*, 1984). Special care must be devoted to control the bulk temperature and to ascertain that the observed effects are those from the irradiation and not from some of its side effects.

VI. CONCLUSION

Hitherto, organic synthesis of fine chemicals has based itself on a majority of homogeneous reactions. This contrasts with the reliance of chemical industry upon heterogeneous catalysis: about 60% of all chemicals are manufactured today through such processes, and more than 90% of the new large-capacity plants use them (Hofmann, 1985). Improved understanding of the molecular ingredients of heterogeneous catalysis now allows rational design of new technologies (Kemball, 1984; Somorjai, 1984). The major gap between industrial processes and supported reagents methodologies arises from the fluid phase of matter used and the range of operating temperatures. Most industrial catalysts work at high temperature in the gaseous state, whereas supported reagents mostly act on liquids at room temperature. Preparative chemistry has much to gain here in looking up to industrial chemistry.

REFERENCES

Adams, J. M., Clapp, T. V., and Clement, D. E. (1983). *Clay Miner.* **18**, 411.
Ando, T., Kawate, T., Yamawaki, J., and Hanafusa, T. (1983). *Synthesis* p. 637.
Ando, T., Sumi, S., Kawate, T., Ichihara (née Yamawaki), J., and Hanafusa, T. (1984). *J.C.S. Chem. Commun.* p. 439.
Balogh, M., Cornélis, A., and Laszlo, P. (1984). *Tetrahedron Lett.* **25**, 3313.
Bergbreiter, D. E., and Chandran, R. (1985). *J. Am. Chem. Soc.* **107**, 4792.
Chalais, S., Cornélis, A., Gerstmans, A., Kołodziejśki, W., Laszlo, P., Mathy, A., and Métra, P. (1985a). *Helv. Chim. Acta* **68**, 1196.
Chalais, S., Laszlo, P., and Mathy, A. (1985b). *Tetrahedron Lett.* **26**, 4453.
Chawla, H. M., and Mittal, R. S. (1985). *Synthesis* p. 70.
Chihara, T., Takagi, Y., Teratani, S., and Ogawa, H. (1982). *Chem. Lett.* p. 1451.
Cohen, Z., Keinan, E., Mazur, Y., and Varkony, T. H. (1975). *J. Org. Chem.* **40**, 2141.
Cornélis, A., and Laszlo, P. (1985). *Synthesis* p. 909.
Cornélis, A., Laszlo, P., and Pennetreau, P. (1984). *Bull. Soc. Chim. Belg.* **93**, 961.
Costa, A., Mestres, R., and Riego, J. M. (1982). *Synth. Commun.* **12**, 1003.
Degl'Innocenti, A., Walton, D. R. M., Seconi, G., Pirazzini, G., and Ricci, A. (1980). *Tetrahedron Lett.* **21**, 3927.
Fadel, A., and Salaün, J. (1985a). *Tetrahedron* **41**, 413.
Fadel, A., and Salaün, J. (1985b). *Tetrahedron* **41**, 1267.
Fétizon, M., and Golfier, M. (1968). *C.R. Hebd. Seances Acad. Sci.*, *Ser. B* **267**, 900.
Foster, M. D. (1951). *Am. Mineral.* **36**, 717.
Gaetano, K., Pagni, R. M., Kabalka, G. W., Bridwell, P., Walsh, E., True, J., and Underwood, M. (1985). *J. Org. Chem.* **50**, 499.
Hofmann, H. (1985). *Appl. Catal.* **15**, 79.
Hojo, M., and Masuda, R. (1975a). *Synth. Commun.* **5**, 169.
Hojo, M., and Masuda, R. (1975b). *Synth. Commun.* **5**, 173.
Keinan, E., and Mazur, Y. (1977). *J. Org. Chem.* **42**, 844.
Kemball, C. (1984). *Chem. Soc. Rev.* **13**, 375.
Kim, B., and Regen, S. L. (1983). *Tetrahedron Lett.* **24**, 689.
Lagaly, G. (1984). *Philos. Trans. R. Soc. London. Ser. A* **311**, 315.
Luche, J.-L, (1982). *Actual. Chim.* Oct, p. 21.
McKillop, A., and Young, D. W. (1979a). *Synthesis* p. 401.
McKillop, A., and Young, D. W. (1979b). *Synthesis* p. 481.
Margulis, M. A. (1976). *Russ. J. Phys. Chem.* **50**, 1.
Mislow, K., Glass, M. A., Hops, H. B., Simon, E., and Wahl, G. H., Jr. (1964). *J. Am. Chem. Soc.* **86**, 1710.
Muzart, J. (1985). *Synth. Commun.* **15**, 285.
Patchornik, A. (1982). *Nouv. J. Chim.* **6**, 639.
Peri, J. B. (1965). *J. Phys. Chem.* **69**, 211.
Petrier, C., Luche, J.-L, and Dupuy, C. (1984). *Tetrahedron Lett.* **25**, 3463.
Pinnavaia, T. J., Raythatha, R., Lee, J. G.-S., Halloran, L. J., and Hoffman, J. F. (1979). *J. Am. Chem. Soc.* **101**, 6891.

Posner, G. H. (1978). *Angew. Chem., Int. Ed. Engl.* **17**, 487.
Posner, G. H., and Gurria, G. M. (1976). *J. Org. Chem.* **41**, 578.
Raythatha, R., and Pinnavaia, T. J. (1983). *J. Catal.* **80**, 47.
Somorjai, G. A. (1984). *Chem. Soc. Rev.* **13**, 321.
Suslick, K. S., and Schubert, P. F. (1983). *J. Am. Chem. Soc.* **105**, 6042.
Swartzen-Allen, S. L., and Matijevic, E. (1974). *Chem. Rev.* **74**, 385.
Taylor, E. C., and Chiang, C. S. (1977). *Synthesis* p. 467.
Taylor, E. C., Chiang, C. S., McKillop, A., and White, J. F. (1976). *J. Am. Chem. Soc.* **98**, 6750.
Tennakoon, D. T. B., Thomas, J. M., Tricker, M. J. and William, J. O. (1974). *J.C.S. Dalton* p. 2207.
Texier-Boullet, F., and Foucaud, A. (1982). *Tetrahedron Lett.* **23**, 4927.
Texier-Boullet, F., Villemin, D., Ricard, M., Moison, H., and Foucaud, A. (1985). *Tetrahedron* **41**, 1259.
Villemin, D. (1985). *Chem. Ind. (London)* p. 166.
Wolfe, D., and Ingold, C. F. (1983). *J. Am. Chem. Soc.* **105**, 7755.
Yamawaki, J., and Ando, T. (1979). *Chem. Lett.* p. 755.
Zarth, M., and de Meijere, A. (1985). *Chem. Ber.* **118**, 2429.

Part II

Physico-Chemical Studies of the Structure of Solid Supports

7
NEW TOOLS FOR THE STUDY
OF SURFACES*

Edward M. Eyring

Department of Chemistry
University of Utah
Salt Lake City, Utah 84112

I. INTRODUCTION

This chapter offers a fairly generous sampling of the currently available new and old surface tools. Techniques useful at solid–liquid interfaces are included along with many methods suitable for solid–gas and solid–high-vacuum interfaces.

To clarify the taxonomic problem let us consider three commonly used methods of investigating solid–high-vacuum interfaces: Auger electron spectroscopy (AES), x-ray photoelectron spectroscopy (XPS), and secondary ion mass spectroscopy (SIMS) (Table I). We find it convenient to lump XPS with AES because in each case an electron ejected from the surface is detected. On the other hand, SIMS has more in common with a general class of ionization techniques called particle-induced desorption (PID) methods that includes fast atom bombardment (FAB) and laser desorption (LD) since in each case an ion (or neutral) ejected from the surface is detected. In general, we will gather together methods on the basis of the common nature of the detected species.

* Financial support by the Department of Energy (Office of Basic Energy Sciences) and illuminating conversations with J. A. Gardella, Jr., M. L. Hunnicutt, and R. C. Yeates are gratefully acknowledged.

TABLE I

Comparison of Characteristics of Three Techniques for Examining
Solid–High-Vacuum Interfaces

	AES	XPS	SIMS
Probe	Electron	X-ray	Ion
Detected emission	Electron	Electron	Ion
Detectable elements	All except hydrogen and helium	All except hydrogen	All
Depth examined (Å)	10–40	40	3–10
Detection limits in atom percent	0.05–5	0.1–5	10^{-7}–10
Lateral resolution (μm)	0.1	15	10
Damages sample	Yes	No	Yes

II. DESCRIPTION OF METHODS

A. Electron Spectroscopies

1. X-Ray Photoelectron Spectroscopy

An electron, ejected from an atomic orbital by a sufficiently energetic x-ray photon (Fig. 1a), has its energy (ranging from tenths of an electron volt to kilovolts) analyzed with a resolution exceeding 0.1%. The Einstein relationship for the photoelectric effect applies. The binding energy of the core electron depends on the chemical environment of its progenitor atom in the surface. While the concept dates back to the early part of the twentieth century, use as an analytical technique only goes back to the work of Siegbahn and others in the late 1950s. The x-ray source is usually an aluminum or magnesium metal anode yielding soft K_α x-rays.

The surface sensitivity of XPS results from the maximum escape depth of a photoelectron of about 2 nm. All elements except hydrogen are detectable by XPS at a surface concentration as low as 10^{-3} at.%. From the measured chemical shifts we can deduce the oxidation state. A disadvantage of XPS has been its comparatively low spatial resolution (of the order of a few square millimeters) arising from the acceptance area of the electron energy analyzer. Recent commercial instruments achieve resolution of areas as small as 200 μm^2.

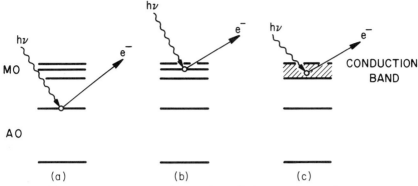

Fig. 1.(a) In XPS a monochromatic x-ray beam displaces an electron from a deep atomic orbital in the sample surface; (b) in UPS a monochromatic ultraviolet light beam displaces an electron from a molecular orbital; and (c) in photoemission an ultraviolet or visible photon ejects an electron from the conduction band of a metal. [Adapted with permission from T. A. Carlson, "Photoelectron and Auger Spectroscopy," p. 4. Plenum, New York, 1975.]

2. Ultraviolet Photoelectron Spectroscopy

Ultraviolet photoelectron spectroscopy formerly meant photoelectron studies carried out with the He I (21.2 eV, $\lambda = 58.43$ nm) and He II (40.8 eV, $\lambda = 30.38$ nm) vacuum UV resonance lines. However, with the advent of synchrotrons and lasers as intense, tunable UV sources UPS can no longer be so narrowly defined. Because UV photons have too little energy to eject most core electrons (Fig. 1b) UPS yields the energies of molecular orbitals and permits their identification in gaseous species. With some solid surfaces the outermost valence shells are no longer well-defined molecular orbitals, and we are dealing instead with band structure (Fig. 1c). Since UPS work can be carried out on the same spectrometer as XPS with only a change in the exciting source the distinction between UPS and XPS is somewhat artificial. The term *electron spectroscopy for chemical analysis* (ESCA) originally coined by Siegbahn and co-workers to describe all areas of electron spectroscopy has generally come to mean XPS. XPS.

3. Auger Electron Spectroscopy

Another type of electron spectroscopy frequently carried out in the same high-vacuum chamber as XPS is Auger electron spectroscopy (AES). This technique uses a focused beam of 1–30-keV electrons or x-rays to cause the emission of secondary electrons (Fig. 2). An incident electron creates a vacancy in an inner shell of an atom by the emission of a secondary electron. This vacancy is filled by an outer level electron

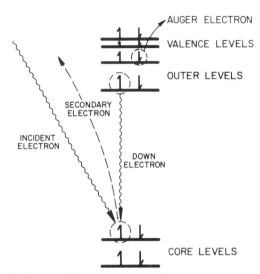

Fig. 2. Schematic diagram of the Auger process. [Adapted with permission from a sketch by Art Hubbard, Snowbird Workshop Notes, University of Utah Chemistry Department, September, 1985.]

("down" electron) dropping into the vacancy. This releases energy that appears either as x-ray fluorescence or in the ejection of an Auger electron with an energy characteristic of the binding energies of the atomic orbitals involved.

Auger electron spectroscopy uses a tightly focused incident electron beam with consequent high spatial resolution (sample spot size as small as 0.03 μm in the diameter). Scanning the electron beam across the sample permits elemental analysis of a large surface area with high resolution at concentrations of detectable species as low as 5×10^{-2} at.%. The Auger electrons can escape from a depth below the surface not exceeding about 2 nm; so AES is definitely a surface technique in the case of solids. Auger electron spectroscopy can depth profile a surface by ion sputtering away the surface and is particularly useful in surface elemental analysis for the first-row elements since x-ray fluorescence from these elements is weak. However, AES is suitable for detecting all the elements except H and He. By varying the angle of incidence of the grazing electron beam we can vary the relative contribution of surface layers and thus scan inhomogeneities in composition of these layers. Thus, AES can detect foreign elements down to about 1% of a monolayer and is well adapted to determine surface cleanliness and segregation of particular elements to the surface through diffusion.

Both AES and XPS yield an elemental analysis of a surface. However, AES measurements can be made more rapidly, which is advantageous when the surface is changing. The AES electron energies are also independent of the exciting source energies. Both techniques require a high-vacuum environment (10^{-8}–10^{-10} torr.)

All three techniques, XPS, UPS, and AES, have been discussed by Carlson (1975) and by Briggs and Seah (1983).

4. Low-Energy Electron Diffraction

Since AES utilizes an electron source and reveals surface elemental composition and cleanliness, it is frequently combined in a high-vacuum system with low-energy electron diffraction (LEED), which permits the determination of the arrangement of surface atoms on a single crystal.

The penetration of low-energy (20–500-eV) electrons into a solid is very small (only a few atomic layers). The wavelength of these electrons ($\lambda \cong (1.5/E)^{1/2}$ nm, where E is in electron volts) is of the same order of magnitude as interatomic distances in the crystalline lattice. This was the basis of the electron diffraction experiment by Davisson and Germer (1927). Scattering of the normal incidence electron beam by the surface atoms yields diffraction maxima at angles α given by the Laue condition $n\lambda = d \sin \alpha$, where n is the diffraction order and d is the surface lattice parameter. (See Fig. 3 for a LEED apparatus schematic.) Electrons diffracted from the surface pass through three grids. With the first grid at ground potential, the electrons are retarded between the second and third grid to eliminate inelastically scattered electrons. Elastically scattered electrons are then accelerated toward a hemispherical fluorescent screen. Sharply defined diffraction spots on the screen indicate a well-ordered surface over distances long compared to the wavelength of the incident electrons. A poorly ordered surface structure yields diffuse diffraction spots. Low-energy electron diffraction data show that there is a contraction of the distances between atomic layers as the surface is approached from the interior of the crystal. For some crystals there is also considerable "reconstruction" at the surface, i.e., surface atoms have different equilibrium positions than the corresponding atoms have in the bulk crystal. Terraces, steps, and kinks in the crystal surface (see Chapter 3) can also be deduced from LEED patterns. The fact that these imperfections can be detected in the diffraction pattern indicates they are not isolated imperfections but rather occur in regular arrays across the crystal surface. Somorjai and Yeates (1984) have reviewed the application of LEED and high-resolution electron energy loss spectroscopy (HREELS) to the study of heterogeneous catalysts.

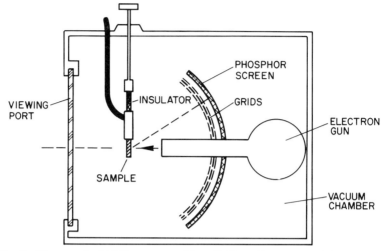

Fig. 3. Diagram of a LEED apparatus. [From P. W. Atkins, "Physical Chemistry," 2nd Ed., p. 1008. Copyright © 1982, by W. H. Freeman and Company San Francisco, California. All rights reserved.]

5. High-Resolution Electron Energy Loss Spectroscopy

In HREELS for studying vibrations, low-energy ($\sim$10-eV) electrons are directed at a surface, and the energy of the scattered electrons is analyzed. Many incident electrons undergo only one inelastic collision at the surface. Accordingly, they experience a definite energy loss ΔE from excitation of a dominant vibrational transition of an adsorbed species. Peaks in a plot of intensity versus energy loss that are independent of surface treatment are attributed to bulk transitions and are neglected. An HREELS apparatus is shown schematically in Fig. 4. Ellis *et al.* (1985) have described refine-

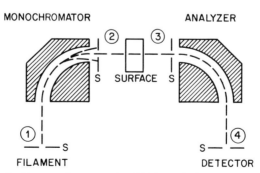

Fig. 4. Schematic diagram of a conventional high-resolution EELS system. [From T. H. Ellis *et al.*, *Science* **230**, 256 (1985). Copyright 1985 by the AAAS.]

ments in the EELS technique that permit measurement of surface vibrational spectra with a signal strength several orders of magnitude larger than is possible with the instrument in Fig. 4. It is now feasible to measure kinetic processes on the surface in real time between 1 msec and 1 h. This time-resolved EELS (TREELS) is the first technique capable of observing the kinetics of the complete chemistry occurring on the surface: adsorption, reaction on the surface, and desorption.

An apparatus for EELS as well as LEED and AES measurements is depicted in Fig. 5. The unusual feature of this experimental system is that it permits high-vacuum HREELS, LEED, and AES and measurements immediately before and after the solid sample is exposed to gas or is otherwise treated at high pressures. Figure 6 depicts a series of EELS spectra for the adsorption of carbon monoxide on the (111) plane of crystalline platinum. As surface coverage by CO increases with longer exposure to CO the major peak corresponding to CO attached perpendicularly to a single Pt atom grows only slightly. The adjoining small peak

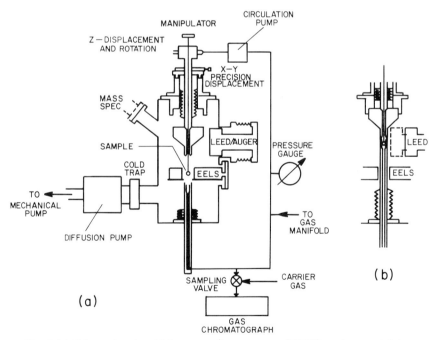

Fig. 5.(a) Schematic of a high-pressure/low-pressure (HPLP) system containing an HREELS apparatus with the high-pressure cell open. (b) Detail with the high-pressure cell closed. [From B. E. Koel, B. E. Bent, and G. A. Somorjai, *Surf. Sci.* **146**, 211 (1984). Copyright 1984 by North-Holland Physics Publ. Co., Amsterdam.]

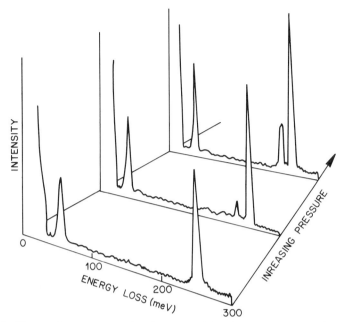

Fig. 6. Electron loss spectrum of carbon monoxide adsorbed on Pt(111). [From P. W. Atkins, "Physical Chemistry," 2nd Ed., p. 1014. Copyright © 1982 by W. H. Freeman and Company, San Francisco, California, 1982. All rights reserved.]

corresponding to CO adsorbed at a bridged site, on the other hand, grows from invisibility to a substantial peak over the same range of increasing pressure. Koel *et al.* (1984) used HREELS to investigate the properties of an adsorbed monolayer of ethylidyne (CCH_3) on a rhodium (111) single crystal surface. Wandass and Gardella (1985a) used HREELS to study Langmuir–Blodgett monolayers of fatty acids.

6. Inelastic Electron Tunneling Spectroscopy

Inelastic electron tunneling spectroscopy (IETS), discovered by Jaklevic and Lambe (1966) and surveyed by DeThomas *et al.* (1985), measures the vibrational spectrum of a few organic molecules in a metal/insulator/sample/metal sandwich. The junction is immersed in liquid helium to increase spectral resolution, and a bias voltage is applied across the junction. (An ac modulation is superimposed on the bias voltage, and a lock-in detector records the second derivative of the response as the bias voltage is varied from 0 to 500 meV to cover the vibrational spectrum from 0 to 4033 cm^{-1}.) When tunneling electrons interact with vibrational modes inelastic tunneling occurs with an increase in tunneling current at each bias

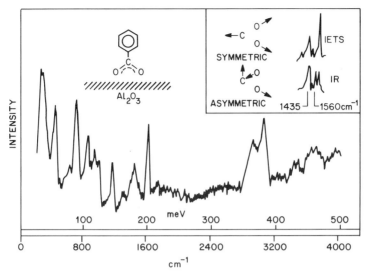

Fig. 7. Inelastic electron tunnelling spectrum of benzoic acid on alumina. The asymmetric carboxylate vibration (1560 cm^{-1}) is weak because of the orientation effect. This is shown via the expanded spectrum depicted in the inset. [From F. A. DeThomas, C. Dybowski, and H. S. Gold, Am. Lab. **17**(11), 23 (1985). Copyright 1985 by International Scientific Communications, Inc.]

voltage that corresponds to a vibrational mode of the sample. A representative spectrum is shown in Fig. 7. Among the advantages of IETS over IR and Raman spectroscopies are the absence of optical selection rules in the IETS and a sensitivity that permits observation of samples constituting less than 1% of a monolayer coverage. Hilliard and Gold (1985) illustrated the utility of IETS for studying supported metal catalysts. They looked at $Os_3(CO)_{12}$ adsorbed on alumina and observed a bridging Os-H-Os bond at 1509 cm^{-1}. For a more general reference on IETS see Wolfram (1978).

B. Photon Spectroscopies

Our focus will be on those optical spectroscopies that elucidate the vibrational spectra of the surface species.

1. Fourier Transform Infrared Spectroscopy

Hair (1980) described transmission Fourier transform infrared (FT-IR) applications. Transmission FT-IR spectroscopy remains one of the most powerful tools for examining supported metal oxides suspended in KBr pellets or otherwise introduced into the sample beam as an optically thin layer of materials. Sample cells described by Hicks *et al.* (1981) and Tagawa and Amenomiya (1985) permit *in-situ* spectroscopic investigation of heterogeneous catalytic reactions at elevated temperatures and pressures.

A new wrinkle in FT-IR is time-resolved IR spectroscopy. There are at least three strategies for securing such data. In the "double-modulation" (DM) method (Nafie and Vidrine, 1982) a sample is modulated between two states at a frequency greatly exceeding the maximum Fourier frequency. Modulation can be effected, for example, by changes in light polarization, in voltage on an electrode, or in tension on a polymer. When the detector output is sent through a lock-in amplifier tuned to the frequency and phase of the sample excitation waveform the ac-modulated signal (the single-beam difference spectrum) that is acquired can be ratioed against the time-averaged single-beam spectrum of the sample so that processes having time constants in the ~5-μsec range are accessible. Liabilities of DM are the requirement that the chemical process should be perfectly reversible, millisecond and longer time-scale processes are inaccessible, and complete IR spectra are not given directly as a function of time.

In the digital time-resolved (DTR) method (Mantz, 1978; Moore *et al.*, 1985) a repetitive voltage, laser, or other perturbing pulse is applied to the sample. Each point in an IR interferogram is acquired over a short time interval (the digitization time of the analog-to-digital converter) a specified time after the start of the forward sweep of the interferometer moving mirror. Points in the interferogram are acquired at evenly spaced (in distance) increments of mirror retardation. For a well-designed rapid-scanning interferometer these points have well-defined times relative to a selected spatially fixed event (the mirror position). These discrete points serve as the time base for the chemical event being monitored. With the relaxation experiment triggered by this time base each interferogram point has a characteristic retardation and time relative to the sample event. By sequentially triggering the sample event at later times in the scan for successive interferogram files we build a complete set of data yielding successive interferograms over an extended time range. Sorting of these data by grouping common timed interferogram points from each interferogram yields new interferograms representative of fixed times in the relaxation experiment. The DTR method yields a true time-varying spectrum and is suitable for studying reactions with time constants as short as 10 μsec or as long as 1.0 sec. The chemical experiment, however, must be exactly repeatable many successive times. Small irreversibilities can totally obscure spectral features.

The third and easiest time-resolved IR technique is rapid-scan dynamic (RSD) FT-IR spectroscopy. In the RSD method each time the interferometer mirror completes a scan, a time-resolved spectrum results. Since time resolution is limited by the time required to collect and store the data from a single scan, high time resolution can only be obtained by sacrificing some spectral resolution since the latter is proportional to the number of data points acquired in each scan. The Nicolet 60SX can collect approximately 50 scans per second with a spectral resolution of 8 cm^{-1} and thus can achieve 20-msec time resolution. While the time resolution of the RSD method is inferior to that of the DM and DTR methods, the RSD technique allows the study of irreversible, single-shot reaction processes and avoids spectral artifacts associated with repetitive scan techniques.

These time-resolved infrared spectroscopic techniques should find many applications in the study of transient intermediates adsorbed on solid supports.

We turn now to reflectance techniques. Let us define some terms. Plane specular reflection occurs from a polished flat surface. Diffused specular reflection describes single reflections from a rough surface. Finally, diffuse reflection denotes multiple reflections and transmissions of each ray of light within a fairly transparent, granular scattering medium.

2. Ellipsometry

A wavelength scanning infrared ellipsometer consists of two fixed polarizers that bracket a rotating polarizer and a flat polished sample surface. A glo-bar IR source is at one end of the system, and a liquid-nitrogen cooled, IR-sensitive, solid-state detector is at the other end. The monochromator can immediately follow the IR source or immediately precede the IR detector.

In ellipsometry (ELL), also called grazing angle reflectance (GAR), the collimated incident beam is at an angle of 5°–10° to the flat sample surface. The spectrum of the reflected radiation with perpendicular polarization resembles the absorption spectrum of the surface. This vibrational spectroscopic technique is insensitive to the same molecules in the gas phase that it detects on the surface at less than a monolayer of coverage. This is a significant advantage over FT-IR photoacoustic spectroscopy (PAS) since in PAS we must purge the sample cell of absorbing gases before we can detect the much weaker photoacoustic signal from molecules adsorbed on the sample surface. Fedyk and Dignam (1980) have described IR ellipsometric spectroscopy (IRES) and its application for distinguishing carbon monoxide in a variety of surface structures when adsorbed on a Ni (100) surface.

3. Diffuse Reflectance Infrared Fourier Transform Spectroscopy

Diffuse reflectance infrared Fourier transform (DRIFT) spectroscopy covers IR, VIS, and UV. Schoonheydt (1984) discussed its use in studying supported catalysts. Fuller and Griffiths (1978) first reported the use of an ellipsoidal collecting mirror interfaced to a commercial FT-IR spectrometer (Fig. 8) to obtain good signal-to-noise ratio DRIFT spectra of powdered solids. Sample absorptivity is related to its reflectivity by the Kubelka–Munk equation

$$f(R_\infty) = (1 - R_\infty)^2/2R_\infty = k/s$$

where R_∞ is the absolute reflectance of the sample layer, s the scattering coefficient, and k the molar absorption coefficient. Since a perfect diffuse reflectance standard has not been found, we must replace R_∞ with

$$R'_\infty = \frac{R'_\infty(\text{sample})}{R'_\infty(\text{standard})}$$

where $R'_\infty(\text{sample})$ is the single-beam reflectance spectrum of the sample and $R'_\infty(\text{standard})$ the single-beam reflectance spectrum of a nonabsorbing material with high diffuse reflectance such as finely ground KCl. For quantitative results the parameter s must be held constant, which requires uniform particle size. Sample dilution with KCl is also required to make spectral subtraction reliable for sample mixtures. Reflectivity of high-absorptivity spectral regions can be dominated by specular reflectivity

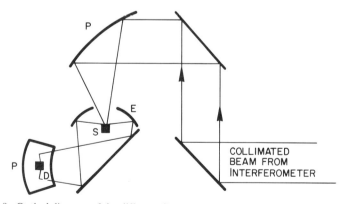

Fig. 8. Optical diagram of the diffuse reflectometer. Radiation from the interferometer is focused on the sample S by the off-axis paraboloid P, and the diffusely reflected radiation is collected by the ellipsoid E and then focused onto the detector D. (Note that the detector is not coplanar with the other optics.) [Reprinted with permission from M. P. Fuller and P. R. Griffiths, *Anal. Chem.* **50**, 1906 (1978). Copyright 1978 American Chemical Society.]

when the rest of the spectrum really constitutes true diffuse reflectance. This can cause "band reversal" in which the absorptivity peak of a band has higher reflectivity than the band wings. Yeboah *et al.* (1984) found that by subjecting powdered samples to pressures of ~12,000 psi packing density becomes more uniform and band intensities become more reproducible.

Comparisons of DRIFT with Fourier transform infrared photoacoustic spectroscopy (FT-IR/PAS) have been made by Freeman *et al.* (1980), Tilgner (1981), Yeboah *et al.* (1981), Delumyea and Mitchell (1983), and Meichenin and Auzel (1983). The consensus is that the techniques are complementary with DRIFT having greater sensitivity but requiring more sample preparation.

4. Internal Reflection Spectroscopy

Attenuated total reflectance (ATR), sometimes also identified as multiple internal reflectance (MIR) or frustrated total reflectance (FTR), is the basis of internal reflectance spectroscopy (IRS), perhaps the most frequently used IR surface spectroscopy. Infrared radiation moving through a high refractive index, IR-transparent ATR crystal reflects off the surface of the crystal from the inside. Materials from which the ATR crystal may be fashioned include KRS-5 (thallium bromoiodide), silicon, germanium, ZnSe, and CaF_2. Assume the angle of reflection θ is more grazing than the critical angle θ_c given by the relation $\sin \theta_c = n_2/n_1$, where n_1 is the refractive index of the ATR crystal and n_2 that of the medium in contact with the crystal at the point of reflection. Then since the electric vector of the light extends beyond the transparent crystal surface at the point of reflection into the lower refractive index absorbing sample medium in contact with the crystal surface, the reflected light will be slightly attenuated. This evanescent wave extends only a few micrometers into the sample medium so only the surface absorptivity of the sample is measured by the ATR technique. The experimental arrangement for ATR spectroscopy is shown schematically in Fig. 9. Multiple reflections enhance the detectability of a weak absorption band.

5. Modulated Specular Reflectance Spectroscopy

At a liquid electrolyte–metal electrode interface (in a battery, for example) the surface chemistry occurs in a boundary layer of liquid several nanometers thick called the diffusion layer. Mechanisms of organic chemical reactions in the electrochemical diffusion layer following electron transfer at a metal electrode (see Chapter 5) are best studied by fast UV–VIS techniques. They discriminate between the reactants,

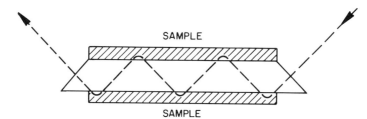

Fig. 9. Schematic representation of attenuated total reflectance in a multireflection element. [From A. Lee Smith, *In* "Treatise on Analytical Chemistry, Part I, Theory and Practice" (P. J. Elving, E. J. Meehan, and I. M. Kolthoff, eds.), 2nd Ed., Vol. 7. Copyright 1981 Wiley Inc., New York.]

intermediates, and products present in the mixture. By rapid modulation of the metal electrode potential, the formation of intermediates and products is also modulated, causing an ac component in the analyzing optical beam reflected through the diffusion layer. This small ac signal may be extracted from the background and amplified. The magnitude of this signal as a function of wavelength yields an absorption spectrum of all components in the diffusion layer.

Differences in the phase angle between the applied potential and the optical signal can be related to the reaction kinetics of the various components in the system. Other signal processing techniques at fixed wavelengths are utilized to record absorbance–time transients of various intermediates, the analysis of which leads to the evaluation of kinetic parameters. This technique, modulated specular reflectance spectroscopy (MSRS), is used to study the kinetics of oxidation of aromatic hydro-carbons, the mechanism of metalloporphyrin oxidation, hydride formation at noble and base metal electrodes, corrosion, and electrocatalysis.

Bewick *et al.* (1984) and Pons *et al.* (1984) have extended it to the IR. Using a reflectance arrangement in an FT-IR spectrometer, they recorded the spectra of both soluble intermediate species in the electrochemical diffusion layer and adsorbed species in monolayer quantities at the electrode surface. Thus, the recording of vibrational spectra of organic radical intermediates, if they can be generated electrochemically, is easy to accomplish.

6. Microwave and Radiowave Magnetic Spectroscopies

Other chapters in this book are devoted to electron spin resonance (ESR) and nuclear magnetic resonance (NMR) spectroscopic studies of surface species.

Opportunities for new surface science discoveries lie in the direction of combining the magnetic resonance spectroscopies with some of the other surface techniques. For instance, in conventional ESR we monitor the reflected microwave power from a resonant cavity containing the sample. Since the photoacoustic effect (discussed in Section II.D.1) can be used to detect absorption of electromagnetic radiation at wavelengths extending from the x-ray to the radiowave regions, it is not surprising that a pulsed microwave acoustic technique has been found suitable for detecting ESR signals. Microphones (Melcher, 1980; Netzelmann *et al.*, 1982), piezoelectric transducers (DuVarney *et al.*, 1981; Netzelmann *et al.*, 1983), and even second sound in superfluid helium (Main and McCann, 1984) have been the means of carrying out such ESR experiments. The method turns out to be particularly suitable for the study of boundary or sandwich structures because of the considerable temperature gradients generated in such systems by absorbed microwave radiation (Netzelmann *et al.*, 1983).

7. Extended X-Ray Absorption Find Structure

When an x-ray is absorbed by an atom an inner electron can be excited, and the wave associated with this excited electron can interfere constructively or destructively with the electrons of neighboring atoms. This interference produces a fine structure or modulations on the high-energy side of an absorption edge as shown in Fig. 10. From these modulations we can deduce coordination numbers and interatomic distances from a given type of absorber atom. Since the extended x-ray absorption fine structure

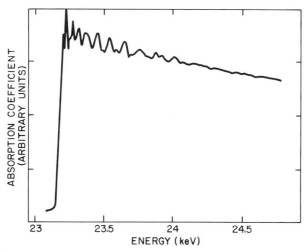

Fig. 10. EXAFS x-ray absorption spectrum of metallic rhodium at 100 K in the region of the K-absorption edge. [Reprinted from J. H. Sinfelt, G. H. Via, and F. W. Lytle, *Catal. Rev.—Sci. Eng.* **26**, 81 (1984). Reprinted by courtesy of Marcel Dekker, Inc.]

(EXAFS) can be determined for each element in a complex sample, we may identify the environments of the different types of atoms present in a supported reagent. A high-intensity wavelength-tunable x-ray source is required. Synchrotron radiation from a particle storage ring has been the usual source. Since synchrotrons are limited in number, development of rotating-anode x-ray emitters will facilitate the execution of EXAFS experiments at many more laboratories.

Sinfelt *et al.* (1984) reviewed the application of EXAFS to bimetallic cluster catalysts. In a shorter report, Boudart *et al.* (1985) described an EXAFS investigation of the structural environment of molybdenum in a CoMoS/γ-alumina catalyst.

8. Fluorescence Spectroscopy

Fluorescence intensities and emission maxima of surface-bound and/or adsorbed molecules can provide information about the relative polarity, distribution, and dynamics of molecules at solid–liquid interfaces. Lochmüller *et al.* (1981a) studied chemically modified surfaces commonly used in reverse-phase liquid chromatography. The microheterogeneity of these surfaces was elucidated by the Stokes shift of fluorescence from surface bound dansylamide molecules as a function of excitation wavelength (Lochmüller *et al.*, 1981b). The probe fluorophore, dansylamide, was covalently attached to a silica surface which simultaneously had bonded to it a methyl, or *n*-octadecyl, phase of interest. Since dansylamide is a polar molecule, the Stokes shift of its fluorescence is a sensitive means of determining the polarity of the molecules in their environment. The spectroscopic data revealed three distinct environments for the probe species corresponding to three sets of band positions and Stokes shifts. There are parts of the surface that are not covered by the *n*-alkyl bonded phase. Also residual reactive silanol groups remain despite chemical treatment of the surface.

Lochmüller *et al.* (1984) also studied the time-dependent luminescence of pyrene silane molecules chemically bound to silica to determine the distribution of surface-bound molecules and their organization in contact with different solvents. The organization and proximity of such molecules were controlled by an inhomogeneous distribution of chemically reactive silanols on the surface. Analysis of the emission decay spectra at several surface concentrations indicated that the bound molecules are clustered predominantly into regions of high density. The resulting picture of the derivatized surface was one of high-density patches or clusters of ligands (see Chapter 3) in contrast to the uniform and relatively ordered blanket envisioned by earlier workers.

Dynamic, solvent-induced, conformational changes of the bound molecules were also observed in the excimer (excited-state dimer) emission decay spectra due to solvophobic interactions between the bonded ligands and solvent. The chemically modified surface is thus pictured as a dynamic "breathing" surface in which the bonded ligands attempt to adjust their total surface area exposed to solvent depending on the solvating properties of the solvent (Lochmüller and Hunnicutt, 1986).

Advantages of the time-dependent fluorescence technique include its applicability to dielectric surfaces, its capacity for detecting submonolayer coverages of adsorbed species, and the sensitivity of fluorescence lifetimes to such factors as polarity, pH, and ligand proximity. Fluorescence depolarization (Bentley *et al.*, 1985) can be used to determine the rotational degrees of freedom of the fluorophore as well as the microviscosity of the interfacial region.

9. Surface Enhanced Raman Spectroscopy

Being unaffected by the presence of liquid water, Raman spectroscopy (RS) is a particularly attractive means of obtaining the vibrational spectra of adsorbed species at a solid–aqueous solution interface. However, the quantum yield from Raman scattering is so low as to make surface RS quite difficult. Fleischmann *et al.* (1974) made the fortunate discovery of surface enhanced Raman spectroscopy (SERS). They noted a very strong Raman spectrum of pyridine adsorbed on a silver metal electrode following the electrochemical anodization of the silver. A variety of theoretical explanations reviewed by Metiu (1982) have since been invoked. Copper and gold metal electrodes also exhibit this enhancement of the Raman effect.

Molecular Raman scattering cross sections can be considerably enhanced by exciting the sample with a laser at a frequency in resonance with an electronic transition of the molecule [resonance Raman (RR) scattering]. Jeanmaire and Van Duyne (1977) first reported surface enhanced Raman spectra for molecules in resonance (SERRS). Morrow (1980) has surveyed Raman spectroscopic studies of surface species at solid–gas and solid–liquid interfaces in the absence of such enhancements.

C. Heavy Particle (Neutron, Cation, and Anion) Spectroscopies

1. Neutron Scattering Spectroscopy

Thermal neutron beams are ideal for probing the structure of a crystalline lattice as well as for determining the vibrational motions of molecules (Fig. 11). Since neutrons are uncharged, they interact very

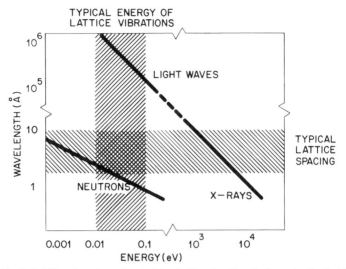

Fig. 11. Suitability of neutrons for studying lattice vibrations is shown by the fact that they have the right wavelength and the right energy. Wavelength is comparable with interatomic spacing; energy is comparable with the energy of lattice vibrations. X-rays have the right wavelength, but energy is 10^5 times too high; light waves have the right energy but wavelength is 10^5 times too long. [From Robert M. Brugger, *Int. Sci. Technol.* Nov., p. 52 (1962).]

weakly with matter and thus do not scatter preferentially from solid surfaces. The exploration of motions and positions of molecules adsorbed on surfaces by neutron scattering therefore becomes possible only on high-surface-area substrates. Furthermore, it is important that the surface be well characterized; one of the crystal planes of the sample solid must be the primary surface on which adsorption occurs. The sample particles also must be large enough (>50 Å) so that the coherence length of the adsorbed film can produce well-defined Bragg reflections. Various forms of graphite have proven particularly suitable even though they have less surface area than the typical heterogeneous catalyst. The superiority of graphite is attributable to the major part of its exposed surface being graphite basal planes of high uniformity.

Although the solid surfaces that can be studied by neutron scattering spectroscopy (NSS) are somewhat limited, NSS has a significant advantage over the electron spectroscopies we have considered previously. Ultrahigh vacuum is not required and adsorbed surface layers can be examined in equilibrium with their own vapor.

The use of neutron diffraction to locate precisely the hydrogen atoms in a crystal structure previously elucidated by x-ray diffraction is familiar to most chemists. Thus, we would expect NSS to be particularly effective in probing the motion and position of hydrogen atoms in submonolayer films of molecules containing hydrogen on a surface. Neutron scattering from adsorbed hydrogen atoms surpasses that from all other surface and nonhydrogen substrate atoms.

Taub (1980) has reviewed the application of NSS to the study of adsorbed monolayers.

2. Secondary Ion Mass Spectrometry

Secondary ion mass spectroscopy (SIMS) is the most expensive high-vacuum technique for examining a sample surface; SIMS bombards the surface with 1–20-keV ions and identifies the ejected ions by quadrupole mass spectrometry (Fig. 12). "Static SIMS" is done at a low current density with minimal damage to the sample but with attendant signal-to-noise ratio difficulties because of the low secondary ion current. "Dynamic SIMS" is carried out at a high current density with the surface layer stripped away in the course of the mesurement and further layers exposed by continued ion bombardment. Thus, "dynamic SIMS" is a valuable means of depth profiling a sample as illustrated by a quantitative analysis of molecular-beam expitaxially grown gallium arsenide thin layers by Chu *et al.* (1982).

We noted previously (Table I) that SIMS has a high lateral resolution in addition to its greater selectivity than AES or XPS for the uppermost surface monolayer. The problems of making SIMS a more quantitative surface technique continue to challenge experimentalists (see, e.g., Galuska and Morrison, 1984). For instance, SIMS has been used to study a cobalt–alumina catalyst (Chin and Hercules, 1982), acrylic polymers

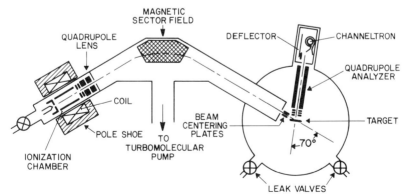

Fig. 12. Schematic of a secondary ion mass spectrometer with mass-selected primary ion beam. [From A. Benninghoven, *Surf. Sci.* **35**, 427 (1973). Copyright 1973 by North-Holland Physics Publ. Co., Amsterdam.]

(Gardella and Hercules, 1981), small hydrocarbons frozen in an argon matrix (Jonkman *et al.*, 1978), and organized molecular layers prepared by Langmuir–Blodgett techniques (Wandass and Gardella, 1985b).

It is only possible to mention here several other surface science techniques loosely related to SIMS. In ion scattering spectrometry (ISS) low-energy ions (<5-keV) scattered from the surface permit the determination of the atomic composition of the surface. In Rutherford backscattering spectroscopy (RBS) high-energy (>1-MeV) incident ions identify heavy impurity atoms in a solid matrix of lower atomic weight. In fast atom bombardment (FAB) the sample is supported in liquid glycerol and is struck with a fast (6–8-keV) beam of neutral argon atoms. The secondary ion beam analyzed by mass spectrometry can have up to ~73500 Dalton sample species in it. Laser desorption (LD) effected with short-duration (<10-nsec) laser pulses is also related to SIMS and is included in a recent survey of particle-induced desorption mass spectrometry techniques by Macfarlane (1982).

D. Thermal and Acoustic Wave Spectroscopies

Photothermal and photoacoustic spectroscopies involve detection of thermal and acoustic waves caused by absorption of electromagnetic radiation. The absorbed radiation creates excited electronic or vibrational energy states in the molecules of the sample. When these excited states undergo radiationless deexcitation thermal and acoustic waves are generated. There are various ways of detecting these waves. A gas in contact with the surface of a cyclically illuminated solid sample can act as a transducer and carry a sound wave to a nearby microphone in what is called photoacoustic spectroscopy (PAS) (see Section II.D.1). When a thermal wave reaches the sample surface it causes a refractive index change in the fluid (gas or liquid) bathing the surface. A laser beam can detect this refractive index change in an experiment called photothermal beam deflection (PBD) spectroscopy.

For lack of space we will not describe in detail several other thermal–acoustic spectroscopies. For instance, for solid samples the increase in temperature of the surface of an illuminated solid can be measured with a remote IR detector as an increase in blackbody radiation. This technique is called photothermal radiometry (PTR) (Tam, 1985; Pekker and Eyring, 1987). A piezoelectric transducer (PZT) attached to the surface of a solid can detect the acoustic waves caused by radiationless decay of excited states in the sample (Farrow *et al.*, 1978). The PZT has the distinct advantage over a microphone of being able to time-resolve events down to times as short as ~10 nsec. The PZT also lends itself to the detection of light-induced sound waves in sample liquids (Tam, 1983). The temperature increase in the surface of a solid sample can also be detected with a pyroelectric polyvinylidene difluoride thin film in intimate contact with the surface (Coufal, 1984). This has been dubbed photopyroelectric spectroscopy (PPES) and may have some significant advantages over PZT detection as, for example, in studying the growth of films on an electrode at an electrode–electrolyte interface (Mandelis, 1984; Mandelis and Zver, 1985).

1. Photoacoustic Spectroscopy

In the mid-1970s Rosencwaig (1980) revived interest in the use of PA techniques for spectral studies of solid surfaces. In a typical PAS experiment (Fig. 13) the sample surface is placed inside a tightly sealed cell with a transparent window that permits illumination of the sample surface by a monochromatic light beam, chopped at an audio frequency f. If the sample surface absorbs this particular wavelength of light, a thermal wave develops in the sample and travels back to the solid–gas interface. There, the thermal wave heats a thin layer of gas in contact with the surface, and this heated layer of gas behaves as a thermal piston that causes a sound wave of the same frequency f to travel to a sensitive microphone. Only that component of the microphonic output that is of frequency f is amplified by a lock-in amplifier. In a dispersive PAS experiment the plot of incident wavelength versus intensity of the acoustic signal will give a PA spectrum that resembles the absorption spectrum obtained on the same sample with

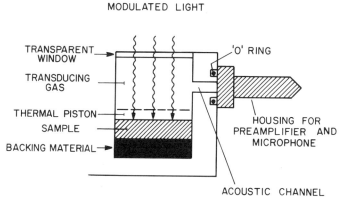

Fig. 13. Schematic of a microphonic photoacoustic sample cell. The diameter of the acoustic channel to the microphone and the height of the transducing gas volume have both been greatly enlarged for visual clarity. Signal strength is inversely proportional to transducing gas volume, so these cell dimensions are kept as small as possible.

a conventional UV–VIS spectrophotometer. A one-dimensional mathematical model devised by Rosencwaig and Gersho (1976), with later embellishments by other authors, has been used with great success.

When the PAS experiment is carried out at IR wavelengths a commercial FT-IR spectrometer is ordinarily used in which the solid-state detector has been replaced by the previously described microphonic cell or by a piezoelectric transducer on the front of which the solid sample has been glued or clamped.

The unique capability of the PA technique is its suitability for depth profiling a sample (see, e.g., Mandelis and Lymer, 1985; Uejima et al., 1985). If the incident light beam is chopped at 10^3 Hz, the PA spectrum will be characteristic of a layer nearer the surface than that giving rise to a spectrum at a chopping frequency of 10^2 Hz. Sampling depths are of the order of micrometers compared to tens of Ångstroms in the case of AES, for example. Photoacoustic spectroscopy is also useful if the solid sample is either so opaque or so extremely transparent that conventional transmission spectroscopy proves impractical. The signal-to-noise ratio in PA experiments is so poor that we will always prefer transmission or reflectance spectroscopies whenever possible. Since scattering and absorption can limit the spectral range accessible to transmission IR spectroscopy of supported metal catalysts to frequencies above 1300 cm^{-1}, it frequently proves valuable to make FT-IR/PAS measurements on such samples at frequencies below 1300 cm^{-1}, in spite of the low signal-to-noise ratio. Helmholtz resonator sample cells have been constructed that permit FT-IR/PAS measurements on supported metal catalysts at temperatures up to 400°C with gas flowing through the cell and covering the full 4000–400 cm^{-1} mid-infrared spectral range (McGovern et al., 1984). Microphonic FT-IR/PAS experiments on this type of sample at both elevated temperatures and elevated pressures have proven more elusive. In spite of such limitations we have found FT-IR/PAS to be a powerful qualitative tool for elucidating adsorbed species on supported metal oxides and other surfaces (see, e.g., Bandyopadhyay et al., 1985; Gardella et al., 1982; McKenna et al., 1984, 1985; Riseman et al., 1985).

2. Photothermal Beam Deflection Spectroscopy

FT-IR photoacoustic beam deflection spectroscopy (PBDS) (Low and Lacroix, 1982; Morterra et al., 1985) based on the mirage effect of Boccara et al. (1980) provides an attractive alternative to microphonic α FT-IR/PAS for studies of supported metal oxides at elevated pressures and temperatures. In FT-IR/PBD spectroscopy (Fig. 14) a solid sample is heated by its absorption of the radiation emerging from the Michelson interferometer of a commercial FT-IR spectrometer. The resulting modulated

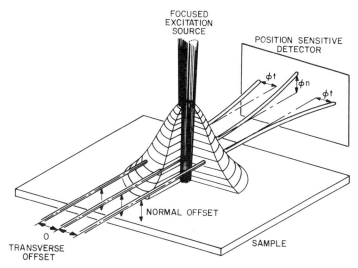

Fig. 14. Diagrammatic representation of the surface temperature profile in a photothermal beam deflection experiment with three different possible probe laser beam trajectories depicted. [From L. C. Aamodt and J. C. Murphy, *J. Appl. Phys.* **54**, 581 (1983). Copyright 1983 American Institute of Physics.]

thermal gradient and consequent refractive index gradient in a layer of fluid (gas or liquid) on top of the heated sample surface is detected by the deflection of a low-power laser beam (propagating parallel to and almost grazing the sample surface) that strikes a position sensor approximately 1 m beyond the sample. Thus, for elevated temperature and pressure experiments the sample cell can be a closed box with an IR transmitting window on the top to admit the output from an interferometer and with laser beam entrance and exit windows on opposite sides of the cell. The key advantage of such an FT-IR/PBD experiment over FT-IR/PAS is the location of the detector outside the sample cell. Contamination of the sample by outgassing of a microphone and possible damage to a microphone by high temperatures and pressures are both avoided. The signal-to-noise ratio in FT-IR/PBD experiments is not large, so it is understandable that dispersive PBD experiments using intense monochromatic light sources (frequently lasers) continue to be reported (see, e.g., Royce *et al.*, 1982; Seager and Land, 1984; Masujima *et al.*, 1984; Field *et al.*, 1985).

E. Relaxation Spectroscopies

Eigen (1954) first described relaxation methods for measuring rates of discrete reaction steps in homogeneous solution. We perturb a chemical equilibrium slightly by a sudden change in a parameter such as temperature, pressure, or electric field intensity and then follow in time the relaxation of the chemical equilibrium to species concentrations appropriate for the new conditions of temperature, pressure, or electric field intensity. Relaxation methods permitted the elucidation of a great many fast reactions in homogeneous liquids (Eigen and De Maeyer, 1963).

Ashida *et al.* (1978) reported the feasibility of measuring the kinetics of proton adsorption on and desorption from TiO_2 (anatase) particles in an aqueous colloidal suspension using the P-jump relaxation technique with electrical conductance detection. In the years since, this research group has measured the kinetics of, for example, adsorption–desorption of OH^- and IO_3^- on colloidal TiO_2, K^+, and OH^- on colloidal zeolite, Pb^{2+} and CrO_4^{2-} on colloidal γ-Al_2O_3, and Cl^- and ClO_4^- on colloidal γ-FeOOH. In the case of colloidal γ-FeOOH Sasaki *et al.* (1983) used the E-jump relaxation technique with conductivity detection and found that anions are free to move on the surface of the colloidal particle but must overcome an activation energy barrier in order to leave the surface.

III. GENERAL CONCLUSIONS

We have scantily described a fraction of all the spectroscopies presently available for studying surfaces. An obvious conclusion to be drawn is that few laboratories can afford to support more than a handful of these rather costly techniques. Thus, chemists in the surface science field of necessity collaborate extensively with workers in other laboratories and regional instrumentation centers.

Another interesting insight suggested by the neutron scattering spectroscopy and relaxation spectroscopy sections is that techniques long thought suitable only for studying homogeneous systems can find powerful ..ew applications in the study of interfaces.

Finally, it is clear that new tools for studying surfaces will continue to proliferate making a survey chapter such as the present one even more impractical to assemble in the future.

REFERENCES

Ashida, M., Sasaki, M., Kan, H., Yasunaga, T., Hachiya, K., and Inoue, T. (1978). *J. Colloid Interface Sci.* **67**, 219.

Bandyopadhyay, S., Massoth, F. E., Pons, S., and Eyring, E. M. (1985). *J. Phys. Chem.* **89**, 2560.

Bentley, K. L., Thompson, L. K., Klebe, R. J., and Horowitz, P. M. (1985). *BioTechniques* **3**, 356.

Bewick, A., Kunimatsu, K., Pons, S., and Russell, J. W. (1984). *J. Electroanal. Chem.* **160**, 47.

Boccara, A. C., Fournier, D., and Badoz, J. (1980). *Appl. Phys. Lett.* **36**, 130.

Boudart, M., Dalla Betta, R. A., Foger, K., Löffler, D. G., and Samant, M. G., (1985). *Science* **228**, 717.

Briggs, D., and Seah, M. P. (1983). "Practical Surface Analysis." Wiley, New York.

Carlson, T. A. (1975). "Photoelectron and Auger Spectroscopy." Plenum, New York.

Chin, R. L., and Hercules, D. M. (1982). *J. Phys. Chem.* **86**, 360.

Chu, P. K., Harris, W. C., Jr., and Morrison, G. H. (1982). *Anal. Chem.* **54**, 2208.

Coufal, H. (1984). *Appl. Phys. Lett.* **44**, 59.

Davisson, C. J., and Germer, L. H. (1927). *Phys. Rev.* **30**, 705.

Delumyea, R. D., and Mitchell, D. (1983). *Anal. Chem.* **55**, 1996.

DeThomas, F. A., Dybowski, C., and Gold, H. S. (1985). *Am. Lab.* **17**(11), 23.

DuVarney, R. C., Garrison, A. K., and Busse, G. (1981). *Appl. Phys. Lett.* **38**, 675.

Eigen, M. (1954). *Discuss. Faraday Soc.* **17**, 194.

Eigen, M., and De Maeyer, L. (1963). *In* "Investigation of Rates and Mechanisms of Reactions" (S. L. Friess, E. S. Lewis, and A. Weissberger, eds.), Technique of Organic Chemistry, 2nd Ed., Vol. 8, Part II, pp. 895–1054. Wiley (Interscience), New York.

Ellis, T. H., Dubois, L. H., Kevan, S. D., and Cardillo, M. J. (1985). *Science* **230**, 256.

Farrow, M. M., Burnham, R. K., Auzanneau, M., Olsen, S. L., Purdie, N., and Eyring, E. M. (1978). *Appl. Opt.* **17**, 1093.

Fedyk, J. D., and Dignam, M. J. (1980). *ACS Symp. Ser.* No. **137**, 75.

Field, R. S., Leyden, D. E., Masujima, T., and Eyring, E. M. (1985). *Appl. Spectrosc.* **39**, 753.

Fleischmann, M., Hendra, P. J., and McQuillan, A. J. (1974). *Chem. Phys. Lett.* **26**, 163.

Freeman, J. J., Friedman, R. M., and Reichard, H. S. (1980). *J. Phys. Chem.* **84**, 315.

Fuller, M. P., and Griffiths, P. R. (1978). *Anal. Chem.* **50**, 1906.

Galuska, A. A., and Morrison, G. H. (1984). *Anal. Chem.* **56**, 74.

Gardella, J. A., Jr., and Hercules, D. M. (1981). *Anal. Chem.* **56**, 74.

Gardella, J. A., Jr., Eyring, E. M., Klein, J. C., and Carvalho, B. B. (1982). *Appl. Spectrosc.* **36**, 570.

Hair, M. L. (1980). *ACS Symp. Ser.* No. 137, 1.

Hicks, R. F., Kellner, C. S., Savatsky, B. J., Hecker, W. C., and Bell, A. (1981). *J. Catal.* **71**, 216.

Hilliard, L. J., and Gold, H. S. (1985). *Appl. Spectrosc.* **39**, 124.

Jaklevic, R. C., and Lambe, J. (1966). *Phys. Rev. Lett.* **17**, 1139.

Jeanmaire, D. L., and Van Duyne, R. P. (1977). *J. Electroanal. Chem.* **84**, 1.

Jonkman, H. T., Michl, J., King, R. N., and Andrade, J. D. (1978). *Anal. Chem.* **50**, 2078.

Koel, B. E., Bent, B. E., and Somorjai, G. A. (1984). *Surf. Sci.* **146**, 211.

Lochmüller, C. H., and Hunnicutt, M. L. (1986). *J. Phys. Chem.* **90**, 4318.

Lochmüller, C. H., Marshall, D. B., and Wilder, D. R. (1981a). *Anal. Chim. Acta* **130**, 31.

Lochmüller, C. H., Marshall, D. B., and Harris, J. M. (1981b). *Anal. Chim. Acta* **131**, 263.

Lochmüller, C. H., Colborn, A. S., Hunnicutt, M. L., and Harris, J. M. (1984). *J. Am. Chem. Soc.* **106**, 4077.

Low, M. J. D., and Lacroix, M. (1982). *Infrared Phys.* **22**, 139.

Macfarlane, R. D. (1982). *Acc. Chem. Res.* **15**, 268.

McGovern, S. J., Royce, B. S. H., and Benziger, J. B. (1984). *Appl. Surf. Sci.* **18**, 401.

McKenna, W. P., Bandyopadhyay, S., and Eyring, E. M. (1984). *Appl. Spectrosc.* **38**, 834.

McKenna, W. P., Higgins, B. E., and Eyring, E. M. (1985). *J. Mol. Catal.* **31**, 199.

Main, P. C., and McCann, J. P. J. (1984). *J. Phys. D.* **17**, 523.

Mandelis, A. (1984). *Chem. Phys. Lett.* **108**, 388.

Mandelis, A., and Lymer, J. D. (1985). *Appl. Spectrosc.* **39**, 473.

Mandelis, A., and Zver, M. M. (1985). *J. Appl. Phys.* **57**, 4421.

Mantz, A. W. (1978). *Appl. Opt.* **17**, 1347.

Masujima, T., Sharda, A. N., Lloyd, L. B., Harris, J. M., and Eyring, E. M. (1984). *Anal. Chem.* **56**, 2975.

Meichenin, D., and Auzel, F. (1983). *J. Phys. (Orsay, Fr.)* **44**, C6-151.

Melcher, R. L. (1980). *Appl. Phys. Lett.* **37**, 895.

Metiu, H. (1982). *In* "Surface Enhanced Raman Scattering" (R. K. Chang and T. E. Furtak, eds.), pp. 1–34. Plenum, New York.

Moore, B. D., Poliakoff, M., Simpson, M. B., and Turner, J. J. (1985). *J. Phys. Chem.* **89**, 850.

Morrow, B. A. (1980). *ACS Symp. Ser.* No. 137, 119.

Morterra, C., Low, M. J. D., and Severdia, A. G. (1985). *Appl. Surf. Sci.* **20**, 317.

Nafie, L. A., and Vidrine, D. W. (1982). *In* "Fourier Transform Infrared Spectroscopy" (J. R. Ferraro and L. J. Basile, eds.), Vol. 3, pp. 83–123. Academic Press, New York.

Netzelmann, U., Goldhammer, E., Pelzl, J., and Vargas, H. (1982). *Appl. Opt.* **21**, 32–34.

Netzelmann, U., Lerchner, H., Pelzl, J., and Sigrist, M. W. (1983). *J. Phys. (Orsay, Fr.)* **44**, C6-221.

Pekker, S., and Eyring, E. M. (1987). *Appl. Spectrosc.* **41**, 260.

Pons, S., Davidson, T., and Bewick, A. (1984). *J. Electroanal. Chem.* **160**, 63.

Riseman, S. M., Bandyopadhyay, S., Massoth, F. E., and Eyring, E. M. (1985). *Appl. Catal.* **16**, 29.

Rosencwaig, A. (1980). "Photoacoustics and Photoacoustic Spectroscopy." Wiley (Interscience), New York.

Rosencwaig, A., and Gersho, A. (1976). *J. Appl. Phys.* **47**, 64.

Royce, B. S. H., Sanchez-Sinencio, F., Goldstein, R., Muratore, R., Williams, R., and Yim, W. M. (1982). *J. Electrochem. Soc.* **129**, 2393.

Sasaki, M., Morlya, M., Yasunaga, T., and Astumian, R. D. (1983). *J. Phys. Chem.* **87**, 1449.

Schoonheydt, R. A. (1984). *In* "Characterization of Heterogeneous Catalysts" (F. Delannay, ed.), pp. 125–160. Dekker, New York.

Seager, C. H., and Land, C. E. (1984). *Appl. Phys. Lett.* **45**, 395.

Sinfelt, J. H., Via, G. H., and Lytle, F. W. (1984). *Catal. Rev.—Sci. Eng.* **26**, 81.

Somorjai, G. A., and Yeates, R. C. (1984). *J. Electrochem. Soc.* **131**, 228C.

Tagawa, T., and Amenomiya, Y. (1985). *Appl. Spectrosc.* **39**, 358.

Tam, A. C. (1983). *In* "Ultrasensitive Spectroscopic Techniques" (D. Kliger, ed.), pp. 1–107. Academic Press, New York.

Tam, A. C. (1985). *Infrared Phys.* **25**, 305.

Taub, H. (1980). *ACS Symp. Ser.* No. 137, 247.

Tilgner, R. (1981). *Appl. Opt.* **20**, 3780.

Uejima, A., Sugitani, Y., and Nagashima, K. (1985). *Anal. Sci.* **1**, 5.

Wandass, J. H., and Gardella, J. A., Jr. (1985a). *Surf. Sci.* **150**, L107.

Wandass, J. H., and Gardella, J. A., Jr. (1985b). *J. Am. Chem. Soc.* **107**, 6192.

Wolfram, T. (1978). "Inelastic Electron Tunneling Spectroscopy." Springer-Verlag, Berlin and New York.

Yeboah, S. A., Griffiths, P. R., Krishnan, K., and Kuehl, D. (1981). *SPIE Semin. Proc.* **289**, 105.

Yeboah, S. A., Wang, S.-H., and Griffiths, P. R. (1984). *Appl. Spectrosc.* **38**, 259.

8

MAGNETIC SPIN RESONANCE METHODS AND APPLICATIONS TO OXIDE SURFACES*

Larry Kevan

Department of Chemistry
University of Houston
Houston, Texas 77004

I. INTRODUCTION

Electron spin resonance (ESR) methods are useful for studying paramagnetic species on solid supports. The paramagnetic species studied fall into two general categories. One category consists of paramagnetic ions, both stable transition metal ions such as cupric ion as well as normally unstable valence states of metal ions produced in oxidations or reductions on solid supports such as silica or zeolites. Examples of unstable valence states stabilized on solid supports are silver (0), palladium(I), and palladium(III). The other general category are free radicals, typically organic, stable free radicals such as nitroxides or normally unstable free radical reaction intermediates that are stabilized on solid supports.

Electron spin resonance is a magnetic resonance spectroscopy that monitors transitions between magnetic energy levels associated with different orientations taken by an electron spin, generally in an external magnetic field. Typical electron magnetic resonance is carried out in a magnetic field of about 0.3 T. This corresponds to an energy absorption frequency of about 9 GHz in the microwave range (Gordy, 1980; Wertz

* This review was supported by the U.S. National Science Foundation and the Robert A. Welch Foundation.

141

PREPARATIVE CHEMISTRY
USING SUPPORTED REAGENTS

and Bolton, 1972). Analysis of the ESR spectrum can give information about the identity of a molecular species, the geometric structure, the electronic structure and the internal or overall rotational or translational motion. The specificity of ESR spectroscopy for species with unpaired electrons is particularly valuable for the study of chemical reaction intermediates or valence state changes of transition metal ions.

As with any spectroscopic technique the useful information is often limited by the resolution of the spectral transitions. This limiting factor is present particularly for paramagnetic species in solids and especially in powder samples typical in studies of solid supports. Double resonance and time-domain electron magnetic resonance methods can then become very useful. The major double resonance method used is electron nuclear double resonance (ENDOR) for which commercial instrumentation is available (Kevan and Kispert, 1976). The time domain technique of electron spin echo (ESE) spectroscopy has been developed for the study of paramagnetic species on solid supports (Kevan and Schwartz, 1979). This technique allows magnetic relaxation times to be measured directly, and more importantly it allows very weak dipolar hyperfine interactions to be measured even in powder samples so that the molecular surroundings of the paramagnetic species can be probed.

Since the reader is probably less likely to be familiar with ESE than with ordinary ESR a few additional words of explanation are in order. Electron spin echoes are generated in pulsed electron spin resonance experiments. The most common type of pulse sequence consists of a 90° focusing pulse followed by a precession time and then a 180° spin flip pulse which causes the spins to refocus within another precession time period. The refocusing produces a burst of microwave energy called an echo. This echo intensity is measured as a function of the time between the two pulses to generate an echo decay envelope with a time constant which gives a transverse magnetic relaxation time (Kevan and Schwartz, 1979). In many cases the echo decay envelope is modulated by weak dipolar hyperfine interactions between the paramagnetic species and neighboring magnetic nuclei, within the range of about 0.2 to 0.6 nm. Analysis of the electron spin echo modulation (ESEM) provides significant structural information which is generally not available from the electron spin resonance spectrum alone, due to resolution limitations. Typically the electron spin echo modulation pattern can be analyzed to yield the number of approximately equivalent interacting magnetic nuclei as well as their distance of interaction together with any small isotropic hyperfine coupling that may be present.

In this brief chapter we first describe the physical parameters obtainable by ESR methods and stress their importance for studies of paramagnetic species on solid supports. Then we present a small selection of current applications to zeolite and silica supports taken mostly from our own current work. Additional information may be found in several reviews of previous work on oxide supports and on zeolites (Che and Tench, 1982; Iton and Turkevich, 1977; Kasai and Bishop, 1976; Lunsford, 1977; Mikheikin *et al.*, 1972).

II. PHYSICAL PARAMETERS MEASURABLE BY ELECTRON SPIN RESONANCE

An observable ESR transition is characterized by its g factor, the electron–nuclear hyperfine splitting, and one or more magnetic relaxation times. The g factor gives the position of the transition at a given magnetic field for a particular microwave frequency and is specifically given as $g = h\nu/\beta H$, where h is Planck's constant, β the Bohr magneton, ν the microwave frequency, and H the magnetic field. For a free electron $g = 2.0023$. For many organic free radicals g is close to 2.00 and does not distinguish well between different radicals. However, for transition metal ions g is often characteristic of a particular ion or of a particular valence state of an ion. In addition, g is typically anisotropic and often shows rhombic or axial symmetry. With sufficient anisotropy the principal components of the g tensor may often be determined from the line shape of the powder ESR spectrum. The g values give information on the symmetry of the species on the solid support and on its ground state, particularly for paramagnetic transition metal ions. This is important since the ground state and the symmetry of the ion depend on its location in the lattice of the solid support.

The hyperfine interaction or splitting can be used to characterize the molecular structure for radical intermediates and can also characterize the number and distance of adsorbate molecules. However, in the latter case ESE modulation spectrometry is generally required for sufficient resolution. If there are magnetic atoms in the solid support lattice itself, such as aluminum for example, ESE modulation spectroscopy can be used to determine the aluminum interactions. In some cases these may even be able to be resolved by ESR.

Magnetic relaxation phenomena are complex, particularly so in solids. In general, a paramagnetic species is characterized by a spin lattice relaxation time T_1. This time characterizes the rate of radiationless transitions between the spin system and the thermal motion of the lattice or surroundings. A second relaxation time is the spin–spin relaxation time denoted by T_2. It is due to energy exchange between spins in the system. Both these relaxation times can be measured by ESE techniques. The transverse relaxation directly measured in powders by ESE is better described as a phase memory time since it is not generally identical to T_2. However, the true T_2 in solids can be obtained by special ESE methods (Kevan and Schwartz, 1979). Little more will be said about either of these types of relaxation times in the applications that follow since they have been used very little for species characterization or to give structural or chemical reaction information. However, this is an area with significant potential for development.

Another effect related to the relaxation times is microwave power saturation. This steady-state continuous-wave phenomenon can give information about the relaxation times but not as directly as pulsed methods. The main use of microwave saturation is differential saturation, often useful for distinguishing between two species with different relaxation times, which consequently saturate to different degrees at the same microwave power.

More details on measurement of all these parameters and ESR spectral analysis can be found in Gordy (1980) and in Wertz and Bolton (1972).

III. APPLICATIONS TO ZEOLITES

The class of aluminosilicates known as zeolites offer useful and rather well-defined solid supports for adsorption and catalysis. Metal ions, typically sodium ions, are present in the zeolite lattices to compensate for the imbalance in charge between the valences of aluminum and silicon. The charge-balancing cation is relatively easily exchanged, and a number of specific transition-metal-cation-exchanged zeolites have been shown to be useful catalysts for a variety of chemical reactions. In many of these the metal cations are considered to be the catalytic sites so it is of considerable importance to elucidate the location of these catalytically active cations and their interaction with adsorbed molecules. Studies of cupric ions, palladium ions, and rhodium ions in zeolites will now be briefly described for their illustrative value.

Divalent cupric ion has been much studied in A-type zeolites by Conesa and Soria (1978), Horser *et al.* (1978), Herman (1980), and others. However, ESR alone has not given unambiguous information about the water coordination or the specific location of cupric ion in the zeolite lattice. More recent ESEM results have answered some of these questions and have shown how to achieve a degree of control of the location and of the adsorbate coordination of the cupric ions in A-type zeolites. Kevan and Narayana (1983) and Anderson and Kevan (1987) have identified hydrated cupric ion in sodium A zeolite as coordinated to three waters in site SII', a β-cage site opposite a hexagonal ring between the α and β cages of the

A zeolite structure. This information was obtained directly by using deuterated water and analyzing the resultant deuterium ESE modulation pattern and by analysis of cesium modulation from known cesium ion positions. From the dipolar hyperfine interaction a cupric ion-to-deuterium distance of 0.27 nm is obtained which corresponds to a cupric ion-to-water oxygen distance of 0.21 nm; this is consistent with direct coordination. This fully hydrated cupric ion species in the sodium A zeolite can be easily dehydrated by brief heating to 50°C to form a cupric ion species coordinated to two water molecules, with a g anisotropy consistent with trigonal bipyramidal symmetry. This suggests that the cupric ion has lost coordination with some of its water molecules and has moved into the hexagonal ring (site SII) between the α and β cages where it is coordinated to three oxygens in the hexagonal ring, to one water molecule in the α cage and to one water molecule in the β cage. Further dehydration causes the loss of all of the water molecules coordinated to the cupric ion, and it then appears to stay close to the hexagonal ring near site SII. If this species is rehydrated it moves into the β cage again and becomes coordinated to three water molecules in the β cage and three oxygens in the hexagonal ring, again being located at site SII'.

The effect on cupric ion of changing the major cocation or changing the silicon-to-aluminum ratio in the zeolite lattice can be investigated by ESEM. The method is quite sensitive to such structural changes and to their effect on the cupric ion location and its water hydration. Results of Anderson and Kevan (1986) have been obtained on ZK4-type zeolites which have the A zeolite structure but in which the silicon-to-aluminum ratio can be varied from near 1 to about 3. Substitution of the potassium ion for the sodium ion causes the cupric ion to decrease its water coordination from three (Na^+) to one (K^+) in site SII'. The other interesting feature as the silicon-to-aluminum ratio is increased from 1.0 to 2.4 is that a new copper species coordinated to two water molecules becomes increasingly dominant. Analysis of the ESEM patterns shows that this cupric ion is located in the α cage. Therefore, by controlling the major cocation and the silicon-to-aluminum ratio the initial hydrated form of cupric ion can be controlled to be in either the α or the β cage and to coordinate with one or three water molecules. This is a good example of the type of structural detail that we can obtain by using the full power of currently available ESR and ESEM methods applied to solid supports. The geometry and coordination of other adsorbates such as methanol and ethylene to cupric ion have also been studied by Ichikawa and Kevan (1981). Analogous studies have also been carried out for cupric ions in X zeolites (Ichikawa and Kevan, 1983a) and in Y zeolites (Ichikawa and Kevan, 1983b).

Other types of ions that have been studied in detail in A-, X-, and Y-type zeolites are silver atoms produced from silver ions by radiolytic reduction (Narayana and Kevan, 1982), and Cd(I) formed by radiolytic reduction from Cd(II) (Narayana and Kevan, 1983).

The generation of two types of paramagnetic palladium ions and their interactions with various organic adsorbates and water have been studied by ESR and ESEM by Michalik *et al.* (1985). The typical activation process for palladium zeolite catalysts of heating to >700 K and then exposure to oxygen at that relatively high temperature followed by evacuation led to characteristic ESR signals of palladium(III) and palladium(I). In zeolites palladium(III) has an isotropic line characterized by $g = 2.23$ whereas palladium(I) has typically an axially symmetric g tensor with $g_\parallel > g_\perp$. Conversely, when the palladium species interacts strongly with water or organic molecules such as benzene, ethylene, and methanol as adsorbates it forms a different ESR spectrum with a reversed axial anisotropy $g_\perp > g_\parallel$. In CuCa–X zeolite evaluation of how the paramagnetic species interacts with various adsorbates including oxygen leads to the conclusion that palladium(III) is located in sites that are accessible to the α cage such as site SII which is in the hexagonal ring between the α and β cages in X zeolite or displaced slightly into the β cage. In contrast palladium(I) is located in a much less accessible site, probably in SI', in the β cage near the hexagonal prism of the X zeolite structure.

A measure of cocation control is described by Michalik *et al.* (1986). When the major cocation is changed from calcium to sodium the relative proportions of palladium(III) and palladium(I) are nearly reversed. Palladium(III) is much more stable in sodium X than in calcium X zeolite and is most probably located in site SI, which is in the hexagonal prism of the X zeolite structure. This is supported by the strong aluminum modulation that is seen in the electron spin echo spectrum.

Similar studies have been carried out by Goldfarb and Kevan (1986) on rhodium(II) species in sodium X zeolite. The rhodium(II) species are generated from rhodium(III) by a thermal activation process. However, two different types of rhodium(II) species are observed depending on the activation temperature. One species, generated at the lower activation temperature, is thought to be located in the β cage whereas the dominant species generated at the higher activation temperature is deduced to be located in the hexagonal prisms of the X zeolite structure. This is based on the interaction distances with adsorbates measured by ESEM and on the broadening effect of oxygen on the ESR lines.

In addition to looking at catalytically active metal ions in zeolites by ESR it is also possible to look at organic radical intermediates produced on zeolites by photolysis, radiolysis, or thermal reactions. For example, the CH_2OH radical formed from the radiolysis or photolysis of methanol adsorbed on zeolites has been studied with the objective of determining the location of such a radical intermediate within the zeolite structure. Ichikawa and Kevan (1980) and Narayana and Kevan (1981) have answered this question with X- and Y-type zeolites as supports. In A zeolite this CH_2OH radical is located in the α cage of the zeolite structure above a six-ring between the α and β cages, which six-ring is a typical site for a sodium or other charge-compensating cation. Furthermore, ESEM data indicate the orientation of the radical with respect to the six-ring axis; the molecular dipole of the radical is oriented toward this axis or in other words toward the cation site. Furthermore, it is found that the radical interacts with the closest neutral molecules in the α cage.

In X- and Y-type zeolites the geometry is found to be similar to that in A zeolite in that the CH_2OH radical is located in the α cage. However, the distances to the closest neutral methanol molecules surrounding it are larger, which is consistent with the larger α cage in the X- and Y-type zeolites compared to A zeolite. This gives an example of the detailed information stemming from combined use of ESR and ESEM techniques.

It is also possible to study molecular packing of adsorbed molecules on surfaces by determining the orientation of an organic radical generated from an adsorbed molecule. This has been carried out for the CH_2OH radical in channel-type zeolites such as mordenite and ZSM-5. This is only possible by using ESEM techniques to measure the distances from the CH_2OH radical to surrounding specifically deuterated adsorbate molecules. For example, Narasimhan et al. (1983) have been able to determine that the methanol molecules are arranged in a wall-to-wall configuration across the channels in mordenite with the OH group alternately oriented on adjacent molecules; any given molecule or hydroxymethyl radical generated from it has two nearest methanol neighbors. The distance between the nearest molecules can also be measured from the weak dipolar hyperfine interaction and is sensitive to the partial pressure of adsorbed methanol. Similar studies have been carried out in ZSM-5 zeolite in which the methanol adsorbate has a high probability of being near the intersections of the two types of channels present in this commercially interesting zeolite.

IV. APPLICATIONS TO SILICA

Silica and alumina are other well-known solid supports for adsorption and catalysis. Some ESR studies have been carried out on alumina, but a greater number have been carried out on silica. Both organic radical species and paramagnetic metal ion species can be investigated along similar lines as already mentioned.

For example, Ichikawa *et al.* (1981, 1982) have used ESR and ESEM to study cupric ions on silica and to measure their geometry and interactions with various adsorbates. Cupric ion interacts with two molecules of adsorbed water and two molecules of adsorbed methanol in a distorted octahedral geometry. It is possible to determine the symmetry of the adsorbate coordination from the g anisotropy of the cupric ion and it is possible to count adsorbed molecules and to measure their interaction distance by measuring weak dipolar hyperfine interactions by ESEM. With ammonia as an adsorbate, the cupric ion also interacts with two ammonia molecules but the symmetry is more nearly square planar than distorted octahedral; the cupric ion interacts with two ammonia molecules and two oxygens of the silica surface to form an approximate square planar configuration.

With larger organic adsorbates such as ethylene a difference is found depending on whether the cupric ion is incorporated into the silica surface by ion exchange or by impregnation. In the first case (ion exchange) Ichikawa *et al.* (1982) find a distorted tetrahedral symmetry in which the cupric ion interacts with only one ethylene molecule at the relatively large distance of 0.38 nm. This distance is too long for direct coordination; thus the distorted tetrahedral geometry points to the cupric ion interacting with four oxygens of the silica surface and more weakly with one ethylene molecule. However Narayana *et al.* (1984) found that for impregnated cupric ion the interaction distance to the ethylene plane decreases to 0.30 nm indicating a much stronger interaction which might be described as a distorted trigonal bipyramid. Zhan *et al.* (1985a) have studied various other adsorbate interactions of impregnated cupric ion on silica.

There have also been a number of studies of paramagnetic vanadium species and paramagnetic molybdenum species on silica by ESR and by ESEM. Research has focused on adsorbate interactions which can only be studied in any detail by measuring weak dipolar hyperfine interactions of the adsorbate molecules by ESEM methods. Studies of this type for vanadium(IV) on silica have been carried out by Narayana *et al.* (1985) and for molybdenum(V) on silica by Zhan *et al.* (1985b). References to extensive earlier ESR work on paramagnetic vanadium and molybdenum species on silica can be found in these recent studies.

REFERENCES

Anderson, M., and Kevan, L. (1986). *J. Phys. Chem.* **90**, 3206.
Anderson, M., and Kevan, L. (1987). *J. Phys. Chem.* **91**, 1850.
Che, M., and Tench, A. J. (1982). *Adv. Catal.* **31**, 77.
Conesa, J. C., and Soria, J. (1978). *J. Phys. Chem.* **82**, 1575, 1847.

Goldfarb, D., and Kevan, L. (1986). *J. Phys. Chem.* **90**, 264.

Gordy, W. (1980). "Theory and Applications of Electron Spin Resonance." Wiley (Interscience), New York.

Herman, R. G. (1980). *Inorg. Chem.* **18**, 995.

Horser, H., Primet, M., and Vedrine, J. C. (1978). *J.C.S. Faraday I* **74**, 335.

Ichikawa, T., and Kevan, L. (1980). *J. Am. Chem. Soc.* **102**, 2650.

Ichikawa, T., and Kevan, L. (1981). *J. Am. Chem. Soc.* **103**, 5355.

Ichikawa, T., and Kevan, L. (1983a). *J. Am. Chem. Soc.* **105**, 402.

Ichikawa, T., and Kevan, L. (1983b). *J. Phys. Chem.* **87**, 4433.

Ichikawa, T., Yoshida, H., and Kevan, L. (1981). *J. Chem. Phys.* **75**, 2485.

Ichikawa, T., Yoshida, H., and Kevan, L. (1982). *J. Phys. Chem.* **86**, 881.

Iton, L. E., and Turkevich, J. (1977). *J. Phys. Chem.* **81**, 435.

Kasai, P. H., and Bishop, R. J. (1976). *In* "Zeolite Chemistry and Catalysis" (J. A. Rabo, ed.), A.C.S. Monograph No. 171, pp. 350–391. Am. Chem. Soc., Washington, D.C.

Kevan, L., and Kispert, L. (1976). "Electron Spin Double Resonance Spectroscopy." Wiley (Interscience), New York.

Kevan, L., and Narayana, M. (1983). *A.C.S. Symp. Ser.* No. 218, 283.

Kevan, L., and Schwartz, R. N. (1979). "Time Domain Electron Spin Resonance." Wiley (Interscience), New York.

Lunsford, J. H. (1977). *In* "Chemistry and Physics of Solid Surfaces" (R. Vanselow and S. Y. Tong, eds.), pp. 255–273. CRC Press, Cleveland, Ohio.

Michalik, J., Narayana, M., and Kevan, L. (1985). *J. Phys. Chem.* **89**, 4553.

Michalik, J., Heming, M., and Kevan, L. (1986). *J. Phys. Chem.* **90**, 2132.

Mikheikin, I. D., Zhidomirov, G. M., and Kazanskii, V. B. (1972). *Russ. Chem. Rev.* **41**, 468.

Narasimhan, C. S., Narayana, M., and Kevan, L. (1983). *J. Phys. Chem.* **87**, 984.

Narayana, M., and Kevan, L. (1981). *J. Am. Chem. Soc.* **103**, 5729.

Narayana, M., and Kevan, L. (1982). *J. Chem. Phys.* **76**, 3999.

Narayana, M., and Kevan, L. (1983). *J. Chem. Phys.* **79**, 5136.

Narayana, M., Zhan, R. Y., and Kevan, L. (1984). *J. Phys. Chem.* **88**, 3990.

Narayana, M., Narasimhan, C. S., and Kevan, L. (1985). *J.C.S. Faraday I* **81**, 137.

Wertz, J. E., and Bolton, J. R. (1972). "Electron Spin Resonance." McGraw-Hill, New York.

Zhan, R. Y., Narayana, M., and Kevan, L. (1985a). *J. Phys. Chem.* **89**, 831.

Zhan, R. Y., Narayana, M., and Kevan, L. (1985b). *J.C.S. Faraday I* **81**, 2083.

9 MAGNETIC RESONANCE METHODS

Harry Pfeifer and Horst Winkler

Sektion Physik der Karl-Marx-Universität Leipzig
DDR-7010 Leipzig, German Democratic Republic

I. NUCLEAR MAGNETIC RESONANCE METHODS

The field of nuclear magnetic resonance (NMR) (Slichter, 1980) has developed very rapidly during the last ten years due to the invention of techniques that allow a measurement of highly resolved NMR spectra in solids.

As is well known, the Hamiltonian operator that determines the properties of a spin system can be written as

$$H_0 = H_Z + H_D + H_\delta + H_J + H_Q + H_{RF} \qquad (1)$$

where H_Z is the Zeeman interaction between the nucleus and the applied static magnetic field B_0, H_D denotes the magnetic dipolar interaction, H_δ the chemical shift interaction, H_J the indirect electron-coupled nuclear spin interaction (J coupling), H_Q the electric quadrupolar interaction which is only present for nuclei with spin $I > \frac{1}{2}$ and H_{RF} the interaction with alternating magnetic fields, mostly applied in the form of short rf pulses.

In nonviscous liquids H_D and H_Q average to zero, which permits the much weaker chemical shift δ and the J coupling to be studied. In the solid state, however, the dipolar and quadrupolar interactions are dominant, which leads to such a broadening of the NMR lines that in general the valuable information contained in the chemical shift and J-coupling [fingerprint system of liquid state NMR (Harris and Mann, 1978)] is hidden.

151

PREPARATIVE CHEMISTRY
USING SUPPORTED REAGENTS

The new techniques of solid-state NMR, namely multiple pulse sequences (Waugh *et al.*, 1968), cross polarization (CP) in combination with high-power decoupling (Pines *et al.*, 1973) and magic angle spinning (MAS) of the sample (Andrew, 1971) eliminate or at least reduce to a certain degree the influence of H_D and H_Q, thus making possible the observation of highly resolved NMR spectra in solids.

More details and a review of the present state of the art can be found in literature (Mehring, 1983; Fyfe, 1983). A review of the application of the more classical NMR methods to surface phenomena is given by Pfeifer (1976) and Pfeifer *et al.* (1983).

In the following, we consider at first porous solids containing Si, O, and, eventually, Al; after that, we consider OH groups and, finally, chemisorbed molecules at the surface of these solids.

II. NUCLEAR MAGNETIC RESONANCE STUDIES OF POROUS SOLIDS

In zeolites (see Chapters 21 and 22), it is possible by means of MAS-^{29}Si-NMR to obtain up to five resonance lines, corresponding to Si in the following five possible building units (Thomas, 1984): $Si(OAl)_4$, $Si(OAl)_3(OSi)$, $Si(OAl)_2(OSi)_2$, $Si(OAl)(OSi)_3$, and $Si(OSi)_4$, an example being shown in Fig. 1. Under the assumption that Loewenstein's rule holds, stating that there are no Al—O—Al links, the Si/Al ratio of the zeolite can be calculated from the relative intensities I of the lines:

$$\frac{Si}{Al} = \sum_{n=0}^{4} I_{Si(OAl)_n(OSi)_{4-n}} \Bigg/ \sum_{n=0}^{4} \frac{n}{4} I_{Si(OAl)_n(OSi)_{4-n}} \qquad (2)$$

The Si/Al ratios for various zeolites as determined by x-ray fluorescence and MAS-^{29}Si-NMR have been compared by Fyfe *et al.* (1983); the agreement is within 5%.

When a zeolite Y is dealuminated, a single resonance peak appears, corresponding to $Si(OSi)_4$ units (Fig. 2) (Thomas, 1984). In fully dealuminated zeolites crystallographically nonequivalent $Si(OSi)_4$ groups produce resolvable signals, the chemical shift of which is correlated to the mean value Θ of the four tetrahedral Si-O-Si angles by the relation (Thomas, 1984)

$$\delta = -25.44 - 0.5793\Theta \qquad (3)$$

where δ is measured in ppm from TMS and Θ in degrees. Corresponding equations also hold for $Si(OAl)_n(OSi)_{4-n}$ building units (Ramdas and Klinowski, 1984). In a highly siliceous zeolite, silicalite, with Si/Al

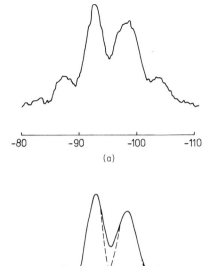

Fig. 1. (a) MAS-^{29}Si-NMR spectrum at 79.80 MHz of zeolite Y with Si/Al ratio of 2.75. The four lines stem from the first four building units mentioned in the text. The resonance shift is given with respect to TMS and is positive to lower field. (b) The computer-simulated spectrum is plotted (Gaussian peaks). [From Thomas (1984).]

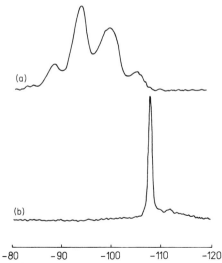

Fig. 2. MAS-^{29}Si-NMR spectrum at 79.80 MHz of (a) zeolite NaY (Si/Al = 2.61) and (b) its dealuminated form. [From Thomas (1984).]

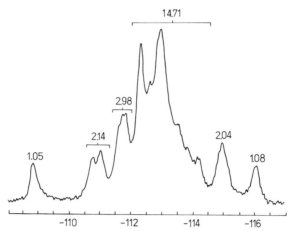

Fig. 3. MAS-^{29}Si-NMR spectrum at 99.3 MHz of silicalite (Si/Al>1000). Relative intensities (normalized to 24) are indicated. [From Thomas (1984).]

ratio >1000, up to 15 resonance lines can be identified, partly as shoulders, as represented in Fig. 3 (Thomas, 1984).

MAS-^{27}Al-NMR performed on zeolites can distinguish tetrahedrally and octahedrally coordinated aluminum (Thomas, 1984). The latter is produced in the process of dealumination, i.e., by Al having escaped from the framework; the corresponding aluminum oxide complexes are more or less mobile, depending on the diameter of the cages in which they are sited. For that reason the framework Al line is wider than the extraframework Al line, as can be seen in Fig. 4. In a thermally treated zeolite of type CaA MAS-^{27}Al-NMR has revealed a third line at 80 ppm (Fig. 5) that must be ascribed to extralattice Al in the form of $Al(OH)_4^-$ anions (Freude *et al.*, 1985).

III. NUCLEAR MAGNETIC RESONANCE STUDIES OF HYDROXYL GROUPS ON SURFACES

From the intensity of ^{1}H-NMR power spectra of the solids considered, the absolute number of OH groups can be determined (whereas problems arise in this respect with IR spectroscopy). Table I shows the OH group density on a silica gel surface of 450 m^2/g for various heat treatment temperatures (Bernstein *et al.*, 1981).

Whether the OH groups are sited at the surface or in the bulk of the adsorbent can be decided by contacting the sample with D_2O vapor (Davydov *et al.*, 1983). Of course, only surface OH groups are accessible to D_2O so that an H–D exchange can occur. The result for two silicas is

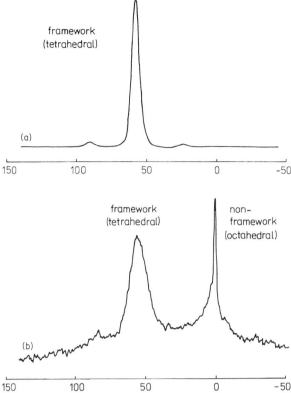

Fig. 4. MAS-^{27}Al-NMR spectrum at 104.2 MHz of zeolite Y (a) before and (b) after dealumination by heating the corresponding NH_4 zeolite. The resonance shift is given with respect to $[Al (H_2O)_6]^{3+}$ and is positive to lower field. [From Thomas (1984).]

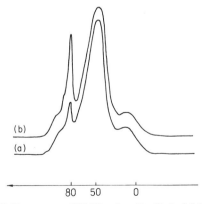

Fig. 5. MAS-^{27}Al-NMR spectra at 70 MHz of zeolite CaA: (a) 200°C deep-bed activated and fully rehydrated and (b) 300°C deep-bed activated and partially rehydrated (6 H_2O per cavity). [From Freude *et al.* (1985).]

TABLE I

Number of OH Groups per nm^2 of Silica Gel 450 (Merck) for Various Heat Treatment Temperatures

Heat treatment temperature (°C)	200	300	400	500	600	700	800
Number of OH groups per nm^2	4.5 ±0.4	3.4 ±0.3	2.4 ±0.3	1.4 ±0.3	1.0 ±0.3	0.7 ±0.3	0.4 ±0.3

TABLE II

Total Number of OH Groups and Number of Nonexchanged OH Groups After D_2O Treatment

Support	Number of OH groups per nm^2	
	Total	Bulk
Silica Gel 450 (Merck)	4.5 ± 0.3	0.5 ± 0.3
Aerosil 300 (DEGUSSA)	20.6 ± 2.0	11.3 ± 0.5

shown in Table II. The specific surface area of Aerosil 300 (DEGUSSA) has been changed by a hydrothermal treatment from 300 to only 50 m^2/g by sintering, so that a considerable portion of the OH groups has become occluded.

In the proton free induction decays (FID) of silica aluminas with an alumina content between about 20 and 90% the rapidly decaying signal of the Al—OH protons can be determined quantitatively as a superposition on the slowly decaying component due to the Si—OH protons (Schreiber and Vaughan, 1975).

The chemical shift anisotropy, obtained from the proton-NMR spectrum with homonuclear decoupling of Silica Gel 450, amounts to (Bernstein *et al.*, 1981) $\Delta\sigma \equiv \sigma_\parallel - \sigma_\perp = 5$–8 ppm. Schreiber and Vaughan (1975) find 6.9 ppm for a different silica gel. It turns out that $\Delta\sigma$ is partly averaged out because of a Brownian rotation of the OH group around the Si—O axis. If the rotation frequency Ω is much greater than the anisotropy, $\Omega \gg \Delta\sigma^{rigid}\gamma B_Z$, then

$$\Delta\sigma = \tfrac{1}{2}\Delta\sigma^{rigid}(3\cos^2\alpha - 1), \tag{4}$$

where α is the Si—O—H angle. With $\Delta\sigma^{rigid} \approx 12$–36 ppm (Rosenberger and Schnabel, 1978) we obtain that α must be in the interval

$132° \leqslant \alpha \leqslant 146°$. *Ab initio* calculations for the model cluster $(HO)_3SiOH$ result in $\alpha = 140°$ or $117°$, if a $4-31G$ or a $6-31G^*$ basis set is used, respectively (Mortier *et al.*, 1984). As the second basis set should yield more accurate values, the obtained too-small angle of $117°$ demonstrates the limits of validity of the underlying model cluster.

The MAS-^{1}H-NMR spectra of an aerosil (Lippmaa *et al.*, 1981) reveal up to four lines that can be attributed to isolated (2.0 ppm) and geminal (2.9 ppm) OH groups, to isolated H_2O molecules (1.1 ppm) and to hydrogen-bonded water (4.3 ppm). If the aerosil is dehydrated and dehydroxylated in part at 730°C for one hour, only isolated and geminal OH groups can be seen in the spectrum. After rehydration in air for 7 days the signals of hydrogen-bonded and isolated water molecules prevail, which is even more the case after rehydration for 100 days.

Adsorption of (deuterated) pyridine, acetone, or benzene on Silica Gel 450 (Merck) shifts the proton line of those OH groups, at which the adsorbed molecules are (hydrogen-) bonded by 7, 4, or (0–1) ppm to lower field, respectively (Rosenberger *et al.*, 1982). In the case of pyridine the intensity of the unshifted line, corresponding to OH groups being not accessible to pyridine, amounts to (0.7 ± 0.3) OH/nm^2 of a total of 2.4 OH/nm^2.

In the MAS-^{1}H-NMR spectra of carefully dehydrated HY zeolites (contained in sealed glass tubes) it was possible to distinguish up to four lines (Fig. 6) and to attribute them (Pfeifer *et al.*, 1985) to nonacidic

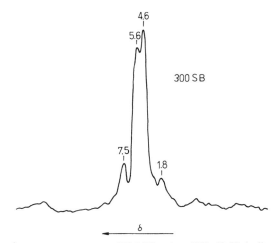

Fig. 6. MAS-^{1}H-NMR spectrum at 270 MHz of an NH$_4$ NaY zeolite treated at 300°C under shallow-bed conditions. The various lines are explained in the text. [From Hunger (1984).]

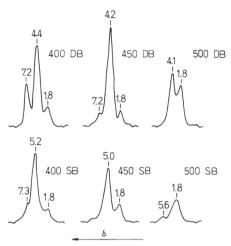

Fig. 7. MAS-^{1}H-NMR spectrum at 270 MHz of an NH$_4$ NaY zeolite for various heat treatments. [From Hunger (1984.)]

Si—OH groups ($\delta = 1.8$ ppm), comparable to those in silica gel, to two kinds of acidic so-called bridging OH groups ($\delta = 3.9$–4.6 ppm and 4.8–6.6 ppm), and to residual NH$_4^+$ cations ($\delta = 7$ ppm).

Figure 7 demonstrates that the four signals shown in Fig. 6, strongly depend on the details of the heat treatment (the two lines at medium field in Fig. 6 corresponding to two kinds of acidic OH groups are not resolved in Fig. 7): at 400, 450, and 500°C under deep-bed (DB) conditions in the upper part and under shallow-bed (SB) conditions in the lower part (Hunger, 1984).

The CP/MAS-^{29}Si-NMR spectrum of silica gel in Fig. 8 allows us to distinguish Si atoms in positions a, b, and c (cf. the inset of Fig. 8), where the signals of Si atoms of type b and c are enhanced through polarization transfer from the OH group protons (Sindorf and Maciel, 1981). The MAS-^{29}Si-NMR spectrum of the same silica gel as in Fig. 8 has also been studied using a conventional high-resolution spectrometer at high magnetic field (^{29}Si resonance at 79.5 MHz) without CP (Fyfe et al., 1985). The spectrum is comparable to that at the much lower field (11.88 MHz) in Fig. 8, i.e., the ^{29}Si chemical shift anisotropy is small enough to be efficiently removed by MAS even at high magnetic fields. However, without using CP all observed resonance lines can now be intepreted quantitatively. Spin lattice relaxation times of the various ^{29}Si nuclei were also measured in order to obtain quantitatively reliable intensities.

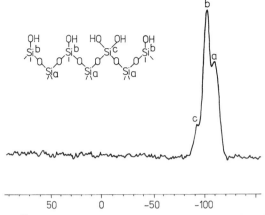

Fig. 8. CP/MAS-^{29}Si-NMR spectrum of Fisher S-157 silica gel at 11.88 MHz. The chemical shift is given with respect to TMS and is positive to lower field. See text for discussion of lines a, b, and c. [Reprinted with permission from Sindorf and Maciel (1981). Copyright 1981 American Chemical Society.]

By combining MAS-^{27}Al-NMR and MAS-^{1}H-NMR measurements, Freude *et al.* (1983) and Pfeifer *et al.* (1985) were able to elucidate the process of dehydroxylation and dealumination of zeolites at high temperatures. In contrast to the well-known model of Uytterhoeven *et al.* (1965) for dehydroxylation, they found (Table III) that (1) the numbers of bridging OH groups and of four-coordinated Al atoms are equal, (2) much less than

TABLE III

Concentration of Bridging OH Groups Corresponding to the Medium ^{1}H Lines in Figs. 6 and 7, and Four-Coordinated Al Atoms of an 90% NH$_4$-Exchanged HY zeolite for Deep-Bed (DB) and Shallow-Bed (SB) Treatment at Various Temperatures

Sample	Bridging OH groups (OH/cavity)	Four-coordinated lattice Al (Al/cavity)
90 HY 300 DB	3.9 ± 0.4	4.7 ± 0.4
400 DB	3.3 ± 0.4	3.7 ± 0.4
500 DB	2.1 ± 0.4	2.7 ± 0.4
90 HY 300 SB	6.2 ± 0.4	5.5 ± 0.4
400 SB	6.0 ± 0.4	5.8 ± 0.4
450 SB	5.8 ± 0.4	5.9 ± 0.4
600 SB	0.0 ± 0.4	1.2 ± 0.4

half of the four-coordinated Al remains when dehydroxylation is complete, and (3) no significant amount of three-coordinated Al exists. This means that dehydroxylation is connected with dealumination.

IV. NUCLEAR MAGNETIC RESONANCE STUDIES OF SURFACE COMPOUNDS (CHEMISORBED MOLECULES)

The chemisorption of molecules results in a modification of the corresponding surface; in this way, especially reactive species can be attached to the surface. The CP/MAS-^{29}Si-NMR spectrum in Fig. 9 reveals the surface modification of the Fisher S-157 silica gel, whose spectrum is plotted in Fig. 8, by trimethylchlorosilane (Sindorf and Maciel, 1981).

In Fig. 10 the influence of atmospheric water vapor on the reaction products of dimethyldichlorosilane and silica gel can be seen (Sindorf and Maciel, 1981). The assignment of lines e, f, and g has been made in comparison with solution data. Line f arises from line g by hydrolysis (in spectrum (a) by contaminating water on the surface). Spectrum (c) reveals an additional surface reaction, namely the condensation of two neighbouring groups of type g and f to give Si of type e'.

Quantitative MAS-^{29}Si-NMR spectra at 79.5 MHz, without using CP,

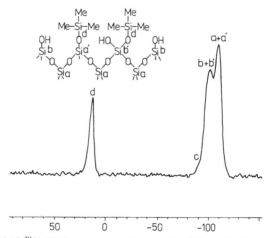

Fig. 9. CP/MAS-^{29}Si-NMR spectrum at 11.88 MHz of Fisher S-157 silica gel (cf. Fig. 8) after reaction with trimethylchlorosilane. The modified surface structure is given in the inset. Line d stems from the Si (CH$_3$)$_3$ groups. Comparison with Fig. 8 shows that intensity from line b has been transferred to line a by the Si atoms of type a'; analogously, an intensity transfer has taken place from line c to b (Si of type b'). [Reprinted with permission from Sindorf and Maciel (1981). Copyright 1981 American Chemical Society.]

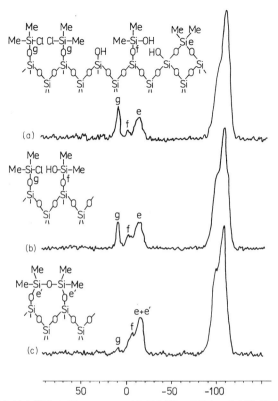

Fig. 10. CP/MAS-^{29}Si-NMR spectra at 11.88 MHz of Fisher S-157 silica gel as modified by reaction with (a) dimethyldichlorosilane and after exposure to air for (b) 12 hrs and (c) 24 hrs. Lines e, f, and g, are discussed in the text. [Reprinted with permission from Sindorf and Maciel (1981). Copyright 1981 American Chemical Society.]

have been obtained for silica gel modified by reaction with trimethylchlorosilane. The intensity of the peak due to the attached silylating reagent (8.9%) is approximately equal to the total change of 9.1% in relative intensities of the lines of the starting silica gel, since by the reaction with $(CH_3)_3SiCl$ some of the $(HO)_2Si(OSi)_2$ groups are converted to $(HO)Si(OSi)_3$ groups (3.7%); and some of the $(HO)Si(OSi)_3$ are converted to $Si(SiO)_4$ groups (5.4%) (Fyfe et al., 1985).

Finally, as an example for the application of CP/MAS-^{13}C-NMR, we mention the study of surface compounds on silica gel prepared by reaction with various monochlorosilanes (Hays et al., 1982). The spectra in Fig. 11 exhibit small linewidths and may be treated in a way similar to liquid spectra. Measurements on these solids using a conventional liquid-state

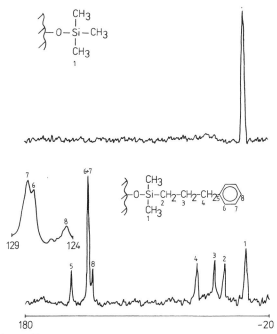

Fig. 11. CP/MAS-^{13}C-NMR spectra at 75.45 MHz of Silica Gel (Merck) whose surface is modified by various monochlorosilanes, as given in the inset. The chemical shift is with respect to TMS and is positive to lower field. [From Hays *et al.* (1982).]

NMR spectrometer, however, show that even a C_{18} chain is not mobile enough to give a well-resolved spectrum, so that we must perform solid-state NMR in the case of these chemisorbed molecules.

REFERENCES

Andrew, E. R. (1971). *Prog. Nucl. Magn. Reson. Spectrosc.* **8**, 1.

Bernstein, T., Ernst, H., Freude, D., Jünger, I., Sauer, J., and Staudte, B. (1981). *Z. Phys. Chem. (Leipzig)* **262**, 1123.

Davydov, V. J., Kiselev, A. V., Pfeifer, H., and Jünger, I. (1983). *Zh. Fiz. Chim.* **57**, 2535.

Freude, D., Fröhlich, T., Hunger, M., Pfeifer, H., and Scheler, G. (1983). *Chem. Phys. Lett.* **98**, 263.

Freude, D., Haase, J., Pfeifer, H., Prager, D., and Scheler, H. (1985). *Chem. Phys. Lett.* **114**, 143.

Fyfe, C. A. (1983). "Solid State NMR for Chemists." C. F. C. Press, Guelph, Ontario.

Fyfe, C. A., Thomas, J. M., Klinowski, J., and Gobbi, G. C. (1983). *Angew. Chem.* **95**, 257.

Fyfe, C. A., Gobbi, G. C., and Kennedy, G. J. (1985). *J. Phys. Chem.* **89**, 277.

Harris, R. K., and Mann, B. E. (1978). "NMR and the Periodic Table." Academic Press, New York.

Hays, G. R., Clague, A. D. H., Huis, R., and Van der Velden, G. (1982). *Appl. Surf. Sci.* **10**, 247.

Hunger, M. (1984). Dissertation. Karl-Marx-Univ., Leipzig.

Lippmaa, E. T., Samoson, A. V., Brer, V. V., and Gorlov. Y. (1981). *Dokl. Akad. Nauk SSSR* **259**, 403.

Mehring, M. (1983). "Principles of High Resolution NMR in Solids." Springer-Verlag, Berlin and New York.

Mortier, W. J., Sauer, J., Lercher, J. A., and Noller, H. (1984). *J. Phys. Chem.* **88**, 905.

Pfeifer, H. (1976). *Phys. Rep.* **26C**, 293.

Pfeifer, H., Meiler, W., and Deininger, D. (1983). *Ann. Rep. NMR Spectrosc.* **15**, 291.

Pfeifer, H., Freude, D., and Hunger, M. (1985). *Zeolites* **5**, 274.

Pines, A., Gibby, M. G., and Waugh, J. S. (1973). *J. Chem. Phys.* **59**, 569.

Ramdas, S., and Klinowski, J. (1984). *Nature* (*London*) **308**, 521.

Rosenberger, H., and Schnabel, B. (1978). *Wiss. Z. Friedrich-Schiller-Univ. Jena, Math-Naturwiss. Reihe* **27**, 257.

Rosenberger, H., Ernst, H., Scheler, G., Jünger, I., and Sonnenberger, R. (1982). *Z. Phys. Chem.* (*Leipzig*) **263**, 846.

Schreiber, L. B., and Vaughan, R. W. (1975). *J. Catal.* **40**, 226.

Sindorf, D. W., and Maciel, G. E. (1981). *J. Am. Chem. Soc.* **103**, 4263.

Slichter, C. P. (1980). "Principles of Magnetic Resonance." Springer-Verlag, Berlin and New York.

Thomas, J. M. (1984). *Proc. Int. Congr. Catal. 8th Berlin* **1**, 31.

Uytterhoeven, J. B., Christner, L. G., and Hall, W. K. (1965). *J. Phys. Chem.* **69**, 2177.

Waugh, J. S., Huber, L. M., and Haeberlen, U. (1968). *Phys. Rev. Lett.* **20**, 180.

10 PHYSICO-CHEMICAL CHARACTERIZATION OF SUPPORTED REAGENTS*

Jack H. Lunsford

Department of Chemistry
Texas A & M University
College Station, Texas 77843

I. INTRODUCTION

Insight into the chemistry of supported reagents is intimately related to the characterization of these systems by a variety of spectroscopic and other physico-chemical techniques. The arsenal of commerically available spectroscopic tools includes infrared (IR) spectroscopy, diffuse reflectance spectroscopy (DRS), photoluminescence spectroscopy, x-ray photoelectron spectroscopy (XPS), Mössbauer spectroscopy, electron paramagnetic resonance (EPR) spectroscopy, and nuclear magnetic resonance (NMR) spectroscopy (see Chapters 7–9). Applications of these spectroscopies to heterogeneous catalysis have been described in some detail (Delgass *et al.*, 1979). In addition a number of thermally related phenomena such as temperature-programmed desorption (TDP) and temperature-programmed reaction (TPR) have been effectively employed (Madix, 1984; Falconer and Schwarz, 1983). Only the diffraction techniques are lacking from this general list, and even here, if the supported metal ions are considered as reagents, considerable application has been made of

* The work carried out in the author's laboratory was supported mainly by the National Science Foundation and the Robert A. Welch Foundation.

**PREPARATIVE CHEMISTRY
USING SUPPORTED REAGENTS**

x-ray diffraction (XRD) and neutron diffraction to define the positions and bonding of these ions in highly ordered systems such as zeolites (Mortier, 1982; Cheetham *et al.*, 1984). Likewise, extended x-ray absorption fine structure (EXAFS) is an emerging new technique that shows exceptional promise for the characterization of supported reagents (Sinfelt *et al.*, 1984; Morrison *et al.*, 1981).

Instead of treating each technique separately in this chapter, several case studies will be considered in order to emphasize the manner in which these methods may be used in concert to resolve the sometimes difficult chemistry of supported reagents. As a class, the most demanding problems usually occur in relation to catalytic cycles because many of the important intermediates are transient and are present only in low concentrations. Nevertheless, considerable progress has been made in unravelling mechanisms with a reasonable degree of confidence. Ideally, we would like to be able to carry out *in-situ* dynamic experiments, and in some cases this has been possible (Iizuka and Lunsford, 1978); but often spectroscopic measurements must be performed under more ideal conditions, with inference being made to the actual reaction conditions.

II. MOLYBDENUM ON SILICA

Molybdenum in a highly dispersed (perhaps monoatomic) state on silica has proven to be a very useful reagent and ultimately a catalyst because of its ability to promote one-electron transfer reactions with nitrous oxide (Kolosov *et al.*, 1975; Ben Taarit and Lunsford, 1973)

$$Mo^V + N_2O \rightarrow Mo^{VI}O^- + N_2 \tag{1}$$

The O^- ion, in turn, is effective in the activation of simple alkanes, including methane, via a hydrogen atom abstraction to form the corresponding alkyl radical. These radicals then react with surface molybdena to form alkoxide ions which in turn either decompose to form aldehydes or react with water to form alcohols. The partial oxidation of methane may be described by the catalytic cycle indicated on the following page. At a CH_4 conversion of 4%, selectivities of 60% to HCHO and 10% to CH_3OH were achieved, and at 1.6% conversion the selectivities were 80% to HCHO and 20% to CH_3OH (Liu *et al.*, 1984).

Spectroscopic evidence has been obtained (1) for the presence of Mo^V and its variation in concentration with respect to the partial pressures of the reaction, (2) for the presence of O^- and its facile reaction with CH_4, (3) for the presence of methyl radicals, and (4) for the formation of methoxide ions and their reaction with H_2O. Molybdenum(V) on silica gives rise to an EPR signal having $g_\perp = 2.020$ and $g_\parallel = 2.004$. The extent of the reaction of Mo^V with N_2O depends on the reduction conditions; however, following reduction of Mo^{VI} at 600°C and reaction with N_2O at 150°C for 4 h the spectrum of Mo^V almost completely disappeared and a spectrum of O^-, reflecting an equivalent number of spins, was detected (Liu *et al.*, 1984). The O^- spectrum is depicted in Fig. 1 where the coordination of the O^- ion to Mo^{VI} is confirmed by the ^{95}Mo and ^{97}Mo superhyperfine splitting (Fig. 1a) (Kolosov *et al.*, 1975; Ben Taarit and Lunsford, 1973). The ^{17}O hyperfine splitting shown in Fig. 1b confirms that the spectrum is indeed that of the O^- ion.

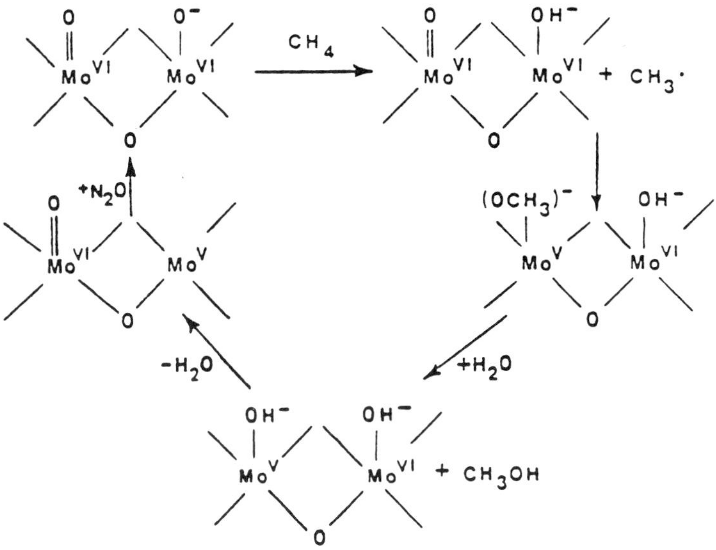

On admission of CH_4 the O^- spectrum rapidly disappeared at 25°C, and a significant fraction reacted over 10 min at -196°C. Although the resulting CH_3 radicals were highly transient, a weak EPR spectrum of them was detected at -196°C (Liu *et al.*, 1984).

At elevated temperatures the CH_3 radicals presumably undergo a reductive addition to $Mo{=}O$ forming surface $Mo^V (OCH_3)^-$ species which give rise to weak infrared bands at 2857, 2928, and 2959 cm^{-1}. It is significant to note that EPR spectroscopy is considerably more sensitive than IR spectroscopy; thus, methyl radicals on the order of 2×10^{18} spins/g, which were easily detectable by the former, were hardly observable as reaction products (methoxide ions) by the latter.

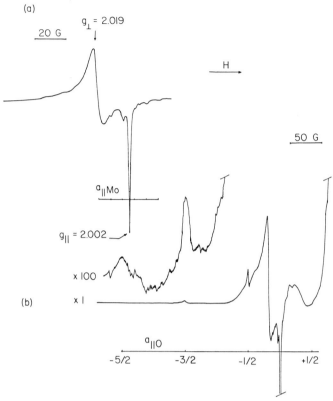

Fig. 1. (a) EPR spectrum of $^{16}O^-$ on Mo supported on silica gel and (b) spectrum of O^- enriched with 51% oxygen-17. [From Ben Taarit and Lunsford (1973).]

III. COPPER AMMINE COMPLEXES IN ZEOLITES

When compared with most other supports, zeolites are a superior medium in which to characterize reagents. This is due in a large part to the regularity of the structure and the presence of metal ions exclusively in ion exchange sites, when strongly coordinating ligands are not present. On adsorption of strongly coordinating ligands, such as ammonia, the metal ions from coordination complexes which, for steric reasons, usually exist in the larger cavities. The synthesis and characterization of numerous complexes in zeolites have been summarized in several review articles (Lunsford, 1975, 1977; Kellerman and Klier, 1975; Seff, 1976).

One of the earliest and best characterized examples of these complexes is tetraamminecopper(II) in zeolite Y (Flentge *et al.*, 1975). This complex is characterized by an EPR spectrum (Fig. 2) having $g_\parallel = 2.236$ and

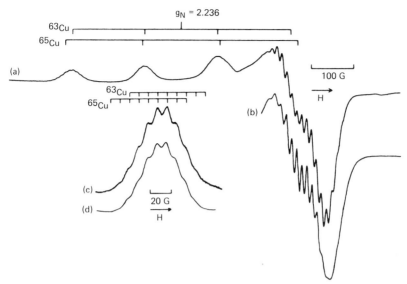

Fig. 2. EPR spectrum of CuY with 350 torr ND$_3$: (a) experimental spectrum, (b) simulated spectrum in the perpendicular region, (c) low-field component with the receiver gain three times the value of spectrum in (a), and (d) simulated spectrum of the low-field component. [From Flentge *et al.* (1975). Copyright 1975 American Chemical Society.]

$g_\perp = 2.035$ with $A_\parallel^{^{63}Cu} = 172$ G, $A_\parallel^{^{65}Cu} = 184$ G, $A_\perp^{Cu} = 20$ G, $A_\perp^{N} = 12.4$ G, and $A_\parallel^{N} = 9$ G. The ^{14}N superhyperfine structure, both in the perpendicular and parallel components, is consistent with four ammine ligands being coordinated to Cu $d_{x^2-y^2}$ orbitals, but on the basis of the EPR spectrum alone we cannot rule out the possible presence of two additional ligands along the d_{z^2} orbital.

By combining EPR spectroscopy with classical adsorption experiments it was found that the EPR spectrum of the ammine complex increased linearly on addition of NH$_3$ up to a level of four NH$_3$ molecules per copper ion, and thereafter the spectrum no longer increased (Flentge *et al.*, 1975). Similar behavior was found in the intensity of the N—H deformation band at 1275 cm^{-1} which is unique to coordinated NH$_3$. This evidence provides strong support for the formation of the $[Cu^{II}(NH_3)_4]^{2+}$ complex in the zeolite.

The complex is a significant reagent because it is an intermediate in a catalytic cycle for the reduction of nitric oxide by ammonia (Williamson and Lunsford, 1976). The reaction has an unusual rate maximum at $\sim110°C$ and a reversible reduction of Cu^{II} to Cu^{I} which begins at slightly lower temperatures as shown in Fig. 3. This reduction was followed by observing the EPR spectrum of Cu^{II} at progressively higher reaction

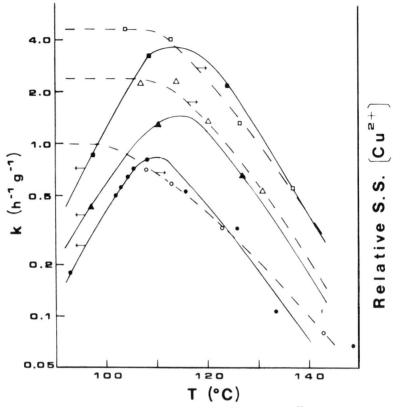

Fig. 3. Rate constants for the reduction of NO by NH_3 over $Cu^{II}Y$ (closed symbols) and the relative steady-state Cu^{II} concentrations (open symbols) as a function of temperature: 6.5% (●,○); 14% (▲, △); and 44% CuY (■, □) [From Williamson and Lunsford (1976). Copyright 1976 American Chemical Society.]

temperatures. After the reaction reached a steady state the sample was evacuated briefly to remove excess gases and cooled to room temperature, at which point the spectrum was recorded. Curiously, the reduction reaction occurs more rapdily in the presence of NH_3 and NO (an oxidant!) then in pure NH_3.

The rate maximum may be understood in terms of competitive reactions in which $[Cu^{II}(NH_3)_4]^{2+}$ is an intermediate in the desired catalytic cycle (i.e., in the reaction of NH_3 with NO to form N_2 and H_2O), but the corresponding Cu^I complex is not an intermediate. Thus, when the Cu^{II} is reduced faster than it can be reformed, the catalytic rate of NO reduction decreases.

IV. RUTHENIUM BIPYRIDINE COMPLEXES
IN ZEOLITES AND CLAYS

Tris (2,2'-bipyridine) ruthenium(II) complexes have played an important role as photoabsorbers in several schemes to achieve the dissociation of water into H_2 and O_2. At one point it was believed that the photoinduced Ru^+ and Ru^{3+} complex ions themselves were capable of reducing and oxidizing water, respectively (Brown *et al.*, 1979; Creutz and Sutin, 1975). The photoefficiency of these processes is often small because of reverse electron transfer and energy transfer reactions. In order to prevent these reverse reactions attempts have been made to immobilize the complexes on surfaces such as zeolites and clays and to study their photochemistry in this state (Ghosh and Bard, 1984; Schoonheydt, 1984; Krenske *et al.*, 1980; De Wilde *et al.*, 1980; Quayle and Lunsford, 1982).

The synthesis of the $[Ru^{II}(bpy)_3)]^{2+}$ complex within zeolite Y is interesting in itself because the ligands are relatively large and the diameter of the complex (10.8 Å) approaches the free internal diameter of the large cavity (13 Å). The complex indeed may be formed within the zeolite first by exchanging in the Ru as the hexaammine complex, $[Ru^{III}(NH_3)_6]^{3+}$, and then by reacting this material with bipyridine at 200°C. Evidence for the formation of the complex in the intracrystalline voids is provided both by XPS and XRD data (Quayle and Lunsford, 1982). Since the escape depth of photoelectrons is 20–40 Å, only atoms near the surface of a zeolite particle are detected by XPS. If enrichment of a Ru ion were to occur in this surface region, then the observed ratio of Ru/Si would be larger than that determined from a bulk analysis. In the $[Ru^{II}(bpy)] - Y$ zeolites the Ru/Si ratio as determined by XPS was 0.08, which may be compared with a value of 0.06 from bulk analysis. Thus, the surface enrichment was small, if it existed at all. Moreover, the XRD patterns showed that the 311 and 220 reflections of the zeolite varied dramatically with the synthesis of the complex, which further supports the presence of the complex within the intracrystalline voids.

The diffuse reflectance spectrum of the orange-colored $[Ru^{II}(bpy)_3] - Y$ zeolite is characterized by a maximum at 460 nm with shoulders at 430 and 540 nm as depicted in Fig. 4a. Two bands in the 425–454-nm region have been reported by several investigators for the $[Ru^{II}(bpy)]^{2+}$ complex in different media (Klassen and Crosby, 1968; Lytle and Hercules, 1969; Van Houten and Watts, 1976). The band at 540 nm has been attributed to an Fe–bpy complex which resulted from iron impurities in the sample.

One of the more interesting aspects of this study was the behavior of the photoluminescence spectrum as a function of the matrix and the extent of loading. The maximum emission in the zeolite at 590 nm is in good

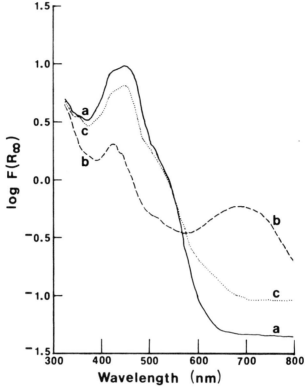

Fig. 4. Visible-region diffuse-reflectance spectra: (a) outgassed 2.1 wt. % Ru, [Ru^{II} (bpy)$_3$]Y; (b) [Ru^{III} (bpy)$_3$Y; and (c) sample (b) after exposure of water. [From Quayle and Lunsford (1982). Copyright 1982 American Chemical Society.]

agreement with the value reported for [Ru^{II}(bpy)$_3$]$^{2+}$ aggregates on the external surface of amorphous zirconium phosphate (Schoonheydt, 1984; Vliers *et al.*, 1985). Emmission maxima for the complex in several clays varied from 620 nm to 602 nm, with complexes in the interlamellar space being more strained and exhibiting greater blue shifts than those on the external surface

A strong *inverse* relationship was observed in the zeolite between emission intensity and concentration, which is characteristic of self-quenching or concentration quenching (De Wilde *et al.*, 1980). This occurs via resonance transfer of electronic excitation energy from an initially absorbing molecule to another identical molecule until the quanta reaches a quenching site. In the zeolite iron impurities probably serve as the quenching sites. This phenomenon is believed to occur in the zeolite because of the unusually large concentration of the complex (typically

1 M) relative, for example, to that found in aqueous solution (typically 10^{-5} M).

Since $[Ru^{III}(bpy)_3]^{3+}$ has been implicated in the oxidation of OH^- ions during the photolytic splitting of water (Creutz and Sutin, 1975) it was of interest to determine whether the reaction

$$[Ru^{III}(bpy)_3]^{3+} + OH^- \rightarrow [Ru^{II}(bpy)_3]^{2+} + \tfrac{1}{4}O_2 + \tfrac{1}{2}H_2O \tag{2}$$

might occur within a zeolite. Instead of carrying out the photolytic oxidation, Cl_2 was used to oxidize the Ru^{II} to Ru^{III} (Quayle and Lunsford, 1982). Oxidation of a dehydrated $[Ru^{II}(bpy)_3] - Y$ zeolite at 65°C yielded a light green sample which contained $[Ru^{III}(bpy)_3]^{3+}$ complexes. Evidence for the latter reagent is found in the DRS of Fig. 4b. Absorption maxima at 685 and 420 nm are in good agreement with literature values of 680 and 418 nm for the Ru^{III} complex. Moreover, it may be deduced from the DRS that approximately 90% of the ruthenium was oxidized. The presence of the $[Ru^{III}(bpy)_3]^{3+}$ complex also was confirmed by an EPR spectrum with $g_\perp = -2.67$ and $g_\parallel = 1.24$.

On addition of water to the $[Ru^{III}(bpy)_3] - Y$ zeolite the color of the sample changed from light green to orange, the DRS changed to that of Fig. 4c, and the EPR spectrum of the Ru^{III} complex disappeared; all of which supports the reduction of Ru^{III} by water to Ru^{II}. However, the formation of O_2, as expected from reaction (2), was not observed. Instead, only a small amount of CO_2 was detected in the gas phase. Clearly, the reduction of $[Ru^{III}(bpy)_3]^{3+}$ in the zeolite is not as straightforward as indicated by reaction (2). If O_2 production is desired from a photocatalytic process in which $[Ru^{II}(bpy)_3]^{3+}$ ions are intended as water-oxidizing agents, immobilization of the complex within zeolites is not a satisfactory approach.

V. COBALT BIPYRIDINE–TERPYRIDINE COMPLEXES IN ZEOLITES

Surface reagents not only have potential in catalytic and photochemical processes but also as reversible adsorbents for separation processes. The mixed-ligand bpy–terpy complexes are particularly attractive in this respect because the fifth coordination site on cobalt is available for forming an oxygen adduct. Moreover, in the absence of water vapor the coordinated oxygen, which is in the form of a superoxide ion, does not attack the other ligands.

The five-coordinated $[Co^{II}(bpy)(terpy)]^{2+}$ complex is characterized by the EPR spectrum of Fig. 5a for which $g_\parallel = 2.012$, $g_\perp = 2.250$,

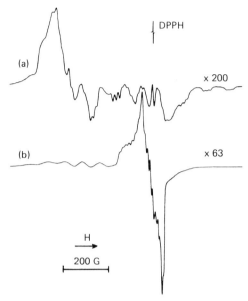

Fig. 5. EPR spectra at 77 K of $Co_{1.2}Na$-Y zeolite after reaction with bipyridine–terpyridine mixture: (a) before exposure to O_2 and (b) after exposure to O_2 [From Mizuno, *et al.* (1984). Copyright 1984 American Chemical Society.]

$A_\parallel^{Co} = 101$ G, $A_\perp^{Co} = 15$ G, and $A_\parallel^N = 15$ G. The presence of ^{14}N superhyperfine splitting in the parallel components supports the five-coordinate complex (Mizuno *et al.*, 1984).

On addition of O_2 to the zeolite the spectrum of this five-coordinate complex reversibly disappeared and the spectrum of a superoxo complex became apparent (Fig. 5b) along with the underlying spectrum of the $[Co^{II}(terpy)_2]^{2+}$ complex. The magnetic parameters of the superoxo complex are $g_x = 1.998$, $g_y = 2.007$, $g_z = 2.063$, $A_x^{Co} = 11.0$ G, $A_y^{Co} = 11.0$ G, and $A_z^{Co} = 15.6$ G, which agree well with other cobalt–oxygen adducts in zeolite–Y (Howe and Lunsford, 1975a,b).

The stability constant for the reaction

$$[Co^{II}(bpy)(terpy)]^{2+} + O_2 \rightleftharpoons [Co^{III}(bpy)(terpy)O_2^-]^{2+} \tag{3}$$

was evaluated from the oxygen pressure at which half of the complexes on the left hand of Eq. (3) were coordinated to oxygen (Imamura and Lunsford, 1985). Under these conditions $K_{O_2} = 1/[O_2]_{1/2}$. The stability constant calculated in this manner was 2.9 torr^{-1} at 298 K, which is large compared with other solid complexes and similar complexes in aqueous media. Moreover, the magnitude of K_{O_2} is such that the separation of O_2

from dry air may be effected in a batch system with separation factors of ~10. The limitation of the technique, however, is the low density of oxygen in the zeolite which is on the order of 10^{-3} g of O_2/g of zeolite.

VI. FUTURE PROSPECTS

As demonstrated by these examples the characterization of surface reagents which have analogs as pure compounds is well in hand, but problems arise when the reagent is uniquely a surface species or, as mentioned earlier, when the reagent of interest is a minor component on the surface. In order to identify reagents which exist only on surfaces it is necessary to obtain structural information on systems which have only short-range or at best two-dimensional order. Currently the most promising technique for attacking such problems is EXAFS, although the methodology has not been extended to multicomponent systems of the type which often are of importance in surface chemistry. For example, with the Mo/SiO_2 catalyst we could have simultaneously isolated Mo ions, polynuclear Mo–oxy complexes, and a two-dimensional oxide phase. In addition, in each of these forms the Mo could be in a variety of oxidation and coordination states. Such complexity indeed seems to be beyond the bounds of EXAFS at the present or for that matter any known technique.

Another exciting new development in the field of surface characterization is solid-state NMR. The wealth of structural information which has been derived from solution-state NMR now appears to be within the grasp of the surface chemist. The potential applications of this technique will be developed more fully in a subsequent chapter.

REFERENCES

Ben Taarit, Y., and Lunsford, J. H. (1973). *Chem. Phys. Lett.* **19**, 348–350.
Brown, G. M., Brunschwig, B. S., Creutz, C., Endicott, J. F., and Sutin, N. (1979). *J. Am. Chem. Soc.* **101**, 1298–1300.
Cheetham, A. K., Eddy, M. M., and Thomas, J. M. (1984). *J. Chem. Commun.* 1337–1338.
Creutz, C., and Sutin, N. (1975). *Proc. Natl. Acad. Sci. U.S.A.*, **72**, 2858–2862
Delgass, W. N., Haller, G. L., Kellerman, R., and Lunsford, J. H. (19779) "Spectroscopy in Heterogeneous Catalysis." Academic Press, New York.
De Wilde, W., Peeters, G., and Lunsford, J. H. (1980). *J. Phys. Chem.* **84**, 2306–2310.
Falconer, J. L., and Schwarz, J. A. (1983). *Catal. Rev.—Sci. Eng.* **25**,141–227.
Flentge, D. R., Lunsford, J. H., Jacobs, P. A., and Uytterhoeven, J. B. (1975). *J. Phys. Chem.* **79**, 354–360.
Ghosh, P. K., and Bard, A. J. (1984) *J. Phys. Chem.* **88**, 5519–5526.
Howe, R. F., and Lunsford, J. H. (1975a). *J. Am. Chem. Soc.* **97**, 5156–5159.
Howe, R. F., and Lunsford, J. H. (1975b). *J. Phys Chem.* **79**, 1836–1842.

Iizuka, T., and Lunsford, J. H. (1978). *J. Am. Chem. Soc.* **100**, 6106–6110.

Imamura, S., and Lunsford, J. H. (1985). *Langmuir* **1**, 326–330.

Kellerman, R., and Klier, K. (1975). *In* "Surface and Defect Properties of Solids," Vol. 4, pp. 1–33. Chem. Soc., London.

Klassen, D. M., and Crosby, G. A. (1968). *J. Chem. Phys.* **48**, 1853–1858.

Kolosov, A. K., Shvets, V.A., and Kazansky, V. B. (1975). *Chem. Phys. Lett.* **34**, 360–361.

Krenske, D., Abdo, S., Van Damme, H., Cruz, M., and Fripiat, J. J. (1980). *J. Phys. Chem.* **84**, 2447.

Liu, H.-F., Liu, R.-,S., Liew, K. Y., Johnson, R. E., and Lunsford, J. H. (1984). *J.Am. Chem. Soc.* **106**, 4117–4121.

Lunsford, J. H. (1975). *Catal. Rev.—Sci. Eng.* **12**, 137–162.

Lunsford, J. H. (1977). *ACS Symp. Ser.* No. **40**, 473–492.

Lytle, F. E., and Hercules, D. M. (1969). *J. Am. Chem. Soc.* **91**, 253–257.

Madix, R. J. (1984). *Catal. Rev.—Sci. Eng.* **28**, 281–297.

Mizuno, K., Imamura, S. I., and Lunsford, J. H. (1984). *Inorg. Chem.* **23**, 3510–3514.

Morrison, T. I., Iton, L. E., Shenoy, G. K., Stucky, D., and Suib, S. L. (1981). *J. Chem. Phys.* **75**, 4086–4089.

Mortier, W. J. (1982). "Compilation of Extra Framework Sites in Zeolites." Butterworth, London.

Quayle, W. H., and Lunsford, J. H. (1982). *Inorg. Chem.* **21**, 97–103.

Schoonheydt, R. A. (1984). *J. Mol. Catal.* **27**, 111–122.

Seff, K. (1976). *Acc. Chem. Res.* **9**, 121–128.

Sinfelt, J. H., Via, G. H., and Lytle, F. W. (1984). *Catal. Rev.—Sci. Eng.* **26**, 81–140.

Van Houten, J., and Watts, R. J. (1976). *J. Am. Chem. Soc.* **98**, 4853–4858.

Vliers, D. P., Schoonheydt, R. A., and de Schrijver, F. C. (1985). *J. C. S. Faraday I* **81**, 2009–2019.

Williamson, W. B., and Lunsford, J. H. (1976). *J. Phys. Chem.* **80**, 2664–2671.

11 X-RAY STUDIES

Jacques Thorez

Claygeology Laboratory
Institute of Mineralogy
Liège State University
Sart-Tilman, B-4000 Liège, Belgium

I. INTRODUCTION

The last three decades in clay sciences have been characterized by a huge amount of work in crystallography, crystallochemistry, mineralogy, geology, pedology, in applied sciences such as agriculture and ceramics, and in industry, particularly in the manufacture of various organic polymers using clays as fillers and as catalytic supports. It is beyond the scope of this chapter to review all the extant analytical data on clay minerals and their complexes with inorganic and/or organic chemicals.

The remarkable wedding of certain clays with inorganic/organic compounds results from the clay properties: their fine-grained (micrometric) sizes, their easy dispersion in water and in other polar liquids (a consequence of their layered structure) their various habits (from nearly perfect hexagonal forms, to fibers, ribbons, tubular, or spheroidal habits) which rule their physical and rheological behavior, and probably the most important property exhibited by some clay minerals: the ability to adsorb at the surface of the layers (and/or between those layers) a great number of foreign compounds. Clays are also the most universal of all the minerals occurring at the surface of the earth. Their abundance makes them cheaply available.

PREPARATIVE CHEMISTRY
USING SUPPORTED REAGENTS

II. THE NATURE AND DEFINITION OF CLAYS

The term *clay* or *clay mineral* is still used in (at least) three different senses: (1) for any unconsolidated (or loose) geological (or pedological) material; (2) to designate a certain particle size (usually below two micrometers); and (3) to ascribe certain remarkable properties or behaviors, such as the swelling when in contact with water.

In fact, any complete definition for those materials must also embody their layered structure and their mineralogy, a hydrous aluminosilicate composition. Any textbook in clay mineralogy provides the classification of the simple clay minerals (kaolinites, micas, chlorites, vermiculites, smectites) but usually ignores the mixed-layer minerals which also share the physical and chemical properties of the simple clay mineral layers that compose these mixed structures. The definition of clay minerals excludes the other components which are often associated to them in variable composition and amounts within the natural clay deposits, such as other silicates, carbonates, sulfates, amorphous compounds and organic material. These "contaminants" must be removed during the industrial treatment, as to produce a "pure" or "purified" clay material. These industrial processes often provoke large damages to the clay minerals themselves (Harward and Theisen, 1961; Harward *et al.*, 1962; Brewster, 1980; Thorez, 1985). Even the "monomineralic" Na-Wyoming montmorillonite (or bentonite), the standard used in a great many intercalations, may have its charge density altered by the industrial treatment (Rengasami *et al.*, 1976). Any clay mineral termed pure and monomineralic may indeed show variable chemical composition within the clay particles themselves and, consequently, within the entire population of the particles (Duplay, 1984).

Phyllosilicates and their related clay minerals are usually classified on the basis of (1) the layer type (1:1 or Te–Oc, and 2:1 or Te–Oc–Te minerals, where Te stands for tetrahedral and Oc for octahedral sheet) (Fig. 1), (2) the layer charge; (3) the type of material filling the interlayer spaces (such as dry, partly or fully hydrated cations, or a hydroxide layer), (4) the nature of the Oc sheet (either dioctahedral or trioctahedral); (5) the polytypism of the minerals (the way the structural units—the layer plus the interlayer—are regularly (or not) stacked along the *c*-dimension; this is usually referred as the *c*-axis); and (6) the chemical composition (mineral species) (Brown, 1984). Figure 1 illustrates the organization of dioctahedral smectites, a subgroup of clays with the best sorption and catalytic reactions, although other clay minerals can also swell but to a lesser degree. The illustration emphasizes the characteristic isomorphous substitution in either the Te or the Oc sheet (Grim, 1968; Brindley and Brown, 1980).

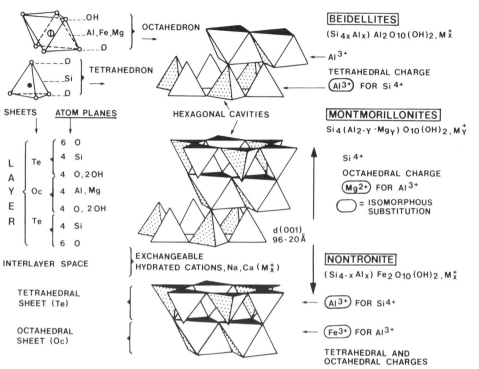

Fig. 1. Crystal structure of dioctahedral smectites, emphasizing the nature and location of the tetrahedrally and octahedrally isomorphous substitutions.

The real structure of the clay minerals is never close to the ideal well-ordered representative components within the groups and subgroups. Tchoubar (1984b) has reviewed the more frequently occurring defects. These include (1) those affecting the layers, i.e., vacant sites in the octahedral sheet, rotation of the tetrahedral around the "hexagonal" cavity, and distribution of the sites for isomorphous substitution; (2) those defects specific to the interlayer spaces, due to the modification in position of the compensating cations or of the inserted molecules; and (3) the defects relative to the stacking mode and arrangement of the layers, i.e., in the mixed layers.

III. X-RAY DIFFRACTION STUDIES OF CLAY–ORGANIC COMPLEXES

Clay minerals, despite their micrometric sizes, are readily identified by x-ray diffraction (XRD) analysis. However, prior to their study, the material must be prepared by removal of contaminants, dispersion and extraction of a specific grain size (usually the fraction less than 2 μm), or preparation as either oriented aggregates or random powders. Strange to say, there is still a lack of standardization for these pretreatments among the clay laboratories (Thorez, 1985). Post-treatments, necessary for proper identification between groups, subgroups, and even species, involve a large set of chemical treatments. Beside the classic air-dried (normal, natural, N) sample (when the material is prepared as oriented aggregates), solvation with polyalcohols (ethylene glycol, EG, or Glycerol, Gl), and heatings (usually to 490–500°C), other identification tests may be applied that concern, for example, cation saturations (with lithium, potassium, magnesium, and sodium chlorides) with subsequent glycolation/glycerolation and heatings, acid attack, and hydrazine saturation. Figure 2 provides an identification key for some swelling minerals (including the smectites) by referring to the behavior of the basal spacing, (001), of these minerals on the latter—cation saturation—tests (Greene-Kelly, 1953; Schultz, 1969; Brindley and Brown, 1980; Thorez, 1975, 1976; Paepe et al., 1978).

Identification by XRD analysis involves recording x-ray patterns (or diffractograms) of the oriented aggregates (particularly in the case of mixtures of clay minerals); the matching of the specific behavior of the basal spacing, $d(001)$, and the other (002) spacings [all these reflections being enhanced in the diffractograms of the oriented aggregates whereas the other, (hk0), reflections (or peaks) are not present]; position and possible variation of the $d(001)$ on the above-quoted tests; measurement of the reflection heights (intensity); and observation of the reflection profiles (shapes). Peak position, expressed as the Bragg angle 2θ, depends on the wavelength of the radiation (ordinarily the CuKa radiation). The angular measurements are converted to d spacings (cf. the c-dimensions in the case of an oriented preparation) by use of the Braggs relation $n\lambda = 2d \sin \theta$. Values of $d/n = 2 \sin \theta/\lambda$ are readily obtained from tables. These values of d/n are called the d spacings but are often submultiples of the true lattice spacings (Brindley and Brown, 1980; Brindley, 1981b). The random powder preparation is better adapted to the study of the a and b dimensions of the lattice, more generally in the case of monomineralic phases, and it indicates if the clay mineral is either dioctahedral or trioctahedral [on the basis of the (060) reflection: 1.49–1.50 Å for the former and 1.52–1.56 Å for the latter].

Swelling occurs in the crystals of smectites, swelling chlorite, some vermiculites, many mixed layers (either regular or randomly interstratified) including "swelling" layers, and even in kaolinites. The mechanism, however, has been intensively studied in smectites. It differentiates swelling from nonswelling clay minerals through intercalation not only of polyalco-

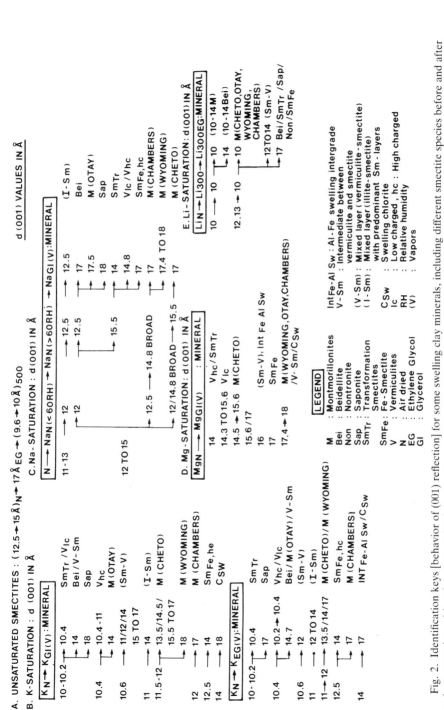

Fig. 2. Identification keys [behavior of (001) reflection] for some swelling clay minerals, including different smectite species before and after cation saturation with subsequent solvation with either ethylene glycol (EG) or glycerol (Gl).

hols (ethylene glycol, and glycerol as a rule), but also of a large variety of other polar liquids.

Swelling is caused by hydration of the exchangeable interlayer cations: Na^+, Ca^{2+}, or both, in natural smectites. It depends mainly on the relative humidity (RH). With Na^+ as the interlayer occupant, smectites hydrate one, two, or three layers of water molecules within the interlayer space. With larger amounts of water, Na smectites become fully expanded and dispersed. The basal spacing at 17–20 Å of the swollen smectites then disappears as a distinct reflection and is replaced by a diffraction halo (Fukushima, 1984). With Ca^{2+} as the interlayer cation, swelling appears to be limited to three water layers with the individual layers separated by a distance of about 9 Å (Brown, 1984). This diverging behavior of Na and Ca smectites, together with the very large expandability of the Na material, are the reasons for which Na-saturated smectites (i.e., Wyoming montmorillonite) are usually selected for the study of adsorption and intercalation of organic compounds. Whatever the nature of the organic compounds, insertion is readily observed in the diffractogram with the $d(001)$ of the (oriented) smectite moving from 12.6–15 Å (air-dried state, normal, N) to 17 Å (EG) or near 18 Å (Gl) (Fig. 2). Comparable values of expansion of the interlayers have been noted with other polar liquids. The literature provides an exploding series of examples. When the smectites are heated (i.e., to 500°C), the interlayers contract ("collapse") to a d spacing of 9.6–10 Å.

However, swelling can be reduced considerably in smectite minerals by application of specific post-treatments, such as cation saturations with subsequent glycolation/glycerolation (Fig. 2). This provides another way to differentiate mineral species. One of the preeminent characteristics of the smectites is the collapse of the interlayers (a phenomenon also shared by vermiculites) by either heating or dehydration. This is due to the very weak bonds between two adjacent layers produced by the intercalation.

A new kind of intercalation agent has been introduced with the manufacture of pillared smectites [CLOS or pillared intercalated clays (PILC)]. These have a high porosity after heating. They are produced by the introduction within the interlayers of polyanionic smectites (Na montmorillonite and Na and F hectorite), of dicationic and polycationic species derived from suitable organic and inorganic compounds (Endo et al., 1980; Occelli and Tindwa, 1983; Pinnavaia et al., 1985; Shabtai et al., 1984; Tokarz and Shabtai, 1985). The originally collapsable smectites have thus become thermally stable on heating to high temperature (700°C) after the pretreatment (intercalation) and do preserve their expanded interlayers at about 17 Å. Products expanded to 27 Å are also produced. The 17-Å basal reflection of the CLOS is remarkably intense, narrow, and symmetric (Fig. 3), which is an indication that the intercalation has worked well globally, the Al–hydroxy pillars being uniformly distributed within all the interlayers and in all the clay particles. This contrasts with the characteristics of the 17-Å peak (ill-defined and exhibiting a diffraction band) of the K 10 (Süd Chemie) clay (Cornelis *et al.*, 1983) (see Fig. 3). The latter product does show a certain pillaring, but it is not homogeneously distributed within the material. Indeed, in the air-dried state (N), and often after glycolation (EG), the basal spacing of the K 10 takes the form of a diffraction band extending from about 10 Å to 17–19 Å, with a plateau inclined toward 17 Å, the latter "reflection" being the most intense. On heating to 500°C, a similar ill-defined "reflection" develops but with the intensity maximum around 10–11 Å, and still with a diffraction band extending to 14–15 Å. This feature indicates an incomplete and heterogeneous collapse of the interlayers.

Natural CLOS do exist in the nature, with incomplete Al–hydroxy layers trapped in the interlayer spaces of former smectites and/or vermiculites. These Al–hydroxy interlayers have been produced in acid conditions of the weathering during pedogenetic processes (Sawhney, 1958; Brydon *et al.*, 1961). Complexes with Fe interlayers have also been reported. (Quigley and Martin, 1963). These materials generally have intermediate properties (i.e., "intergradients") between smectites or vermiculties and chlorites. The occurrence of the "islands" of Al–hydroxy interlayers prevents the complete collapse of the basal spacing to 10 Å after heating. Instead, a diffraction band from 10 to 14 Å (or a poorly defined and broad 14-Å peak) develops after heating. If these "islands" coalesce within the interlayer spaces, a chlorite-like structure replaces the former Al smectite or Al vermiculite. On the other hand, as observed in many Quaternary paleosoils of Belgium (Thorez, in Sommé *et al.*, 1978), the vermiculitic or smectitic component of rather complex clay assemblages can reach the state of a naturally swollen 17-Å structure through the pedogenetic interlayering of Al–hydroxy material (Fig. 3). The latter 17-Å structure is stable on glycolation (with no further expansion), and after heating (to 500°C) just collapses to 15.5–16 Å. This interlayered material is readily withdrawn when the sample is treated with either KOH or NH_4F. However, this 17-Å component is only a minor clay mineral among others (illite, kaolinite, smectite, vermiculite, and degraded chlorite) in these paleosoils.

As briefly discussed here and schematically illustrated in Fig. 2 and 3, XRD analysis is a nice way to show the existence of intercalated swelling clay minerals. However, one question still arises in the way an x-ray pattern (which specifically concerns the part of the sample that is less than two micrometers) can reflect an intercalation process that affects the whole sample, i.e., during industrial manufacture. The answer to this question may be provided within/by the concept of the "clay integron."

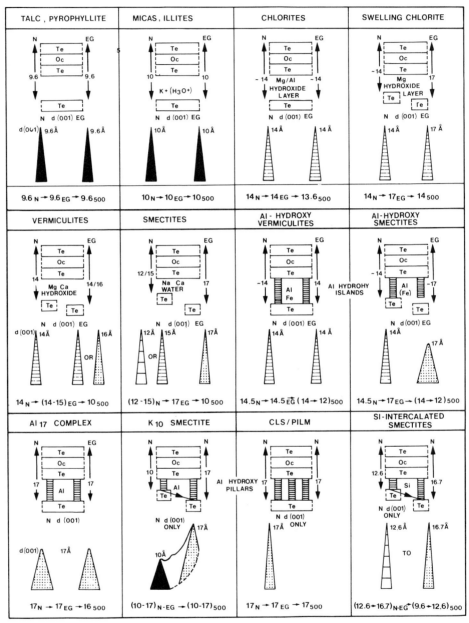

Fig. 3. Schematic representation of the structural arrangement (particularly the interlayer component) of some nonswelling and swelling clay minerals before and after natural or artificial intercalation. Comparison is made between the *d* spacing, (001), at the natural (N), glycolation (EG), and heating states (cf. the cartouche contents), and at the level of the (001) shape.

IV. A PHILOSOPHIC APPROACH OF CLAY SCIENCE:
THE CLAY INTEGRON

The clay integron is a conceptual model of the structural architecture and the physical (spatial) organization and dimensional variation of any clay sample; it is valid from the angstrom scale (characteristic of the layer structure) to a macroscopic cubic centimeter sample (as treated in the laboratory), and to the huge volumes (cubic kilometers) of clay deposits possibly present in the landscape (Fig. 4).

This model bridges units of specific sizes and internal structure and builds these up through a series of integrations more or less in the way of a "box-into-box" system. The clay integron depicted here is based on analytical data provided by different clay scientists or by researches dealing with clays who analyze these materials at different scales of observation. Among these workers, we can cite Pedro (1976; Tessier and Pedro, 1976; Pedro and Tessier, 1985) who described the structural organization of illite (mica), kaolinite, and smectite within the range of tens to hundreds of angstroms. On the other hand, soil mechanical and pedological (micromorphological) studies have provided data at larger scales (from micrometers to meters or beyond) with the scanning electron microscope and the polarizing microscope.

Pedro and Tessier (1985) have distinguished four integrated levels in smectites:

(a) The first level concerns the structural unit, which thus characterizes both the layer and the interlayer.

(b) The second level describes the crystallite, or tactoid, and embodies monocrystalline particles composed by the stacking of several layers, perpendicularly to the c dimension (cf. the "primary particle" of Mering, and Oberlin, 1971).

(c) The third level corresponds to the quasi-crystal in the case of smectites, as it results from the overlapping of a number of crystallites following the basal plane, as to form units of great extension (Aylmore and Quirk, 1971).

(d) The fourth level concerns the assemblage of quasi-crystals into a tridimensional network which gives the smectites their sponge-like habit and property. It is due to this very level that the number of layers increases when the material dries up (as in during the preparation of an oriented aggregate) or diminishes with addition of large amounts of water.

From this organization within the clay integron (Fig. 4), it can be easily demonstrated that any diffraction peak [i.e., the basal reflection (001), which precisely measures the thickness between two adjacent layers] in reality is likewise a statistical integration, which applies to the different

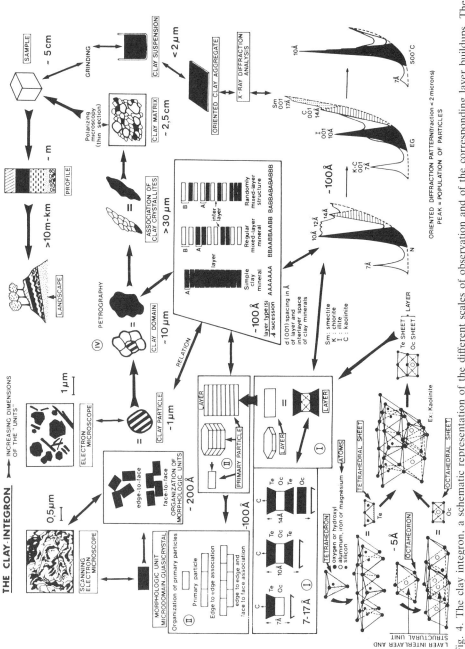

Fig. 4. The clay integron, a schematic representation of the different scales of observation and of the corresponding layer buildups. The relationship between the clay integron and the x-ray pattern (oriented aggregate) is shown.

but integrated units (as shown by the double arrows in Fig. 4). As a consequence, the (001) reflection is a kind of gauge to measure d spacings relative to an entire population of micrometric clay particles (the upper limit size of which is imposed by the preparation of the sample prior to the XRD analysis); the d spacing in the x-ray pattern refers to this population in terms of homogeneity [i.e., the occurrence of narrow intense and symmetric (001) peaks], whereas asymmetric profiles or enlarged peaks reflect a certain inhomogeneity not only at the scale of the whole clay particle population, but probably at that of each particle, in the way the compositional layers are orderly or not stacked one on the other.

V. THE ADVANTAGES OF THE XRD METHOD

Beside the old (but still intensively used) XRD method, the last decade has seen the birth and the remarkable growth of several new techniques for the investigation of the clay minerals, whether intercalated or not with organic or inorganic compounds. All these new and somewhat highly sophisticated methods are assuredly suitable for the investigation of clay minerals down to a single clay particle (i.e., for the study of the ordered stacking of the layers, or even deeper, for the investigation of an atom within the sheet or in the interlayer space).

For instance, electron diffraction provides structural data about a peculiar atom in relation with its immediate neighbors; transmission and scanning electron microscopes, while revealing the size and the shape of the clay particle or the mode of organization of several clay particles, seldom provide direct information about the mineralogical composition. Mössbauer spectroscopy, as applied to clay minerals so far, gives information on the nature of the iron, NMR on the sorbed water, EPR on the transitional element, electron probe on the location and the distribution of elements at the surface of the clay particles. All these highly specialized techniques present a definitive risk, that of examining a "tree", or even a "leaf" of that tree, but forgetting to take in simultaneity whole the "forest." Thomas (1984) is partly in the right when he states that XRD analysis is no longer universally suited for the study of the clay minerals and of their intercalates. However, it all hinges on the scale of the observation and the concerns of the researcher! As for the chemist, who is interested in the reactivity of the clay supports and their intercalates, he is mostly concerned about the overall positive and maximum intercalation of the clay sample. For this very purpose, XRD analysis remains the most adequate, rapid, and accurate investigational and control tool, even if the method operates "blindly" to an extent. Inspection of the d spacings, intensity, and shape of the particular basal reflection of the untreated and then of the intercalated clay substance gives readily the answer at the scale of the whole sample. By implicit reference to the clay integron, the chemist (or more generally the clay scientist) understands the very meaning of any basal reflection. Accordingly, XRD will remain a necessary companion to the other techniques, all of them enabling a zooming prospection of the intercalation mechanism.

VI. CONCLUDING REMARKS

Smectites, but also some vermiculites, exhibit a sorption capacity for organic and inorganic compounds, which both may be made to replace the inorganic counterions present in the interlayer spaces of the natural clay material. Such exchanges take place on the external surfaces of the clay particles and layers and in between the layers. They lead to clay–organic complexes. The intercalates provoke swelling by opening the interlayers. After intercalation, these interlayers are generally collapsable on heating, excepting in the case of the pillared smectites, the CLOS. Such complexes are of considerable economic importance in the petroleum industry, agriculture, and organic chemistry.

The XRD method is a very easy, reproducible technique for the study of intercalation and of its results. The method is considerably easier and cheaper than the previous methods which all allow characterization of a very restricted volume of the clay material (at the scale of the micrometric particle) or of an inner molecule or atom. In fact, to gather complete data about the components and the anatomy of the intercalated clay material, the whole set of these techniques is required. XRD analysis still appears of paramount importance for checking that intercalation has occurred; even if the data have to be considered to be "blindly" obtained because they are integrated over the real inhomogeneous structural and compositional characteristics of a "pure" clay material. As emphasized by Duplay (1984): any monomineralic clay material is in fact an entire population of clay particles with different chemical compositions. A population of clay particles belonging to a single species may include particles whose cations and charge distribution belong to another closely related species.

REFERENCES

Aylmore, L. A. G., and Quirk, J. P. (1971). *Soil Sci. Soc. Amr. Proc.* **35**, 652–654.
Bades, J. M. (1984). *Clays Clay Miner.* **32**, 49–57.
Brewster, G. R. (1980). *Clays Clay Miner.* **28**, 303–310.
Brindley, G. W. (1981a). *In* "Short Course and the Resource Geologist" (F. J. Longstaffe, ed.), pp. 22–38. Mineral. Assoc. Can., Toronto.

Brindley, G. W. (1981b). *In* "Short Course and the Resource Geologist" (F. J. Longstaffe, ed.), pp. 1–21. Mineral. Assoc. Can., Toronto.

Brindley, G. W., and Brown, G., eds. (1980). "Crystal Structures of Clay Minerals and their X-Ray Diffraction," Monogr. 5. Miner. Soc., London.

Brown, G. (1984). *Philos. Trans. R. Soc. London, Ser. A* **311**, 221–240.

Brydon, J. E., Clark, J. S., and Osborne, V. (1961). *Can. Mineral.* **6**, 595–609.

Cornelis, A., Laszlo, P., and Pennetreau, P. (1983). *Clay Miner.* **18**, 437–445.

Dixon, J. B., and Jackson, M. L. (1962). *Soil Sci. Soc. Am. Proc.* **26**, 358–365.

Duplay, J. (1984). *Bull. Soc. Geol. Fr.* **37**, 307–317.

Endo, T., Mortland, M. M., and Pinnavaia, T. J. (1980). *Clays Clay Miner.* **28**, 105–110.

Fukushima, Y. (1984). *Clays Clay Miner.* **32**, 320–326.

Greene-Kelly, R. (1953). *J. Soil Sci.* **4**, 233–237.

Grim, R. E. (1962). "Applied Clay Mineralogy." McGraw-Hill, New York.

Grim, R. E. (1968). "Clay Mineralogy." McGraw-Hill, New York.

Grim, R. E., and Güven, N. (1978). Bentonites: geology, properties and uses. *Dev. Sedimentol.* **24**.

Grim, R. E., and Kulbicki, G. (1961). *Am. Mineral.* **46**, 1329–1369.

Harward, M. E., and Theisen, A. A. (1961). *Soil Sci. Soc. Am. Proc.* **26**, 335–340.

Harward, M. E., Theisen, A. A., and Evans, D. D. (1962). *Soil Sci. Soc. Am. Proc.* **26**, 535–541.

Mering, J., and Oberlin, A. (1971). "The Electron-Optical Investigation of Clays." Miner. Soc., London.

Norrish, K. (1954a). *Discuss. Faraday Soc.* **18**, 126–134.

Norrish, K. (1954b). *Nature (London)* **173**, 256–257.

Occelli, M. L., and Tindwa, R. M. (1983). *Clays Clay Miner.* **31**, 22–28.

Pedro, G. (1976). *Bull. Assoc. Fr. Etude Sol* **2**, 69–84.

Pedro, G., and Tessier, D. (1985). *Eur. Clay Group Meet., 5th, Prague, 1983* pp. 417–428.

Pinnavaia, T J., Tzou, M. S., and Landau, S. D. (1985). *J. Am. Chem. Soc.* **107**, 4783–4785.

Quigley, R. M., and Martin, R. T. (1963). *Clays Clay Miner.* **10**, 107–116.

Rengasami, P., Van Assche, J. B., and Uytterhoeven, J. B. (1976). *J. C. S. Faraday, I* **72**, 376–381.

Robert, M., and Barshad, I. (1972). *C. R. Acad. Sci. (Paris)* **275**, 1465–1469.

Sawhney, B. L. (1958). *Nature (London)* **182**, 1595–1596.

Schultz, L. G. (1969). *Clays Clay Miner.* **17**, 115–149.

Shabtai, J., Rossel, M., and Tokarz, M. (1984). *Clays Clay Miner.* **32**, 95–107.

Sommé, J., Paepe, R., Baeteman, C., Beyens, L., Cunat, N., Geezaerts, R., Hardy, A. F., Hus. J., Juvigné, E., Mathieu, L., Thorez, J., and Vanhoorne, R. (1978). *Bull. Assoc. Fr. Etude Quat.*, 1–3, 81, 1–149.

Tchoubar, C. (1984a). *Philos. Trans. R. Soc. London, Ser. A.* **311**, 259–269.

Tchoubar, C. (1984b). *Clays Clay Miner. Ital.–Span. Congr., 1st, Amalfi, Italy* Abstr., pp. 10–11.

Tessier, D., and Pedro, G. (1976). *Bull. Assoc. Fr. Etude Sol* **2**, 85–99.

Thomas, J. M. (1984). *Philos. Trans. R. Soc. London, Ser. A* **311**, 271–285.

Thorez, J. (1975). *In* "Phyllosilicates and Clay Minerals—A Laboratory Handbook for their X-Ray Studies" (G. Lelotte, ed.). Dison, Belgium.

Thorez, J. (1976). *In* "Practical Identification of Clay Minerals" (G. Lelotte, ed.). Dison, Belgium.

Thorez, J. (1985). *Eur. Clay Group Meet., 5th, Prague, 1983* pp. 383–389.

Tokarz, M., and Shabtai, J. (1985). *Clays Clay Miner.* **33**, 89–98.

Part III

Polymer-Supported Reagents

12 POLYMER-SUPPORTED REAGENTS, POLYMER-SUPPORTED CATALYSTS, AND POLYMER-SUPPORTED COUPLING REACTIONS

Warren T. Ford

Department of Chemistry
Oklahoma State University
Stillwater, Oklahoma 74078

Erich C. Blossey

Department of Chemistry
Rollins College
Winter Park, Florida 32789

I. THE CHEMISTRY OF POLYMER SUPPORTS

A. Advantages and Disadvantages of Polymer Supports

The first uses of ion exchange resins as catalysts and Merrifield's first peptide syntheses (Merrifield, 1963) demonstrated the ease of separation of insoluble polymeric species from reaction mixtures. Large excesses of coupling reagents in peptide syntheses provide higher conversion than achieved in solution syntheses (Barany and Merrifield, 1979). Supported reagents and catalysts are recycled easily and can be used in flow reactors and automated processes. Supported analogs of toxic and odorous reagents are safer to use.

The features that make polymers different from inorganic supports are the variety of structures and morphologies and higher capacities, often up to 4 mmol/g. Most polymer-supported reagents and catalysts are used as gels. The support restricts motion of polymer-bound species but allows permeation of low-molecular-weight species. The gel can be manipulated so as to either promote or inhibit contact between polymer-bound species. The variety of polymer structures makes it possible to conduct reactions in many different environments.

PREPARATIVE CHEMISTRY
USING SUPPORTED REAGENTS

The disadvantages of insoluble polymer-supported reagents and catalysts are higher costs and sometimes slower or incomplete reactions due to diffusional limitations.

The aim of this chapter is to describe selected examples of polymeric species whose chemistry provides some distinct advantage over analogous species in solution. For more thorough surveys see Hodge and Sherrington (1980), Mathur *et al.* (1980), Ford (1986a) and the reviews of Akelah and Sherrington (1981, 1983). For catalysts see also Manecke and Storck (1978).

We shall use the following symbols:

$\textcircled{P}$ = the polystyrene repeat unit,

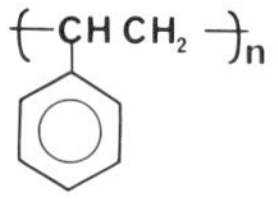

$$\left(\!\!\begin{array}{c} CH\,CH_2 \\ \mid \\ \bigcirc \end{array}\!\!\right)_n$$

DF = degree of functionalization, the fraction of repeat units functionalized.
% CL = percent cross-linked, either wt. % or mol. % of the cross-linking
monomer used in the original polymerization.

B. The Gel State of Matter

A gel is a macroscopic solid and a microscopic liquid comprised of polymer and solvent (Tanaka, 1981; Flory, 1953). The polymer prevents flow, and the solvent provides molecular mobility. Most polymers used as supports are cross-linked, such as copolymers of styrene and divinylbenzene. Polymer-supported reagents and catalysts are at least slightly solvent-swollen for use. This makes them different from reagents supported on silica gel and alumina, which are porous solids with chemical species bound to the internal surfaces of their pores. The active species in polymer gels can be distributed throughout the polymer network, and should be described as in, not on, the support. Solvent swelling creates micropores in the polymer network. The principle "like dissolves like" may be extended to "like swells like" to find a swelling solvent.

C. Macroporous Polymers

Polymers that retain porosity and have high internal surface area in the dry state can be prepared by copolymerizations with cross-linking monomers in the presence of solvent from which the polymer precipitates as the network forms (Seidl *et al.*, 1967; Guyot and Bartholin, 1982). Subsequent removal of the solvent leaves pores if the polymer is rigid enough to retain its shape. The % CL and the amount and nature of the inert solvent control the porosity. A macroporous polymer also is a gel. It can swell in good solvents, although the degree of swelling may be low because of high % CL. Macroporous polystyrenes are used as column packings for size exclusion (gel permeation) chromatography of organic polymers and as active supports in high-pressure liquid chromatography (HPLC) columns.

D. Kinetics of Reactions in Polymer Supports

Frequently chemical reactions of polymer-supported reagents and catalysts are slower than their counterparts in solution due to slow diffusion of a reactant through the gel or slow mass transfer of a reagent from bulk liquid to the particle surface (Satterfield, 1970). The functional sites within the polymer have variable environments. Their reactions do not follow a simple rate law and often slow down at high conversion or stop short of complete conversion (Pan and Morawetz, 1980). However, average conversions in the coupling steps of Merrifield peptide syntheses exceed 99% (Sarin *et al.*, 1984).

Initial functionalization of the polymer can be performed by copolymerization of a functional monomer with a cross-linking monomer or by chemical transformation of a preformed polymer. In a copolymerization the functional monomer can be incorporated randomly or alternately with another monomer, depending on the copolymer reactivities of the monomers (Odian, 1981).

Functionalization of a preformed network polymer may proceed by either of two limiting mechanisms:

(1) If reaction is much slower than diffusion of reagent into the polymer, functional sites are formed uniformly throughout the particle. Less reactive sites in the network remain unchanged if functionalization is incomplete.

(2) If chemical reaction is much faster than diffusion of the reagent into the polymer, functionalization proceeds by a shell-diffusive mechanism (Schmuckler and Goldstein, 1977). Use of a deficient amount of the reagent, or stopping the reaction before completion, gives a polymer functionalized only in its most accessible sites.

By either mechanism, efficient use of the functional sites in subsequent reactions is favored by incomplete functionalization of the polymer. We must be careful, however, that unreacted functional groups from one synthetic step do not become reactive in a later step with a more aggressive reagent or a better swelling solvent.

E. Stability of Polymer-Supported Reagents and Catalysts

Polystyrene has been used as support most often for both historical and practical reasons. Polystyrene-based ion exchange resins have been used commercially since about 1950 (Helfferich, 1962), and gel polystyrenes have been highly successful for Merrifield peptide synthesis. Polystyrene is easily functionalized by ring substitution reactions, and it has no other reactive sites under most conditions. It is stable for days at 200°C but degrades at >250°C. Polystyrene beads may be crushed to a powder by a magnetic stirring bar against the inner wall of a glass flask. Osmotic forces during rapid swelling or deswelling can shatter polymer beads to powders. Shaking or mechanical stirring and slow swelling and deswelling of the polymer are recommended.

F. Analysis of Polymer-Supported Reagents and Catalysts

In the literature polymeric reagents often have been inadequately described. The polymer structure, the method of functionalization and DF, the method of cross-linking and % CL, and the particle size range should be reported.

The repeat unit structure is usually known from the method of synthesis. Compositions of copolymers often can be determined by elemental analysis. Infrared and UV/VIS spectra can be obtained from KBr pellets and solvent mulls. Liquid-state single-pulse Fourier transform ^{13}C NMR spectra can be obtained with highly swollen gels (Ford et al., 1984). Less swellable polymers can be analyzed by cross-polarization and magic angle sample spinning (CPMAS) solid-state NMR spectroscopy (Fyfe, 1983). Specific functional groups can be analyzed by the same wet chemical methods used in solutions.

G. Available Polymers

Many polymeric reagents and catalysts are commercial, and many more parent polymers are available for functionalization. New polymers and copolymers can usually be prepared by modification of standard procedures in the literature (Sorenson and Campbell, 1968; Hodge and Sherrington, 1980; Odian, 1981; Ford, 1986a). The most common initially functionalized polystyrenes are (1) *p*-brominated polystyrene and (2) chloromethyl-polystyrene, which is para by alkylation with a chloromethyl ether (Pepper *et al.*, 1953), or a 70/30 *m/p* mixture by copolymerization of commercial chloromethylstyrenes (Balakrishnan and Ford, 1982). *Warning*: Chloromethyl methyl ether and bis(chloromethyl) ether are carcinogenic, and other chloromethyl ethers may be also. For methods of modification of polymers into reagents and catalysts see Frechet (1981).

II. STOICHIOMETRIC REACTIONS

A. Carbon–Carbon Bond Formation

1. Ester Enolate Reactions

Polystyrenes more highly cross-linked than Merrifield resins enabled 73–87% isolated yields of ketones such as **II** from acylations and alkylations of the enolate of 3-phenylpropanoic ester **I** at room temperature (Chang and Ford, 1981). The keys to success were the use of 10–20% CL polystyrenes, formation of the polymer enolate and introduction of the alkylating or acylating reagent as rapidly as possible, and incomplete conversion of the chloromethyl sites of the starting polymer to esters. Normally alkylations and acylations of esters in solution must be performed at dry ice temperature to avoid self-condensation product **III** [58% in solution, Eq. (1)]. The polymer-supported method avoids the need for low temperature, but it still requires an expensive organolithium base.

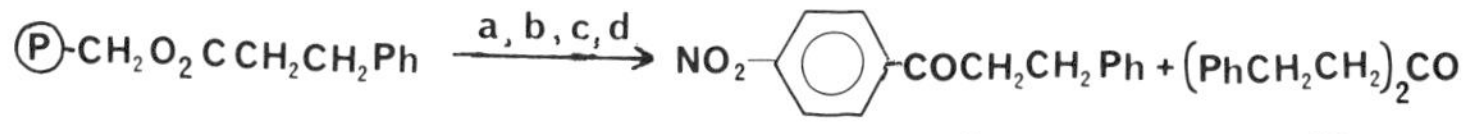

$$\text{\textcircled{P}-CH}_2\text{O}_2\text{C CH}_2\text{CH}_2\text{Ph} \quad \xrightarrow{\text{a, b, c, d}} \quad \text{NO}_2\text{-}\langle\bigcirc\rangle\text{-COCH}_2\text{CH}_2\text{Ph} + (\text{PhCH}_2\text{CH}_2)_2\text{CO}$$

$$\textbf{II}\textbf{III}(1)$$

(a) $\text{Ph}_3\text{CLi, THF}$; (b) $\text{p-NO}_2\text{C}_6\text{H}_4\text{COCl}$; (c) OH^-; (d) HCl

2. Sulfur Ylides

Reactions of a dimethyl(polystyryl)sulfonium fluorosulfonate (**IV**, DF 0.32) with benzaldehyde, acetophenone, and benzophenone under phase-transfer catalysis conditions [Eq. (3)] produced epoxides in nearly quantitative yields (Farral *et al.*, 1979). Analogous phase-transfer catalyzed epoxidations in solution gave lower yields from ketones. The polymeric reagent could be recycled five times as in Eq. (2) with no loss of activity. The polymer-bound sulfide by-product has no odor.

$$\text{(P)}-S-CH_3 \xrightarrow{\quad a \quad} \text{(P)}-\overset{+}{S}(CH_3)_2 \ FSO_3^- \qquad (2)$$
$$\text{IV} \qquad\qquad\qquad\qquad \text{V}$$

$$\text{V} + RCOR' \xrightarrow{\quad b \quad} \text{IV} + RR'C\underset{O}{-}CH_2 \qquad (3)$$

$$(a) \ FSO_3CH_3 \ ; \ (b) \ CH_2Cl_2 \ , \ aq. \ NaOH, \ n-Bu_4NOH$$

3. Wittig Reactions

Wittig reagents bound to polymer supports allow easy separation and recycling of the polymeric phosphine oxide by-product. For a review see Ford (1986b). Yields of alkenes from 2% CL, DF 0.3–0.4 polystyrene-bound benzyl and methylphosphonium salts are as high as yields from analogous solution Wittig reactions [Eqs. (4)–(7)]. Yields better than 90% were obtained from the methylidenephosphorane even with bulky ketones such as 10-nonadecanone and cholest-4-en-3-one (Bernard and Ford, 1983). The improved yields are due to (1) limitation of the functional groups to accessible sites in the polymer by the phosphination step, and (2) use of solvent/base systems that both swell polystyrene and solvate the phosphonium ion sites.

$$\text{(P)}-H \xrightarrow{\quad a, b \quad} \text{(P)}-P \ Ph_2 \qquad (4)$$

$$\text{(P)}-P \ Ph_2 + RCH_2X \longrightarrow \text{(P)}-\overset{+}{P}(Ph)_2CH_2R, \ X^- \qquad (5)$$
$$\text{VI} : R = H, \ X = I$$
$$\text{VII} : R = Ph, X = Br$$

$$\text{VI} + RCOR' \xrightarrow{\quad c \quad} RR'C=CH_2 + \text{(P)}-P(O) \ Ph_2 \qquad (6)$$

$$\text{VII} + RCHO \xrightarrow{\quad d \quad} RCH=CHPh + \text{(P)}-P(O)Ph_2 \qquad (7)$$

$$(a) \ Br_2, Tl(OAc)_3 \ , \ CCl_4 \ ; \ (b) \ Li \ PPh_2 \ , THF \ ;$$

$$(c) \ NaCH_2S(O)CH_3, (CH_3)_2SO, \ THF; \ (d) \ NaOEt, EtOH, THF$$

Heitz and Michels (1973) found highly stereoselective alkene formation from polymeric Wittig reagents under Li-salt and Li-free conditions similar to results in solution.

Allyl and benzyl phosphonium salts give high yields of alkenes from aldehydes with solid potassium carbonate in THF (Hodge et al., 1985).

Polymer-supported phosphonates can be used in the Horner–Wittig synthesis of alkenes (Cainelli *et al.*, 1980). Percolation of a dilute ethereal solution of the diethyl phosphonate through a column of Amberlyst A-26 in OH$^-$ form gives the polymer-supported carbanion [Eq. (8)]. The phosphonate **VIII** and carbonyl compounds gave 83–97% yields of α,β-unsaturated nitriles in 1–2 h [Eq. (9)]. Phosphonate ester **IX** reacted only with aldehydes [Eq. (10)]. Unlike many other polymer-supported syntheses, the Horner–Wittig reaction is independent of solvent, succeeding in hexane, methanol, benzene, or THF–water (9:1). Sequential reactions were performed in one flask with mixtures of two polymer-supported reagents. Dioxolanes were treated with a mixture of Amberlyst 15 [Eq. (11)], and the aldehyde or ketone was converted to alkene with the polymer–phosphonate **VIII** [Eq. (9)]. For example, 2-phenyl-1,3-dioxolane gave cinnamonitrile in 98% yield.

$$\textcircled{P}\text{-CH}_2\overset{+}{\text{N}}(\text{CH}_3)_3\text{OH}^- + (\text{EtO})_2\text{P(O)CH}_2\text{X} \longrightarrow \textcircled{P}\text{-CH}_2\overset{+}{\text{N}}(\text{CH}_3)_3\ (\text{EtO})_2\text{P(O)}\overset{-}{\text{C}}\text{HX} \quad (8)$$

$$\text{VIII}: \text{X} = \text{CN}$$
$$\text{IX}\ : \text{X} = \text{CO}_2\text{Et}$$

$$\text{VIII} + \text{RCOR}' \longrightarrow \text{RR}'\text{C}=\text{CHCN} \quad (9)$$

$$\text{IX} + \text{RCHO} \longrightarrow \text{RCH}=\text{CHCO}_2\text{Et} \quad (10)$$

$$\textcircled{P}\text{-SO}_2\text{OH} + \text{Ph}\diagup \longrightarrow \text{PhCHO} \quad (11)$$

B. Functional Group Conversions

1. Substitution at Saturated Carbon

Quaternary ammonium anion exchange resins are readily available. The bound anions have been used as nucleophiles, bases, oxidizing agents, and reducing agents (Cainelli, 1981). Simple ion exchange in a column can convert one of the common resins in the Cl$^-$ or OH$^-$ form to any resin with the desired anion. The synthesis in Fig. 1 of the β-adrenergic receptor antagonist propranolol uses three different reagents based on the macroporous resin Amberlyst A-26 (Cardillo *et al.*, 1986).

Triphenylphosphine and carbon tetrachloride convert alcohols to alkyl chlorides. Polystyryldiphenylphosphine (1% CL, DF 0.70) reacts up to 20 times faster than triphenylphosphine and allows easy filtration of the

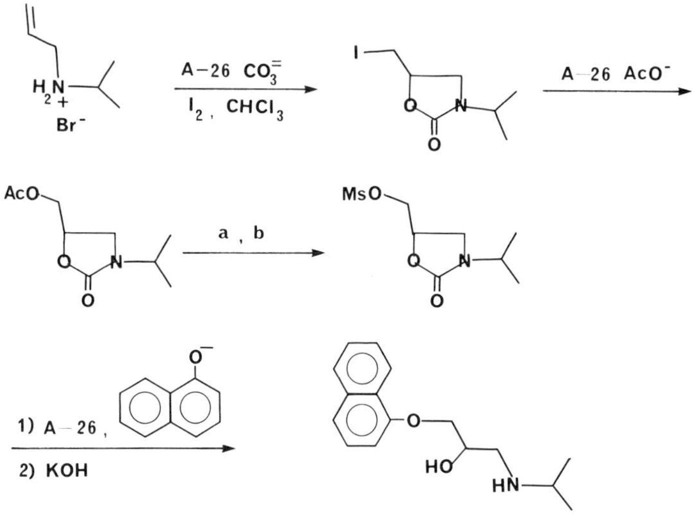

(a) K_2CO_3, EtOH; (b) CH_3SO_2Cl, Et_3N,CH_2Cl_2

Fig. 1. Synthesis of propranolol.

phosphine oxide by-product (Harrison et al., 1983). It gives primary alkyl chlorides from alcohols in >90% yields. With secondary alcohols, 40–70% yields result along with some elimination products. Even the bulky 5-β-cholan-24-ol and 4,4,4-triphenylbutan-l-ol gave 98% and 94% yields of alkyl chlorides. The polymer cannot be recycled due to concurrent formation of a dichloromethylidenephosphorane.

A polymeric tosyl azide converted the acidic methylene groups of β-diketones and β-ketoesters to diazo groups in 68–97% yields (Roush et al., 1974). Unlike the monomeric tosyl azide, the polymeric reagent showed no tendency to detonate with shock.

2. Substitution at Acyl Carbon

The Patchornik group (Shai et al., 1985) has developed a flow system in which two different polymers are involved in acyl transfers for peptide synthesis by means of a mediator [Eqs. (12) and (13)]. A polymeric o-nitrophenyl ester of phenylalanine (DF = 0.36) was prepared in a 3% CL macroporous polystyrene and placed in a column. A 1% CL polystyrene containing 0.3 mmol/g of carboxyl-bound leucine was loaded into a second column. A solution of imidazole as mediator and ethyldiisopropylamine in dichloromethane was circulated through the two columns, with on-line monitoring of solution composition by UV spectrophotometry. After 24 h the dipeptide was cleaved from the resin and deprotected to give a 99.8% yield of phenylalanylleucine. Continued coupling of other amino acid residues is possible by simply changing the polymer in column 1 after the deblocking step. This method requires much longer time for the coupling step than Merrifield's peptide synthesis.

$$ \text{(P)}-\underset{\text{column 1}}{\underset{\underset{\text{CH}_2\text{Ph}}{|}}{\overset{\text{NO}_2}{\bigcirc}-\text{O}_2\text{CCHNH}-\text{BOC}}} \quad \xrightarrow{\text{ImH}} \quad \underset{X}{\underset{\underset{\text{CH}_2\text{Ph}}{|}}{\text{N}\diagdown\text{N}-\overset{\overset{\text{O}}{\|}}{\text{C}}\text{CH NH}-\text{BOC}}} \tag{12}$$

$$ \underset{\text{column 2}}{\text{(P)}-\text{CH}_2\text{O}_2\overset{\underset{i-\text{Bu}}{|}}{\text{C CHNH}_2}} \quad X \quad \xrightarrow{-\text{ImH}} \quad \text{(P)}-\text{CH}_2\text{O}_2\overset{\underset{i-\text{Bu}}{|}}{\text{CCH}}\text{ NH}\overset{\overset{\text{O}}{\|}}{\text{C}}\overset{\underset{\text{CH}_2\text{Ph}}{|}}{\text{CHNH}}-\text{BOC} \tag{13}$$

Soluble nitriles, ketones, and amides can be acylated in sequential reactions involving trityllithium bound to one polymer to generate the enolate, and an active ester bound to a second polymer [Eqs. (14) and (15)] (Cohen *et al.*, 1981). No self-condensation of the acetophenone occurs. The corresponding solution reaction gave dibenzoylmethane in only 48% yield.

$$ \text{(P)}-\text{C}(\text{Ph})_2\text{Li} + \text{CH}_3\text{CO Ph} \quad \longrightarrow \quad \underset{XI}{\text{LiCH}_2\text{CO Ph}} \tag{14}$$

$$ XI + \text{(P)}-\text{CH}_2-\overset{\text{NO}_2}{\bigcirc}-\text{O}_2\text{C Ph} \quad \longrightarrow \quad \text{PhCO CH}_2\text{CO Ph} \tag{15}$$

3. Protecting Groups

Functionalized insoluble polymers can remove compounds containing specific functional groups from mixtures. They can also serve to protect a given functional group in a polyfunctional compound (Table 1). Low DF polymers are required to minimize double binding of difunctional substrates. Polymeric protecting groups have been used in syntheses such as that of the sex attractant of the fall army worm moth [Eqs. (16)–(18)] (Leznoff *et al.*, 1977).

TABLE I

Polymer-Supported Protecting Groups[a]

Protecting group[b]	DF	Substrate	% Bound	Reaction	Product	% Yield[c]	Reference
Acid							
—CH$_2$OH	0.17	ClCO(CH$_2$)$_4$COCl	71	i, PhMgI ii, K$_2$CO$_3$ iii, CH$_2$N$_2$	CH$_3$OCO(CH$_2$)$_4$C(OH)Ph$_2$	85	Leznoff and Yedidia (1980)
Alcohols							
—B(OH)$_2$	0.06	CH$_2$OHCHOHCH$_2$OH	—	PhCOCl	CH$_2$OHCHOHCH$_2$OCOPh	93	Frechet and Seymour (1978)
Aldehydes and ketones							
—CH$_2$OCH$_2$CHOH $\quad\quad\quad$ | $\quad\quad\quad$ CH$_2$OH	0.16	OCH—Ph—CHO—p	56	i, PhMgBr ii, 4 M HCl–dioxane	OCH—Ph—C=CH$_2$ $\quad\quad\quad\quad\quad$ | $\quad\quad\quad\quad\quad$ Ph	95	Xu et al. (1983)
Amines							
—CH$_2$OC—O—Ph—NO$_2$—p $\quad$ || $\quad$ O	0.30	NH$_2$(CH$_2$)$_8$NH$_2$	100	i, PhCOCl ii, CF$_3$COOH	CF$_3$CONH(CH$_2$)$_8$NHCOPh	81	Dixit and Leznoff (1978)

[a] All polymers were cross-linked polystyrenes.
[b] Indicates the attachment site on the polymer.
[c] Refers to yield from polymer.

$$\text{\textcircled{P}-C(Ph)}_2\text{Cl} \xrightarrow{\text{a,b,c}} \text{\textcircled{P}-C(Ph)}_2\text{O(CH}_2)_8\text{C}{\equiv}\text{C(CH}_2)_3\text{CH}_3 \tag{16}$$
$$\text{XII}$$

$$\text{XII} \xrightarrow{\text{d}} \text{HO(CH}_2)_8\text{C}{\equiv}\text{C(CH}_2)_3\text{CH}_3 \tag{17}$$
$$\text{XIII}$$

$$\text{XIII} \xrightarrow{\text{e,f}} \text{AcO(CH}_2)_8\diagdown\diagup\text{(CH}_2)_3\text{CH}_3 \tag{18}$$

overall yield 36% (16% in solution)

(a) $HO(CH_2)_8OH$; (b) CH_3SO_2Cl, pyridine; (c) $Li C{\equiv}C(CH_2)_3CH_3$;

(d) HCl, dioxane; (e) H_2/Pd / $CaCO_3$; (f) Ac_2O

Chapman and Walker (1975) used a high capacity (DF = 0.63) diazomethylene polymer [$\text{\textcircled{P}}$—$C(N_2)Ph$] for the protection of carboxylic acid end groups of some penicillins and cephalosporins.

C. Inorganic Syntheses

Burlitch and Winterton (1975) synthesized bis(di-*n*-butylchlorotin)-tetracarbonylosmium (**XVII**) [Eqs. (19)–(22)] on a macroporous 20% CL polystyrene using a very low DF of 0.008 to minimize intrapolymer reactions. The key step was formation of the tin–osmium bond [Eq. (20)] without competing formation of an osmium dimer (the only product observed in a solution synthesis).

$$\text{\textcircled{P}-Hg O}_2\text{CCF}_3 \xrightarrow{\text{a,b,c}} \text{\textcircled{P}-Sn(Bu)}_2\text{Cl} \tag{19}$$
$$\text{XIV}$$

$$\text{XIV} \xrightarrow{\text{d}} \text{\textcircled{P}-Sn(Bu)}_2\text{OsH(CO)}_4 \tag{20}$$
$$\text{XV}$$

$$\text{XV} \xrightarrow{\text{e}} \text{\textcircled{P}-Sn(Bu)}_2\text{Os(CO)}_4\text{Sn(Bu)}_2\text{Cl} \tag{21}$$
$$\text{XVI}$$

$$\text{XVI} \xrightarrow{\text{f}} \text{Cl Sn(Bu)}_2\text{Os(CO)}_4\text{Sn(Bu)}_2\text{Cl} \tag{22}$$
$$\text{XVII}$$

(a) $(CH_3)_4NCl$; (b) n–BuLi ; (c) n–Bu_2SnCl_2 ;

(d) $H_2Os(CO)_4$, Et_3N ; (e) n–Bu_2SnCl_2, Et_3N ; (f) HCl

III. POLYMER-SUPPORTED CATALYSTS

A. Experimental Design

The faster the intrinsic chemical reaction, the more likely mass transfer of a reactant from bulk liquid to the particle surface and intraparticle diffusion to the active site will limit the rate (Satterfield, 1970). Hence, reaction mixtures should be agitated vigorously. A surface-functionalized macroporous polymeric support or microporous polymer particles as small as practical should be used. The times required for mass transport to the particle surface and to the active site are proportional to the external surface area per unit volume of the catalyst. (Surface area per unit weight is proportional to the inverse radius of a sphere.) Polystyrene beads are commonly available in 200/400 mesh (37–74-μm) size. Still smaller beads can be difficult to filter with conventional glass frits and filter papers.

B. Transition Metal Catalysts

For reviews see Pittman (1983), Bailey and Langer (1981), Holy (1983), and Chauvin *et al.* (1977). See also Chapters 13 and 14 of this volume.

Mass transfer and intraparticle diffusion in reactions catalyzed by transition metals on polymer supports are discussed by Ekerdt (1986). When kinetics are limited by intraparticle diffusion, greater retardation of the diffusion of larger molecules causes competing reactions to occur with selectivity for smaller product molecules (Grubbs and Kroll, 1971).

Polymer supports may concentrate polymer-bound ligands around the metal or cause aggregation of metal species to a level not possible in solution because of low solubility. Polymer supports may also isolate metal sites from one another. "Concentration" of polymeric ligands or of a metal species is favored by a high DF and a high degree of swelling of the polymer. "Site isolation" is favored by low DF, high % CL, and low swelling. (For elaboration of these principles see Ford, 1986c.)

Titanocene catalysts on a 20% CL macroporous polystyrene were 25–120 times more active than the corresponding soluble catalysts, an increase attributed to a monomeric titanium species (Grubbs *et al.*, 1977). A model in which (1) the Ti sites were assumed to be randomly distributed over the 90 m^2/g surface area of the polymer, (2) the area of one titanocene was assumed to be 22 Å^2, and (3) only titanocenes with no neighbor in an adjacent site were active, fit the hydrogenation rate dependence on mmol Ti/g amazingly well. Use of nonswelling solvents limited active sites to the internal surface of the polymer and prevented their redistribution to the interior.

Cooperation of polymer-bound ligands is inferred from normal/branched aldehyde ratios of >10 in the hydroformylation of 1-pentene with a 1% CL polystyrene-supported Wilkinson catalyst having DF = 0.4 and a P/Rh ratio of 19 (Pittman and Hanes, 1976). The soluble analog $RhH(CO)(PPh_3)_3$ gave n/b ratios of about 3.

The dimerization–alkoxylation of butadiene [Eq. (23)] is catalyzed by soluble $Pd(PPh_3)_4$ and by the analogous 1% CL polystyryldiphenylphosphine catalyst (Pittman and Ng, 1978). Higher rates with the polymeric catalyst are attributed to a higher degree of coordination and less aggregation of the Pd in the polymer matrix.

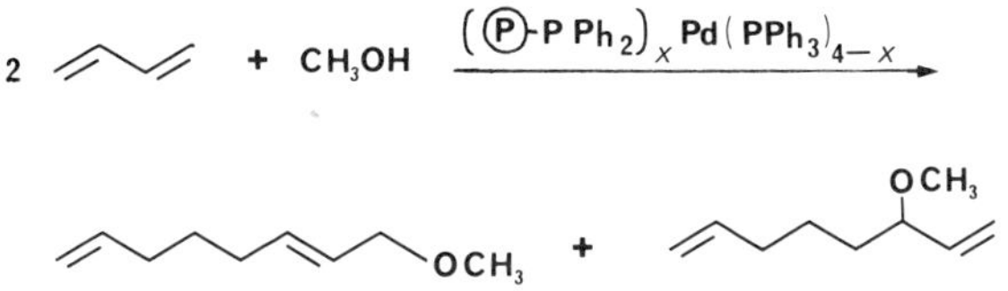

$$2 \, \diagdown\!\!\diagup + CH_3OH \xrightarrow{\left(\text{P}\text{-}P\,Ph_2\right)_x Pd\left(PPh_3\right)_{4-x}}$$

$$\diagdown\!\!\diagup\!\!\diagdown\!\!\diagup\!\!\diagdown\!\!\diagup OCH_3 \; + \; \diagdown\!\!\diagup\!\!\diagdown\!\!\diagup\overset{OCH_3}{\diagdown}\!\!\diagup \tag{23}$$

Polymer-bound tetrakis(triarylphosphine)palladium(0) catalyzed cyclization of a carbanion to a π-allylpalladium [Eq. (24)] with 0.7 M substrate, the highest solution concentration known for high-yield synthesis of large rings (Trost and Warner, 1983). Large ring syntheses normally require "high dilution," typically leading to final product concentrations of below centimolar.

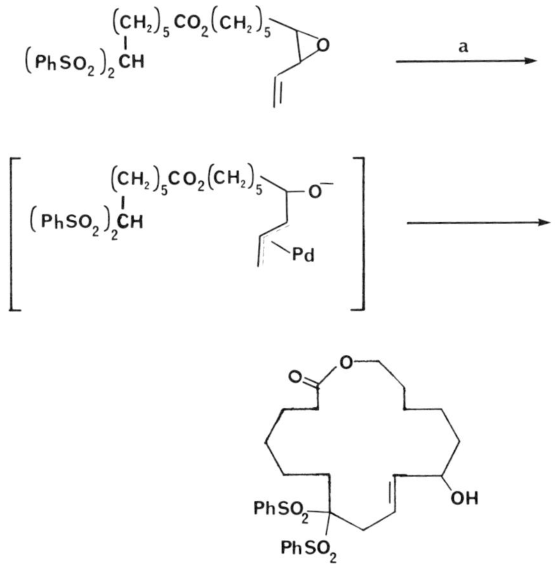

$$\overset{(CH_2)_5 CO_2 (CH_2)_5}{\underset{(PhSO_2)_2 CH}{\diagdown}} \!\!\!\!\diagdown\!\!O \xrightarrow{\;a\;}$$

$$\left[\overset{(CH_2)_5 CO_2 (CH_2)_5}{\underset{(PhSO_2)_2 CH}{\diagdown}} \!\!\!\!\diagdown\!\!-O^- \underset{Pd}{\diagdown} \right] \longrightarrow$$

$$\underset{\underset{PhSO_2}{PhSO_2}}{\diagdown}\!\!\!\diagdown\!\!OH \tag{24}$$

(a) $\left(\text{P}\text{-}P\,Ph_2\right)_x Pd\left(PPh_3\right)_{4-x}$

Most polymer-bound transition metal catalysts suffer from leaching of the metal or from chemical decomposition that would prevent continuous use for months (Garrou, 1986). There are many examples of recycling and of use of such catalysts under flow conditions without loss of activity over a few days (Allum *et al.*, 1976a,b). Thus, the most promising industrial applications of polymer-supported transition metal catalysts are for fine and specialty chemicals. Polymer-supported catalysts also may be used in the form of fibres, films, or membranes (Lieto *et al.*, 1983).

C. Phase-Transfer Catalysts

Reactions between water-insoluble organic compounds and water-soluble salts require a common phase, such as an insoluble gel polymer with quaternary ammonium or phosphonium salts (**XVIII, XIX**), crown ethers (**XX**), cryptands, or poly(ethylene glycol) ethers (**XXI**) [Eqs. (25)–(26)] (Regen, 1979; Chiellini *et al.* 1981; Ford and Tomoi, 1984).

$$\text{(P)-CH}_2\overset{+}{\text{N}}(\text{CH}_3)_3\text{ Cl}^- \qquad \text{(P)-CH}_2\overset{+}{\text{P}}(n\text{-Bu})_3\text{ Cl}^-$$

$$\textbf{XVIII} \qquad\qquad\qquad \textbf{XIX}$$

$$\text{(P)-(CH}_2)_7\text{-N} \qquad\qquad \text{(P)-CH}_2(\text{OCH}_2\text{CH}_2)_8\text{O CH}_3$$

$$\textbf{XX} \qquad\qquad\qquad \textbf{XXI}$$

$$n\text{-C}_8\text{H}_{17}\text{Br} + \text{NaCN}_{(aq)} \xrightarrow{\textbf{XIX}} n\text{-C}_8\text{H}_{17}\text{CN} \tag{25}$$

$$\text{PhCH}_2\text{CN} + n\text{-BuBr} + \text{NaOH}_{(aq)} \xrightarrow{\textbf{XVIII}} \text{PhCH}(\text{Bu})\text{CN} \tag{26}$$

Nucleophilic displacement reactions with hard anions such as cyanide, acetate, and chloride are catalyzed most effectively by DF 0.1–0.2 polystyryl quaternary ammonium and phosphonium ions [for example, Eq. (25)] (Tomoi and Ford, 1981). Fast reaction rates are promoted by rapid mechanical stirring, small catalyst particles, low % CL, lipophilic active sites, spacer chains between active site and polymer backbone, good swelling solvents, and high concentration of the inorganic reactant in the aqueous phase (Ford and Tomoi, 1984).

Commercial quaternary ammonium anion exchange resins (**XVIII**) in the OH$^-$, OAc$^-$, CN$^-$, and other forms are catalysts for aldol condensations, Michael reactions, Knoevenagel condensations, cyanoethylations, and cyanohydrin formation (Chiellini *et al.*, 1981; Ford and Tomoi, 1984).

Displacement reactions can be carried out by solid/solid/liquid phase transfer catalysis using solid potassium phenoxide (MacKenzie and Sherrington, 1980) and solid sodium iodide (Yanagida et al., 1979). The insoluble 2% CL polystyrenes with bound PEG ethers (**XXI**) can be reused.

In contrast to the lipophilic quaternary onium ion-substituted polystyrenes that are active for most nucleophilic displacement reactions, highly hydrated commercial anion exchange resins (**XVIII**) with DF $\geqslant 0.5$ are most effective for phase-transfer reactions that require generation of a carbanion with aqueous sodium or potassium hydroxide, such as the alkylation of phenylacetonitrile [Eq. (26)] (Komeili-Zadeh et al., 1978; Balakrishnan and Ford, 1983). The selectivity for monoalkylation is as high as has ever attained with a soluble phase-transfer catalyst. The strongly alkaline conditions, however, degrade the catalyst. Poly(ethylene glycol) methyl ether (**XXI**) grafted to 1% cross-linked polystyrene is highly active for alkylation of phenylacetonitrile and benzyl phenyl ketone, and for Williamson ether synthesis, with a longer lifetime than quaternary onium ion catalysts under strongly basic conditions (Kimura et al., 1983).

Anion exchange resins (**XVIII**) are easily separated catalysts for dichlorocyclopropane syntheses from chloroform and solid sodium hydroxide [Eq. (27)] and for dehydration of amides to nitriles [Eq. (28)] (Yanagida et al., 1979).

$$\text{(cyclohexene)} + \text{CHCl}_3 + \text{NaOH}_{(s)} \xrightarrow{\textbf{XVIII}} \text{(bicyclic)} \underset{\text{Cl}}{\overset{\text{Cl}}{<}} \qquad (27)$$

$$\text{PhCONH}_2 + \text{CHCl}_3 + \text{NaOH}_{(s)} \longrightarrow \text{PhCN} \qquad (28)$$

Two immiscible liquid phases present special problems in continuous flow reactions in packed beds. Reaction of 1-bromooctane in o-dichlorobenzene with aqueous KI catalyzed with a polystyrene-bound quaternary phosphonium ion in an upflow column reactor has been demonstrated on a small scale with activity as high as 0.4 times that achieved with the same catalyst in a stirred reaction mixture (Ragaini et al., 1986).

D. Amine Catalysts

Polymeric amine catalysts may function as proton acceptors, acyl transfer agents, or ligands for metal ions. Most readily available are the 2- and 4-isomers of poly(vinylpyridine) and tertiary amine anion exchange resins. The poly(vinylpyridine)s (PVPs) have been used in stoichiometric amounts for preparation of esters from acid chlorides and alcohols and for preparation of trimethylsilyl ethers and trimethylsilylamines from chloro-trimethylsilane and alcohols or amines (Frechet and de Meftahi, 1984). The acid salts of the PVP are easily converted back to the free base form for reuse.

Several different polymers bearing 4-(N,N-dialkylamino)pyridine functionality, **XXII–XXIII**, have been prepared as analogs of N,N-dimethylaminopyridine and used as catalysts for esterification of tertiary alcohols [Eq. (29)], hydrolyses of active esters, and other substitutions at acyl carbon (Shinkai et al., 1981; Tomoi et al., 1982).

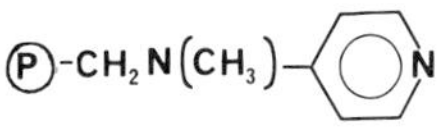

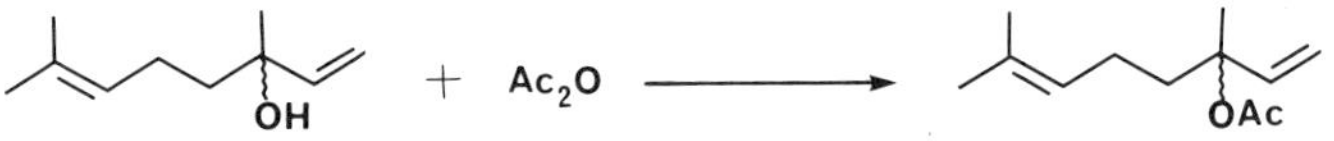

$$
\text{(29)}
$$

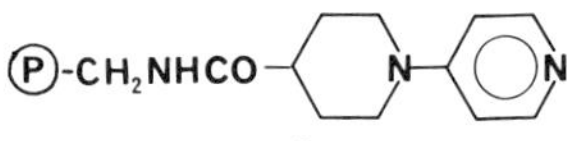

Catalyst **XXIII** acted as an acyl transfer reagent in a circulating fixed bed reactor to give, for example, p-nitrophenyl thiolbenzoate in quantitative yield [Eq. (30)] (Shai et al., 1985). Other acyl donors used in this column method were acetyl chloride, isobutyl chloroformate, p-toluenesulfonyl chloride, and isopropyl o-chlorophenyl phosphoryl chloride. Nucleophiles included alcohols to form esters, carboxylic acids to form anhydrides, HF to form a carboxylic acid fluoride, and amines to form carboxamides, sulfonamides, and phosphoramides.

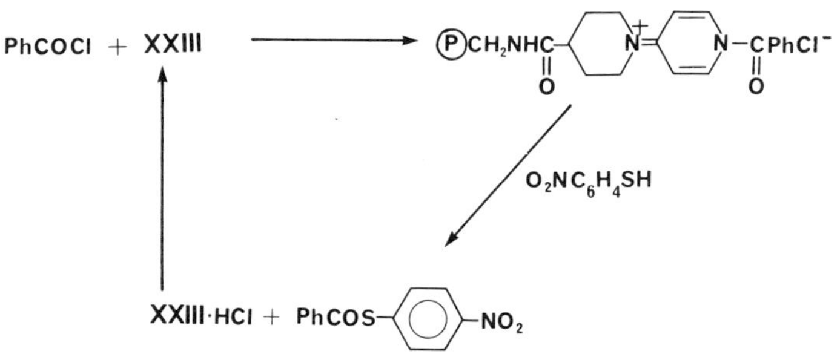

$$
\text{(30)}
$$

Tertiary amine ion exchange resins are catalysts for aldol condensations, Knoevenagel condensations, Perkin reactions, cyanohydrin formation, and redistributions of chlorosilanes (Chiellini et al., 1981; Ford and Tomoi, 1984).

The *Cinchona* alkaloids, quinine, quinidine, cinchonine, and cinchonidine, bound to polymers are catalysts for asymmetric Michael additions to α,β-unsaturated ketones, esters, and nitro compounds [Eqs. (31) and

(32)]. Chemical activity and stereoselectivity were obtained only by attachment of the alkaloids to the polymer through the remote vinyl group, either by copolymerization with acrylonitrile (Kobayashi, 1984) or by addition of a Ⓟ—CH₂SH to the double bond (Hodge *et al.*, 1985). Although the enantiomeric excesses with the polymeric catalysts are somewhat lower than with the soluble alkaloids, ease of recovery and reuse is a distinct advantage.

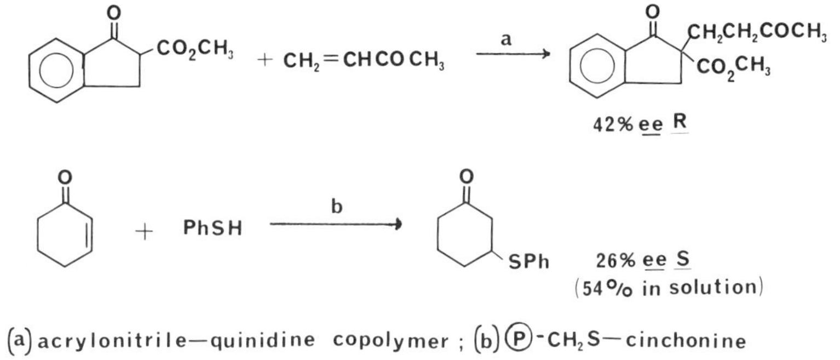

(a) acrylonitrile—quinidine copolymer ; (b) Ⓟ⁻CH₂S—cinchonine

E. Acidic Catalysts

Poly(styrenesulfonic acid) ion exchange resins and a perfluoroalkanesulfonic acid polymer (Nafion) (Waller, 1986) are noncorrosive, recoverable analogs of *p*-toluenesulfonic acid and of super acids. Commercial poly(styrenesulfonic acid) resins catalyze hydrolyses of carboxylic acid derivatives, esterifications, acetal and ketal formation, condensation, alkylations, and rearrangements (Akelah and Sherrington, 1981, 1983). In many cases, such as the dimerization of α-methylstyrene [Eq. (33)], higher yields and cleaner products have been obtained than with *p*-toluenesulfonic acid or sulfuric acid as the catalyst (Duncan *et al.*, 1975).

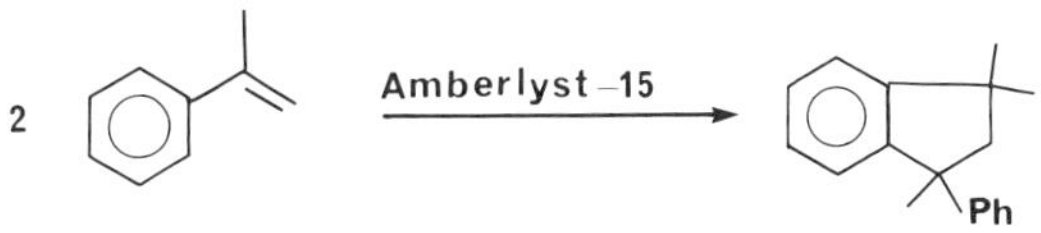

The gasoline anti-knock additive methyl *t*-butyl ether is manufactured in at least 30 plants worldwide by addition of methanol to isobutene catalyzed by a macroporous poly(styrenesulfonic acid) (Ancillotti *et al.*, 1977; Widdecke, 1984). Other industrial processes employing sulfonic acid ion exchange resin catalysts are alkylations of phenol, hydration of propene to isopropyl alcohol, alkylations of benzene, and the 2:1 condensation of phenol with acetone to produce bisphenol-A (Widdecke, 1984).

A polystyrene-supported triarylmethyl cation is an efficient and gentle catalyst for aldol condensations of trimethylsilyl enol ethers with acetals or aldehydes [Eq. (34)] by batch or flow methods (Mukaiyama and Iwakiri, 1985).

$$\underset{}{\text{OSi(CH}_3)_3} \quad + \quad \text{Ph CH}(\text{O CH}_3)_2 \quad \xrightarrow{\text{a}} \quad \underset{}{\text{O}} \quad \underset{}{\text{OCH}_3} \quad \text{Ph} \qquad (34)$$

$$\text{(a)} \quad CH_2Cl_2, \, -78°\,C., \, 2\text{--}3 \text{ mol. }\% \quad \textcircled{P}\text{-}\overset{+}{C}Ph_2 \; ClO_4^-$$

IV. CONCLUSIONS

A wide variety of preparative reactions can be carried out with polymer-bound reagents and catalysts. Many are commercially available. Others may be prepared in one or two steps from preformed polymers or from monomers using techniques almost the same as those of small-molecule preparative chemistry. In the future polymeric reagents and catalysts will be more common in both research laboratories and manufacturing plants because they are safer and more economical to use.

REFERENCES

Akelah, A., and Sherrington, D. C. (1981). *Chem. Rev.* **81**, 557.
Akelah, A., and Sherrington, D. C. (1983). *Polymer* **24**, 1369.
Allum, K. G., Hancock, R. D., Howell, I. V., Pitkethyl, R. C., and Robinson, P. J. (1976a) *J. Catal.* **43**, 322.
Allum, K. G., Hancock, R. D. Howell, I. V., Lester, T. E., McKenzie, S., Pitkethly, R. C., and Robinson, P. J. (1976b). *J. Catal.* **43**, 331.
Ancillotti, F., Mauri, M. M., and Pescarollo, E. (1977). *J. Catal.* **46**, 49.
Bailey, D. C., and Langer, S. H. (1981). *Chem. Rev.* **81**, 109.
Balakrishnan, T., and Ford, W. T. (1982). *J. Appl. Polym. Sci.* **27**, 133.
Balakrishnan, T., and Ford, W. T. (1983). *J. Org. Chem.* **48**, 1029.
Barany, G., and Merrifield, R. B. (1979). *In* "The Peptides" (E. Gross and J. Meienhofer, eds.), Vol. 2, Part A, Chap. 1. Academic Press, New York.
Bernard, M., and Ford, W. T. (1983). *J. Org. Chem.* **48**, 326.
Burlitch, J. M., and Winterton, R. C. (1975). *J. Am. Chem. Soc.* **97**, 5605.
Cainelli, G., Contento, M., Manescalchi, R., and Regnoli, R. (1980). *J.C.S. Perkin I*, p. 2516. Hutchinson, eds.) Pergamon, New York, pp. 19–28.
Cainelli, G., Contento, M., Manescalchi, F., and Regnoli, R. (1980). *J.C.S. Perkin I*, p. 2516.
Cardillo, G., Orena, M., and Sandri, S. (1986). *J. Org. Chem.* **51**, 713.
Chang, Y. H., and Ford, W. T. (1981). *J. Org. Chem.* **46**, 5364.
Chapman, P. H., and Walker, D. (1975). *J.C.S. Chem. Commun.* p. 690.
Chauvin, Y., Commereuc, D., and Dawans, F. (1977). *Prog. Polym. Sci.* **5**, 95.

Chiellini, E., Solaro, R., and D'Antone, S. (1981). *Makromol. Chem., Suppl.* **5**, 82.

Cohen, B. J., Kraus,, M. A., and Patchornik, A. (1981). *J. Am. Chem. Soc.* **103**, 7620.

Dixit, D. M., and Leznoff, C. C. (1978). *Isr. J. Chem.* **17**, 248.

Duncan, W. P., Eisenbraun, E. J., Taylor, A. R., and Keen, G. W. (1975). *Org. Prep. Proced, Int.* **7**, 225.

Ekerdt, J. (1986). *A.C.S., Symp. Ser.* No. 308, 68.

Farrall, M. J., Durst, T., and Frechet, J. M. J. (1979). *Tetrahedron Lett.* 203.

Flory, P. J. (1953). "Principles of Polymer Chemistry." Cornell Univ. Press, Ithaca, New York.

Ford, W. T., ed. (1986a). "Polymeric Reagents and Catalysts," *A.C.S. Symp. Ser.* No. 308.

Ford, W. T. (1986b). *A.C.S. Symp. Ser.* No. 308, 155.

Ford, W. T. (1986c). *A.C.S. Symp. Ser.* No. 308, 247.

Ford, W. T., and Tomoi, M. (1984). *Adv. Polym. Sci.* **55**, 49.

Ford, W. T., Mohanraj, S., and Periyasamy, M. (1984). *Br. Polym. J.* **16**, 179.

Frechet, J. M. J. (1981). *Tetrahedron* **37**, 663.

Frechet, J. M. J., and de Meftahi, M. V. (1984). *Br. Polym. J.* **16**, 193.

Frechet, J. M. J., and Seymour, E. (1978). *Isr. J. Chem.* **17**, 253.

Fyfe, C. A. (1983). "Solid State NMR for Chemists," C. F. C. Press, Guelph, Ontario.

Garrou, P. (1986). *A.C.S. Symp. Ser.* No. 308, 84.

Grubbs, R. H., and Kroll, L. C. (1971). *J. Am. Chem. Soc.* **93**, 3062.

Grubbs, R.H., Lau, C.P., Cukier, R., and Brubaker, C., Jr. (1977). *J. Am. Chem. Soc.* **99**, 4517.

Guyot, A., and Bartholin, M. (1982). *Prog. Polym. Sci.* **8**, 277.

Harrison, C. R., Hodge, P., Hunt, B. J., Khoshdel, E., and Richardson, R. (1983), *J. Org. Chem.* **48**, 3721.

Heitz, W., and Michels, R. (1973). *Liebig's Ann. Chem.* p. 227.

Helfferich, F. (1962). "Ion Exchange." McGraw-Hill, New York.

Hodge, P., and Sherrington, D. C., eds. (1980). "Polymer-Supported Reactions in Organic Synthesis." Wiley (Interscience), New York.

Hodge, P., Khoshdel, E., Waterhouse, J., and Frechet, J. M. J. (1985). *J.C.S. Perkin I* p. 2327.

Holy, N. L. (1983). *In* "Homogeneous Catalysis by Metal Phosphine Complexes" (L. H. Pignolet, ed.) pp. 443–484. Plenum, New York.

Kimura, Y., Kirszensztejn, P., and Regen, S. L. (1983). *J. Org. Chem.* **48**, 385.

Kobayashi, N. (1984). *Br. Polym. J.* **16**, 205.

Komeili-Zadeh, H., Dou, H. J.-M., and Metzger, J. (1978). *J. Org. Chem.* **43**, 156.

Leznoff, C. C., and Yedidia, V. (1980). *Can. J. Chem.* **58**, 287.

Leznoff, C. C., Fyles, T. M., and Weatherston, J. (1977). *Can. J. Chem.* **55**, 1143.

Lieto, J., Milstein, D., Albright, R. L., Minkiewicz, J. V., and Gates, B. C. (1983). *CHEMTECH* **13**, 46.

MacKenzie, W. M., and Sherrington, D. C. (1980). *Polymer* **21**, 791.

Manecke, G., and Storck, W. (1978). *Angew. Chem.* **90**, 691.

Mathur, N. K., Narang, C. K., and Williams, R. E. (1980). "Polymers As Aids in Organic Chemistry." Academic Press, New York.

Merrifield, R. B. (1963). *J. Am. Chem. Soc.* **85**, 2149.

Mukaiyama, T., and Iwakiri, H. (1985). *Chem. Lett.* p. 1363.

Odian, G., (1981). "Principles of Polymerization," 2nd ed. Wiley (Interscience), New York.

Pan, S., and Morawetz, H. (1980). *Macromolecules* **13**, 1157.

Pepper, K. W., Paisley, H. M., and Young, M. A. (1953). *J. Chem. Soc.* p. 4093.

Pittman, C. U., Jr. (1983). *In* "Comprehensive Organometallic Chemistry" (G. Wilkinson,

F. G. A. Stone, and E. W. Abel, eds.), Vol. 8, pp. 553–611. Pergamon, Oxford.

Pittman, C. U., Jr., and Hanes, R. M. (1976). *J. Am. Chem. Soc.* **98**, 5402.

Pittman, C. U., Jr., and Ng, Q. (1978). *J. Organomet. Chem.* **153**, 85.

Ragaini, V., Verzella, G., Ghignone, A., and Colombo, G. (1986). *Ind. Eng. Chem. Process Des. Dev.* **25**, 878.

Regen. S. L. (1979). *Angew. Chem., Int. Ed. Engl.* **18**, 421.

Roush, W. R., Feitler, D., and Rebek, J. (1974). *Tetrahedron Lett.* p. 1391.

Sarin, V. K., Kent, S. B. H., Mitchell, A. R., and Merrifield, R. B. (1984). *J. Am. Chem. Soc.* **106**, 7845.

Satterfield, C. N. (1970). "Mass Transfer in Heterogeoeous Catalysis." M.I.T. Press, Cambridge, Massachusetts.

Schmuckler, G., and Goldstein, S. (1977). *In* "Ion Exchange and Solvent Extraction" (J. A. Marinsky and Y. Marcus, eds.), Vol. 7, pp. 1–28. Dekker, New York.

Seidl, J., Malinsky, J., Dusek, K., and Heitz, W. (1967). *Adv. Polym. Sci.* **5**, 113.

Shai, Y. Jacobson, K. A., and Patchornik, A. (1985). *J. Am. Chem. Soc.* **107**, 4249.

Shinkai, S., Tsuji, H., Hara, Y., and Manabe, O. (1981). *Bull. Chem. Soc. Jpn.* **54**, 631.

Sorenson, W. R., and Campbell, T. W. (1968). "Preparative Methods of Polymer Chemistry," 2nd Ed. Wiley, New York.

Tanaka, T. (1981). *Sci. Am.* **244**, 124.

Tomoi, M., and Ford, W. T. (1981). *J. Am. Chem. Soc.* **103**, 3821.

Tomoi, M., Akada, Y., and Kakiuchi, H. (1982). *Makromol. Chem., Rapid Commun.* **3**, 527.

Trost, B. M., and Warner, R. W. (1983). *J. Am. Chem. Soc.* **105**, 5940.

Waller, F. J. (1986). *A.C.S. Symp. Ser.* No. 308, 42.

Widdecke, H. (1984). *Br. Polym. J.* **16**, 188.

Xu, Z.-H., McArthur, C. R., and Leznoff, C. C. (1983). *Can. J. Chem.* **61**, 1405.

Yanagida, S., Takahashi, K., and Okahara, M. (1979). *J. Org. Chem.* **44**, 1099.

13 POLYMER-SUPPORTED OXIDATIONS

Georges Gelbard

Laboratoire des Matériaux Organiques
Centre National de la Recherche Scientifique
69390 Vernaison, France

I. INTRODUCTION

A. Scope of Oxidation and Halogenation

In this chapter, oxidation is taken in a broad sense to mean

(1) the removal of two hydrogen atoms and subsequent formation of additional bonds (insaturation, coupling),
(2) the introduction of oxygen or halogen atoms by substitution or addition reactions, and
(3) the introduction of oxygen atoms with bond breaking.

A book series (Wiberg and Trahanowsky, 1965–1982) describes the most common oxidizing reagents; helpful table (Hudlicky, 1977) overviews oxidizing and reducing agents and their relevant substrates; a short review on polymer-bound oxidizing agents is given by Taylor (1986).

B. Functional Groups Involved as Substrates

According to available or already described reagents, the possible modifications of functional groups by oxidation or halogenation are reported in Table I.

PREPARATIVE CHEMISTRY
USING SUPPORTED REAGENTS

TABLE I

Functional Transformations by Oxidation

Substrate	Product	Reagent	Chapter section
R—CH$_2$CH$_2$COX	RCH=CHCOX	Hg(AcO)$_2$ quinones	II.A
R—C≡C—R	R—C—C—R (O)	Peracids	II.F
R—C≡C—R	R—CO—CO—R	Cr(VI)	III.B.3
R—C≡C—R	RCO$_2$H	MnO$_4^-$	
—CH$_2$—	—CH(OAc)—	Pb(AcO)$_4$ CrO$_2$(AcO)$_2$	
R$_2$CHOH	R$_2$C=O	$\overset{+}{S}$—Cl	II.C
R$_2$CHOH	R$_2$C=O	$\overset{+}{S}$—O$^-$	II.D
R$_2$CHOH	R$_2$C=O	$\overset{+}{N}$—O$^-$	II.E
R$_2$CHOH	R$_2$C=O	Cr(VI)	II.G
R$_2$CHOH	R$_2$C=O	Cr(VI)	III.B.3
R$_2$CHOH	R$_2$C=O	MnO$_4^-$	
—C(OH)—C(OH)—	C=O	Pb(AcO)$_4$ IO$_4^-$	III.B.1
RCHO	RCO$_2$H	X$_2$ Cr(VI) MnO$_4^-$	III.B.3
Ketone	Ester	Peracid	II.F
RCOCH$_2$R′	RCOCH(OAc)R′	Pb(AcO)$_4$	
C—COCH$_3$	C—COCO$_2$H	K$_3$Fe(CN)$_6$	
RCOR′	RCO$_2$H, R′CO$_2$H	MnO$_4^-$	
R—NH$_2$	RCHO		
	RCN	Pb(AcO)$_4$	
	R—NO	MnO$_4^-$	
	R—NO$_2$		
R—CH$_2$NHR	R—CH=N—R	Hg(AcO)$_2$	
≥N	≥N—O	Peracids	II.F

(*continues*)

TABLE I (*Continued*)

Substrate	Product	Reagent	Chapter section
R—NH—NH—R	R—N=N—R	Cu^{2+}	
		Cr(VI)	
R—SH	R—SS—R	Cr(VI)	
		MnO_4	
		$N—Br_2$	II.B
		$K_3Fe(CN)_6$	
		Se=O	II.D
RSH	RSO_3H	MnO_4^-	
RSR'	RSOR'	MnO_4^-	
	RSO_2R'	Peracids	II.F
		N—Cl	II.B

R—C=C—R'	R—C—C—R'
	   | |
	   X X
		X_3^-	III.B.2

$RCH_2CO—R'$	R—CH—CO—R'
	   |
	   X
		X_3	III.B.2

The use of the relevant reagent is indicated in the corresponding section of this chapter; we report here reagents which are not yet described as supported species but which appear as potentially useful.

The choice of a reagent for a given substrate may be helped by the knowledge of the redox potential of both substrate and reagent. Unfortunately many values are scattered in the literature and determined using experimental conditions somewhat different from those required in an organic chemistry experiment (Clark, 1960).

There are two main groups:

(1) neutral species in which the reactive function is covalently bound to the polymer and
(2) ionic species fixed on ion exchange resins.

With the exception of *N*-halogenated amides (Section II.B) all the supported reagents decribed here are cross-linked with divinylbenzene. To avoid any damage to the polymer, reagents able to react with the benzene ring or at benzylic position must be discarded.

II. COVALENTLY BOUND REAGENTS

All these supported species share the common feature that after oxidation, the final polymer is again in the primitive form; hence it can be regenerated by an auxiliary inorganic oxidant:

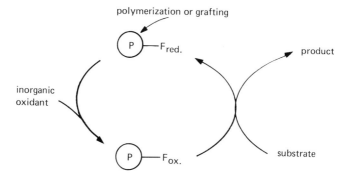

This important characteristic, if often mentioned, is hardly ever performed.

A. Quinones

Manecke *et al.* (1959) and Cassidy and Kun (1965) did pioneering work on oxidation–reduction polymers; the functional resins are obtained by (co)polymerization of vinyl monomers or by chemical modification of cross-linked styrene–divinylbenzene copolymers.

All materials thus obtained have been examined in detail for their physicochemical aspects, some analytical applications are also given (Cassidy, 1972), but useful procedures for organic chemistry are limited. The most used systems are reported here.

A cross-linked resin obtained by polycondensation of a mixture of

I

hydroquinone, phenol and formol is prepared by Manecke *et al.* (1964); the following oxidations are possible at acceptable rates:

ascorbic acid $\longrightarrow$ dehydroascorbic acid

cysteine $\longrightarrow$ cystine

tetraline $\longrightarrow$ naphtalene

amino acids $\longrightarrow$ aldehyde $+ CO_2 + NH_3$ (Strecker degradation)

A copolymer made with 2-isopropenylanthraquinone **II** was, at one

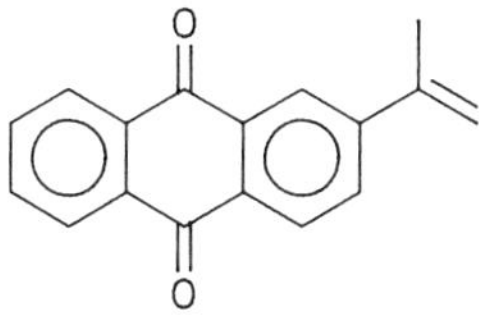

II

time, marketed by E. Merck (Manecke and Creuzburg, 1966) and exhibited the same properties as polymer **I**. A similar polymer obtained by copolymerization of 2-vinylanthraquinone, styrene and divinylbenzene is able to oxidize 1,4 dihydro- and 1,2,3,4-tetrahydrocarbazole into carbazole (Manecke, 1969).

B. N-Halogenated Amines, Amides, and Imides

Directly inspired from pyridine perbromide C_6H_5N,Br_2 (Mc Elvain and Morris, 1951), several copolymers of 2- or 4-vinylpyridine are brominated and the resulting polymers used as oxidizing or brominating reagents. Linear 2-polyvinylpyridine forms a $1:1$ complex when in contact with bromine vapor; the orange polymer thus obtained converts aliphatic, aromatic, and acid thiols into their corresponding disulfides (Christensen and Heacock, 1978) (Table II). The yields (not optimized) are above 80%.

Optically active linear and cross-linked copolymers of 4-vinylpyridine with $(-)$-menthylacrylate or $(-)$-menthylmethacrylate with different composition are complexed with bromine in solution (Chiellini et al., 1981).

TABLE II

Oxidation of Thiols to Disulfides
$2RSH + 2\text{-PVP, } Br_2 \rightarrow$
$RS\text{—}SR + 2\text{-PVP, HBr} + HBr$

R	R
$p\text{-}CH_3C_6H_5$—	$o\text{-}NO_2\text{—}C_6H_4$—
$C_6H_5CH_2$—	C_6H_5CO—
$p\text{-}ClC_6H_5$—	$n\text{-}C_3H_7$—
C_6H_5—	$HOCH_2CH_2$—
	$HOCO\text{—}CH\text{—}CH_2$—
	$\quad\quad\quad |$
	$\quad\quad\quad NH_2$

The resulting polymer **III** is used for the bromination of several racemic

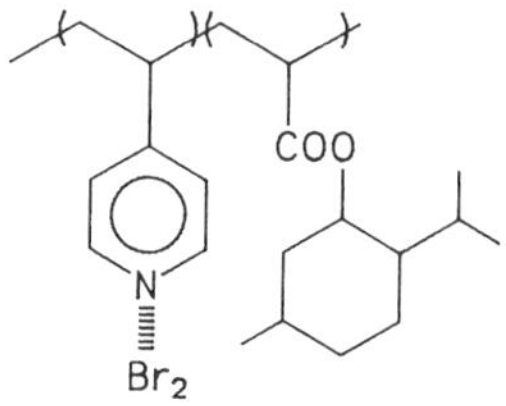

III

and prochiral alkenes:

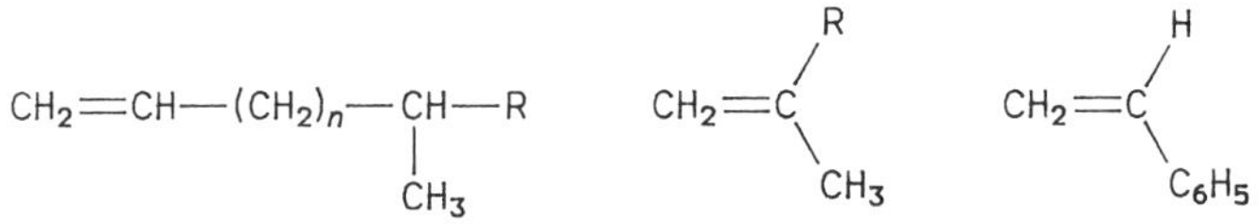

The yields of dibromo adducts exceed 80% after 2 h at room temperature (Chiellini *et al.*, 1985). Despite the fact that neither kinetic resolution nor asymmetric induction appear, polymer effect is evidenced in the regioselectivity and in the diastereoselectivity of bromination of several dienes including 4-vinylcylohexane, 1,3-pentadiene, and 1,4-hexadiene.

Several alkylsubstituted benzene and napthalene compounds are brominated on the side chain with the bromine complexes of cross-linked copolymers of styrene with 2- or 4-vinylpyridies (Sket and Zupan, 1986).

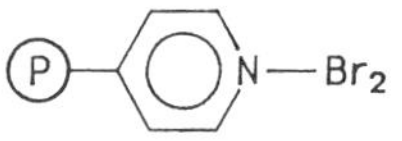

IV 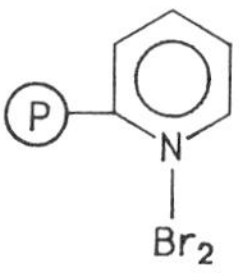   **V**

Because of obvious steric interactions with the polymer backbone, **IV** is more reactive than **V**; in both cases a radical initiator (dibenzoylperoxyde) is required. Some typical results are reported in Table III.

The selectivity in the monobromination of dimethylbenzenes is difficult to manage and results in mixtures of mono and dibromo derivatives.

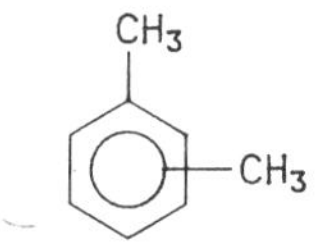

TABLE III

Side Chain Bromination of Alkyl-Substituted Molecules with **IV** under Reflux in CCl_4[a]

Substrate	Product	Yield(%)
Methylbenzene	(Bromomethyl)benzene	78
1-Methylnaphtalene	1-(Bromomethyl)naphtalene	63
2-Methylnaphtalene	2-(Bromomethyl)naphtalene	79
Ethylbenzene	1-Phenyl-1-bromoethane	81
1,2-Dimethylbenzene	1,2-Bis(bromomethyl)benzene	85
1,3-Dimethylbenzene	1,3-Bis(bromomethyl)benzene	75
1,4-Dimethylbenzene	1,4-Bis(bromomethyl)benzene	82
2,6-Dimethylpyridine	2,6-Bis(bromomethyl)pyridine	66
Hexamethylbenzene	1-(Bromomethyl)pentamethylbenzene	75

[a] From Sket and Zupan (1986). Reproduced with permission. Copyright 1986 American Chemical Society.

Allylic and benzylic bromination is performed by a polymeric N-bromo(phenyl-4-vinylphenyl)methane amine **VI** obtained after a whole cascade of transformations (Manecke and Stärk, 1975).

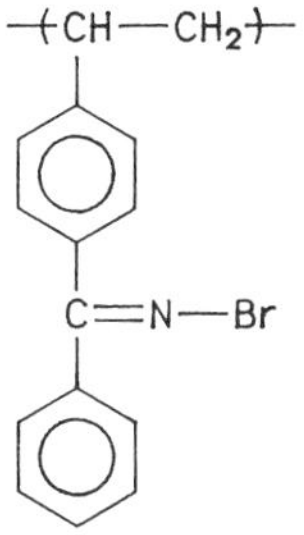

VI

Cyclohexene, toluene, and p-chlorotoluene are substituted with moderate yield after several hours in refluxing benzene. Though regeneration is possible, the preparation of the polymer requires a multistep procedure and the use of the pyridine–bromine complex as brominating agent is unwieldy.

More interesting are the N-halogenated polyamides obtained very simply from powdered commercially available Nylon 6-6, Nylon 6, and Nylon 3. They are used for the oxidation of alcohols to aldehydes or ketones or for the oxidation of sulfides to sulfoxides.

When treating Nylon 6-6 with t-butyl hypochlorite, extensive N-halogenation occurs giving polymer **VII** ; secondary alcohols give readily the

$$-NH-(CH_2)_6-NH-CO-(CH_2)_4CO- \longrightarrow \quad -N-(CH_2)_6-N-CO-(CH_2)_4-CO-$$
$$\begin{array}{cc} \quad\;\; | & \;\;\; | \\ \quad\;\; Cl & \;\;\; Cl \end{array}$$

VII

corresponding ketones whereas primary alcohols give less satisfactory results (Schuttenberg and Schultz, 1971) (Table IV).

Sulfides give mixtures of sulfoxides and starting material (Schuttenberg *et al.*, 1972, 1973); however, excellent results in oxidation of sulfide to sulfoxides are obtained with reagent **VII** in benzene–methanol mixtures (Sato *et al.*, 1972) (Table V).

Among the different polyamides examined, Nylon 3, Nylon 6, and Nylon 6-6, only the latter is chlorinated at the higher level regardless of the chlorinating agent being used (t-BuOCl, Cl_2O, or HClO) (Schuttenberg *et al.*, 1972).

Depending on the solvent used for the reaction, these chlorinated nylons, which are not cross-linked, dissolve in the reaction mixture but the dechlorinated reaction product precipitates, so the insoluble starting material can be separated.

Polymeric analogs of N-bromosuccinimide **IX** and N-chlorosuccinimide **X** (Filler, 1963) are prepared by halogenation of polymers of maleimide **VIII**: polymaleimide, maleimide–styrene soluble polymer, and maleimide–sytrene–divinylbenzene insoluble copolymer (Yanagisawa *et al.*, 1969). Benzylation of phenylmethanes: Φ-CH_3, Φ_2CH_2, and Φ_3CH in refluxing CCl_4 occurs with lower yield than that obtained with N-bromosuccinimide (50% instead of 80%); no allylic bromination is observed with cyclohexene but cyclohexanol can be oxidized to some extent in aqueous dioxane.

TABLE IV

Oxidation of Primary and Secondary Alcohols by Polymer **VII** in Benzene at 35°C[a]

Alcohol	Ketone	Aldehyde	Ester
Borneol	97		
Cyclohexane	82		
$C_6H_5CH_2CH(OH)CH_3$	90		
$C_6H_5(CH_2)_2CH(OH)CH_3$	95		
$C_6H_5CH(OH)CH_3$	95		
$C_6H_5CH_2OH$		95	0
$CH_3(CH_2)_6CH_2OH$		5	80
$C_6H_5-CH=CH-CH_2OH$		10	

[a] From Schuttenberg *et al.* (1971). Reproduced with permission. Copyright 1971 Hütig Verlag.

TABLE V

Oxidation of Sulfides to Sulfoxides with Polymer **VII** in
Benzene–Methanol at 35°C[a]

Sulfide	Reaction time (h)	Yield of sulfoxide (%)
$C_6H_5SCH_3$	0.5	97
$C_6H_5CH_2SCH_2C_6H_5$	0.5	100
$C_6H_5CH_2SC_6H_5$	0.5	100
$C_6H_5SC_6H_5$	15	97
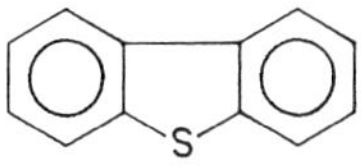	21	100
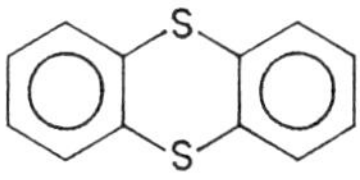	22	99
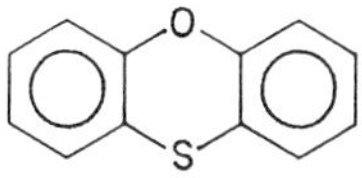	24	99

[a] From Sato *et al.* (1972). Reproduced with permission. Copyright
1972 Chemical Society of Japan.

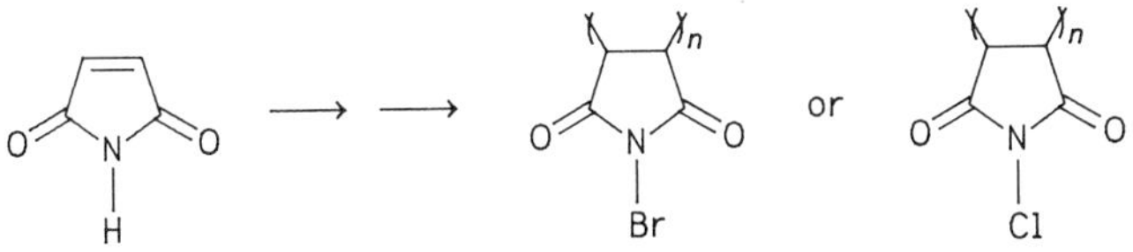

Some interesting polymer effects are shown by reagent **IX** in the
bromination of cumene (Yaroslawsky *et al.*, 1970);

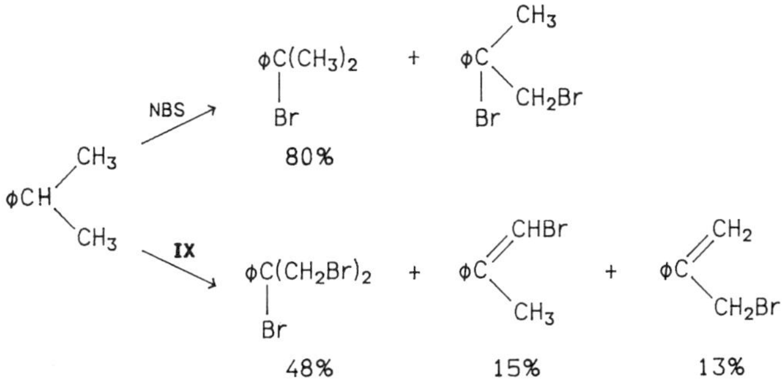

Several products are obtained whereas monomeric N-bromosuccinimide (NBS) gives mainly the monobrominated product. The polarity of the polymer medium (unlike that of CCl_4) causes dehydrobromination of the primary reaction product.

The difference is more striking with polymeric N-chlorosuccinimide (NCS): chlorination of ethylbenzene affords 100% of aromatic substitution whereas with monomeric NCS, only chain substitution occurs (Yaroslawsky and Katchalsky, 1972).

C. Sulfonium Halides

The oxidation of primary alcohols to aldehydes is achieved by a polymeric analog of thioanisole dichloride described previously (Corey and Kim, 1972). Macroreticular polystyrene is first brominated then treated with an excess of n-butyllithium followed by treatment with methyl disulfide to give the poly (p-methylmercaptostyrene) **XI**

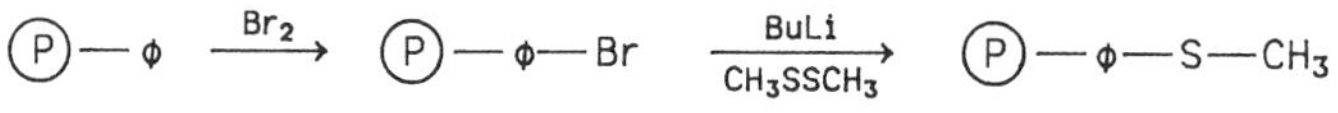

$$\text{(P)}-\phi \xrightarrow{Br_2} \text{(P)}-\phi-Br \xrightarrow[CH_3SSCH_3]{BuLi} \text{(P)}-\phi-S-CH_3$$

XI

$$\textbf{XI} + Cl_2 \longrightarrow \text{(P)}-\phi-\overset{CH_3}{\underset{Cl}{\overset{|}{S}{+}}} \quad Cl^-$$

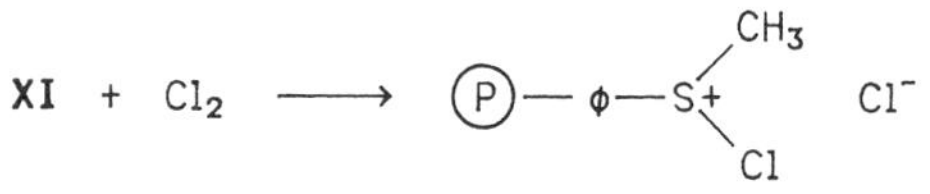

XII

The oxidizing species **XII** is generated *in situ* with a solution of chlorine in dichloromethane containing the alcohol to be oxidized (Crosby *et al.*, 1975). The results of the oxidation of several primary alcohols are given in Table VI.

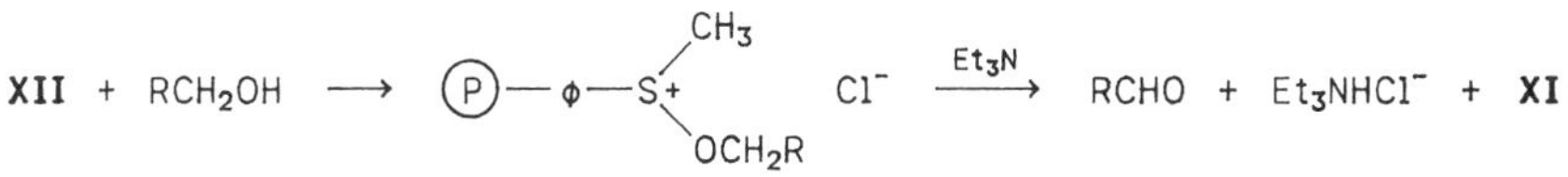

$$\textbf{XII} + RCH_2OH \longrightarrow \text{(P)}-\phi-\overset{CH_3}{\underset{OCH_2R}{\overset{|}{S}{+}}} \quad Cl^- \xrightarrow{Et_3N} RCHO + Et_3NHCl^- + \textbf{XI}$$

The reported yields are excellent but attempts at selective monooxidation of 1,7-heptanediol failed even with stoichiometric amounts of reagent.

D. Sulfur and Selenium Oxides

Sulfoxides, essentially DMSO, are used for the oxidation of halides and, to a smaller extent, of alcohols to the corresponding aldehydes. To avoid

TABLE VI

Oxidation of Alcohols with the Polymeric Sulfide Reagent **XII** and Chlorine[a]

Alcohol	Sulfide (equiv.)	Reaction time (h)	Yield (%)
1-Octanol	3.56	3	95
4-Phenylcyclohexanol	3.56	4	90
Benzyl alcohol	3.30	5	53
3-(*p*-Methoxyphenyl)propan-1-ol	5.0	4	94

[a] From Crosby *et al.* (1975). Reproduced with permission. Copyright 1975 American Chemical Society.

formation of foul-smelling volatile sulfide, polymer-bound sulfoxides are prepared in several ways.

Linear chloromethylated polystyrene is treated by ethylmercaptide to give the polymer sulfide **XIIa**;

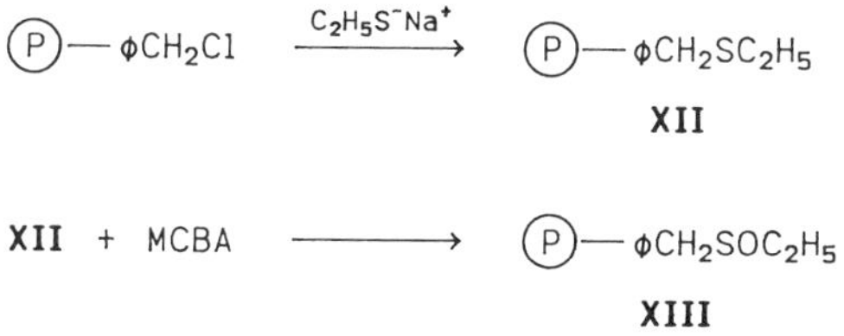
P — ϕCH2Cl
C2H5S⁻Na⁺
P — ϕCH2SC2H5
XII
XII + MCBA
P — ϕCH2SOC2H5
XIII

Oxidation with *m*-chloroperorybenzoic acid (MCBA) affords the sulfoxide **XIII** which converts benzyl chloride into benzaldehyde; the yield is 55% (Davies and Sood, 1983), and the soluble sulfide **XIIa** is recovered by precipitation.

Cross-linked polystyrenes with the pendant group —CH_2SOCH_3 (Janout *et al.*, 1984) are obtained by the following sequences:

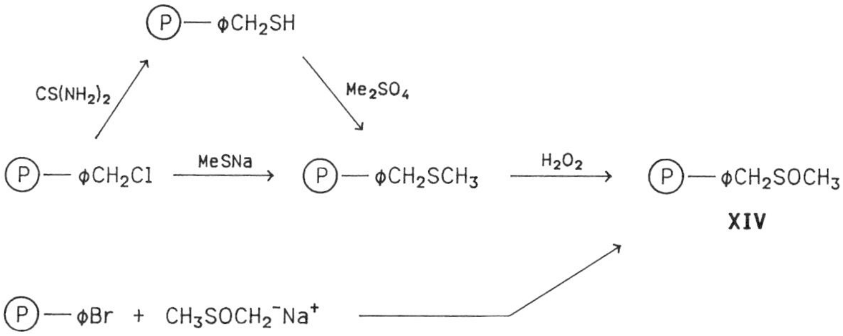
P — ϕCH2SH
CS(NH2)2
Me2SO4
P — ϕCH2Cl
MeSNa
P — ϕCH2SCH3
H2O2
P — ϕCH2SOCH3
XIV
P — ϕBr + CH3SOCH2⁻Na⁺

but used only as "solid solvents" in triphase catalysis.

Oxidation of iodooctane and benzyl alcohol by methylsulfoxide **XIV** occurs with yields of 66% (Foureys *et al.*, 1985). The polymer is obtained by a two-step modification of cross-linked chloromethylated polystyrene

using phase-transfer catalysis

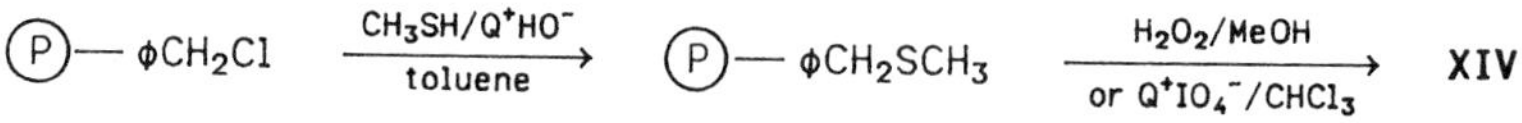

Selenium reagents are prepared from vinylselenide by copolymerization with DVB (Michels *et al.*, 1976).

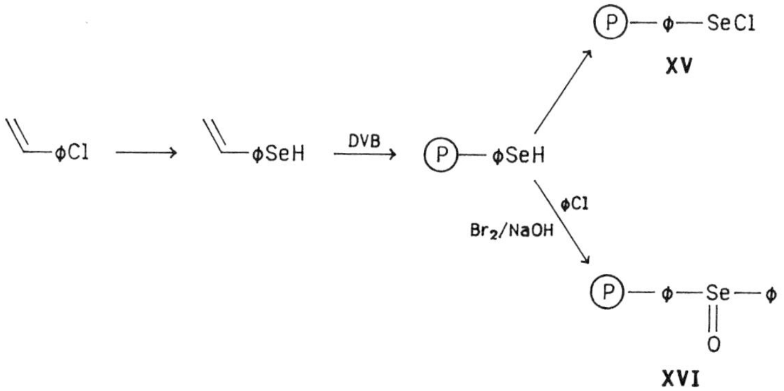

4-Methylcyclohexanone is converted into its α,β-unsaturated derivative by the chloride **XV**. The yield is 91%; the selenoxide **XVI** allows the transformation of 2-methyl-2-heptene into the corresponding glycol.

In connection with the versatile oxidizing abilities of bis(*p*-methoxy-phenyl) selenoxide, the polystyrene bound selenoxide **XVIII** is prepared from lithiated cross-linked polystyrene (Hu *et al.*, 1985) through selenide **XVII**.

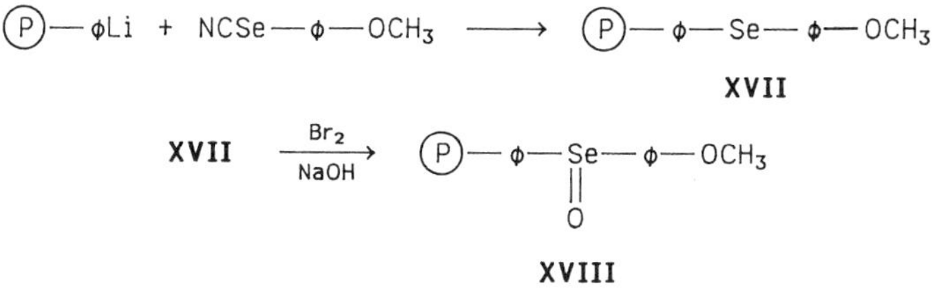

It readily oxidizes thiol to disulfide, sulfide to sulfoxide, phosphine to phosphine oxide, and hydroquinone to quinone under very mild conditions. The results are summarized in Table VII.

E. Amine Oxides

Oxidation of primary and secondary halides to aldehydes and ketones, respectively, is achieved by tertiary amine oxides (Frechet *et al.*, 1980a) formed by oxidation of polymer bound tertiary amines.

TABLE VII

Oxidations of Thiol, Sulfide, Phosphine, and Hydroquinone with Polystyrene-Bound Selenoxide **II** at Room Temperature[a]

Substrate	Solvent	Time (h)	Product	Yield[b] (%)
C_6H_5SH	CH_2Cl_2	1.5	$(C_6H_5S)_2$	95
$C_6H_5CH_2SH$	CH_2Cl_2	1.5	$(C_6H_5CH_2S)_2$	93
$n\text{-}C_{16}H_{33}SH$	CH_2Cl_2	3	$(n\text{-}C_{16}H_{33}S)_2$	97
$NH_2CH_2CH_2SH$	CH_2Cl_2	1.5	$(NH_2CH_2CH_2S)_2$	100
$HOCH_2CH_2SH$	CH_2Cl_2	1.5	$(HOCH_2CH_2S)_2$	97
$(C_6H_5CH_2)_2S$	CH_3COOH	3.5	$(C_6H_5CH_2)_2SO$	98
$(C_6H_5)_3P$	CH_2Cl_2	3.5	$(C_6H_5)_3PO$	87
p-Hydroquinone	CH_3COOH	12	p-Benzoquinone	63
3,5-Di-t-butyl catechol	CH_3COOH	4.5	3,5-Di-t-butyl o-Benzoquinone	96

[a] From Hu et al. (1985). Reproduced with permission. Copyright 1985 Chemical Society of Japan.
[b] Isolated yield.

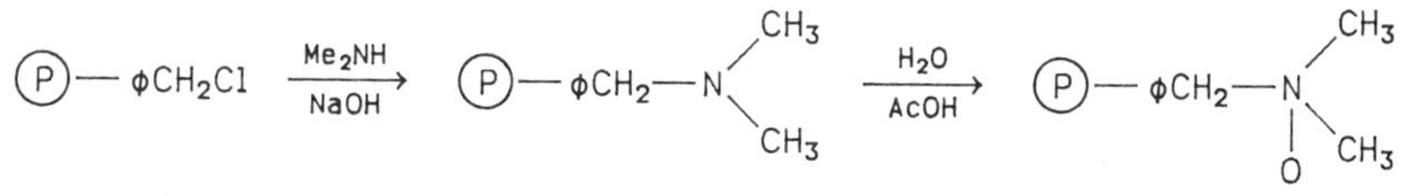

XIX   **XX**

The reactivity sequence is: primary iodide > secondary iodide ≈ primary bromide > secondary bromide; tosylates react at approximately the same rate as bromides (Table VIII).

TABLE VIII

Oxidation of Alkyl Halides and Tosylates with Amine Oxide[a]

Starting material	C=O (%)	C=C (%)	C—OH (%)
1-Iodoheptane	95	1	3
1-Bromoheptane	87	1	8
2-Bromoheptane	96	—	4
4-Tosylheptane	85	2	7
Benzyl chloride	95	—	2
1-Iodotetradecane	80	3	—
(1-Bromoethyl)benzene	77	20	1
(2-Bromoethyl)benzene	—	86	—
Bromocyclopentane	87	6	7

[a] Frechet et al. (1980a). Reproduced with permission. Copyright 1980 American Chemical Society.

The onium-type reagents mentioned in Sections II.C, II.D, and II.E may involve an alkoxy onium ion **XXI** as an intermediate:

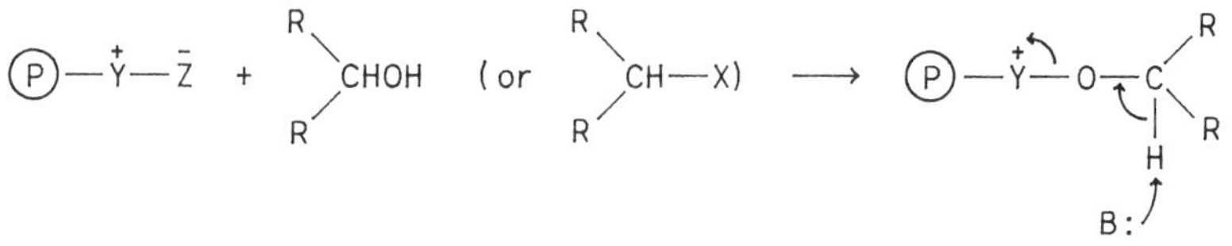

XXI

which undergoes proton abstraction to give the carbonyl derivative and the reduced form of the reagent (**XI**, **XII**, **XVII**, or **XIX**) which can be regenerated.

F. Peracids

Peracids are valuable reagents for the epoxiditation or hydroxylation of compounds with olefinic double bounds. Polyacrylic resins are transformed into the peroxy form with hydrogen peroxyde only when arenesulfonic groups are present in the polymer (Helfferich and Luten, 1964); copolymers of styrene and ethylacrylate are thus saponified and sulfonated.

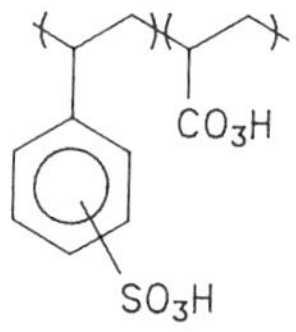

The percarboxylic groups are able to convert cyclohexene and butene into the corresponding glycols with moderate yields. Commercially available cross-linked polymethacrylic acid (Amberlite XE 69) can be oxidized directly with hydrogen peroxide (Takagi, 1967); satisfactory yields are obtained with several olefins (Table IX). Other substrates, sulfoxides, sulfides, and ketones, are also oxidized in sulfones, esters, and lactones (Takagi, 1975).

The more stable perbenzoic acids have been the subject of larger investigations than the percarboxylic acids. Cross-linked chloromethylated polystyrene is the common starting material of several perbenzoic resins **XXIII**;

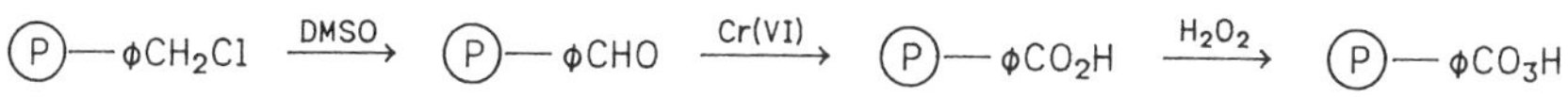

TABLE IX

Epoxidation[a] of Olefins with
Peracid Type Resins[b]

Substrate	Yield (%)
Cyclohexene	85
1-Pentene	44
2-Pentene	85
Undecylenate	44
Oleate	73
Styrene	30
trans-Stilbene	5

[a] Epoxidation carried out with 2–3 equivalents of peracid in dioxanne at 40°C.

[b] From Takagi (1967). Reproduced with permission. Copyright 1967 John Wiley and Sons, Inc.

Oxidation with DMSO and chromic acid gives the benzoic acid resin which is oxidized with hydrogen peroxide in the presence of methane- or toluenesulfonic acids (Frechet and Haque, 1975).

The polymer bound perbenzoic acid can be obtained also from the aldehydic form **XXII** directly with hydrogen peroxide (Harrison and Hodge, 1976); epoxidation of cyclohexene occurs with 86% yield. Swellable resins exhibit better mechanical resistance though their capacity drops after several reduction–oxidation cycles (Frechet and Haque, 1975) but with high contents of perbenzoic acid, the epoxidation of olefins can proceed readily at 40°C (Harrison and Hodge, 1976). The reactivity parallels that of homogeneous reagents and yields are fair with di-substituted and trisubstituted olefins. Some results are reported in Table IX.

Various sulfides, tetrahydrothiophene, methionine, penicillins, and cephalosporins, are oxidized to sulfoxides and sulfones in high yields (Harrison and Hodge, 1976); with highly cross-linked polymers the oxidation of sulfides occurs to a smaller extent (Greig *et al.*, 1980).

A persulfonic acid, derived from Amberlyst A-15 treated with hydrogen peroxyde, epoxidizes quinols (Jefford *et al.*, 1985).

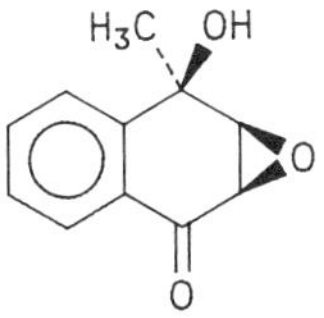

G. Chromium Trioxide

Dark red resins are obtained by fixing chromium trioxide on various copolymers of styrene and 4-vinylpyridine (Brunelet and Gelbard, 1983)

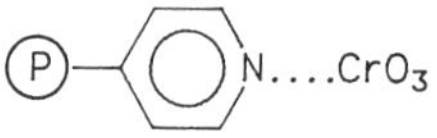

$$\text{(P)}-\langle\bigcirc\rangle\text{N} \ldots \text{CrO}_3$$

1-Octanol, 2-octanol, cyclohexanol, and benzyl alcohol are oxidized with good to excellent yields to the parent carbonyl compounds in refluxing cyclohexane. Macroporous polymers give the most efficient reagents.

III. IONIC SPECIES ON ION EXCHANGE RESINS

Ion exchange resins are the first polymer beads marketed in a large scale for water treatment (Helfferich, 1962). They are well designed to fix a great number of inorganic anions and cations widely used as oxidizing salts. Table X presents the redox potentials of several metallic cations, inorganic anions and organic dyes.

The latter, which are redox indicators, might be used as reagents in their oxidized form, or, as they are reversible, in a catalytic way by coupling with an external oxidant.

TABLE X

Redox Potentials of Significant Metallic Cations, Organic Anions, and Dyes[a]

Tl^{3+}	2.2	Ferroine	1.14
FeO_4^{2-}	1.9	p-Nitrodiphenylamine	1.06
MnO_4^-	1.7	Porphyrexide	0.97
Pb^{4+}	1.46	ClO^-	0.90
Ce^{4+}	1.4	Porphyrindine	0.81
Porphyrexide	1.34	Ag^+	0.79
$Cr_2O_7^{2-}$	1.3	Fe^{3+}	0.77
Terpyridine	1.25	BrO_3^-	0.61
Nitroferroine	1.25	Hg^{2+}	0.55
$HCrO_4^-$	1.2	$Fe(CN)_6^{3-}$	0.46
ClO_3^-	1.2	Cu^{2+}	0.16
IO_3^-	1.2		

[a] Related to hydrogen electrode; compiled from "CRC Handbook of Chemistry and Physics," 58th Ed. CRC Press, Boca Raton, Florida, 1978.

Only some of these ionic species are reported as supported reagents; the purpose of this table is to emphasize a wider range of use, especially since the polymers are commercially available.

A. Metallic Cations

The only known example concerns the oxidation of ascorbic acid with a strong acidic resin in the Fe^{3+} form (Inczedy, 1962).

B. Inorganic Anions

1. Iodate, Periodate, and Mesodiperiodate

The iodate and periodate form of commercial macroporous ion exchange resins

$$\text{(P)} - \phi CH_2 \overset{+}{N}(CH_3)_3 \ X^-, \qquad \begin{array}{l} X^- = IO_3^- \\ \\ X^- = IO_4^- \end{array}$$

are prepared by standard ion exchange procedures (Harrison and Hodge, 1982). 1,2-diols are cleaved into aldehydes; quinols and sulfoxides are oxidized to quinones and sulfones, respectively. Typical results are given in Table XI.

The cleavage of diols, that of the ribose ring of some nucleosides in particular is achieved by similar reagents (Gibietis et al., 1983, 1984). The

TABLE XI

Oxidation with Polymer-Supported Iodate and Periodate[a]

Substrate	Refluxing solvent	Product (% yield)
Quinol	$CHCl_3$	Quinone (86)
2-Methoxyquinol	$CHCl_3$	Quinone (93)
2-Chloroquinol	$CHCl_3$	Quinone (96)
4-t-Butylcatechol	CH_2Cl_2	Quinone (99)
1,2-Dihydroxynapthalene	$CHCl_3$	Quinone (96)
Cyclohexane-1,2-diol	$CHCl_3$	Adipaldehyde (90)
2-Phenylpropane-1,2-diol	$CHCl_3$	Acetophenone (86)
1-Phenylethane-1,2-diol	$CHCl_3$	Benzaldehyde (97)
Dibenzylsulfide	MeOH	Sulfoxide (99)
Thioanisole	MeOH	Sulfoxide (81)
Triphenylphosphine	$CHCl_3$	Phosphine oxide (100)

[a] From Harrison et al. (1982). Reproduced with permission. Copyright 1982 The Royal Society of Chemistry.

cleavage of 1,2-diphenylethane-1,2-diol in benzaldehyde is quantitative with the supported mesodiperiodate anion $I_2O_9^{4-}$ (Villemin and Ricard, 1982).

2. Perhalides

Beside covalent N-halogenated compounds (Section II.B), ionic perhalides X_3^- are found useful for halogenation of olefinic double bounds and carbonyl compounds. Polypyridinium tribromide **XXIV** is prepared from a cross-linked copolymer of 4-vinylpyridine (Frechet et al., 1977; Zupan and Sket, 1982).

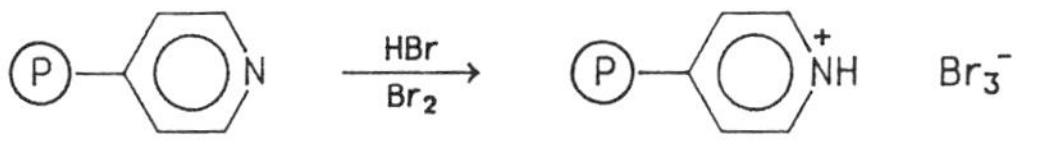

XXIV

Quaternary ammonium tribromide **XXV** is obtained from Amberlyst A-26 in the Br^- form (Cacchi et al., 1979; Bongini et al., 1980).

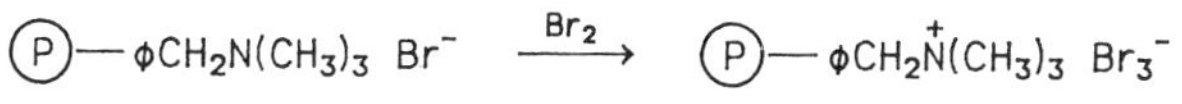

XXV

Quaternary phosphonium tribromide **XXVI** derives from bromomethylated polystyrene (Akelah et al., 1983, 1984).

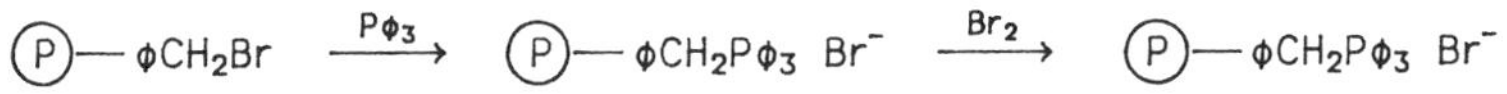

XXVI

Olefins, styrene, cis-stilbene, and 1,7-octadiene, are brominated in excellent yields at room temperature with polymer **XXIV** (Frechet et al., 1977).

The bromination of aldehydes and ketones occurs in the α position, selected examples of the efficiency of the different supported tribromides are reported in Table XII. The reaction requires generally mild conditions (room temperature or refluxing methylene chloride).

3. Chromates and Complex Chromates

Several Cr(VI) compounds in the form of chromate, dichromate, and complex mono- and dichromates are fixed on various types of polymeric beads. Polyvinylpyridine resins are impregnated by percolation or by soaking with an acidic solution of chromium trioxide; quaternary ammonium resins (mostly Amberlyst A-26) are impregnated by ion exchange procedures. The resins thus prepared are shown in Table XIII.

TABLE XII

Bromination of Aldehydes and Ketones with Tribromides

Substrate	Reagent	Product	Yield (%)
$\phi COCH_3$	**XIV**	$\phi COCH_2Br$	99
$\phi COCH_3$	**XV**	$\phi COCH_2Br$	55
$\phi COCH_3$	**XVI**	$\phi COCH_2Br$	75
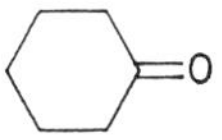 cyclohexanone =O	**XIV**	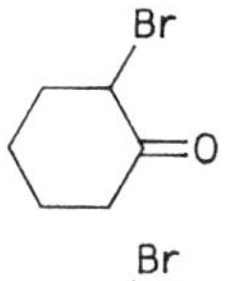 2-bromocyclohexanone	100
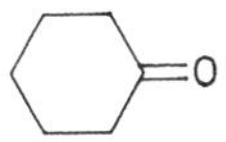 cyclohexanone =O	**XV**	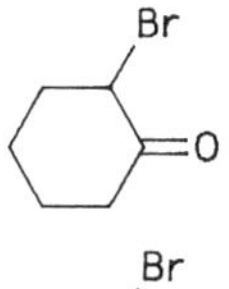 2-bromocyclohexanone	90
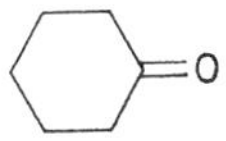 cyclohexanone =O	**XVI**	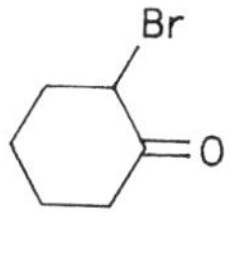 2-bromocyclohexanone	90
$C_4H_9CH_2CHO$	**XV**	$C_4H_9CHBrCHO$	95
$C_5H_{11}CH_2COCH_3$	**XV**	$C_5H_{11}CHBrCOCH_3$	70
CH_3CH_2CHO	**XVI**	$CH_3CHBrCHO$	75

TABLE XIII

Supported Chromates and Complex Chromates

Polymer-supported chromate			References
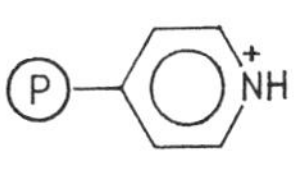 (P)—⟨O⟩NH⁺	$ClCrO_3^-$	**XXVII**	Frechet (1978)
	$(Cr_2O_7^{2-})1/2$	**XXVIII**	Frechet (1980b)
	$CF_3CO_2CrO_3^-$	**XXIX**	Brunelet and Gelbard (1986)
(P)—ϕ—$CH_2\overset{+}{N}Me_3$	$HCrO_4^-$	**XXX**	Cainelli (1976)
	$Cl(CrO_3)_n^-$	**XXXI**	Brunelet and Gelbard (1986)
	$CF_3CO_2CrO_3^-$	**XXXII**	Brunelet and Gelbard (1986)

The reagents are remarkably effective in oxidizing primary and secondary alcohols to aldehydes and ketones. The oxidation is clean and no trace of acids or by-products is detected in the reaction mixtures, but the reaction is rather slow and an excess of reagent is required for completion. Some comparative results are reported in Table XIV. Allylic and benzylic halides are also converted into carbonyls with polymer **XXX**.

TABLE XIV

Oxidation of Acohols by Supported
Chromate Anions

Substrate	Reagent	Yields (%)
Octan-1-ol	XXX	94
	XXXI	80
Octan-2-ol	XXIX	96
	XXXI	92
	XXXII	90
Cyclohexanol	XXVII	94
	XXVIII	76
	XXX	77
	XXXI	90
Benzylalcohol	XXVII	95
	XXVIII	96
	XXX	96
	XXXI	73
Cinnamylalcohol	XXVII	100
	XXVIII	98
	XXX	95
Menthol	XXX	86
	XXXI	98

IV. PROSPECTIVE CONSIDERATIONS

As most of the aforementioned supported reagents are closely related to monomeric soluble species, their advantages lie not only in the case of separation of excess reagent, spent reagent and product: unexpected selectivities are observed due to polymer effects yet to be explained. Supported oxidizing reagents appear as species with controlled reactivity; efficiency can be managed by both the structure of the polymer and the nature of the bound function.

According to the scheme (see p. 216) the external inorganic oxidant is transformed into a supported reagent which is *less* reactive but exhibits a *higher selectivity*. This precludes the use of the polymer as a catalyst involved in a cycle derived from this scheme unless a triphase catalysis is possible with some anions.

Discontinuous batch processes are proned to be more widely exploited provided that reagents with high redox potentials are used, i.e., halogens, Cr(VI) which are then put in a non-noxious state. This allows, simultaneously, safe handling, high reactivity and good selectivity.

REFERENCES

Akelah, A., Hassanein, M., and Abidel-Galil, I. (1983). *Polym. Prepr., Am. Chem. Soc., Div. Polym. Chem.* **24**, 467–468.

Akelah, A., Hassanein, M., and Abidel-Galil, I. (1984). *Eur. Polym. J.* **20**, 221–223.

Bongini, A., Cainelli, G., Contento, F., and Manescalchi, F. (1980). *Synthesis* 143–146.

Brunelet, T., and Gelbard, G. (1983). *Nouv. J. Chim.* **7**, 483–490.

Brunelet, T., Jouitteau, C., and Gelbard, G. (1986). *J. Org. Chem.* **51**, 4016.

Cacchi, S., Caglioti, L., and Cernia, E. (1979). *Synthesis* 64–66.

Cainelli, G., Cardillo, G., Orena, M., and Panunzio, M. (1976). *J. Am. Chem. Soc.* **98**, 6737–6738.

Cassidy, H. G. (1972). *J. Polym. Sci., Part D* **6**, 1–58.

Cassidy, H. G., and Kun, K. A. (1965). "Oxidation-Reduction Polymers," Wiley (Interscience), New York.

Chiellini, E., Solaro, R., and D'Antone, S. (1981). *Chim. Ind. (Milan)* **63**, 512–515.

Chiellini, E., Callaioli, A., and Solaro, R. (1985). *React. Polym.* **3**, 357–368.

Christensen, L. W., and Heacock, D. J. (1978). *Synthesis* 50–51.

Clark, W. M. (1960). "Oxidation Reduction Potentials of Organic Species," Williams & Wilkins, Baltimore, Maryland.

Corey, E. J., and Kim, C. U. (1972). *J. Am. Chem. Soc.* **94**, 7586.

"CRC Handbook of Chemistry and Physics," 58th ed. CRC Press, Boca Raton, Florida, 1978.

Crosby, G. A., Weinshenker, N. M., and Uh, H. S. (1975). *J. Am. Chem. Soc.* **97**, 2232.

Davies, J. A., and Sood, A. (1983). *Makromol. Chem. Rapid Commun.* **4**, 777–782.

Filler, R. (1963). *Chem. Rev.* **63**, 21.

Foureys, J. L., Hamaide, T., Yaacoub, E., and Le Perchec, P. (1985). *Eur. Polym. J.* **21**, 221.

Frechet, J. M. J., and Haque, K. E. (1975). *Macromolecules* **8**, 130–134.

Frechet, J. M. J., Farral, M. J., and Nuyens, L. J. (1977). *J. Macromol. Sci., Chem.* **A11**, 507–514.

Frechet, J. M. J., Warnock, J., and Farral, M. J. (1978). *J. Org. Chem.* **43**, 2618.

Frechet, J. M. J., Darling, G., and Farral, M. J. (1980a). *Polym. Prepr., Am. Chem. Soc., Div. Polym. Chem.* **21**, 270–271.

Frechet, J. M. J., Darling, P., and Farral, M. J. (1980b). *Polym. Prepr., Am. Chem. Soc., Div. Polym. Chem.* **21**, 272–273.

Gibietis, J., Zicmanis, A., Spince, B., and Berzina, A. (1983). *Latv. PSR Zinat. Akad. Vestis* 104–110 [C.A. **98**, 179815t (1983)].

Gibietis, J., Zicmanis, A., Lipsberg, I., and Spince, B. (1984). *Latv. PSP Zinat. Akad. Vestis* 233–238 [C.A. **100**, 130661F (1984)].

Greig, J. A., Hancock, R. D., and Sherrington, D.C. (1980). *Eur. Polym. J.* **16**, 293–298.

Harrison, C. R., and Hodge, P. (1976). *J.C.S. Perkin I* 605–609; 2252–2254.

Harrison, C. R., and Hodge, P. (1982). *J.C.S. Perkin I* 509–511.

Helfferich, F. (1962). "Ion Exchange." McGraw-Hill, New York.

Helfferich, F., and Luten, D. B. (1964). *J. Appl. Polym. Sci.* **8**, 2899–2908.

Hu, N. X., Aso, Y., Otsubo, T., and Ogura, F. (1985). *Chem. Lett.* 603–606.

Hudlicky, M. (1977). *J. Chem. Educ.* **54**, 100–106.

Inczedy, J. (1962). *Z. Chem.* **2**, 302–305.

Janout, V., Kahovec, J., Hrudovká, H., Svéc, F., and Cefelin, P. (1984). *Polym. Bull.* **11**, 215–221.

Jefford, C. W., Bernardinelli, G., Rossier, J. C., Kohmoto, S., and Boukavalas, J. (1985). *Tetrahedron Lett.* **26**, 615–618.

Mc Elvain, S. M., and Morris, L. R. (1951). *J. Am. Chem. Soc.* **73**, 206.

Manecke, G., and Creuzburg, K. (1966). *Makromol. Chem.* **93**, 271.

Manecke, G., and Stärk, M. (1975). *Makromol. Chem.* **176**, 285–297.

Manecke, G., Bahr, C., and Reich, C. (1959). *Angew. Chem.* **71**, 646–650.

Manecke, G., Kossmehl, G., and Singer, S. (1964). *Naturwissenschaften* **51**, 336.

Manecke, G., Kossmehl, G., Gawlik, R., and Hartmich, G. (1969). *Angew. Makromol. Chem.* **6**, 89.

Michels, R., Kato, M., and Heitz, W. (1976). *Makromol. Chem.* **177**, 2311–2320.

Sato, Y., Kunieda, N., and Kinoshita, M. (1972). *Chem. Lett.* 1023–1026.

Schuttenberg, H., and Schultz, R. C. (1971). *Angew. Makromol. Chem.* **18**, 175–182.

Schuttenberg, H., Klump, G., Kaczmar, U., Turner, R., and Schultz, R. C. (1972). *Polym. Prepr., Am. Chem. Soc., Div. Polym. Chem.* **13**, 866–871.

Schuttenberg, H., Klump, G., Kaczmar, V., Turner, R., and Schuttz, R. C. (1973). *J. Macromol. Sci., Chem.* **A7**, 1085–1095.

Sket, B., and Zupan, M. (1986). *J. Org. Chem.* **51**, 929.

Takagi, T. (1967). *Polym. Lett.* **5**, 1031–1035.

Takagi, T. (1975). *J. Appl. Polym. Sci.* **19**, 1649–1662.

Taylor, R. T. (1986). *ACS Symp. Ser.* No. 308, 132–154.

Villemin, D., and Ricard, M. (1982). *Nouv. J. Chim.* **6**, 605–607.

Wiberg, K. B., and Trahanowsky, W. S. (1965–1982). "Oxidation in Organic Chemistry," Vols. 5A–5D. Academic Press. New York.

Yanagisawa, Y., Akiyama, M., and Okawara, M. (1969). *Kogyo Kagaku Zasshi* **72**, 1399–1402 [*C.A.* **71**, 113410t (1969)].

Yaroslawsky, C., and Katchalsky, E. (1972). *Tetrahedron Lett.* 5173–5174.

Yaroslawsky, C., Patchornik, A., and Katchalsky, E. (1970). *Tetrahedron Lett.* 3629–3632.

Zupan, Y. J., and Sket, B. (1982). *J. Chem. Soc. Perkin Trans.*, p. 2059.

14 POLYMER-SUPPORTED REDUCTIONS

Joseph Lieto and Hmaïd Marrakchi

Ecole Supérieure d'Ingéniérie
  de Pétroléochimie et de Synthèse Organique Industrielle
Centre de Saint-Jérôme
13397 Marseille Cédex 13, France

I. INTRODUCTION

The literature dealing with polymer-supported reductions is not very abundant as compared to that dealing with polymer-supported oxidations (see Chapter 13). Nevertheless, it demonstrates the usefulness of polymers as supports for reduction reactions. This chapter is not intended to be a comprehensive review of reductions activated by polymer-supported reagents and/or catalysts. General reviews, available in the literature, provide such information (Allum *et al.*, 1975; Chauvin *et al.*, 1977; Gates, 1980; Gates and Lieto, 1980, 1985; Hodge and Sherrington, 1980; Chiellini and Solaro, 1981; Kanego and Tsuchida, 1981; Mathur *et al.*, 1980; Pittman, 1982; Lieto *et al.*, 1983). This chapter concentrates instead on the activity–polymer matrix relationship through selected examples. The second section presents a general comparison of polymer-supported and soluble reductions. The third section is devoted to a brief review of the syntheses and characterization methods for polymeric materials. Polymer-bound reducing agents and polymer-bound catalysts are then presented in two different sections. Finally, some future opportunities offered by polymer-supported reductions are discussed.

235

II. POLYMER-SUPPORTED REDUCTION VERSUS
SOLUBLE REDUCTION

An optimal industrial process is one in which all reactants are transformed into desired products with 100% yield and 100% selectivity using an economical and durable equipment. Often the transformation of reactants into products is not complete. Furthermore, the products are contaminated with side products. The chemical nature of reactants and products can also lead to security and corrosion problems.

A process development engineer must take into account all these problems and can take advantage, today, of the opportunities offered by supported reagents and catalysts to solve them.

A. Technological Arguments

Some processes involve very expensive metallic catalysts: it is absolutely necessary to recover the precious metal. With a soluble catalyst, metal recovery might imply energy-consuming distillation and/or evaporation. It becomes unwarranted if the precious metal is anchored on a solid, i.e., if the catalyst belongs to a separate phase from the mixture of reactants and products. Thus, using a support makes separations considerably easier.

With easier recovery of the catalyst, its regeneration is simplified. Furthermore, the reactor design can be simplified. Very simple reactor configurations such as fixed-bed reactors can be applied; smaller equipment can often be used because there is no longer a need for solvent volume for catalyst dilution. Dispersion of the active species is ensured by the support itself. Very often soluble catalyst reactions are developed at laboratory scale in a stirred reactor. When the process is extrapolated, the power input per unit volume of the laboratory-stirred reactor may require expensive engines. Process feasibility can be achieved using supported catalysts.

Corrosion problems can also be partly solved by use of a support. Corrosion results from contact between the chemicals (reactants or products) and the equipment walls. Such contacts are decreased by fixation of the corrosive species on a support. Likewise, toxicity is less when toxic species are supported.

B. Chemical Arguments

The stability of a reactive species can be increased on fixation. Some chemical species have the tendency to react with themselves. Such degradation may be stopped by anchoring the active species on supports.

A chemical reaction results from contact between reactants. With supported reactants, the active species are embedded in a nearly inert polymeric matrix. The mutual accessibility of the supported species is therefore limited (Challa, 1983). There are two possible consequences:

(1) The support might become an obstacle to mass transfer. Chemical kinetics are masked by physical kinetics (diffusion), and the reaction rate is limited by the mass transfer rate.

(2) The support acts as a filter for the reactants. Only the molecules with the right size and shape have access to the supported reactants. This applies mostly to asymmetric reactions, to which we shall return.

Anchoring species onto a support may lead to special selectivity associated with the physical structure of the support which forces a definite geometry to be respected by the active species. Stabilization of reaction intermediates and their *in-situ* identification are eased with polymer-supported reactions. This stabilization facilitates the trapping of labile species, which may shed light on some reaction mechanisms.

For completeness sake, disadvantages of polymer-supported reactions should be mentioned. Besides introduction of diffusional problems, the lack of thermal and mechanical stability is the main drawback of polymeric supports. Polymeric beads can be obtained which can sustain the weight of packed columns; however, these beads cannot resist the mechanical stress of the stirrer, and they are crushed into fine powder which can cause clogging. The thermal stability is limited to that of classical organic compounds, polytetrafluoroethylene being an exception.

An important problem with polymer-supported reactions is active species leaching. It is of utmost importance to check the catalyst mass balance and the stability of the bonding between support and active species.

III. POLYMERIC MATERIALS USED FOR POLYMER-SUPPORTED REDUCTIONS

The most usual support for polymer-supported reduction (more generally for polymer-supported reactions) is poly(styrene–divinylbenzene) because of its industrial use. Functionalized polymers are prepared by copolymerization of functionalized monomers or by functionalization of poly(styrene–codivinylbenzene). Figures 1 and 2 summarize the most common methods for introducing functional groups onto the aromatic ring.

Polymers with different forms and with different distributions of a variety of ligands can be obtained. Polymers can be prepared as macroporous beads, gels, or membranes. Each type can incorporate ligands with a block or a random distribution. These properties can be used as a basis for catalyst design; block ligand distribution can induce multisubstitution of

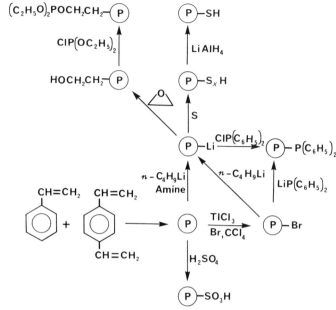

Fig. 1. Synthetic routes for functionalization of poly(styrene–divinylbenzene). [Reprinted with permission from Lieto *et al.* (1983). Copyright 1983 American Chemical Society.]

the catalyst. A random ligand distribution is associated with a monosubstitution of the active species.

A number of organic polymers have been used as catalyst supports, including polybutadiene, polyvinyl alcohol, polyvinylchloride, polyvinylpyridine, and polyamides. Some of these polymers are listed in Table I.

A. Catalytic Group Attachment

Several methods are available for introducing catalytic metal species in polymers. They require an appropriate swelling agent for the support which dissolves the reagents and allows access of the reagent to the functional group inside the polymer.

Ligand exchange or association reactions can be carried out. Occasionally a satisfactory alternative is *in-situ* synthesis of the supported complex based on the synthesis of the molecular analog (Lieto *et al.*, 1980). These metal fixation reactions can be schematized as follows:

$$\text{complex } L + \text{(P)}\!-\!L' \longrightarrow \text{(P)}\!-\!L' \text{ complex} + L$$

$$\text{unsaturated complex} + \text{(P)}\!-\!L' \longrightarrow \text{(P)}\!-\!L' \text{ complex}$$

$$\text{complex precursor} + \text{(P)}\!-\!L' \longrightarrow \text{(P)}\!-\!L' \text{ complex}$$

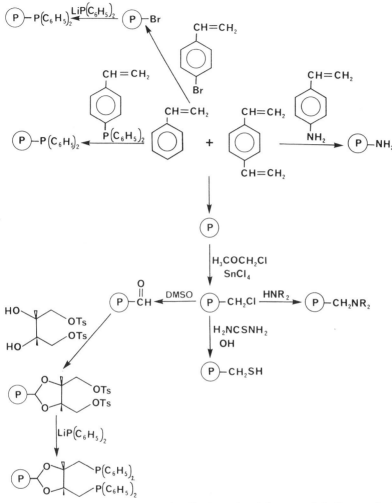

Fig. 2. Synthetic routes for functionalization of poly(styrene–divinylbenzene). [Reprinted with permission from Lieto *et al.* (1983). Copyright 1983 American Chemical Society.]

B. Catalyst Characterization

When the active moities are embedded in the polymer matrix, it is important to determine whether the supported species are indeed the desired ones. Most identification methods are based on the stoichiometry of the synthesis reaction and on comparison of spectra of the supported species with those of known compounds. The most commonly used method

TABLE I

Some Organic Polymers Used as Reduction Catalyst Supports[a]

Polymer	Ligand	Metal bonded to the ligand	Reference
Polyvinyl alcohol	$-O-PR_2$	Co, Rh, Re	Allum and Hancock (1972)
Polymetacrylate	$-\overset{\displaystyle \mid}{\underset{\displaystyle O-O}{C}}-NH$	Pd	Petrov et al. (1968)
Polyamide	$-\overset{\displaystyle \mid}{\underset{\displaystyle -O}{C}}-NH$	Rh, Pd, Pt	Bernard et al. (1975)
Polyacrylonitrile	$-CN$	Pd	Tyurenkova and Chimarova (1970)
Polyvinylpyridine	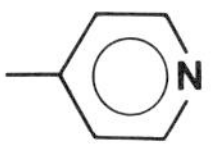 (pyridine ring) N	Fe, Co	Agnew (1976)
Polyvinylferrocene	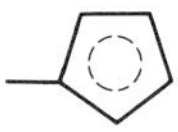 (cyclopentadienyl ring)	Fe	George and Hayes (1973)

[a] Adapted from Chauvin et al., 1977.

is infrared spectroscopy. Fourier transform spectrophotometers allow rapid multiple scanning and substraction of absorption associated with the support. Extended x-ray absorption fine structure (EXAFS), solid-phase NMR, and XPS (x-ray) photoelectron spectroscopy) also serve to characterize polymer-supported species (Lieto et al., 1983) (see Chapters 7–9).

Characterization of the active species in the working state is important. Therefore, identification methods should offer the opportunity of monitoring in-situ the transformation of the catalytic species.

IV. POLYMER-BOUND REDUCING REAGENTS

A. Ionic Reducing Agents

These agents are readily available from ion exchange resins widely produced commercially. Quaternary ammonium groups can be incorporated in poly(styrene–divinylbenzene) matrix producing strong basic anions exchange resins **I**.

Resin functionalized with hydroxy groups can also be used to produce, through reaction with the conjugated acid corresponding to the anion, an anion exchange resin **II**.

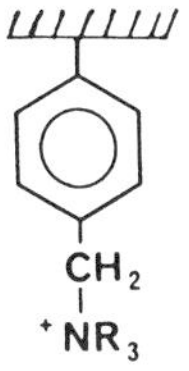

Common cation exchange resins present either sulfonic acid or carboxylic groups. Titanous or chromous forms of cation exchange resins have been used to reduce organic substrates by passing the reactant solubilized in an appropriate solvent through a column packed with the resin. For example, supported Ti^{3+} ions reduced 1,4-naphtoquinone (in ethanol) into 1,4-dihydroxynaphtalene with a 90% yield at 60°C (Dahbor, 1958).

Metal ions can also be supported on polymers. One example involves an anionic triiron cluster $HFe_3(CO)_{11}^-$ supported on polymer incorporating ammonium groups. This supported cluster is able to reduce nitrobenzene into aniline under very mild conditions ($T = 20$°C, $P = 1$ atm) (Nguini Effa et al., 1983).

Sodium borohydride has been attached on ion exchange resins (Gibson and Bailey, 1977). Beads were made with an ionic bond between quaternary cations and BH_4^-. The insoluble beads readily reduced aldehydes but were less active than the soluble $NaBH_4$. This type of supported reducing agent is now commercially available (Ventron Corporation).

Ion-exchange-resin bound cyanoborohydride provides a convenient reagent for a variety of reductions. Reductive amination, amine dimethylation, reduction of conjugated enones to allylic alcohols, deshalogenations, and conversion of pyridinium ions into tetrahydro derivatives have been performed (Hutchins et al., 1978). The supported reductions were slower than their soluble analogs, and higher temperatures were necessary to obtain adequate conversions. The resin may be reused by rinsing the spent material with dilute acid followed by regeneration.

Polymer-supported cyanide and thiocyanate were used to prepare nitriles, thiocyanates, and isothiocyanates. The functionalized polymers were prepared by treating the chloride forms of a resin incorporating quaternary ammonium groups with aqueous KCN or KSCN (Harrison and Hodge, 1980). Sodium dithionite immobilized on ion exchange macroporous resins has been used as reducing agent for various aldehydes and ketones (Chalmin Louis-André, 1985). However, the poor stability of the supported ammonium groups led to some extent of degradation of the supported reagents.

B. Covalently Bound Reagents

1. Reduction of Aldehydes and Ketones

Metal hydrides can be embedded in polymeric matrices to produce solids that reduce carbonyl compounds. Preparation of such a supported reducing agent is illustrated in the following scheme

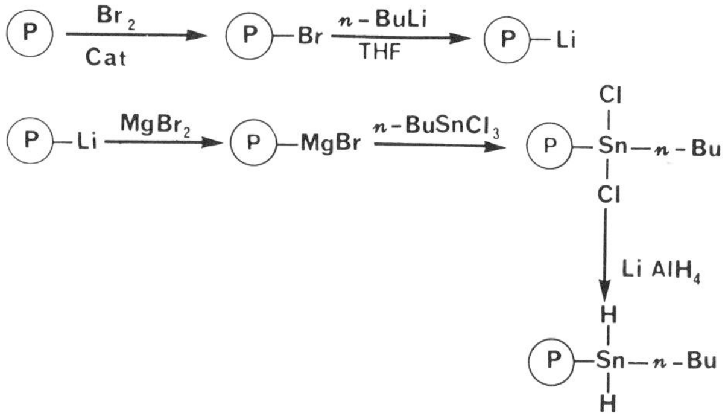

Reaction of the organotin dihydride polymer with an iodoalkyl can be used to determine the hydride content of the solid. The supported organotin dihydride, easily recovered and regenerated, was nearly odorless and nontoxic. However, it slowly lost its activity when stored at 20°C. Carbonyl compounds (acetophenone, benzaldehyde, and benzophenone) were efficiently reduced by this tin compound, but long reaction times were required (Cassidy, 1972). When dicarbonyl compounds were reacted with supported organotin dihydride (Weinshenker et $al.$, 1975) monoalcohol hydroxycarbonyl compounds were obtained with good selectivity (6:1). The following intermediate has been proposed:

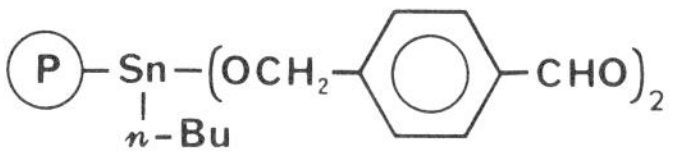

which shows that the second carbonyl group has restricted accessibility.

Polymer-supported borane complexes were also used to reduce aldehydes and ketones (Hallensleben, 1974). The preparation of the reducing agent supported on poly(4-vinylpyridine) is illustrated in the following scheme

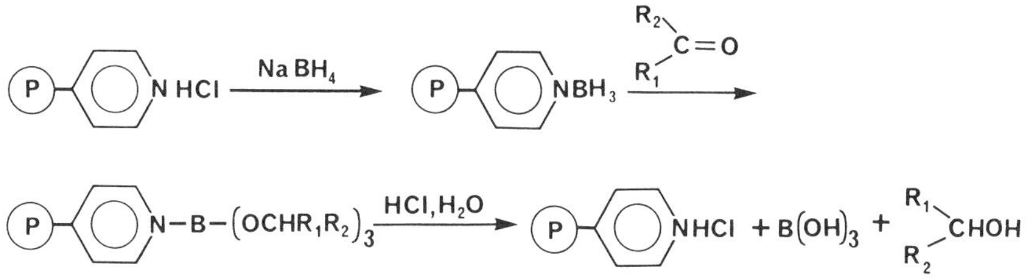

Benzaldehyde, *p*-nitrobenzaldehyde, *p*-chlorobenzaldehyde, cyclopentanone, and benzophenone were reduced by these polymer-supported boranes. Two types of poly(4-vinylpyridine) served as supports. When the support was not reticulated, the supported reducing agent placed in dry refluxing benzene was not very efficient. When the support was reticulated, it could be placed in a column, the substrate in hot benzene (10°C) being passed down the column at $T = 79°C$. The polymeric hydride could then be easily regenerated.

Reductions of amines and aldehydes to nitriles by polymer-supported phosphide in carbon tetrachloride were described (Harrison *et al.*, 1977). The phosphine resin was prepared by introducing diphenylphosphine groups into polystyrene cross-linked with 1% divinylbenzene. High yields of products were generally obtained, and in many instances removal of the resin at the end of the reaction period left a solution of nearly pure product.

The reduction of trifluoroacetophenone to the corresponding alcohol could be effected with a regenerable dihydrinicotinamide fixed on poly-(styrene–divinylbenzene) (Eling *et al.*, 1983). The reaction is illustrated in the following scheme

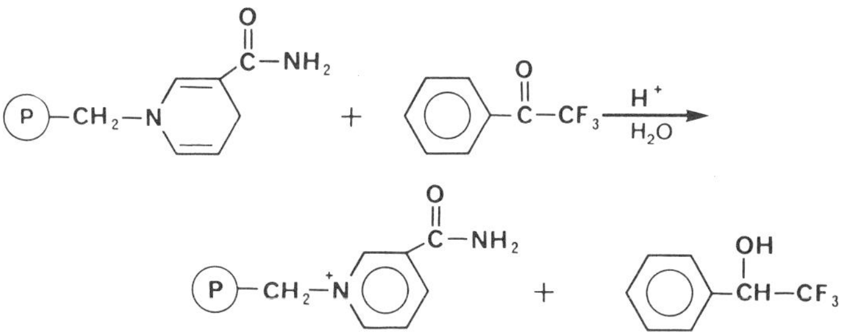

The second-order rate constant for the polymer-supported reagent is significantly lower than that of the soluble analog. However, it was

possible to reduce side reactions to 10–15% and, therefore, to reutilize the supported reagent.

Trisubstituted phosphine dichloride reagents were prepared by reaction of phosphine with a polymer incorporating phosphine oxide reagents. The reagents were good reducing agents transforming amide into nitrile (Relles and Schluenz, 1974).

An analogous supported borane complex was obtained by replacing poly(4-vinylpyridine) with the polymeric version of $Me_2S:BH_3$. The odorless and easily handled polymer:

$$(P)-CH_2-MeS \rightarrow BH_3$$

reduced acetophenone quantitatively (Crosby, 1976).

2. Reduction of Halides

The organotin dihydride supported complex was successfully used to replace iodo and bromo groups by hydrogen (Weinshenker *et al.*, 1975). Experimental conditions were mild, and the yields were generally high. Filtration of the resins at the end of the reaction allowed recovery of a nearly pure product.

3. Reduction of Disulfides to Thiols

Lipoic acid was attached covalently via different aminoethyl groups. Subsequent reduction of the disulfide-containing polymers by $NaBH_4$ produced supported diols which were in turn used to reduce disulfide bridges in peptides and proteins (Gorecki and Patchornick, 1973).

V. POLYMER-BOUND CATALYSTS

Methods for preparing supports, anchoring metal species, and characterizing them were briefly discussed in Section III. We present here some examples dealing with hydrogenation, hydroformylation, and hydrosilylation. Triphase reduction catalysis (Chiellini and Solaro, 1981) and biochemistry-related reduction catalysis (Kijima *et al.*, 1984) have been covered elsewhere.

A. Multiple C—C Bond Hydrogenation

Wilkinson's catalyst has been anchored on polymer incorporating PPh_2 groups and used for a variety of alkene hydrogenations (Grubbs *et al.*,

1973). Alkene size was an important parameter for the reaction. Cyclohexene hydrogenation was about six times faster than that of cyclooctene. It was concluded that the active sites were mostly on the surface of the support.

The literature provides only a few quantitative comparisons of catalysis by transition metal complexes in solution and on polymers. Alkene isomerization and hydrogenation catalyzed by $H_2Os_3(CO)_9PPh_3$, and its polymer-supported analogs have been reported (Freeman *et al.*, 1982; Nguini Effa *et al.*, 1982); they had approximately equivalent activities per catalytic groups. These studies showed that the performance of metal complexes catalysts is rather similar in solution and in polymers highly swelled, e.g., by chlorinated solvents, THF, or aromatic hydrocarbons.

A work published in 1975 provides an interesting example of stabilization by the support (Bonds *et al.*, 1975). Supported titanocene compounds were able to activate alkene hydrogenation whereas the soluble metal analog was inactive for the same reaction. Catalysts were formed by functionalization of a highly cross-linked polystyrene.

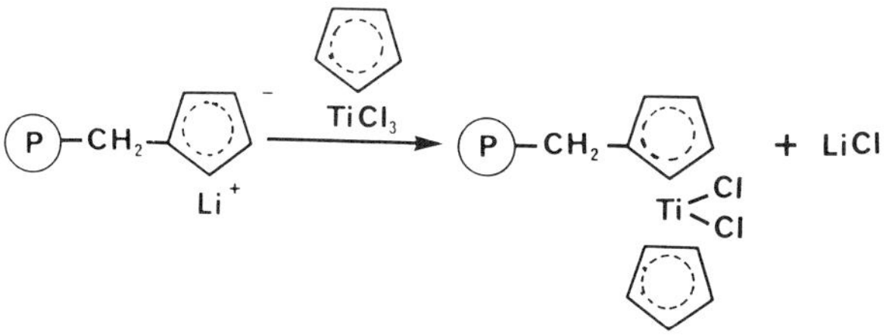

$$\text{(P)}-CH_2-\text{Cp}^- Li^+ \quad \xrightarrow{\ TiCl_3\ } \quad \text{(P)}-CH_2-Cp\,Ti(Cp)Cl_2 \quad + \quad LiCl$$

On reduction with butyllithium, a catalyst was formed presumably with the following structure

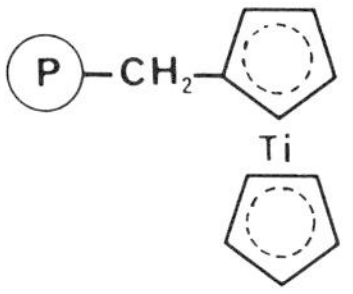

$$\text{(P)}-CH_2-Cp\,Ti\,Cp$$

The sandwich complex of Ti has coordinative insaturation to allow bonding of reactant hydrogen and alkene: hydrogenation can be performed. In contrast, the soluble titanocene analog cannot be isolated since, in solution, Ti—Ti bonds form leading to oligomer or polymer lacking coordinative insaturation; a self-inhibition of the soluble catalyst is observed which was prevented by the rigid polymeric ligand bonded to the Ti. A list of some of the polymer–bond transition metal complexes used for alkene hydrogenation is given in Table II.

TABLE II

Selection of Polymer–Bond Transition Metal Complexes Catalysts for
Alkene Hydrogenation

Catalyst	Alkene	Reference
(P)—PPh$_2$)$_x$PdCl$_2$	Polyene	Bruner and Bailor (1973)
(P)—DIOP RhCl(C$_2$H$_4$)$_2$	Methylstyrene	Dumont $et\ al.$ (1973)
(P)—PPh$_2$)$_2$Fe$_2$Pt(CO)$_8$	Ethylene	Pierantozzi $et\ al.$ (1979)
(P)—PPh$_2$Ir$_4$(CO)$_{11}$	Ethylene	Lieto $et\ al.$ (1980)
(P)—CH$_2$—NEt$_2$/RuCl$_2$(SPPh$_2$)$_2$	Crotonic acid	Joò and Beck (1984)

B. Alkene Hydroformylation

This reaction illustrates the different support effects on the catalytic activity of supported transition metal complexes. The reaction can be schematized as follows:

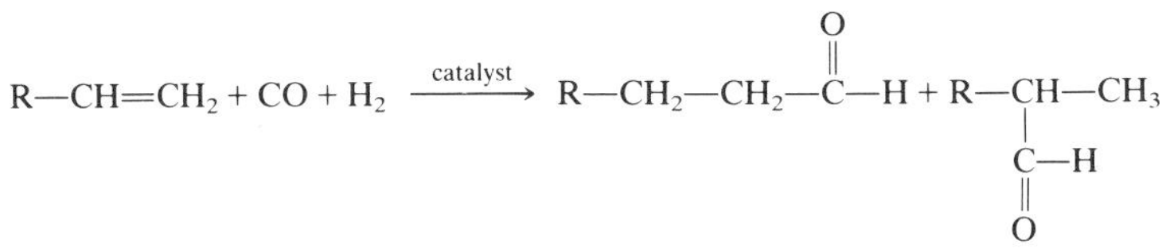

$$R-CH=CH_2 + CO + H_2 \xrightarrow{\text{catalyst}} R-CH_2-CH_2-\overset{\overset{O}{\|}}{C}-H + R-\underset{\underset{O}{\overset{\|}{C}-H}}{\overset{}{CH}}-CH_3$$

normal   branched

The aldehydes finding industrial application are normal aldehydes. Therefore, the normal–branched ratio is an important parameter. Several studies have demonstrated that differences in selectivity can be achieved using supported catalysts instead of their soluble analogs. Co$_2$CO$_8$ has been anchored on poly(styrene–divinylbenzene) incorporating —PPh$_2$ groups. The resulting solid is a good catalyst for pent-1-ene hydroformylation at 150°C and 68 atm CO:H$_2$, producing complete conversion to aldehyde with only a small amount of n-pentane (less than 4%) (Evans et $al.$, 1974). The normal–branched ratio is about 2:1 for the supported and soluble catalysts under these experimental conditions. However, on heating above 155°C, the reaction produces alcohol and the normal–branched ratio decreases to 1.4:1 for the solid catalyst whereas the soluble analog gives a 3:1 ratio. Some metal leaching occurred as indicated by the IR spectrum of the recovered catalyst which exhibits CO frequencies indicative of ⓟ —PPh$_2$Co$_2$(CO)$_7$ instead of the initial ⓟ —PPh$_2$Co$_2$(CO)$_6$. Indeed, in all the experiments, a small amount of Co$_2$(CO)$_8$ may leach out of the solid.

Soluble and polymer-supported rhodium complexes have been used as alkene hydroformylation catalysts. The supported analogs of RhCl(CO)(PPh$_3$)$_2$ transformed pent-1-ene under mild conditions (70°C, 30 atm CO/H$_2$) exclusively to aldehydes with a normal–branched product ratio of about 3 : 1. No metal leaching was observed (Evans *et al.*, 1974).

A strong support effect was observed with RhH(CO)(PPh$_3$)$_3$ anchored on poly(styrene–divinylbenzene) functionalized with phosphine groups, and used as pent-1-ene hydroformylation catalyst (Pittman and Hanes, 1976). On successive recycling, a replacement of the free phosphine ligand by the bound phosphine group occurs according to the following scheme:

$$(ⓟ—PPh_2)_2RhH(CO)(PPh_3) \longrightarrow (ⓟ—PPh_2)_3RhH(CO)$$

At higher temperatures the internal mobility of the polymer increases, and bound phosphine groups contact more rhodium atoms decreasing the content of coordinatively unsaturated rhodium.

The complex Rh$_2$Cl$_2$(CO)$_4$ has been attached on poly(styrene–divinylbenzene) incorporating a variety of ligands such as: —PBu$_2$, —CH$_2$—PPh$_2$, —NMe$_2$, —CH$_2$—NMe$_2$, and —SH, with little influence on selectivity (N/B aldehydes) of the soluble or supported catalysts (Jurewicz *et al.*, 1974).

The complex Rh(CO)$_2$ (acac) has been fixed through —PPh$_2$ groups on poly(styrene–divinylbenzene) and used for hex-1-ene hydroformylation (Allum *et al.*, 1975). With a pilot plant scale apparatus, 97% of hex-1-ene is transformed into aldehydes (with 95% selectivity) at 80°C and under 40 atm CO : H$_2$ = 1 : 1. The N/B ratio for aldehydes is 2–2.5/1 which is indicative of formation of a P—Rh bond during catalysis.

Soluble and polymer-supported homo- and heterotrimetallic anionic clusters have been used for alkene hydroformylation (Marrakchi *et al.*, 1985b, 1986). The support strongly influences the activity and selectivity of the anion clusters. Hydroformylation and isomerization of hex-1-ene occurred simultaneously with triiron and triosmium clusters. The triiron clusters experienced decomposition under the catalytic experimental conditions. In the case of osmium, decomposition was observed for the soluble system whereas the supported cluster anion could be recovered almost quantitatively under anhydrous conditions. High normal–branched ratios (15 : 1) were obtained with this system.

C. Alkene Hydrosilylation

A hydrosilylation produces a C—Si bond by reacting an alkene with an organosilicon hydride. Carbon–silicon bond formation is important for composite material and organosilicon industry:

$$R—CH=CH_2 + HSi(OEt)_3 \xrightarrow[\text{catalyst}]{} R—CH—CH_2Si(OEt)_3$$

Rhodium and platinum complexes supported on poly(styrene–divinylbenzene) incorporating —PPh$_2$ groups are catalyst for hydrosilylation reactions (Capka *et al.*, 1971, 1972; Svoboda *et al.*, 1972) at ambient temperature and pressure.

Rhodium complexes with phosphinoalkylorganosilicon ligands were used as hydrosilylation catalysts (Marcinec *et al.*, 1982). The support did not affect the kinetic parameters and mechanism of hex-1-ene hydrosilylation.

VI. PROSPECTS

A. Bifunctional and Multifunctional Catalysis

Intricate catalytic reactions and reaction sequences can be carried out by introducing the right combination of catalytic groups in polymers. Natural catalysts such as enzymes are good examples of such combinations. The same polymer may support two or more catalysts anchored on the support through identical or different ligands.

Several successful examples of multifunctional catalytic polymers are described in the literature. The sequential butadiene cyclooligomerization and hydrogenation of the cyclooligomeres has been carried out (Pittman *et al.*, 1975). Complexes, equivalent to Ni(CO)$_2$(PPh$_3$)$_2$ and RhCl(PPh$_3$)$_3$ were attached to poly(styrene–divinylbenzene) resin presenting —PPh$_2$ groups. The Ni complex induced butadiene cyclooligomerization leading to mainly cycloocta-1,5-diene (60%), vinylcyclohexene (30%), and cyclododeca-1,5,9-triene (10%). The unsaturated compounds were quantitatively hydrogenated in the presence of H$_2$ and the Rh complex. When the Wilkinson-type catalyst is replaced by supported complexes analogous to (PPh$_3$)$_3$RhHCO or (PPh$_3$)$_3$RhCO, a selective hydrogenation of the cyclooligomers to monomers on the selective terminal double bond hydroformylation are obtained, respectively. In each of these examples, the catalytic groups were recognized as effective for the soluble reactions.

The Aldox process in which propene is converted to a 2-ethyl hexanol provides a good example of a three-step sequential reaction carried out in a "one-pot" reaction. A polymer incorporating rhodium complexes cata-

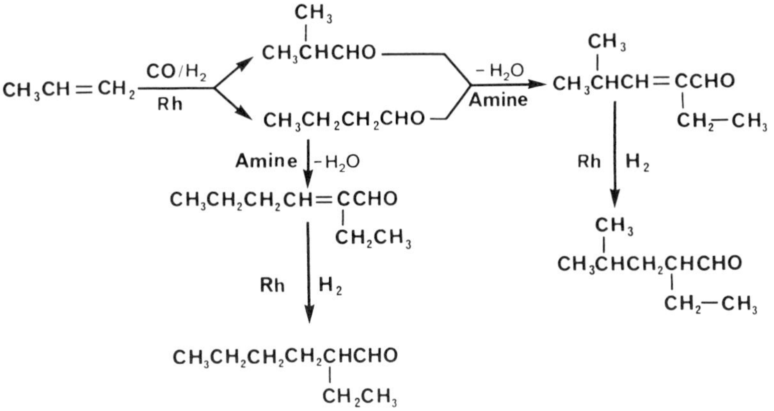

Fig. 3. The Aldox reaction network catalyzed by a styrene–divinylbenzene copolymer incorporating a rhodium complex and amine groups.

lyzes hydroformylation whereas a polymer incorporating amine groups catalyzes aldol condensation. A polymer incorporating both groups catalyzes sequentially the transformation of propene into 2-ethylhexanol (Batchelder et al., 1977). The sequences of reactions are shown in Fig. 3.

The selected examples discussed here demonstrate that polymer support offers the opportunity of designing multifunctional catalysts with different ligand–metal ratio and different metal.

B. Asymmetric Synthesis

Chiral ligands can be introduced on polymers to provide the necessary chirality for asymmetrical catalysis. Groups analogous to 2,3-o-isopropyl-idene-2,3-dihydroxy-1,4-bis(diphenylphosphine)butane (DIOP) have been supported on poly(styrene–divinylbenzene).

In one method, the support is functionalized with an appropriate set of reactions as illustrated in Fig. 2. The supported DIOP group is used as a bidentate ligand for Rh and Ru (Dumont et al., 1973; Balavoine et al., 1980), and asymmetric hydrogenation, hydrosilylation and hydroformyla-tion are activated with the supported complexes. In another preparation, the monomer is functionalized in the same way and then polymerized (Takaishi et al., 1978; Masuda and Stille, 1978).

If polar solvents are necessary for the asymmetric reaction to occur, poly(styrene–divinylbenzene) supports are no longer acceptable. Chiral polymers, obtained by copolymerizing N-acryloyl (2S-4S-(4-diphenyl-phosphino)-2-(diphenylphosphino)methyl) pyrolidine with hydroxy-methylmethacrylate and ethylene dimethacrylate, are swelled in ethanol

(Stille and Parrinello, 1983). Such chiral polymers associated with Pd salts catalyzed, with high optical yield, alkene hydroformylation in the presence of a stannous chloride cocatalyst.

Aromatic ketones were reduced with a polymer-bound chiral aminoalcohol reagent. The reaction is represented in the following scheme

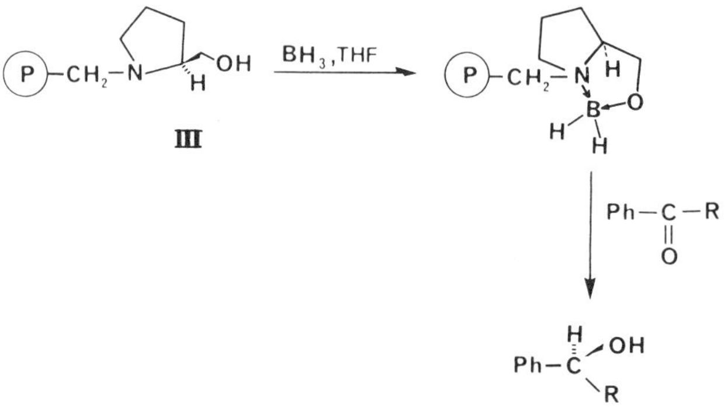

The reactant **III** could be regenerated by a simple hydrolysis and reused several times (Itsuno *et al.*, 1982).

We briefly discussed the catalytic properties of metal complex associated to support through chiral ligand. More details will be found in a review by Stille (1985). Another type of asymmetric catalyst involves polymer supports that are inherently chiral as a consequence of their secondary structure. This type includes cellulose, natural and synthetic polypeptides, and chiral stereoregular polymers.

C. Identification of Labile Species

Polymer supports can help the understanding of reaction mechanisms. Recent published works illustrate this use of polymer-supported reagents. Transformation of nitrobenzene into phenylisocyanate in the presence of carbon monoxide and a triiron anionic cluster was known. However, the mechanism of such transformation was not elucidated, a nonisolated nitrene complex being proposed as an intermediate. Use of a supported triiron complex allowed the isolation and identification of the nitrene species, with a spectroscopic reaction and vapor-phase reactants (Nguini Effa *et al.*, 1983; Marrakchi *et al.*, 1985a).

Surface catalysis involving metal aggregates dispersed on inorganic oxides are industrially very important. However, the catalytic mechanism involved with these catalysts is not well understood. Some authors proposed the use of metal clusters supported on poly(styrene–divinylbenzene) to model the behavior of metal aggregates on inorganic oxides. This type of work represents an attempt to analyze separately and progressively the factors that influence the activity and selectivity of surface catalysts (Marrakchi *et al.*, 1986).

REFERENCES

Agnew, N. H. (1976). *J. Polym. Sci., Polym. Chem. Ed.* **14**, 2819.

Allum, K. G., and Hancock, R. D. (1972). Br. Pat. 1, 287, 566.

Allum, K. G., Hancock, R. D., Howell, I. V., Pitkethly, R. C., and Robinson, J. P. (1975). *J. Organomet. Chem.* **87**, 189.

Balavoine, G., Dang, T. P., Eskenaji, C., and Kagan, H. B. (1980). *J. Mol. Catal.* **7**, 531.

Batchelder, R. F., Gates, B. C., and Kuilpers, F. P. J. (1977). *Proc. Int. Congr. Catal., 6th* **1**, 499.

Bernard, G., Huang-Van, G., and Teichner, S. J. (1975). *J. Chim. Phys.* **72**, 729; **73**, 1217.

Bonds, W. D., Jr., Brubaker, C. H., Jr., Chandrasekaran, E. S., Gibbons, C., Grubbs, R. H., and Kroll, L. C. (1975). *J. Am. Chem. Soc.* **97**, 2128.

Bruner, H. S., and Bailor, J. C. (1973). *Inorg. Chem.* **12**, 1465.

Capka, M., Svoboda, P., Cerny, M., and Hetflejs, J. (1971). *Tetrahedron Lett.* p. 4787.

Capka, M., Svoboda, P., Kraus, M., and Hetflejs, J. (1972). *Chem. Ind. (London)* p. 650.

Cassidy, H. G. (1972). *J. Polym. Sci.* **6**, 1.

Challa, G. (1983). *J. Mol. Catal.* **21**, 1.

Chalmin Louis-André, O. (1985). Thesis dissertation, University of Lyon (France).

Chauvin, Y., Commereuc, D., and Dawans, F. (1977). *Prog. Polym. Sci.* **5**, 95.

Chiellini, E., and Solaro, S. (1981). *Makromol. Chem., Suppl.* **5**, 81.

Crosby, G. A. (1976). U.S. Pat. 3,928,293 [*C.A.* **84**, 1064994 (1976)].

Dahbor, H. (1958). *Chem. Ber.* **91**, 1955.

Dumont, W., Poulin, J. C., Dang, T. P., and Kagan, H. B. (1973). *J. Am. Chem. Soc.* **95**, 8295.

Eling, B., Challa, G., and Paudit, U. K. (1983). *J. Polym. Sci., Polym. Chem. Ed.* **21**, 1125.

Evans, G. O., Pittman, C. U., McMillan, P., Beach, R. T., and Jones, R. (1974). *J. Organomet. Chem.* **67**, 296.

Freeman, M. B., Patrick, M. A., and Gates, B. C. (1982). *J. Catal.* **73**, 82.

Gates, B. C., and Lieto, J. (1980). CHEMTECH pp. 196, 248.

Gates, B. C., and Lieto, J. (1985). *In* "Encyclopedia of Polymer Science and Engineering" (N. M. Bikales, H. F. Mark, and N. G. Gaylord, eds.), Vol. 2, p. 708. Wiley, New York.

Gates, B. C. (1980). *In* "Chemistry and Chemical Engineering of Catalytic Processes" (Prins, R., and Schuit, G. C.A., eds.) p. 437. Sijthoff & Noordhoff, The Hague.

George, N. H., and Hayes, G. F. (1973). *J. Polym. Sci., Polym. Chem. Ed.* **11**, 471.

Gibson, H. W., and Bailey, F. C. (1977). *J.C.S. Chem. Commun.* p. 815.

Gorecki, M., and Patchornick, A. (1973). *Biochim. Biophys. Acta* **303**, 36.

Grubbs, R. H., Kroll, L. C., and Sweet, E. M. (1973). *J. Macromol. Sci., Chem.* **A7**, 1047.

Hallensleben, M. L. (1974). *J. Polym. Sci., Polym. Symp.* **47**, 1.

Harrison, C. R., and Hodge, P. (1980). *Synthesis* p. 299.

Harrison, C. R., Hodge, P., and Rogers, W. J. (1977). *Synthesis* p. 41.

Hodge, P., and Sherrington, D. C., eds. (1980). "Polymer-Supported Reactions in Organic Synthesis." Wiley, New York.

Hutchins, R. O., Natale, N. R., and Taffer, I. M. (1978). *J.C.S. Chem. Commun.* p. 1088.

Itsuno, S., Hirao, A., and Nakahuma, S. (1982). *Makromol. Chem., Rapid Commun.* **3**, 673.

Joò, F., and Beck, M. T. (1984). *J. Mol. Catal.* **24**, 135.

Jurewicz, A. T., Rollman, L. D., and Whitehurst, D. D. (1974). "Homogeneous Catalysis," Vol. II, Advances in Chemistry Series, No. 132, p. 240. *Am. Chem. Soc.*, Washington, D. C.

Kanego, M., and Tsuchida, E. (1981). *J. Polym. Sci., Macromol. Rev.* **16**, 397.

Kijima, M., Namba, Y., Eudo, T., and Okawara, M. (1984). *J. Polym. Sci., Polym. Chem. Ed.* **22**, 821.

Lieto, J., Rafalko, J. J., and Gates, B. C. (1980). *J. Catal.* **62**, 149.

Lieto, J., Milstein, D., Albright, R. L., Minkiewicz, J. V., and Gates, B. C. (1983). *CHEMTECH* p. 46.

Marcinec, B., Urbaniak, W., and Pawlak, P. (1982). *J. Mol. Catal.* **14**, 323.

Marrakchi, H., Nguini Effa, J. B., Haimeur, M., Lieto, J., and Aune, J. P. (1985a). *Bull. Soc. Chim. Fr.* **3**, 390.

Marrakchi, H., Nguini Effa, J. B., Haimeur, M., Lieto, J., and Aune, J. P. (1985b). *J. Mol. Catal.* **30**, 101.

Marrakchi, H., Haimeur, M., Escalant, P., Lieto, J., and Aune, J. P. (1986). *Nouv. J. Chim.* **10**, 159.

Masuda, T., and Stille, J. K. (1978). *J. Am. Chem. Soc.* **100**, 268.

Mathur, N. K., Narang, C. K., and Williams, R. E. (1980). "Polymers as Aids in Organic Chemistry." Academic Press, New York.

Nguini Effa, J. B., Lieto, J., and Aune, J. P. (1982). *J. Mol. Catal.* **15**, 367.

Nguini Effa, J. B., Djebailli, S., Lieto, J., and Aune, J. P. (1983). *J.C.S. Chem. Commun.* p. 408.

Petrov, Y. I., Klabunovski, E. I., and Baladin, A. A. (1968). *Kinat. Katal.* **8**, 1175.

Pierantozzi, R., McQuade, K. J., Gates, B. C., Wolf, M., Knözinger, H., and Ruhmann, W. (1979). *J. Am. Chem. Soc.* **101**, 5436.

Pittman, C. U., Jr. (1982). *In* "Comprehensive Organometallic Chemistry" (G. Wilkinson, F. G. A. Stone, and E. W. Abel. eds.), Vol. 8, p. 553. Pergamon, Elmsford, New York.

Pittman, C. U., Jr., and Hanes, R. M. (1976). *J. Am. Chem. Soc.* **98**, 5402.

Pittman, C. U., Jr., Smith, L. R., and Hanes, R. M. (1975). *J. Am. Chem. Soc.* **97**, 1742.

Relles, H. M., and Schluenz, R. W. (1974). *J. Am. Chem. Soc.* **96**, 6469.

Stille, J. K. (1985). *In* "Chem. Met.—Carbon Bond" (F. R. Hartley and S. Pataï, eds.), Vol. 2, pp. 625–787. Wiley, New York.

Stille, J. K., and Parrinello, G. (1983). *J. Mol. Catal.* **21**, 203.

Svoboda, P., Capka, M., Chvalovsky, V., Bazant, V., Hetyleys, J., Jahr, H., and Pracejus, H. (1972). *Angew. Chem. Int. Ed. Engl.* **12**, 153.

Takaishi, N., Imai, H., Bertelo, C. A., and Stille, J. K. (1978). *J. Am. Chem. Soc.* **100**, 264.

Tyurenkova, O. A., and Chimarova, L. A. (1970). *Russ. J. Phys. Chem.* **44**, 1289.

Weinshenker, N. M., Crosby, G. A., and Wong, J. Y. (1975). *J. Org. Chem.* **40**, 1966.

Part **IV**

Graphite Intercalates

15 INTERCALATION COMPOUNDS OF GRAPHITE AND THEIR REACTIONS

Ralph Setton

Centre de Recherche sur les Solides à Organisation Cristalline Imparfaite
Centre National de la Recherche Scientifique
F-45071 Orléans Cédex 2, France

The literature on graphite intercalation compounds is now extensive. The following reviews are good starting points: for general chemical properties: Selig and Ebert (1980), Hérold (1979); for physical properties: Solin (1982), Dresselhaus and Dresselhaus (1981), Fischer (1979); for use as catalysts: Ebert (1982), Boersma (1974, 1977), Golé et al. (1982).

Owing to the severely limited space available, only a selection of references (mainly centered on the years 1974–1984) will be presented, and reaction mechanisms will seldom be discussed; the original references should be consulted for this information when it is needed.

The following standard abbreviations will be used: Ac = acetyl $CH_3CO—$, Bu = butyl $C_4H_9—$, Bz = benzoyl $C_6H_5CO—$, Et = ethyl $C_2H_5—$, Me = methyl $CH_3—$, Ph = phenyl $C_6H_5—$, Pr = propyl $C_3H_7—$, R^1, R^2, R^3 = radicals, c = cis, gem = geminated, i = iso, n = normal, p = para, sec = secondary, t = trans, ter = tertiary, vic = vicinal, and G = graphite.

I. GENERAL FEATURES OF THE GRAPHITE INTERCALATION COMPOUNDS

Graphite has stacked parallel sp^2-hybridized (aromatic) carbon sheets separated by about 3.3 Å, with each carbon atom bearing a π electron which plays a vital role in the formation of the graphite intercalation compounds (GICs), with charge transfers occuring between the intercalate

255

and the host. When the charge transfer is much less than 1 electron per carbon atom, no covalent bond exists and a true GIC is formed with the intercalate: donor or acceptor GICs are formed when the intercalate gives an electron to or receives an electron from the C atoms, respectively. Intercalation may yield a periodic sequence of filled and empty spaces. When this sequence is regular (this may not always be the case), the stage of the GIC is defined as the ratio of host layers to guest layers. In a first-stage GIC, every interlayer space is occupied while, for instance, only every third space is occupied in a third-stage GIC.

The formation of a donor GIC is usually a fairly reversible process

$$M + nC \rightleftharpoons MC_n \quad \text{or rather} \quad M^+C_n^- \tag{1}$$

(where it is understood that the charge transfer may be incomplete).

Acceptor GICs need an electrical or chemical oxidant for their formation

$$nC + AlCl_3 + \epsilon\tfrac{1}{2}Cl_2 \longrightarrow C_nAlCl_{3+\epsilon} \tag{2}$$

so that the final reaction is now no longer fully reversible because of the formation of a stable $AlCl_4^-$ ion in Eq. (2) by the excess chlorine. In fact, the "residue" GICs which remain after attempts to remove all the intercalate may well be the ionic portion of the intercalated material. Note that all GICs are nonstoichiometric compounds and that ideal formulas are, at best, convenient mnemonics.

II. THE GRAPHITE INTERCALATION COMPOUND AS A REAGENT

When a GIC reacts with a chemical species, the following can occur:

(1) The intercalate diffuses out and reacts outside the solid phase; the reaction is then little or not affected by the graphite unless one or both of the reacting species are adsorbed on it.

(2) The nonintercalated species diffuses into the GIC and reacts with the intercalate *in situ*; in this case, steric parameters and local potentials are of paramount importance in directing the course of the reaction.

(3) The graphite participates chemically in the reaction.

All three types of behavior are met, singly or combined.

A. The Preparation of Activated Metal Species

It has been claimed that some of the transition metals can be made to form GICs by *in-situ* reduction of their easily intercalated halide

$$C_mMX_n + nR \longrightarrow M \text{ (in G?)} + nRX$$

where M is the transition metal, X a halogen, and R a reductant. The GIC (usually containing a chloride) is treated with a solution of Na in liquid NH_3 or hexamethyl phosphotriamide (HMPT) (halides of Fe, Mn, Co, Ni, Cu, Mo, and W: Vol'pin *et al.*, 1975; Bewer *et al.*, 1977), or the solution in tetrahydrofuran (THF) of the complex formed by an alkali metal with a polycyclic aromatic hydrocarbon (same halides as above: Vol'pin *et al.*, 1975; Schlögl *et al.*, 1983; intercalated mixed halides: Nefed'ev *et al.*, 1980), or H_2 to eliminate gaseous HCl (same halides as above: Vol'pin *et al.*, 1975; Parkash *et al.*, 1977, 1978; Parkash and Hooley, 1980; Kikuchi, 1978; Kikuchi *et al.*, 1979, 1980, 1982; Ino *et al.*, 1980; Kalucki *et al.*, 1981; Jung *et al.*, 1982; intercalated mixed halides: Kikuchi, 1978; Xu *et al.*, 1983), or even K vapor (same halides as above, also halides of Cr, Ru, Pd, Os, Pt: Mashinskii *et al.*, 1976; Bewer *et al.*, 1977; Nefed'ev *et al.*, 1977; Vol'pin *et al.*, 1977; Chernenko *et al.*, 1983). "Activation" of the transition metal was obtained by deliberately using an excess of reductant.

Activated metal species have also been prepared through reduction of the halide, dissolved or suspended in THF, by a potassium GIC. The anhydrous halides of Mg, Hg, Cd, Zn, Ni, and Fe were thus reduced by Ungurenasu and Palie (1977). Later, Braga *et al.* (1978, 1979) also reduced $Ti(i\text{-PrO})_4$, $MnCl_2 \cdot 4H_2O$, $FeCl_3$, $CoCl_2 \cdot 6H_2O$, $CuCl_2 \cdot 2H_2O$, and $ZnCl_2$. The exact nature of the species formed has been questioned (Böhm *et al.*, 1985), and THF is retained in the solid (Inagaki *et al.*, 1984).

B. Halogen Chemistry

1. Formation of a Bond with Fluorine

Rabinowitz *et al.* (1977a,b) and Agranat *et al.* (1977) showed the advantages of using GICs of $XeOF_4$, XeF_6, and XeF_4 rather than the neat substances: handling of less corrosive and solid (rather than gaseous) products, absence of explosive by-products. The preparation of these GICs is rather specialized (Selig and Gani, 1972; Selig *et al.*, 1976), but their use presents no major difficulty (Rabinowitz *et al.*, 1977b). The reactions are usually run in CH_2Cl_2, in standard glass or plastic ware, in a fume cupboard if necessary. The main results are presented in Table I. Even if the yields are, on the whole, fairly modest, the simplicity and the ease of handling are a decided advantage over other methods.

The GIC $C_{19}XeF_6$, inactive toward simple ketones, is efficient for the fluorination of β-diketones (Yemul *et al.*, 1980), halogenation of the methylene group taking precedence over that of an aromatic ring (Table II). The 90% yield of the pharmacologically important 5-fluorouracil is the highest reported yet.

TABLE I

Fluorination of Aromatic Hydrocarbons[a,b]

Fluorinating agent	II	III	V	VI	VII	IX	X	XI
XeF$_2$	0	0	45	9	15	30	36	4
XeF$_2$ + AHF[c]	75	3						
C$_{26}$XeF$_4$	0	0						
C$_{26}$XeF$_4$ + AHF	26	11						
C$_{19}$XeF$_6$	37	19	12	5	20	20	18	10
C$_9$XeOF$_4$	0	0	23	4	2	6	10[d]	1
C$_9$XeOF$_4$ + AHF	22	0						

(continues)

TABLE I (*continued*)

Fluorinating agent	XIII	XIV	XVI	XVIII	XIX	XX	XXII	XXIII
XeF_2	33	21?	31	22	14	1	22	?
XeF_2 + AHF[c]	36	18						
$C_{26}XeF_4$	22	19						
$C_{26}XeF_4$ + AHF								
$C_{19}XeF_6$	34	9	22	20	7		17	
C_xXeOF_4	22	10		10[d]	2			
C_xXeOF_4 + AHF								

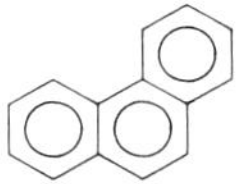

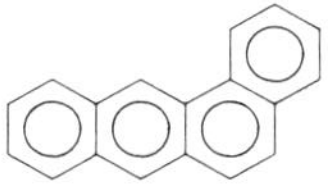

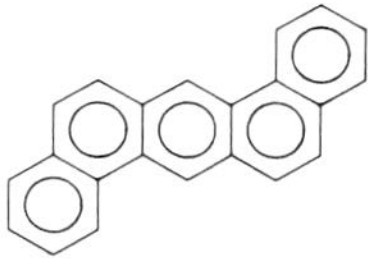

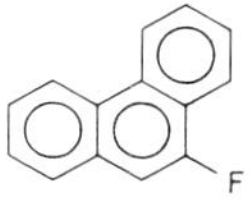

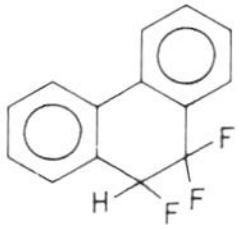

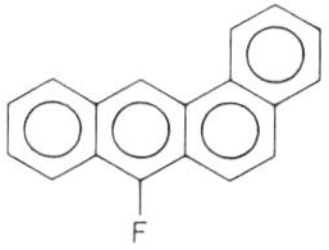

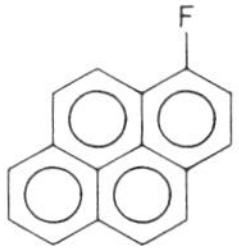

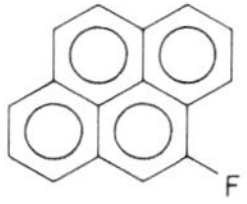

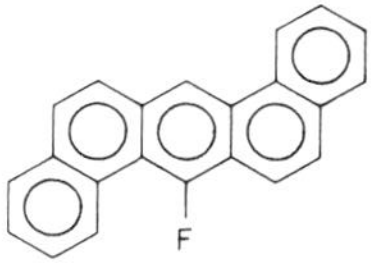

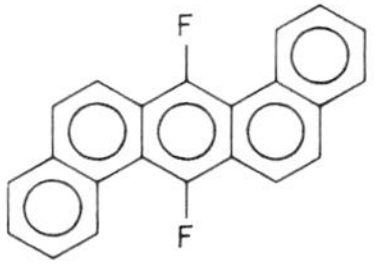

[a] The hydrocarbon and the compounds formed are given in the first and second line, respectively. The first column indicates the fluorinating agent. In the other columns: proportion (%) of the compound in the resulting mixture.

[b] Information for **I** and **XIII** from Rabinowitz *et al.* (1977a); **IV, VIII, XIV,** and **XVI** from Rabinowitz *et al.* (1977b); and **XVII** and **XXI** from Agranat *et al.* (1977).

[c] AHF = anhydrous HF.

[d] Also 24% anthraquinone.

259

TABLE II

Fluorination of β-Diketones[a][b]

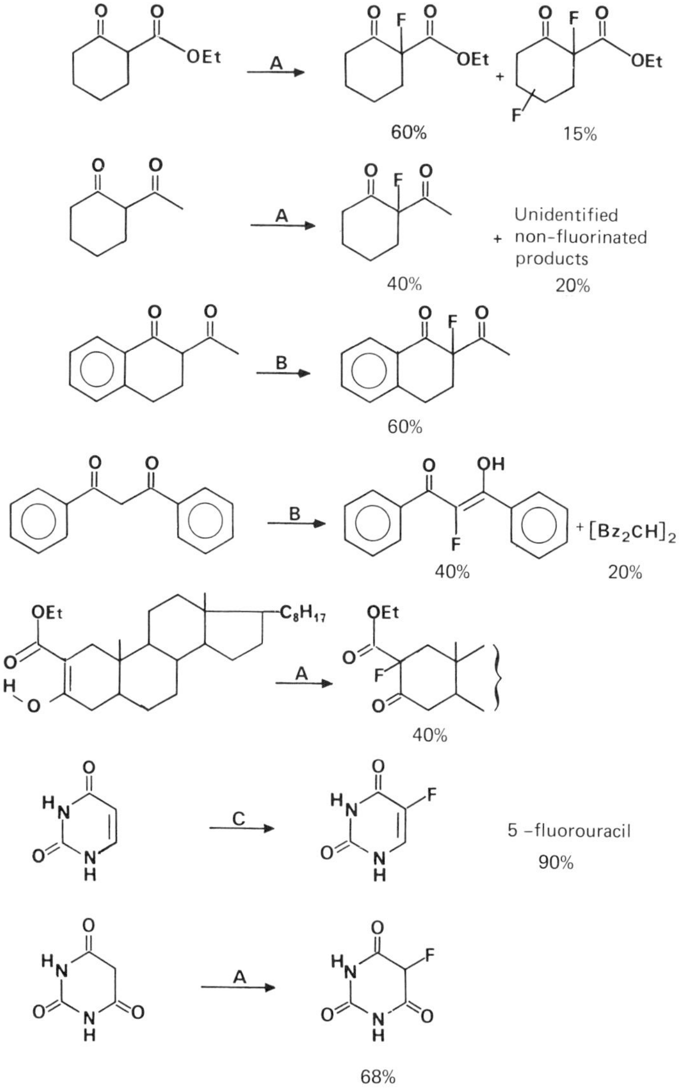

[a] Reprinted with permission from Yemul et al. (1980). Copyright 1980, Pergamon Press, Ltd.

[b] 1 mmole organic compound in 20 mL CH_2Cl_2; 1 mmole $C_9X_eF_6$ added as follows: A = addition at room temperature, stirring 24 h; B = addition at −196, stirring 24 h with gradual warming up to room temperature; and C = addition at room temperature to 2 mmole C_9XeF_6 in CH_2Cl_2, stirring 24 hr.

TABLE III

Formation of Mono- and Difluorosilicon Derivatives by Fluorination with a GIC of SbF_5[a,b]

Organosilane	A	Solvent	Time/ temperature (h/°C)	Product	Yield (%)
Ph(α-Np)MeSiAll	0.5	Et₂O	24/25	Ph(α-Np)MeSiF	80
Ph₃SiAll	0.5	Et₂O	10/25	Ph₃SiF	70
Me₂PhSiOMe	0.21	Pe	20/25	Me₂PhSiF	81
Ph(α-Np)MeSiOMe	0.3	Pe or Et₂O	20/25	Ph(α-Np)MeSiF	79
Me₂PhSiCl	0.25	Pe	10/25	Me₂PhSiF	85
Ph(α-Np)MeSiCl	0.23	Pe or Et₂O	10/25	Ph(α-Np)MeSiF	80
Ph(α-Np)Si(H)Cl	0.39	Pe	2/25	Ph(α-Np)Si(H)F	52
Ph(α-Np)Si(H)OMe	0.47	Pe	25/25	Ph(α-Np)Si(H)F	50[c]
				Ph(α-Np)Si(F)OMe	35[c]
Ph(α-Np)Si(H)OMen	0.42	Pe	24/25	Ph(α-Np)Si(H)F	95[c]
Me₂PhSiH	0.22	Pe	24/25	No reaction	
Me₂PhSiSPh	0.21	Pe	100/25	No reaction	
Ph₂(OEt)SiAll	0.5	Et₂O	24/25	Ph₂SiF₂	100
Ph(α-Np)(OEt)SiAll	0.5	Et₂O	24/25	Ph(α-Np)SiF₂	100
Ph(α-Np)Si(H)(OEph)	0.52	Pe	72/25	Ph(α-Np)SiF₂	100
Ph(α-Np)Si(OEt)₂	0.3	Et₂O	17/25	Ph(α-Np)SiF₂	90
(m-CF₃Ph)₂Si(OMe)₂				(m-CF₃Ph)₂SiF₂	80
PhCH₂(PhCH₂CH₂)SiOMe				PhCH₂(PhCH₂CH₂)SiF₂	80
Ph₂Si(H)OMe				Ph₂SiF₂	75
Ph(α-Np)Si(OMe)₂				Ph(α-Np)SiF₂	73
Ph(α-Np)Si(H)Cl	0.39	Pe	2/25	Ph(α-Np)SiF₂	48[c]
Ph(α-Np)Si(H)OMe	0.47	Pe	25/25	Ph(α-Np)SiF₂	15[c]

[a] From Setton et al. (1982).
[b] A = SbF_5/silane, Pe = pentane, Ph = phenyl, All = allyl, α-Np = α-naphthyl, Me = methyl, OEt = ethoxy, OMe = methoxy, OEph = ephedrinoxy, OMen = menthyloxy.
[c] In mixture.

Gem-difluorination of monofunctional ketones can be performed by $C_{20}MoF_6$ in the presence of CF_3COOH as catalyst (Setton et al., 1982). Bond formation between fluorine and Ge or Si was obtained by Corriu et al. (1980) using the SbF_5 GIC (Table III).

2. Formation of a Bond with Chlorine or Bromine

Heineman and Latscha (1981) used $C_{23}SbCl_5$ to chlorinate p-benzoquinone and were thus able to isolate or identify all the intermediates leading to 2,3,5,6-tetrachloroquinone.

Page Lécuyer and co-workers (1973) compared the results obtained on various substrates by using a "residue" GIC containing 4–15% Br_2, or Br_2 itself, or a Br_2–thallium acetate reagent; better yields and selectivity were found for the GIC in the bromination of the sterically hindered 1,1'-binaphtyl and its tetramethyl homologs.

3. Dehalogenation Reactions

Münch and Selig (1980) obtained (90%) $t\text{-}N_2F_2$ from N_2F_4 by using $C_{10}AsF_{5+x}$. The reaction is carried out at room temperature and involves fluorine abstraction by the graphite, as was later proved by Setton (1984) who showed that the driving force of the reaction is the exothermic formation of the G—F bonds; the intercalated species is therefore the catalyst for the reaction between G and fluorine.

Dehalogenations have been obtained with donor GICs. Thus, the technologically important dechlorination of polychlorinated biphenyl is obtained with KC_8 or NaC_{16} using 3 K (or Na) per Cl atom (Oku et al., 1980; Okawara et al., 1984). Likewise, vic-dibromo derivatives are debrominated by KC_8 at room temperature in benzene (Rabinowitz and Tamarkin, 1984), and even the active Fe/G prepared by Braga's method will give the corresponding alkenes with little loss of geometrical purity and in excellent yields (Savoia et al., 1982).

Alkyl halides are also reduced by KC_8 (Bergbreiter and Killough, 1978). The best results (80–100%) are obtained with a 3- to 13-fold excess of the GIC reacting on n-chloro compounds in Et_2O or on n-iodo compounds in pentane since n-bromo or iodo compounds in THF also give extensive Wurtz coupling, and sec halogen derivatives (also in THF) give increasing quantities (15–88%) of the corresponding alkene as the halogen gets heavier. Chlorine or bromine on aromatic cycles are cleanly reduced (50%) with no formation of by-products.

4. Halogen Exchange Reactions

The GIC of $SbCl_5$ is a prime reagent for halogen exchange on non-aromatic carbons (Bertin et al., 1974b). Best results are obtained when the reagent is opposed to Br or I on sec-C atoms (Table IV). Control experiments with neat $SbCl_5$ prove that graphite plays a specific role since the products formed are quite different: free $SbCl_5$ chlorinates the aromatic cycle of $PhCH_2CH_2Br$ or $PhCHBrCH_3$ whereas the GIC forces a halogen exchange exclusively.

C. Formation of a C—C Bond

1. Grignard-Type Reagents

The activated Mg/G obtained by reducing MgI_2 by KC_8 reacts even in benzene with alkyl, aryl, or allyl halides to form Grignard reagents which

TABLE IV[a]

Halogen Exchange Reactions with the $SbCl_5$ Graphite Intercalation Compound

R—X substrate	Monohalide R—Cl	R—X recovered	Dihalide
$n\text{-}C_6H_{13}CH(Br)\text{—}CH_3$	$n\text{-}C_6H_{13}\text{—}CH(Cl)\text{—}CH_3$ 98%	2%	–
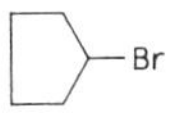 cyclopentyl–Br	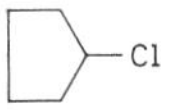 cyclopentyl–Cl 60%	5%	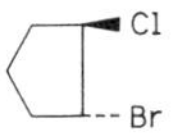 cyclopentane (Cl, Br) 35%
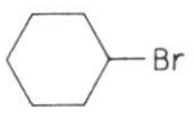 cyclohexyl–Br	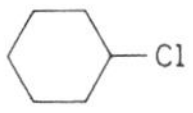 cyclohexyl–Cl 62%	20%	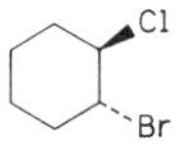 cyclohexane (Cl, Br) 18%
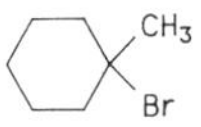 1-methylcyclohexyl–Br	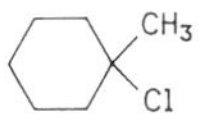 1-methylcyclohexyl (CH₃, Cl) 25%	–	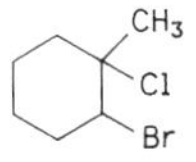 cyclohexane (CH₃, Cl, Br) 75%
$C_6H_5\text{—}CH_2\text{—}CH_2\text{—}Br$	$C_6H_5\text{—}CH_2\text{—}CH_2\text{—}Cl$ 30%	70%	–
$C_6H_5\text{—}CH(Br)\text{—}CH_3$	$C_6H_5\text{—}CH(Cl)\text{—}CH_3$ 98%	2%	–
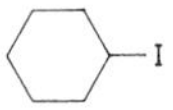 cyclohexyl–I	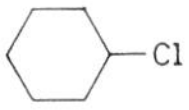 cyclohexyl–Cl 86%	14%	–
$n\text{-}C_6H_{13}CH(OTs)\text{—}CH_3$	$n\text{-}C_6H_{13}CH(Cl)\text{—}CH_3$ 15%	85%	–
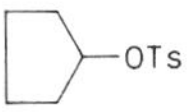 cyclopentyl–OTs	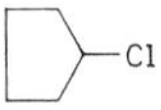 cyclopentyl–Cl 5%	90%	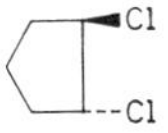 cyclopentane (Cl, Cl)
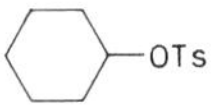 cyclohexyl–OTs	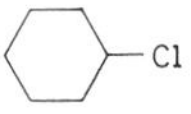 cyclohexyl–Cl 22%	78%	–

[a] Reprinted with permission from Bertin et al. (1974b). Copyright 1974, Pergamon Press, Ltd.

TABLE V

Condensation of Organozinc Derivatives with Carbonyl Compounds[a,b]

Bromo derivative	Carbonyl compound	Product	Yield (%)
Br~CO₂Et	(nonanal)	(structure) CO₂Et	90
Br~CO₂Et	(cyclohexanone)	(structure) CO₂Et	88
Br~CO₂Et	(6-methyl-5-hepten-2-one)	(structure) CO₂Et	87
Br~CO₂Et	(acetophenone)	(structure) CO₂Et	89
Br~CO₂SiMe₃	(cyclohexanone)	(structure) CO₂H	75
(structure) Br~CO₂Et	(structure) CO₂Et	EtO₂C (structure)	86
(structure) Br lactone	(fluorenone)	(structure)	88
(structure) Br lactone	CH₂=N⁺Me₂ I⁻	(structure)	55
Br~~CO₂Me	CHO (benzaldehyde)	(structure) CO₂Me	85
(allyl bromide)	(heptanone)	(structure) OH	94
(prenyl bromide)		(structure) OH	94

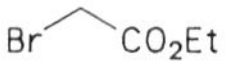

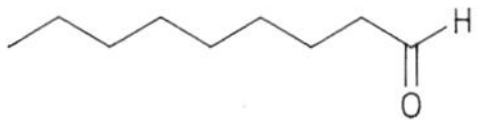

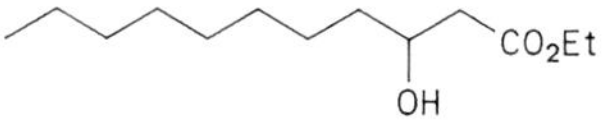

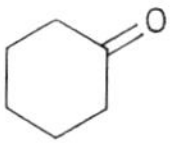

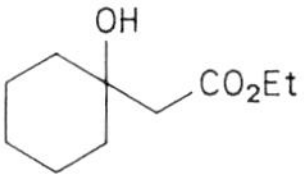

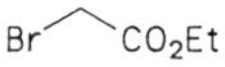

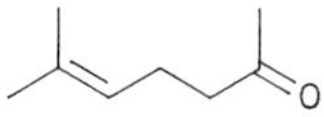

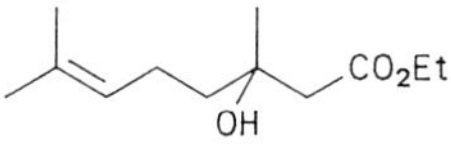

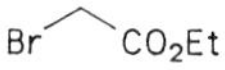

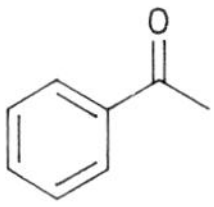

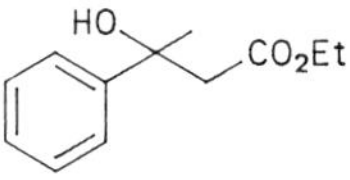

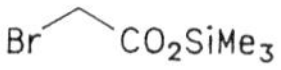

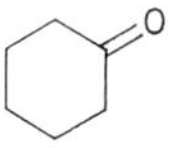

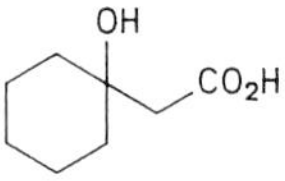

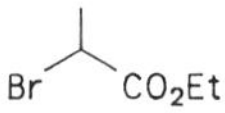

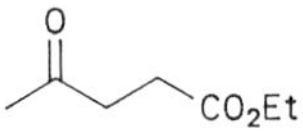

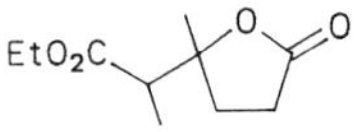

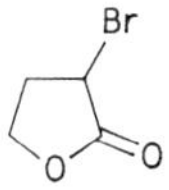

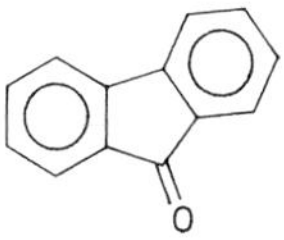

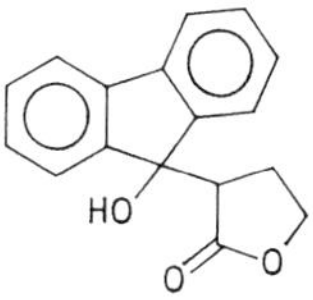

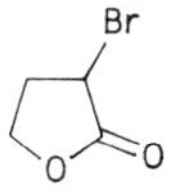

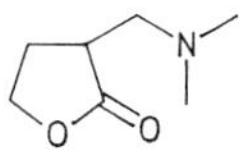

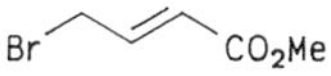

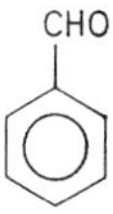

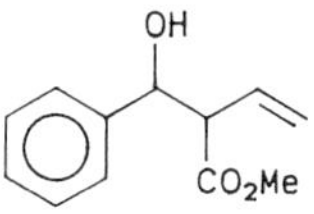

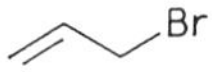

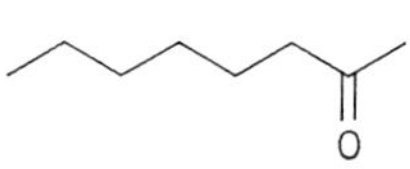

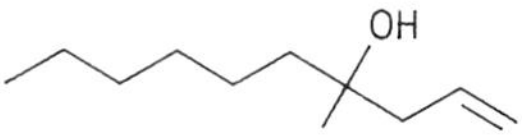

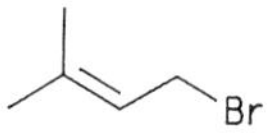

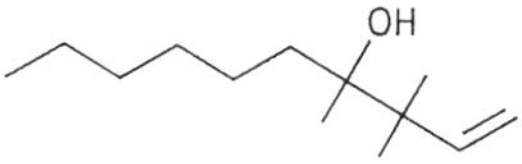

(*continues*)

TABLE V (*continued*)

Bromo derivative	Carbonyl compd	Product	Yield (%)
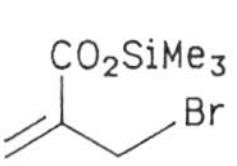	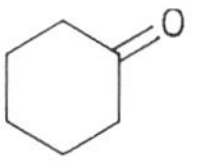	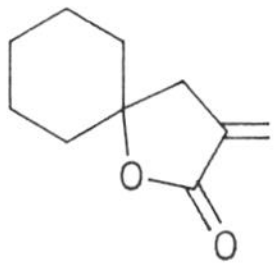	90
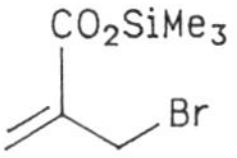	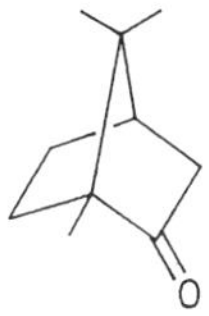	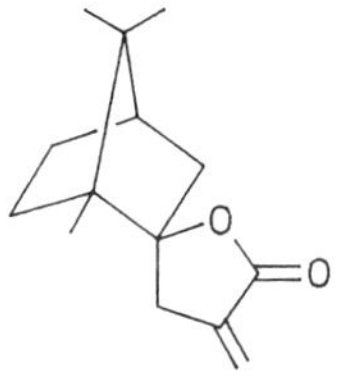	65

[a] Reprinted with permission from Boldrini *et al.* (1983). Copyright 1983 American Chemical Society.

[b] All reactions were carried out with ratios Zn/G : bromo derivative : carbonyl compound = 15 : 12 : 10 mmole. The first seven reactions and the ninth reaction were carried out at 0°C, all others at 20 ± 2°C.

react with CO_2, H_2O, or Ph_2CO to give the expected products (87–96%). After preparation, the Grignard reagent—which is held within the solid matrix—can be filtered off and dry-stored several months with no visible loss of activity (Ungurenasu and Palie, 1977). Similarly, activated Zn/G has been used (Boldrini *et al.*, 1983) for Reformatsky reactions between α-activated bromine derivatives and carbonyl compounds to form hydroxy esters or cyclic lactones. The yields are excellent (Table V) with negligible side reactions.

The activated Fe/G which gives alkenes from *vic*-dibromides can also lead to condensation reactions (Table VI) (Savoia *et al.*, 1982); the preparation of 2-oxallyl cations from α,α'-dibromo derivatives is particularly interesting in view of their use in the synthesis of cyclic compounds.

2. Coupling Reactions

A surprising feature of some GICs is their propensity to promote coupling of aromatic rings. A first example of this was the formation of 3,3'-biphenanthryl (or the 1,1' isomer) from phenanthrene in the presence of the AsF_5 GIC (Lin *et al.*, 1975). The donor GIC KC_8 promoted condensation of benzene to biphenyl (Lalancette and Roussel, 1976; Béguin and Setton, 1976), and it was later proved that the metal moiety of

TABLE VI

Syntheses with Fe/G

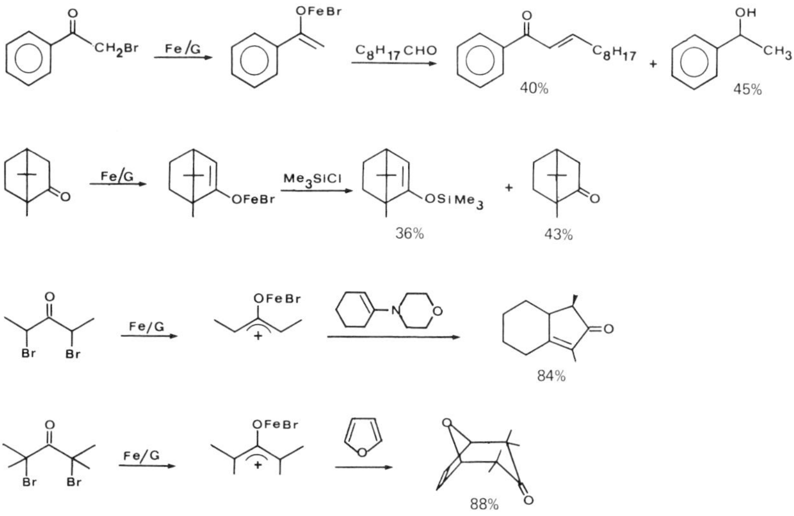

the GIC and the biphenyl which is formed are both needed to fix the hydrogen liberated by the formation of the C—C bond (Setton, 1984). Another mechanism must be involved in the formation of phenan-threnequinones from benzils in the presence of KC_8 with loss of gaseous H_2 (Tamarkin et al., 1984).

KC_8 promotes coupling of Ph_2CO to $Ph_2C(OH)C(OH)Ph_2$ (Tamarkin and Rabinowitz, 1984) in aprotic solvents, while proton-donating solvents lead to monomolecular reduction.

3. Alkylation Reactions

KC_8 is a convenient agent for metallation; even if the amount of metal extracted from it depends on the electron affinity and/or the redox potential of the molecule to be metallated (Rashkov et al., 1979, 1980), the metallation can always be forced to completion by increasing the relative ratio of moles of KC_8 to moles of substrate.

Savoia and co-workers (1977) alkylated nitriles and esters at $-60°C$:

$$RCH_2Y \xrightarrow{KC_8} (RCHY)^- \xrightarrow{R^1X} RCHR^1Y; \qquad Y = CN; R = H, Et, Ph;$$
$$Y = COOEt; R = Ph$$

With the molar ratios $RCH_2Y : KC_8 : R^1X = 1:2:2$, best yields (40–70%) were obtained when R^1 was n-alkyl, dropping to about one-half these values when R^1 was sec-alkyl. No dialkylated derivatives were found, in spite of the excess of halide provided to counteract the Wurtz coupling which always occurred.

KC_8 was also used (Savoia *et al.*, 1978) in the synthesis of aldehydes and ketones

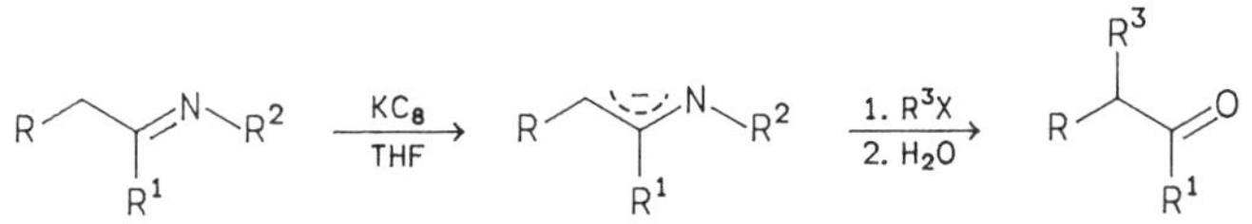

with R = H, alkyl or aryl; R^1 = H or alkyl; R^2 = ter-Bu or cyclohexyl; R^3 = alkyl. Here too, the best yields (50–70%) were obtained with the ratios imine : $KC_8 : R^3X = 1:2:2$, but they were halved when the reaction was carried out in hexane.

The metallation of 2,4,4,6-tetramethyl-5, 6-dihydro-1,3-oxazine **XXIV** also led to the aldehydes **XXV** by supplying the two C atoms other than those in R:

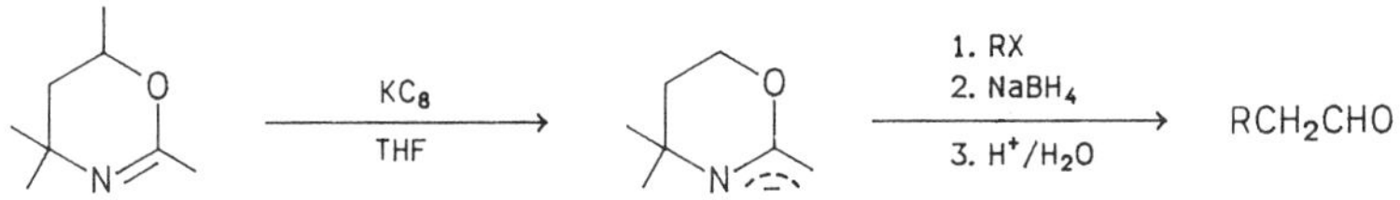

XXIV   **XXV**

D. C—H Bond Chemistry

Contento *et al.* (1979) have studied the reduction by KC_8 of activated double bonds in α,β-unsaturated ketones, α,β-unsaturated acids and Schiff bases. The reactions are run at 25°C, and the yields are excellent (Table VII). The reduction of compounds **XXVI–XXIX** were run in the presence of hexamethylsilazane $[(CH_3)_3Si]_2NH$ which considerably betters the yield for a yet unknown reason.

E. C—O Bond Chemistry

(See Section II C.1 for the synthesis of alcohols using Grignard reagents prepared from Mg/G, and Bergbreiter and Killough (1978) for the reduction of tosylates to alcohols or phenols by KC_8.)

TABLE VII

Reduction of Activated Double Bonds by KC_8[a,b]

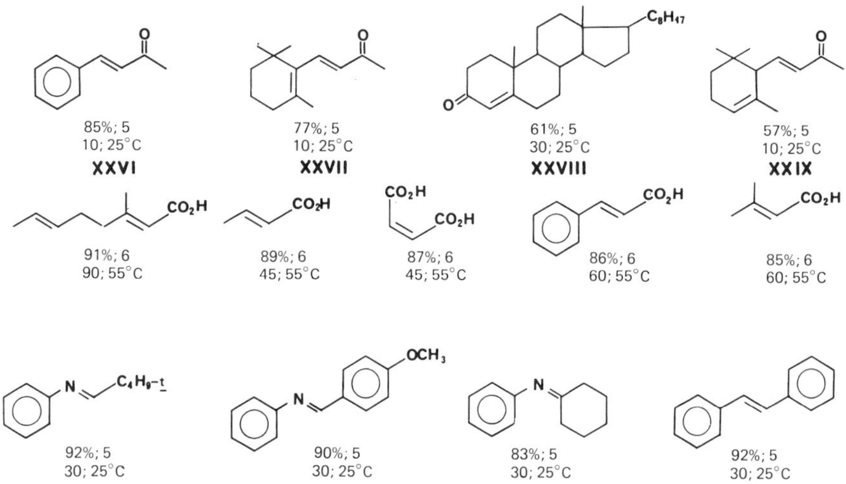

85%; 5 10; 25°C **XXVI**	77%; 5 10; 25°C **XXVII**	61%; 5 30; 25°C **XXVIII**	57%; 5 10; 25°C **XXIX**
91%; 6 90; 55°C	89%; 6 45; 55°C	87%; 6 45; 55°C	86%; 6 60; 55°C	85%; 6 60; 55°C
92%; 5 30; 25°C	90%; 5 30; 25°C	83%; 5 30; 25°C	92%; 5 30; 25°C

[a] From Setton et al. (1982).

[b] Under each compound are given: the yield (%); the ratio KC_8/reducible component; duration of stirring (min); and the temperature (°C).

The commercially available "Seloxcette" (purported to be the GIC of CrO_3) will successfully oxidize primary alcohols to aldehydes (Lalancette et al., 1972), but it is not a true GIC (Ebert and Matty, 1982), while $C_{12}HNO_3$ smoothly oxidizes most benzoins to benzils (Alazard et al., 1977; Setton, 1979).

Tetrahydrofuran co-intercalated with K in the ternary $K(THF)_{2.2} C_{24}$ undergoes a number of chemical changes leading to the formation of gaseous H_2 and ethylene and, after hydrolysis of the solid phase, to n-butanol and ethyl acetate. The formation of these compounds, occurring within the solid matrix, can be explained fairly easily (Setton, 1983); other reactions must be involved in the formation of, among other compounds, cyclopropane and butyraldehyde when THF is refluxed over KC_8 or KC_{24} for prolonged periods of time (Schlögl and Böhm, 1984).

F. Chiral Synthesis

A synthetic graphite was partially oxidized to carry carboxylic groups while retaining good mechanical stability (Kagan et al., 1978), and the following sequence of transformations was performed:

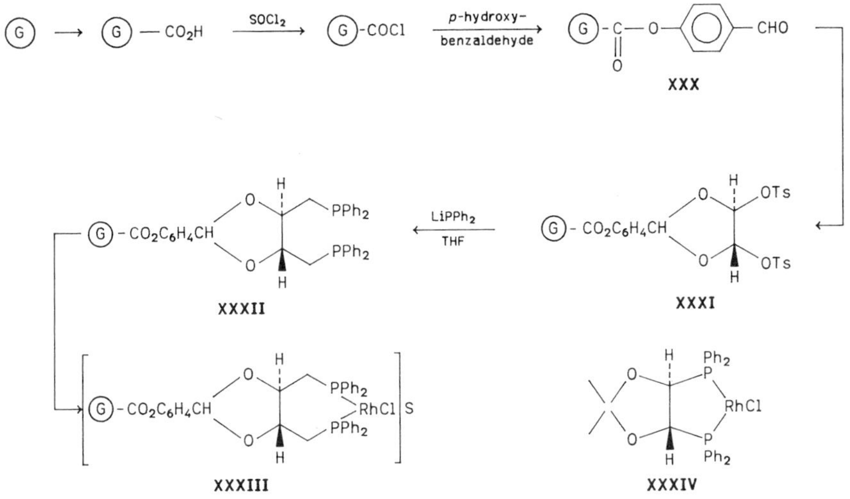

As $(-)$-1,4-ditosylthreitol was used to acetalize the supported aldehyde **XXX**, compounds **XXXI**, **XXXII**, and **XXXIII** were optically active. The fixation of Rh on **XXXII** was obtained by ligand exchange using, for instance, [Rh $(C_2H_4)_2Cl]_2$ so that **XXXIII** contained a molecule of the solvent or an olefin, here noted S. The complex **XXXIII** catalyzed the hydrosilylation of acetophenone, but the reduction of α-acetyl-aminocinnamic acid was very slow. As compared with the results of homogeneous reactions with **XXXIV** as catalyst, the G-supported catalyst always led to smaller optical yields, but the most surprising result was that the optical configuration in nearly all cases was opposite to that observed in the homogeneous reaction.

G. Miscellaneous Reactions

KC_8 has been used, again as a metallation reagent

(1) in the condensation of Ph_3SnBr to $(Ph_3Sn)_2$ (Glockling and Kingston, 1961);

(2) in the formation, at room temperature, of carbonyl complexes (Ungurenasu and Palie, 1975)

$$2 [M (CO)_x] + 2KC_8 \longrightarrow K_2[M_2 (CO)_{2x-2}] + 2 CO + G \quad (M = Cr, Mo, W; x = 6$$

$$M = Fe; x = 5)$$

$$[Co_2 (CO)_8] + 2KC_8 \longrightarrow 2K[Co (CO)_4] + G$$

(3) in the dehalogenation of MeBBr$_2$ to MeB= and its condensation with acetylenes RC≡CR^1 to give diborcyclohexadienes **XXXVI** by dimerization of the borinene **XXXV** and other heterocyclic boron compounds (Van der Kerk *et al.*, 1984):

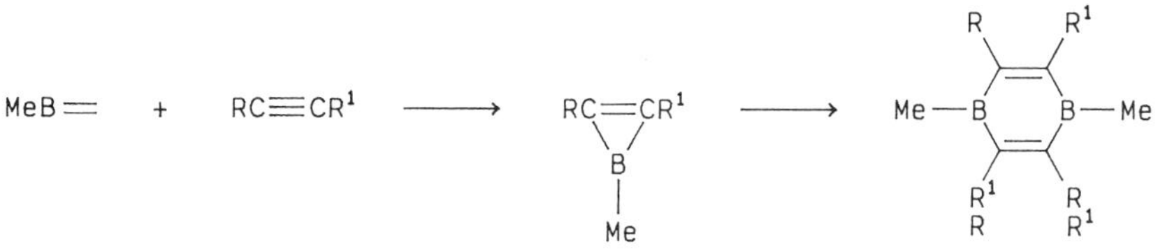

XXXV   **XXXVI**

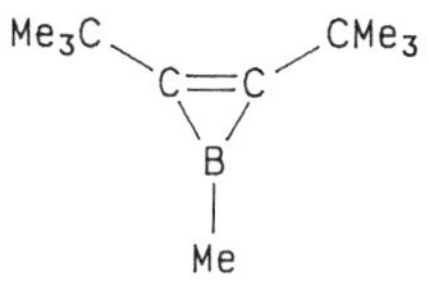

XXXVII

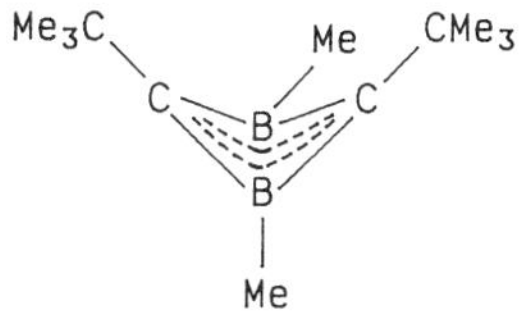

XXXVIII

III. THE GRAPHITE INTERCALATION COMPOUND AS A CATALYST

A. Isotope Effects

Since the elimination of the alkali metal M in a GIC such as MC$_8$ involves diffusion out of the interlamellar space, we can expect isotope effects. Thus, Billaud and co-workers (1973) found that RbC$_8$ was enriched in the heavier isotope [87]Rb by 1.2% after four passages, which corresponds to the value 1.003 for the enrichment factor K defined by

$$(^{87}Rb/^{85}Rb)_{compound} = K\ (^{87}Rb/^{85}Rb)_{metal}$$

Much more spectacular are the values obtained with the isotopes of hydrogen. We must first distinguish between the chemisorption of H$_2$ in KC$_8$, which results in the formation of a new GIC KH$_{2/3}$C$_8$ (Hérold and Saehr, 1965) and the reversible low-temperature physisorption in MC$_{24}$ (with M = K, Rb, or Cs) in which up to 2 moles of H$_2$ are probably retained in vacancies of the intercalated layer. The quantity of H$_2$ physisorbed by the GIC is related to the presence of structural defects in the carbon used for the preparation of the GIC and Terai and Takahashi (1981, 1983) found that the carbons best suited to this purpose were the ones which had never been heated above 1500°C, with enough defects to stabilize first-stage preparations formulated KC_{12}. Lagrange and co-workers (1976a) characterized the efficiency of separation of two isotopes such as D_2 and H_2 by a partition coefficient α defined as

$$\alpha = \frac{\text{proportion of } D_2 \text{ to } H_2 \text{ in the solid phase}}{\text{proportion of } D_2 \text{ to } H_2 \text{ in the gas phase}}$$

which is a function of the relative concentration of D_2 in the gas since it must be 1 for pure D_2. For G-derived KC_{24}, $\alpha = 4.5$ at 77 K and for vanishingly small concentrations of D_2, which should be compared with $\alpha = 10.2$ obtained at the same temperature with KC_{12} (Terai et al., 1982) and $\alpha = 5.0$ for HT/H_2 mixtures, also on KC_{12} at 77 K (Terai and Takahashi, 1984).

In contrast, the room-temperature chemisorption in KC_8 involves the formation of a new species akin to the hydride K^+H^-. This results in values of α which are quite different, dropping from 1 for pure D_2 to 0.48, for low concentrations of D_2 in H_2 (Lagrange et al., 1976b): the solid phase is now richer in the lighter isotope, probably because of the difference in the energy of dissociation of KH and KD (Setton, 1984). By using physisorption or chemisorption, the gas can be enriched at will in the lighter or the heavier isotope.

B. Catalyzed C—C Bond Chemistry

1. Isomerization

Lalancette and Roussel (1976) found that mixtures of KC_{24} and KC_8 quantitatively transformed cis-stilbene dissolved in benzene into the trans isomer and that 2-alkynes were isomerized into a mixture of the 1-yne isomer and the 1,2-dienic (i.e., allenic) derivative. Since, in the latter case, yields were improved when cyclohexane (which does not form a ternary compound with either of the K GICs) was used instead of benzene, it is highly probable that the reactions took place principally in KC_{24}, especially since Béguin et al. (1980) showed that benzene is only readily absorbed by KC_{24} but not by KC_8.

The results of attempts at isomerization of n-hydrocarbons to "high octane" (i.e., branched) paraffins (see also Chapters 22 and 23) over KC_8 or KC_{24} can be summarized as follows:

(1) 1-olefins easily isomerize to the cis compound over KC_8: after 5 min onstream at 25°C, 1-butene gives 45.5% of cis-2-butene, with 81.5% selectivity (Harris and Johnson, 1979);

(2) KC_{24} behaves essentially like KC_8: in general, the reactivity of *n*-olefins increases with the length of the chain and branched olefins are less reactive than the *n*-isomer, each Me group on the chain or on the C atoms of the double bond decreasing the overall reactivity (Tsuchiya *et al.*, 1982, 1983, 1984).

Thermodynamic equilibrium is not reached with either of the above GICs, neither does one get skeletal isomerization; this is more easily obtained using GICs of Brønsted or Lewis acids: *i*-pentane is formed when pentane is passed at 292 K over the GICs of sulfuric or nitric acid, or of $AlCl_3$, $AlBr_3$, $FeCl_3$, or SbF_5 (Imamura and Tsuchiya, 1981). Isomerization occurs at low temperature, with few unfavorable side reactions such as cracking or polycondensation. The first-stage GIC of SbF_5 is particularly effective (Laali *et al.*, 1980; Le Normand *et al.*, 1982b,c), with a very high, temperature-dependent selectivity for isomerization or cracking even at high conversions. The isomerizations are thermodynamically controlled in batch processes, but they can be kinetically controlled in on-stream processes; even then, for long enough residence times, there is a tendency to reach the thermodynamically predictable values. Longer life is obtained for the catalyst when the SbF_5 GIC is formed from partially fluorinated G (Heinerman and Gaaf, 1981) or when the intercalation of the Lewis acid (BF_3, PF_5, or SbF_5) is performed in the presence of gaseous F_2 (Rodewald, 1977; Cohen, 1978). However, slight modifications of the catalyst can result in profound differences in the products formed (Yoneda *et al.*, 1981). Catalysts containing SbF_5 (or $AlCl_3$) and chloroplatinic acid have also been claimed to be useful in the hydroisomerization of cyclohexane to methylpentane and of *n*-hexane to C_6 isomers (Rodewald, 1976).

2. Catalyzed Transalkylations and Cracking

The first-stage SbF_5 GIC efficiently promotes the formation of Et_2Ph isomers and benzene from EtPh by simple mixing at room temperature, with an overall yield of about 30%, while cumene disproportionates (50%) to $(i\text{-}Pr)_2Ph$ and benzene (Laali and Sommer, 1981). However, unlike the Et or *i*-Pr groups, Me groups only disproportionate slowly. All the reactions run under thermodynamic control.

The same reagent catalyzes cracking reactions in gaseous mixtures of C_6 hydrocarbons and H_2 (Le Normand *et al.*, 1982a). The cracking process (which yields *i*-butane and *i*-pentane) competes successfully with the isomerization process, especially when the starting hydrocarbon is 2-methyl-pentane. No cracking occurs with *n*-hexane or 2,2-dimethylbutane which have no tertiary hydrogen atoms.

3. Catalyzed Alkylation Reactions

Wilkes (1982) has confirmed that C_3-C_6 olefins give better yields of linear olefinic products with n-hydrocarbons when the alkylation is performed over K GICs than when performed over simple dispersions of the metal: thus, ethylene and 2-pentene give 57% of 3-heptene and 37% of 3-ethyl-1-pentene, as against 19% and 54%, respectively, over K dispersions.

The $AlCl_3$ GIC is a milder catalyst than pure $AlCl_3$, which promotes polyalkylation in batch processes (Lalancette et al., 1974). The main problem with these catalysts in on-stream processes is, apparently, leaching out of the active species (Olah et al., 1977) and buildup of nonvolatile products (Olah and Kaspi, 1977). In spite of these shortcomings, we should note the alkylation over SbF_5/G of phenols by a number of substituted butenes and pentenes (Kvatcheva et al., 1984), the alkylation of benzene by olefins or alkyl bromides, and its acylation (Furin et al., 1981). In this last case, the oxidative properties of the catalyst were also used to obtain (by reaction with C_6H_6) C_6F_5SePh from $(C_6F_5Se)_2$, $PhHgF$ from HgO, PhSCN from Pb $(SCN)_2$ and $p\text{-}ClC_6F_4SPh$ from $p\text{-}ClC_6F_4SH$.

This same SbF_5/G catalyst was one among many other combinations patented to promote the single-pass transformation of mixtures of natural gas and olefins into C_4 to C_6 hydrocarbons, with high overall conversion yield (Olah, 1983).

C. Catalyzed C—H Bond Chemistry

1. C—H Bond Formation

Early work on the use of the K GICs as catalysts for the hydrogenation of aromatics has been re-examined, and apparent differences in behavior and results explained in terms of the necessity of formation of a ternary GIC (of K and H), which is the true hydrogenating agent (Setton, 1984).

Novikov and co-workers (1973) found that a Pd–G catalyst permitted the hydrogenation of 1-hexene but was completely deactivated for olefin hydrogenation by treatment with an alkali metal, displaying then a high selectivity for the formation of olefins from alkynes. This should be compared with the results of Savoia et al. (1981): reduction of alkynes mainly to alkenes, and quantitative reduction of 2-octene, 3-nitrotoluene, and 4-nitrobenzoic acid, respectively, to n-octane, 3-aminotoluene, and 4-aminobenzoic acid. Similarly, Jones et al. (1984) used a catalyst obtained by partial deintercalation of C_9FeCl_3 by K naphthalenide to convert a 3 $H_2/1$ CO mixture to C_2H_2 with 95% selectivity and 20% efficiency.

2. C—H Bond Cleavage

Novikov *et al.* (1976) claim satisfactory performances from the reduction products of the NiCl$_2$ and CoCl$_2$ GICs, used as such or after oxidation of the metal moiety, in the vapor-phase dehydrogenation of *n*-alcohols to aldehydes and of *sec*-alcohols to ketones.

D. Catalyzed C—O Bond Chemistry

1. Oxidations

The oxyhydrolysis of a third-stage MoCl$_5$ GIC forms a catalyst well suited to the important oxidation of propylene to acrolein (Volta *et al.*, 1979); there is a narrow range of temperature (460–470°C) in which the proportion of (100) faces in the microcrystals of MoO$_3$ (on which the catalysis occurs), and the yield are both maximized (Volta, 1982).

The liquid- and vapor-phase oxidation of hydrocarbons by molecular oxygen over some transition metal GICs (especially of CoCl$_2$ and CuCl$_2$) was investigated by Belousov *et al.* (1980a,b). The main products of the oxidation of EtPh were the hydroperoxide PhCHMeOOH (**XXXIX**) and lesser quantities each of the ketone PhCOMe and the alcohol PhCHOHMe; cumene gave the corresponding hydroperoxide and tertiary alcohol. Later studies on the oxidation of cyclohexene over similar catalysts revealed that appreciable quantities of the epoxide could be obtained over the MoCl$_5$ GIC (Kovtyukhova *et al.*, 1983a). It is then easy to understand why the oxidation of mixtures of EtPh and allyl chloride yields 1-chloro-2,3-epoxypropane (Kovtyukhova *et al.*, 1983b): the epoxyde must be obtained through the hydroperoxide **XXXIX** formed by the oxidation of the ethylbenzene.

The 1-alkene epoxides $\overline{\text{RCH—CH}_2\text{O}}$ isomerize to the corresponding aldehyde RCH$_2$CHO (~70% yield) when passed over Cr or W oxide GICs at 150–500°C (Volta and Cognion, 1977); the same epoxides, and the 1,2-disubstituted homologs, form the corresponding 1,4-dioxanes when in contact with the strong Brønsted acid C$_9$FSO$_3$H **XL** (Setton *et al.*, 1982):

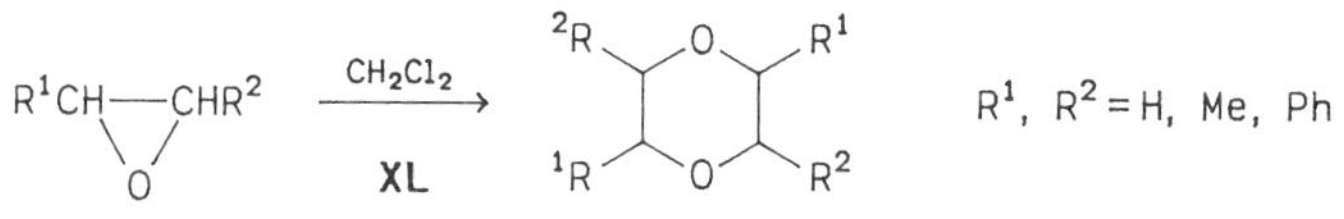

$$R^1CH\!\!-\!\!CHR^2 \quad \xrightarrow{\text{CH}_2\text{Cl}_2} \quad \begin{array}{c} {}^2R \diagup O \diagdown R^1 \\ {}^1R \diagdown O \diagup R^2 \end{array} \qquad R^1,\ R^2 = H,\ Me,\ Ph$$

2. Esterification

Enolizable ketones form enol acetates or benzoates in the presence of **XL** and an acetylating agent [acetic anhydride Ac$_2$O **XLI** or isopropenyl acetate $CH_3C(=CH_2)OAc$ **XLII**] or a benzoylating agent [isopropenyl benzoate $CH_3C\ (=CH_2)\ OBz$ **XLIII**] (Lanconi, 1980):

$$XLI: R = Ac,\ R^1 = MeCO-;\qquad HOR^1 = MeCOOH \tag{3}$$

$$XLII: R = Ac,\ R^1 = MeC(=CH_2)-;\ HOR^1 = MeC(=CH_2)OH \longrightarrow Me_2CO \tag{4}$$

$$XLIII: R = Bz,\ R^1 = MeC(=CH_2)-,\ HOR^1 = MeC(=CH_2)OH \longrightarrow Me_2CO \tag{5}$$

As seen in Table VIII, the yields are equal to or better than the best in the literature. Depending on the nature of R^2, R^3, and R^4 in **XLIV**, the ester **XLV** is a single product or a mixture of two isomers.

Graphite derivatives of strong acids such as $C_{24}{}^{+}HSO_4{}^{-}\cdot 2H_2SO_4$ catalyze the room-temperature synthesis of esters from aliphatic acids and alcohols (Table IX) (Bertin $et\ al.$, 1974a). The ratios acid: alcohol: GIC = 1:1:0.06 often lead to quantitative results, and optical configuration is retained. When ethyl esters are required, EtOH can be replaced by ethyl orthoformate, when even lesser quantities of the catalyst can be used. The fact that this last esterification is not obtained with equivalent quantities of free sulfuric acid is proof that the acid is not simply leached out of the GIC. The same catalyst activates the formation of ketals from ketones and ethyl orthoformate (Alazard $et\ al.$, 1977).

E. Fischer–Tropsch Synthesis of Hydrocarbons

Earlier work on the use of GIC's as catalysts in the Fischer–Tropsch process (i.e., the synthesis of hydrocarbons from mixtures of CO and H_2) has been reviewed by Setton and co-workers (1982) and Ebert (1982).

Morita and co-workers (1978) have patented the use of an H_2-reduced FeCl$_3$ GIC in the production of a high calorific value fuel gas enriched with CH_4 and C_2H_6. Work by Kikuchi and co-workers (1979, 1982) and by Ino and co-workers (1980) confirmed the beneficial effect of pressure: under 5–20 atm. and between 300–400°C, parasitic formation of CO_2 is negligible and the proportion of CH_4 (62–86%) and C_2 hydrocarbons (15–6%) quite important, at least for rates of conversion less than 20%. There is a distinct difference of behavior with similar catalysts made from Fe $supported\ on$—rather than $intercalated\ in$—graphite so that the catalytic activity was associated by the authors "with highly dispersed Fe in the

TABLE VIII

Formation of Enol Esters[a]

Ketone	Reaction[b]	Q^c	Yield Exp.	Yield Lit.	Products formed
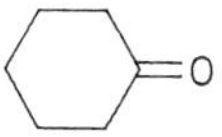	3A	37	88	68	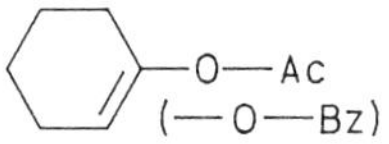
	4A	55	86		
	5A	22	64	30	
	4B	55	60		
	5B	22	30	30	
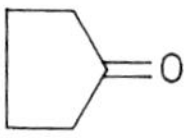	4A	55	86 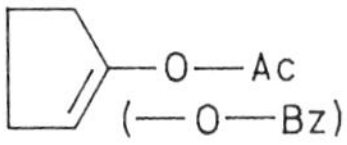		
	4B	22	82		
	5A	22	71	35	
	3A	44	22	12	
	3B	37	18	12	
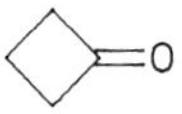	5A	13	20	25	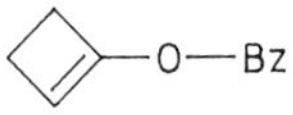

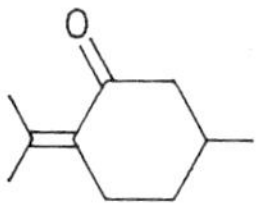

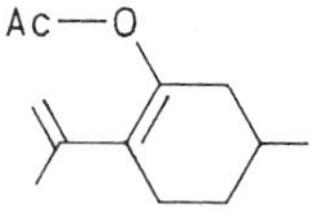

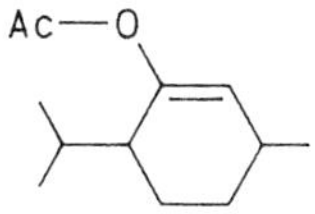

Ketone	Reaction[b]	Q^c	Yield Exp.	Products formed	
	4A	22	85	35%	65%
	4B	22	71	15%	85%

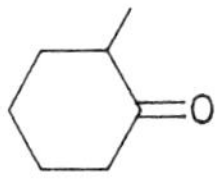 Ketone	Reaction[b]	Q^c	Yield Exp.	Yield Lit.	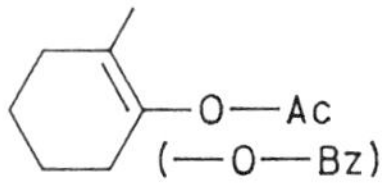	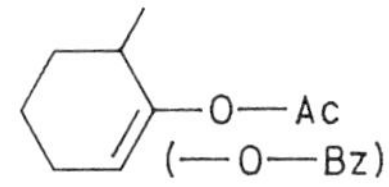
	4B	22	85	50	80%	20%
	5A	44	50	30	70%	30%
	5B	44	30		85%	15%

[a] From Setton et al., (1982).

[b] A is the reaction carried out at P = 240 mm Hg, with continuous distillation and elimination of the more volatile acetic acid or acetone and B at room temperature and atmospheric pressure. The numbers refer to reactions (3), (4) and (5) in text.

[c] Q is the molar ration **LXIV** C_9FSO_3H.

intercalated layers of graphite," but arguments have been presented (Parkash and Hooley, 1980) contending that the catalytically active species is that fraction of the halide which has diffused out from between the G layers during the initial treatment. Recent investigations (Jung et al., 1982; Vaishnava and Montano, 1982, 1983) are still not conclusive.

TABLE IX

Esterification of Aliphatic Acids,[a]

Acid	Alcohol	Ester[b] (%)	Time (hr)
HCOOH	$CH_3CH_2CH_2CH_2OH$	98	1
	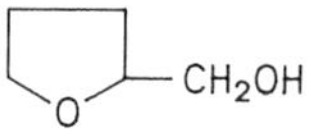 —CH_2OH (tetrahydrofurfuryl)	98	1
	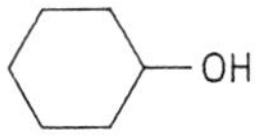 (cyclohexyl)—OH	98	17
	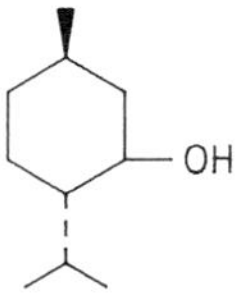 (menthyl)—OH (−)	98[c]	17
CH_3COOH	$C_6H_5CH_2CH_2OH$	97	22
	$CH_3CH_2CH(CH_3)CH_2OH$	98	17
	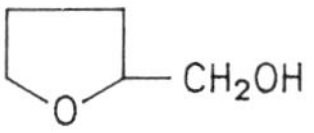 —CH_2OH	94	17
	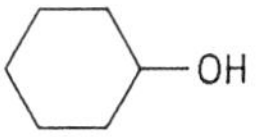 (cyclohexyl)—OH	87	17
	$C_6H_5CH(OH)CH_3$	96	0.5
$C_6H_5CH_2COOH$	$CH_3CH_2CH(CH_3)CH_2OH$	95	26
	$CH_3CH_2\overset{*}{C}H(OH)CH_3(+)$	74[c]	50
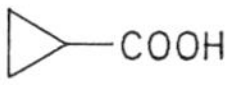 ▷—COOH	$CH_3CH_2CH(CH_3)CH_2OH$	53	17
$HOCOCH_2COOH$	$C_6H_5CH_2OH$(2 equiv)	94	17
$HOCO(CH_2)_3COOH$	$C_6H_5CH_2OH$(2 equiv)	96	17
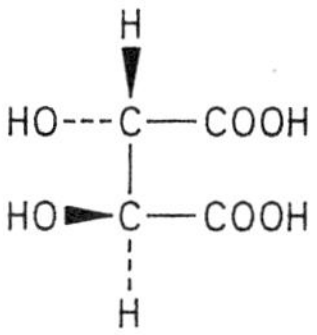 (tartaric acid) (−)	CH_3OH(2 equiv)	99[c]	60
$C_6H_5COCOOH$	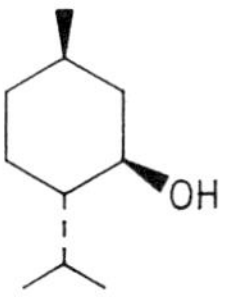 (menthyl)—OH (−)	50[c]	17

[a] Reprinted with permission from Bertin et al. (1974a). Copyright 1974 American Chemical Society.

[b] Identifications were made by ir, nmr, and comparison with authentic samples. Vpc analyses were run on a Perkin-Elmer F11 chromatograph (2 m column, Carbowax 20M 15%). Yields are calculated from the isolated ester.

[c] Esterification occurs with retention of configuration.

F. Synthesis of Ammonia

The potassium-activated transition metal GICs have come into their own in the synthesis of ammonia but the yields obtained are characterized by extreme dispersion: 90% at 350°C with a catalyst $G:FeCl_3: K = 2:0.2:2$ (batch process, Ichikawa et al., 1972). 0.2% at 350°C with $C_{10.4}RuCl_3K_{15}$, and 0.22% with C_{23} $(FeCl_3)_2OsCl_3K_{27}$ at 400°C (on-stream, Vol'pin et al., 1977; Postnikov et al., 1975; Novikov and Vol'pin, 1981). Note that in the last two catalysts the $K:C$ ratio is greater than 1, but lesser values lower the yields.

Recent publications and patents often refer to highly complex mixtures and stress the length of duration of the catalyst. Both Xu et al. (1983) and Ungurenasu et al. (1978) use a $FeCl_3/AlCl_3/G$ catalyst prereduced by H_2 then treated with K, the latter group claiming about 12% conversion even after 300 h; similar durations are claimed by Kalucki et al. (1981) and by Kepinsky et al. (1982, 1983) who also used activation by Cu or other transition metals.

G. Polymerizations

1. Polymerization of Organocyclosiloxanes

The polymerization of organocyclosilaxane monomers with Li or K GICs as initiators has been studied. One of the monomers was an eight-membered ring $(—SiMe_2O—)_4$ while the others were six-membered rings, namely $(—SiMe_2O—)_3$ and three homologs obtained by replacing one of the two Me groups by a Ph, $CF_3C_2H_4$, or CF_3 group. Important differences of behavior were found between the Li and K GICs, while the molecular weights of the polymers ranged between 4 and 10×10^5, depending on the temperature and the initiator. At no time was any metal (from the GIC) found in the polymers (Vol'pin et al., 1982; Kakuliya et al., 1980, 1982; Novikov et al., 1979).

2. Polymerization of Lactones

The GIC KC_{24} has been found to be a prime agent for the anionic polymerization of many lactones; high molecular weights $(1–5 \times 10^5)$ are generally obtained and many of the polymers (e.g., from caprolactone, propiolactone, and pivalolactone) are often semicrystalline (Mazier et al., 1980). The molecular weight of the polymers obtained depends on the temperature at which the polymerization is performed and whether it is carried out in THF or xylene (Rashkov et al., 1983).

3. Polymerization at C=C Double Bonds

All the monomers found to be polymerizable by anionic initiators have given positive results with the alkali metal GICs, but the most interesting were obtained in the polymerization of dienes such as isoprene. Detailed studies (Loria, 1979; Loria *et al.*, 1981; Golé *et al.*, 1982) showed that the microstructure of the polymer is a function of the rate of completion of the reaction. The initial proportion of 1,4-cis linkages is over 90% but, when the end of the reaction is reached, there are about equal proportions both of 1,4-cis and 1,4-trans linkages.

As in the Fischer–Tropsch synthesis, combinations of transition and alkali metals have also been found to be useful in the polymerization of isoprene (Chernenko *et al.*, 1983). The most stereospecific catalysts, such as $C_{9.5}CoK$ or $C_{9.5}CoK_2$ gave 95% of 3,4-addition, in contrast to the results discussed earlier.

[Considerable promise accrues from the use of GICs, some handsome results have already been achieved and were focussed on in this chapter. Future progress hinges, it would seem, on improved structural and mechanistic understanding.—ED.]

REFERENCES

Agranat, I., Rabinowitz, M., Selig, H., and Lin, C.-H. (1977). *Synthesis* p. 267.

Alazard, J. P., Kagan, H. B., and Setton, R. (1977). *Bull. Soc. Chim. Fr.* p. 499.

Béguin, F., and Setton, R. (1976). *J.C.S. Chem. Commun.* p. 611.

Béguin, F., Setton, R., Facchini, L., Legrand, A. P., Merle, G., and Mai, C. (1980). *Synth. Met.* **2**, 161.

Belousov, V. M., Kovtyukhova, N. I., Mikhalovskii, S. V., Novikov, Y. N., and Vol'pin, M. E. (1980a). *Dokl. Akad. Nauk SSSR* **253**, 346.

Belousov, V. M., Pal'chevskaya, T. A., Zaitsev, Y. P., Zazhigalov, V. A., and Novikov, Y. N. (1980b). *Katal. Katal.* **18**, 43.

Bergbreiter, D. E., and Killough, J. M. (1978). *J. Am. Chem. Soc.* **100**, 2127.

Bertin, J., Kagan, H. B., Luche, J.L., and Setton, R. (1974a). *J. Am. Chem. Soc.* **96**, 8113.

Bertin, J., Luche, J. L., Kagan, H. B., and Setton, R. (1974b). *Tetrahedron Lett.* p. 763.

Bewer, G. Wichmann, N., and Böhm, H. P. (1977). *Mater. Sci. Eng.* **31**, 73.

Billaud, D., Hérold, A., and Leutwein, F. (1973). *C. R. Hebd. Seances Acad. Sci., Ser. C* **277**, 419.

Böhm, H .P., Ko, Y.-S., Ruisinger, B., and Schögl, R. (1985). *In* "Chemical Reactions in Organic and Inorganic Constrained Systems" (R. Setton, ed.), p. 429. Reidel Publ., Dordrecht, The Netherlands.

Boersma, M. A. M. (1974). *Catal. Rev.—Sci. Eng.* **10**, 243.

Boersma, M. A. M. (1977). *In* "Advanced Materials in Catalysis(J. J. Burton and R. L. Garten, eds.), pp. 67–69. Academic Press, New York.

Boldrini, G. P., Savoia, D., Tagliavini, E., Trombini, C., and Umani-Ronchi, A. (1983). *J. Org. Chem.* **48**, 4108.

Braga, D., Ripamonti, A., Savoia, D., Trombini, C., and Umanu-Ronchi, A. (1978). *J.C.S. Chem. Commun.* p. 927.

Braga, D., Ripamonti, A., Savoia, D., Trombini, C., and Umani-Ronchi, A. (1979). *J.C.S. Dalton* p. 2026.

Chernenko, G. M., Tinyakova, E. I., Kakuliya, T. V., Khananashvili, L. M., Novikov, Y. N., and Vol'pin, M. E. (1983). *Vysokomol. Soedin., Ser. B* **25**, 219.

Cohen, A. D., (1978). U.S. Pat. 4,128,499.

Contento, M., Savoia, D., Trombini, C., and Umani-Ronchi, A. (1979). *Synthesis* p. 30.

Corriu, R. J. P., Fernandez, J. M., and Guerin, C. (1980). *J. Organomet. Chem.* **192**, 347.

Dresselhaus, M. S., and Dresselhaus, G. (1981). *Adv. Phys.* **30**. 139.

Ebert, L. B. (1982). *J. Mol. Catal.* **15**, 275.

Ebert, L. B., and Matty, L., Jr. (1982). *Synth. Met.* **4**, 345.

Fischer, J. E. (1979). In "Intercalated Layered Materials" (F. Levy, ed.), Physics and Chemistry of Materials with Layered Structures, Vol. 6, pp. 481–532. Reidel, Dordrecht, The Netherlands.

Furin, G. G., Avramenko, A. A., Nikonorov, Y. I., and Yakobson, G. (1981). *Zh. Org. Khim.* **17**, 1505.

Glockling, F., and Kingston, D. (1961). *Chem. Ind. (London)* p. 1037.

Golé, J., Merle, G., and Pascault, J. P. (1982). *Synth. Met.* **4**, 269.

Harris, J. R., and Johnson, M. M. (1979). U.S. Pat. 4,150,065.

Heineman, M. G., and Latscha, H. P. (1981). *Chem.-Ztg.* **105**, 255.

Heinerman, J. J. L., and Gaaf, J. (1981). *J. Mol. Catal.* **11**, 215.

Hérold, A. (1979). *In* "Intercalated Layered Materials" (F. Levy, ed.), Physics and Chemistry of Materials with Layered Structures, Vol. 6, pp. 321–422. Reidel, Dordrecht, The Netherlands.

Hérold, A., and Saehr, D. (1965). *Bull. Soc. Chim. Fr.* p. 3130.

Ichikawa, M., Kondo, I., Kawase, K., Sudo, M., Onishi, T., and Tamaru, K. (1972). *J.C.S. Chem. Commun.* p. 176.

Imamura, H., and Tsuchiya, S. (1981). *Z. Phys. Chem. (Wiesbaden)* **125**, 251.

Inagaki, M., Shiwachi, Y., and Maeda, Y. (1984). *J. Chim, Phys.* **81**, 847.

Ino, T., Ito, N., Kikuchi, E., and Morita, Y. (1980). *Sekiyu Gakkaishi* **23**, 408.

Jones, W., Schlögl, R., and Thomas, J.M. (1984). *J.C.S. Chem. Commun.* p. 464.

Jung, H. J., Walker, P. L., Jr., and Vannice, M. A. (1982). *J. Catal.* **75**, 416.

Kagan, H. B., Yamagishi, T., Motte, J. C., and Setton, R. (1978). *Isr. J. Chem.* **17**, 274

Kakuliya, T. V., Khananashvili, L. M., Dneprovskaya, L. S., Novikov, Y. N., and Vol'pin M. E. (1980). *Vysokomol. Soedin., Ser. A* **22**, 1690.

Kakuliya, T. V., Khananashvili, L. M., Novikov, Y. N., and Vol'pin, M. F. (1982). *Izv. Akad. Nauk Gruz. SSSR, Ser. Khim.* **8**, 151.

Kalucki, K., Morawski, W., and Arabczyk, W. (1981). *Stud. Surf. Sci. Catal.* **7**, 1496.

Kepinsky, J., Morawski, W., Kalucki, K., and Kalenczuk, R. (1982). Pol. Pat. 116, 777.

Kepinsky, J., Kalucki, K., Morawski, W., and Arabczyk, W. (1983). Pol. Pat. 119, 912.

Kikuchi, E. (1978). *Kenkyu Hokoku—Asahi Garasu Kogyo Gijutsu Shoreikai* **32**, 276.

Kikuchi, E., Ino, T., and Morita, Y. (1979). *J. Catal.* **57**, 27.

Kikuchi, E., Ino, T., and Morita, Y. (1980). *J. Catal.* **62**, 189.

Kikuchi, E., Inoue, S., and Morita, Y. (1982). *Nippon Kagaku Kaishi* p. 185.

Kovtyukhova, N. I., Belousov, V. M., Mikhalovskii, S. V., Kvacheva, L. D., Novikov, Y. N., and Vol'pin, M. E. (1983a). *Izv. Akad. Nauk SSSR, Ser. Khim.* p. 25.

Kovtyukhova, N. I., Belousov, V., Novikov, Y. N., and Vol'pin, M. E. (1983b). *Izv. Akad. Nauk SSSR, Ser. Khim.* p. 1728.

Kvatcheva, L. D., Klimakhina, S. T., Grigor'eva, L. T., Podberezina, A. S., Novikov, Y. N., Kheifits, L. A., and Vol'pin, M. E. (1984). *Izv. Akad. Nauk SSSR, Ser. Khim.* p. 1907.

Laali, K., and Sommer, J. (1981). *Nouv. J. Chim.* **5**, 469.

Laali, K., Muller, M., and Sommer, J. (1980). *J.C.S. Chem. Commun.* p. 1088.

Lagrange, P., Portman, M. H., and Hérold, A. (1976a). *C. R. Hebd. Séances Acad. Sci., Ser. C* **283**, 511.

Lagrange, P. Portmann, M. H., and Hérold, A. (1976b). *C. R. Hebd. Séances Acad. Sci., Ser. C* **283**, 557.

Lalancette, J. M., and Roussel, R. (1976). *Can. J. Chem.* **54**, 2110.

Lalancette, J. M., Rollin, G., and Dumas, P. (1972). *Can. J. Chem.* **50**, 3058.

Lalancette, J. M., Fournier-Bréault, M. J., and Thiffault, R. (1974). *Can. J. Chem.* **52**, 589.

Lanconi, A. (1980). D. E. A. Chim. Org., Univ. Lille.

Le Normand, F., Fajula, F., and Sommer, J. (1982a). *Nouv. J. Chim.* **6**, 291.

Le Normand, F., Fajula, F., Gault, F., and Sommer, J. (1982b). *Nouv. J. Chim.* **6**, 411.

Le Normand, F., Fajula, F., Gault, F., and Sommer, J. (1982c). *Nouv. J. Chim.* **6**, 417.

Lin, C. H., Selig, H., Rabinowitz, M., Agranat, I., and Sarig, S. (1975). *Inorg. Nucl. Chem. Lett.* **11**, 601.

Loria, E. (1979). Thesis, Univ. Lyon.

Loria, E., Merle, G., Pascault, J. P., and Rashkov, I. B. (1981). *Polymer* **22**, 95.

Mashinskii, V. L., Postnikov, V. A., Novikov, Y. N., Lapidus, A. L., Vol'pin, M. E., and Eidus Y.T. (1976). *Izv. Akad. Nauk SSSR, Ser. Khim.* p. 2018.

Mazier, C., Douillard, C., Merle, G., and Pascault, J. P. (1980). *Eur. Polym. J.* **16**, 773.

Morita, Y., Kikuchi, E., and Ino, T. (1978). Jpn. Pat. 74,502.

Münch, V., and Selig, H. (1980). *J. Fluorine Chem.* **15**, 253.

Nefed'ev, A. V., Stukan, R. H., Alekseev, V. P., Postnikov, V. A., Shur, V. B., Novikov, Y. N., and Vol'pin, M. E. (1977). *Proc.—Int. Conf. Moessbauer Spectrosc., Bucharest* **1**, 259.

Nefed'ev, A. V., Stukan, R. A., Makarov, V. A., Kondarov, V. A., Shuvaev, A. T., Lapkina, N. D., Novikov, Y. N, and Vol'pin, M. E. (1980). *Zh. Strukt. Khim.* **21**, 68.

Novikov, Y. N., and Vol'pin, M. E. (1981). *Physica B + C (Amsterdam)* **105, B + C**, 471.

Novikov, Y. N., Kakuliya, T. V., Khananashvili, L. M., Kopylov, V. M., and Vol'pin M. E. (1979). *Dokl. Nauk. SSSR* **245**, 848.

Novikov, Y. N., Postnikov, V. A., Salyn, Y. V., Nefed'ev, V. I., and Vol'pin, M. E. (1973). *Izv. Akad. Nauk SSSR, Ser. Khim.* p. 131.

Novikov, Y. N., Lapkina, N. D., and Vol'pin, M. E. (1976). *Kinet. Katal.*, **17**, 1537.

Okawara, Y., Hirosaki, N., and Kurosawa, O. (1984). *Kenkyu Hokoku—Kanto Gakuin Daigaku Kogakubu* **27**, 25.

Oku, A., Ueda, H., and Tamatani, H. (1980). *Nippon Kagaku Kaishi* p. 1903.

Olah, G. A. (1983). Eur. Pat. Appl. EP 73673.

Olah, G. A., and Kaspi, J. (1977). *J. Org. Chem.* **42**, 3046.

Olah, G. A., Kaspi, J., and Bukala, J. (1977). *J. Org. Chem.* **42**, 4187.

Page Lécuyer, A., Luche, J. L., Kagan, H. B., Colin, G., and Mazières, C. (1973). *Bull. Soc. Chim. Fra.* p. 1690.

Parkash, S., and Hooley, J. G. (1980). *J. Catal.* **62**, 187.

Parkash, S., Chakrabartty, S. K., and Hooley, J. G. (1977). *Carbon* **15**, 307.

Parkash, S., Chakrabartty, S. K., and Hooley, J. G. (1978). *Carbon* **16**, 231.

Postnikov, V. A., Dimitrienko, L. M., Ivanova, R. F., Dobrolyubova, N. I., Golubeva, M. A., Gapeeva, T. I., Novikov, Y. N., Shur, V. B., and Vol'pin, M. E. (1975). *Izv, Akad. Nauk SSSR, Ser. Khim.* p. 2642.

Rabinowitz, M., and Tamarkin, D. (1984). *Synth. Commun* **14**, 377.

Rabinowitz, M., Agranat, L., Lin, C. H., and Ebert, L. B. (1977a). *J. Am. Chem. Soc.* **99**, 953.

Rabinowitz, M., Agranat, I., Selig, H., Lin, C. H., and Ebert, L. B. (1977b). *J. Chem. Res.*, *Miniprint* p. 2350.

Rashkov, I. B., Panayotov, I. M., and Shishkova, V. C. (1979). *Carbon* **17**, 103.

Rashkov, I., Shiskova, V., Panayotov, I., Merle, G., and Letoffé, J. M. (1980) *Proc. 3rd, Baden-Baden Int. Carbon Conf.*, p. 96.

Rashkov, I., Gitsov, I., and Panayotov, I. (1983). *J. Polym. Sci., Polym. Chem. Ed.* **21**, 923.

Rodewald, P. G. (1976). U.S. Pat. 3,976,714.

Rodewald, P. G. (1977). U.S. Pat. 4,035,434.

Savoia, D., Trombini, C., and Umani-Ronchi, A. (1977). *Tetrahedron Lett.* p. 653.

Savoia, D., Trombini, C., and Umani-Ronchi, A. (1978). *J. Org. Chem.* **43**, 2957.

Savoia, D., Trombini, C., Umani-Ronchi, A., and Verardo, G. (1981). *J.C.S. Chem. Commun.* p. 540.

Savoia, D., Tagliavini, E., Trombini, C., and Umani-Ronchi, A. (1982). *J. Org. Chem.* **47** 876.

Schlögl, R., and Böhm, H. P. (1984). *Carbon* **22**, 341.

Schlögl, R., Jones, W., Böhm, H. P., and Breimeir, W. (1983). *Synth. Met.* **8**, 323.

Selig, H., and Ebert, L. B. (1980). *Inorg. Chem. Radiochem.* **23**, 281.

Selig, H., and Gani, O. (1972). *Inorg. Nucl. Chem. Lett.* **11**, 75.

Selig, H., Rabinowitz, M., Agranat, I., Lin, C. H., and Ebert, L. (1976). *J. Am. Chem. Soc.* **98**, 1601.

Setton, R. (1979). Unpublished results.

Setton, R. (1983). *Tanso* p. 180.

Setton, R. (1984). *J. Mol. Catal.* **27**, 263.

Setton, R., Béguin, F., and Piroelle, S. (1982). *Synth. Met.* **4**, 1982.

Solin, S. A. (1982). *Adv. Chem. Phys.* **49**, 455.

Tamarkin, D., and Rabinowitz, M. (1984). *Synth. Met.* **9**, 125.

Tamarkin, D., Benny, D., and Rabinowitz, M. (1984). *Angew. Chem., Int. Ed. Engl.* **8**, 642.

Terai, T., and Takahashi, Y. (1981). *J. Nucl. Sci. Technol.* **18**, 643.

Terai, T., and Takahashi, Y. (1983). *Synth. Met.* **7**, 49.

Terai, T., and Takahashi, Y. (1984). *Carbon* **22**, 91.

Terai, T., Inoue, A., and Takahashi, Y. (1982). *Proc. Int. Symp. Carbon, Toyohashi, Jpn.* p. 220.

Tsuchiya, S., Yamamoto, S., Ohuye, N., and Imamura, H. (1982). *Proc. Int. Carbon, Toyohashi, Jpn.* p. 224.

Tsuchiya, S., Yamamoto, S., and Imamura, H. (1983). *Synth. Met.*, **6**, 1.

Tsuchiya, S., Yamomoto, S., Ohuye, N., and Imamura, H. (1984). *J. Catal.* **88**, 225.

Ungurenasu, C., and Palie, M. (1975). *J.C.S. Chem. Commun.* p. 388.

Ungurenasu, C., and Palie, M. (1977). *Synth. React. Inorg. Met. Org. Chem.* **7**, 581.

Ungurenasu, C. Streba, E., Ionescu, C., Barbinta, V., Popa, T. V., and Palie, M. (1978). Rom. Pat. 63,826.

Vaishnava, P. P., and Montano, P. A. (1982). *J. Phys. Chem. Solids* **43**, 809.

Vaishnava, P. P., and Montano, P. A. (1983). *Proc. Mater. Res. Soc. Symp., Boston, Mass.* **20**, 397.

Van der Kerk, S. M., Budzelaar, P. H. M., Van Eekeren, A. L. M., and Van der Kerk, G. J. M. (1984). *Polyhedron* **3**, 271.

Vol'pin, M. E., Novikov, Y. N., Lapkina, N. D., Kasatochkin, V. I., Struchkov, Y. I., Kasakov, M. E., Stukan, R. A., Povitskij, V. A., Karimov, Y. S., and Zvarikina, A. V. (1975). *J. Am. Chem. Soc.* **97**, 3367.

Vol'pin, M. E., Novikov, Y. N., Postnikov, V. A., Bayerl, B., Kaden, L., Wahren, M.,

Dmitrienko, L. M., Stukan, R. A., and Nefed'ev, A. V. (1977). *Z. Anorg. Allg. Chem.* **428**, 231.

Vol'pin, M. E., Novikov, Y.N., Kopylov, V. M., Khananashvili, L. M., and Kakuliya, T. B. (1982). *Synth. Met.* **4**, 331.

Volta, J. C. (1982). *Synth. Met.* **4**, 319.

Volta, J. C., and Cognion, J. M. (1977). Fr. Pat. Appl. 2,338,920.

Volta, J. C., Desquesnes, W., Moraweck, B., and Coudurier, G. (1979). *React. Kinet. Catal. Lett.* **12**, 241.

Wilkes, J. B. (1982). *Ind. Eng. Chem., Prod. Res. Dev.* **21**, 585.

Xu, J., Liu, X., Zhang, Z., Xu, L., Huang Q., Liu, X., Chen, Y., Niu, S., and Yang, Z. (1983). *Chi Lin Ta Hsueh, Tzu Jan K'o Hsueh Hsueh Pao* p. 123.

Yemul, S. S., Kagan, H. B., and Setton, R. (1980). *Tetrahedron Lett.* **21**, 277.

Yoneda, N., Fukuhara, T., Abe, T., and Suzuki, A. (1981). *Chem. Lett.* p. 1485.

Part **V**

Alumina-Supported Reagents

16 ALUMINA AND ALUMINA-SUPPORTED REAGENTS

Gary H. Posner

Department of Chemistry
The Johns Hopkins University
Baltimore, Maryland 21218

I. INTRODUCTION

This chapter is an update dealing with organic reactions promoted by chromatographic γ-alumina (Posner, 1978). Because an accompanying chapter by Foucaud and Bram in this volume deals with alumina-promoted carbon–carbon bond-forming reactions, this chapter is limited mainly to functional group interconversions promoted by alumina.

More and more, the general chemical community is realizing that many porous solids [e.g., alumina and silica, (McKillop and Young, 1979)] are useful not only for industrial operations (e.g., cracking of petroleum) but also for many small, laboratory-scale, heterogeneous reactions often involving milder condition ($\leq 100°C$), more specific (chemospecific, regiospecific, and stereospecific) transformations, and easier product isolation than in the corresponding homogeneous reactions. The many examples provided in this chapter illustrate the growing popularity of alumina for diverse organic reactions and attest to the significant advantages of heterogeneous reactions in a wide variety of chemical operations.

Although this chapter emphasizes uses of alumina in small laboratory-scale organic syntheses, industrial use of alumina on a large scale often as a solid support for normally homogeneous catalysts is quite important. Several spectroscopic studies [IR, (Evans and Gracey, 1980; Lavalley *et al.*, 1980; Riseman *et al.*, 1982, NMR (Dawson *et al.*, 1981, 1982; Ripmeester, 1983), inelastic electron tunneling (Evans and Weinberg,

**PREPARATIVE CHEMISTRY
USING SUPPORTED REAGENTS**

1980) and transmission electron microscopy (Miller and Majda, 1985)] of the surface properties of γ-alumina have revealed interesting structural details and a more accurate picture of the interaction between acidic and basic sites on the surface (Misono and Okuhara, 1980; Stair, 1982). Such studies help characterize those surface features that are essential for alumina's ability to promote various organic reactions, help produce aluminas having consistent and reliable surface properties, and thus help change the use of alumina from an art to a science. To standardize reporting, all researchers using alumina are *urged* to specify in their publications the source, the activity, and the acidity–basicity of the alumina they have used (e.g., Woelm–Act. I–Neutral).

II. ADDITION REACTIONS

A. Intramolecular Addition Reactions

Several examples have appeared since 1978 involving intramolecular Michael-type additions of hydroxyl groups across activated carbon–carbon double bonds (RT = room temperature) [Eq. (1) (Caine and Smith, 1980); Eq. (2) (Danishefsky and Pearson, 1983), and Eq. (3) (Kan-Fan and Husson, 1980)]. Equation (2) represents successful alumina-promoted spiroketal cyclizations that are unsuccessful with strong acids. Note that Eqs. (1)–(3) involve structurally complex, multifunctional organic molecules that react chemospecifically on alumina.

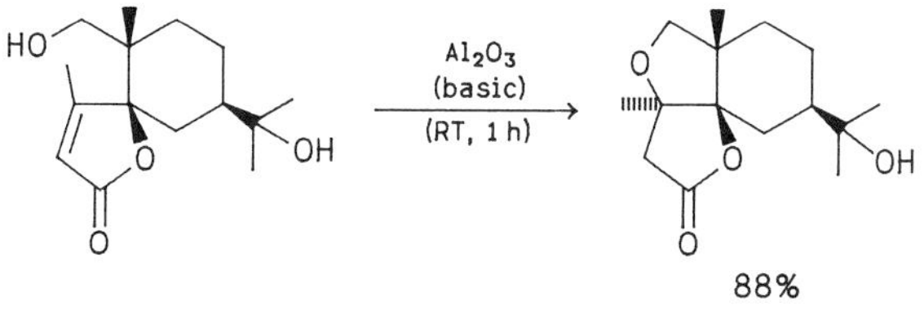

$$\text{(1)}$$

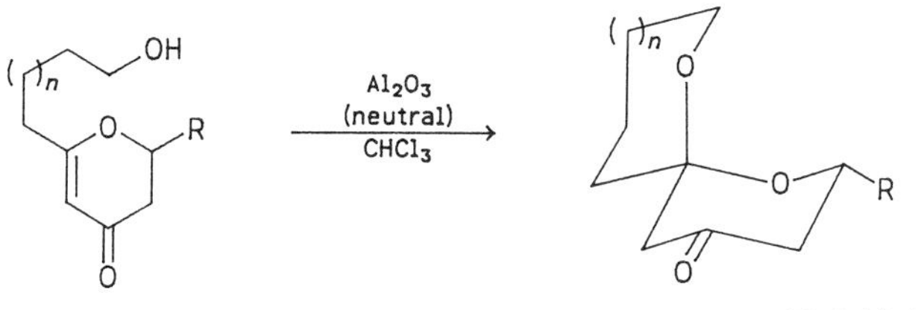

88%

$$\text{(2)}$$

Yield (%)

$n = 0$,	R = Et	80
$n = 0$,	R = Ph	82
$n = 1$,	R = Me	56

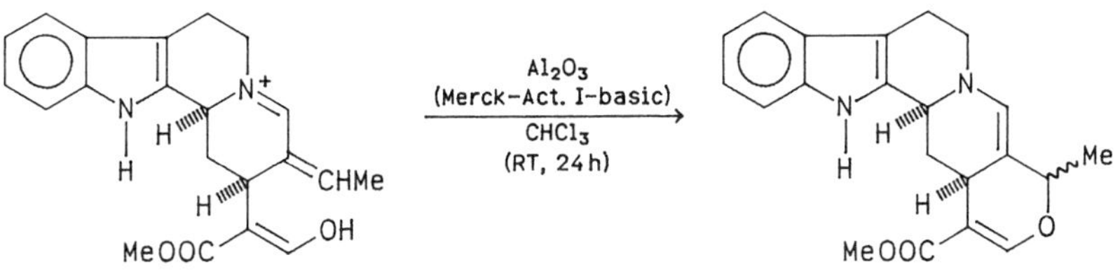

$$\text{(3)}$$

Epoxides carrying nucleophilic groups in the same molecule can be activated by alumina to undergo intramolecular opening at room temperature [Eq. (4) (Boeckman *et al.*, 1978), Eq. (5) (Tanis and Herrinton, 1983), and Eq. (6) (Niwa *et al.*, 1979; Yamamura *et al.*, 1982)].

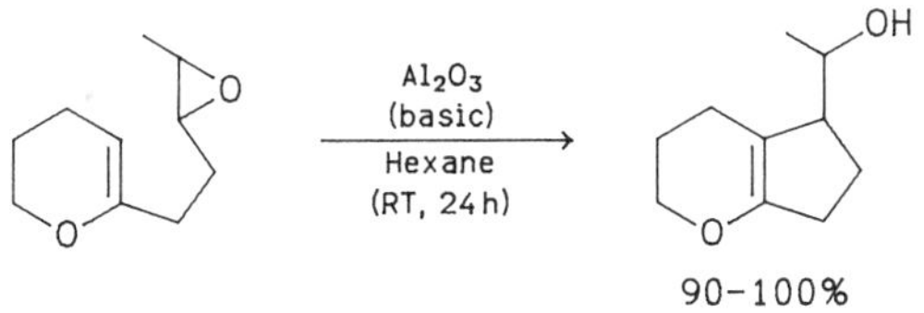

$$\xrightarrow[\substack{\text{Hexane} \\ \text{(RT, 24 h)}}]{\substack{\text{Al}_2\text{O}_3 \\ \text{(basic)}}} \qquad \text{90--100\%} \qquad \text{(4)}$$

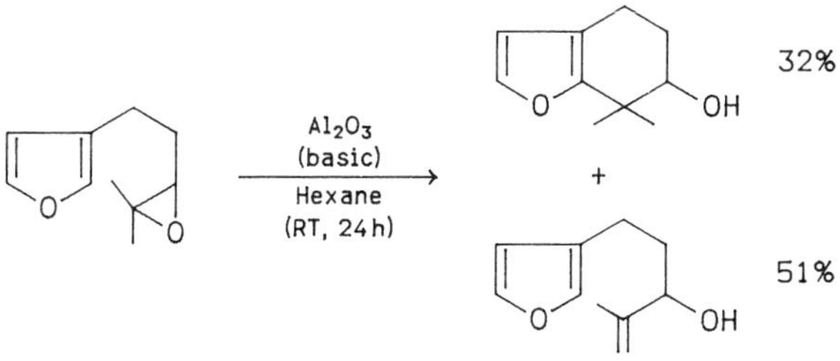

$$\xrightarrow[\substack{\text{Hexane} \\ \text{(RT, 24 h)}}]{\substack{\text{Al}_2\text{O}_3 \\ \text{(basic)}}} \qquad \begin{array}{l} 32\% \\ + \\ 51\% \end{array} \qquad \text{(5)}$$

$$\xrightarrow[\substack{\text{Hexane} \\ \text{(Rt, 2.5 h)}}]{\substack{\text{Al}_2\text{O}_3 \\ \text{(basic)}}} \qquad \begin{array}{l} 11\% \\ + \\ 10\% \end{array} \qquad \text{(6)}$$

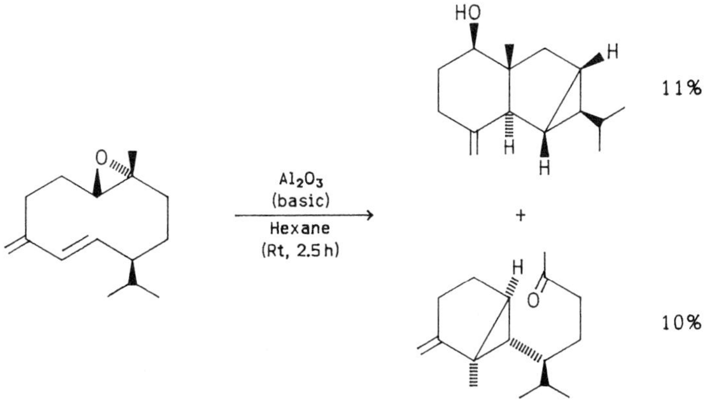

B. Intermolecular Addition Reactions

1. Carbonyl–Oxygen Bonds

  a. Epoxides.   Opening of epoxides is usually achieved homogeneously by treating oxiranes with strong acids or strong bases. Commercially available chromatographic γ-alumina promotes highly efficient heterogeneous opening of a wide range of epoxides by a variety of nucleophiles (e.g., alcohols, acetic acid, amines, mercaptans, and selenols) under extremely mild (25°C) and neutral conditions; this procedure is the method of choice for clean nucleophilic opening of medium-ring epoxides (Posner and Rogers, 1977). Oxiranes and oxetanes and also opened by alkyl hydroperoxides on alumina at room temperature (Kropf and Torkler, 1985a,b,).

  Medium-ring epoxides are not only less reactive than cyclohexene or acyclic alkene oxides, but they are also more prone to transannular reactions. *cis*-Cyclooctene oxide, as a typical example, undergoes solvolysis at 60°C with acetic acid to give 54% of transannular reaction product(s); even with sodium acetate as a buffer, 24% of transannular product(s) is formed. Under the best previous homogeneous conditions, *trans*-2-acetoxtcyclooctanol has been isolated in only 22% yield from *cis*-cyclooctene oxide (Cope *et al.*, 1959). In sharp contrast to these homogeneous reactions, which produce a mixture of products requiring careful chromatography for purification, heterogeneous opening of medium-ring epoxides on neutral alumina at 25°C proceeds stereospecifically, regiospecifically, and on a preparatively useful gram-scale to yield 2-substituted cycloalkanols which are purified simply and conveniently by distillation. Table I provides some examples of this alumina-promoted opening of medium-ring epoxides (Posner *et al.*, 1981b).

  We have reported that catalytic amounts of Woelm alumina at 100°C cause stereospecific trans opening of a representative olefine epoxide, cyclohexene oxide, by alcohols, acetic acid, and aniline (Posner and Hulce, 1980). As shown in Eq. (7), 3 g of Woelm–200–neutral alumina effects clean opening of 30 g of cyclohexene oxide by methanol, allyl alcohol, acetic acid, and aniline. 1,2-Glycol monoethers and monoesters and 1,2-amino alcohols are valuable industrial intermediates. Alcoholysis usually has been achieved homogeneously by treating epoxides with alcohols and strong acids or strong bases, and homogeneous aminolysis of epoxides often requires vigorous reaction conditions and often leads to a mixture of products that are difficult to separate. In contrast, Eq. (7) represents a reproducible, convenient, simple, efficient, neutral, and nonaqueous new procedure for preparation of multigram quantities of

TABLE I

Woelm–200–Neutral Alumina-Promoted Opening of Medium
Ring Epoxides by Acetic Acid at 25°C

Epoxide	Product	% Yield (after distillation)
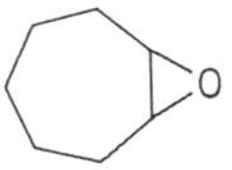	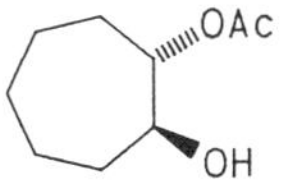	92
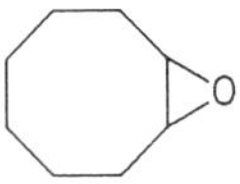	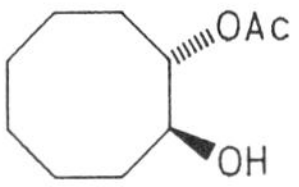	77–78
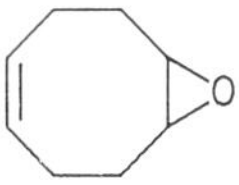	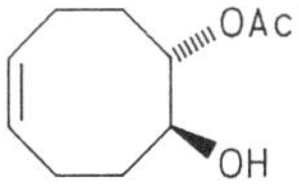	65
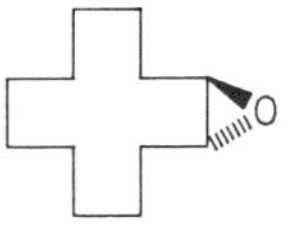	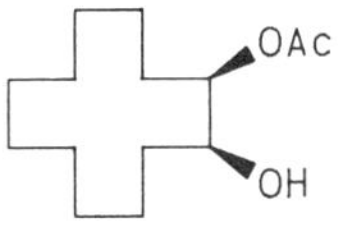	69

trans-1,2-glycol and *trans*-1,2-amino alcohol derivaties. This alumina catalysis (i.e., catalytic conversion of 100 mmol of epoxide per 1 g of alumina)
compares favorably with zeolite catalysts used for this type of process
(Gerveny and Ruzicka, 1975). Indeed these alumina-catalyzed epoxide
openings have some distinct advantages over the zeolite reaction: (1) much
higher conversion of epoxide to product and (2) the ability to tolerate large
(e.g., steroidal) epoxides and large nucleophiles (e.g., aniline). For a study
of effective zeolite-promoted opening of epoxides by amines on a
millimolar scale see Onaka *et al.* (1985).

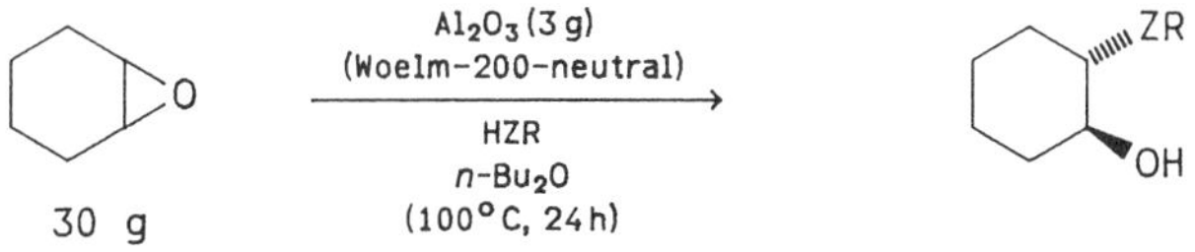

HZR	Yield (%)
$HOCH_2CH = CH_2$	86
$HOCH_3$	87
$HOOCCH_3$	71
H_2NPh	91

(7)

b. Aldehydes and Ketones. Alumina promotes reaction between ethylene glycol and various aliphatic and aromatic aldehydes to produce the corresponding acetals cleanly and in high yield [Eq. (8) (Kamitori *et al.*, 1985a]. The alumina may be facilitating the initial carbonyl addition step and/or the subsequent dehydration step. Kamitori *et al.* (1985b) show that this acetalization process is very selective for aldehydes in preference to ketones [Eq. (9)] and that alumina is more generally effective than silica.

$$
\underset{\text{(5 mmol)}}{\underset{\substack{\| \\ \text{RCH}}}{\text{O}}} \xrightarrow[\substack{\text{HO} \\ \text{HO} \\ \text{CCl}_4 \\ \text{Reflux} \\ (24\,\text{h})}]{\substack{\text{Al}_2\text{O}_3\,(0.25\,\text{g}) \\ \text{(Woelm–TLC–acidic)}}} \underset{87-100\%}{\text{RCH}\langle{}^{\text{O}}_{\text{O}}\rangle} \tag{8}
$$

$$
\text{PhC}\overset{\text{O}}{\underset{\|}{-}}\langle\bigcirc\rangle\text{—CH}\overset{\text{O}}{\underset{\|}{}} \xrightarrow[\substack{\text{HO} \\ \text{HO} \\ \text{CCl}_4 \\ \text{Reflux} \\ (24\,\text{h})}]{\substack{\text{Al}_2\text{O}_3\,(0.25\,\text{g}) \\ \text{(Woelm–TLC–acidic)}}} \underset{100\%}{\text{PhC}\overset{\text{O}}{\underset{\|}{-}}\langle\bigcirc\rangle\text{—CH}\langle{}^{\text{O}}_{\text{O}}\rangle} \tag{9}
$$

Alumina also promotes reaction between amines and ketones (Texier-Boullet, 1985). For examples, a sulfonylhydrazine reacts with acyclic ketones, including acid-sensitive β-keto esters, to produce the corresponding sulfonylhydrazone condensation products [Eq. (10) (Vinczer *et al.*, 1984)], and morpholine reacts with unsymmetrical cyclic ketones in the presence of titanium tetrachloride on alumina to give mixtures of regioisomeric enamines [Eq. (11) (Chou and Chu, 1984)].

$$
\underset{R^2}{\overset{R^1}{>}}{=}\text{O} \xrightarrow[\substack{\text{H}_2\text{NNTs} \\ \text{EtOH, Reflux} \\ (0.5-2\,\text{h})}]{\substack{\text{Al}_2\text{O}_3 \\ \text{(Act. I)}}} \underset{\underset{40-80\%}{R^2}}{\overset{R^1}{>}}{=}\text{NNTs} \tag{10}
$$

$$
R^1 = \text{alkyl}; \quad R^2 = \text{ethyl alkanoate}
$$

$$
\langle\text{cyclohexanone}\rangle{=}\text{O} + \text{HN}\langle\text{O}\rangle \xrightarrow[\substack{\text{TiCl}_4 \\ \text{Reflux} \\ (3\,\text{h})}]{\substack{\text{Al}_2\text{O}_3 \\ \text{(Neutral)}}} \underset{94\%}{\langle\text{enamine}\rangle\text{—N}\langle\text{O}\rangle} \tag{11}
$$

Addition of the cyanide anion to an α-bromoketone is facilitated by alumina and leads stereoselectively to an epoxide [Eq. (12) (Takahashi *et al.*, 1981)].

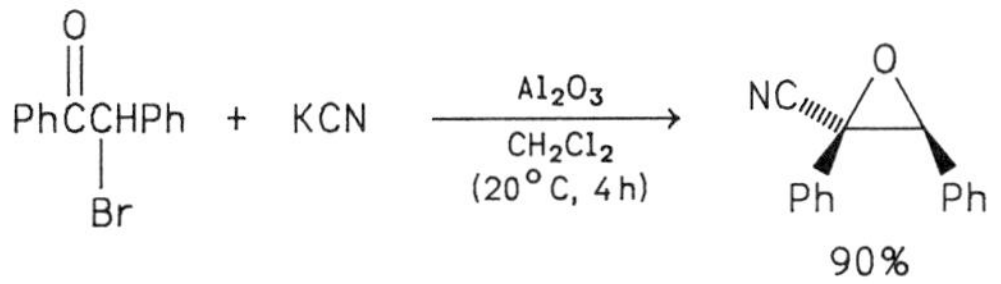

$$
\underset{\substack{|\\ Br}}{PhCCHPh} + KCN \xrightarrow[\substack{CH_2Cl_2 \\ (20°C,\,4h)}]{Al_2O_3} \quad \overset{O}{\underset{\substack{Ph \quad\quad Ph \\ 90\%}}{NC_{\prime\prime\prime\prime}}} \tag{12}
$$

2. Carbon–Nitrogen Multiple Bonds

Alumina-promoted addition of water to the carbon–nitrogen double bond of some *p*-toluenesulfonylhydrazones leads to the mild and neutral hydrolysis of these hydrazones into the parent carbonyl compounds [Eq. (13) (Coutts *et al.*, 1980)], a reversal of alumina-promoted condensation of hydrazines and carbonyl compounds [Eq. (10)]. Steroidal aldimines also can be hydrolyzed simply by chromatography on activity-IV alumina [Eq. (14) (Paradisi and Zecchini, 1982)]. Clearly whether alumina promotes hydrazone formation or hydrazone hydrolysis depends critically on the reaction conditions and especially on the degree of hydration of the alumina; generally, activity-II–IV alumina promotes hydrolysis of C=N linkages.

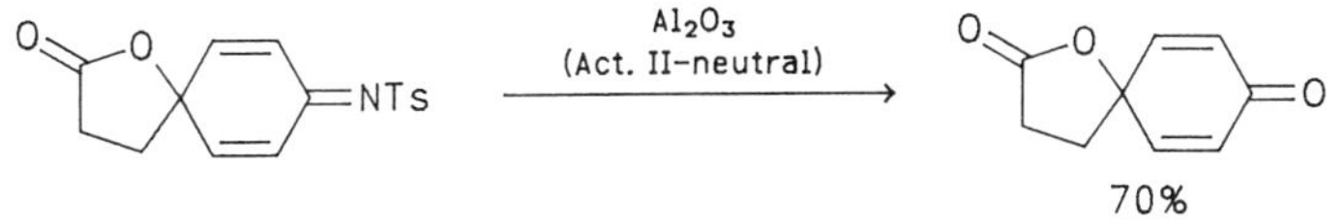

$$\tag{13}$$

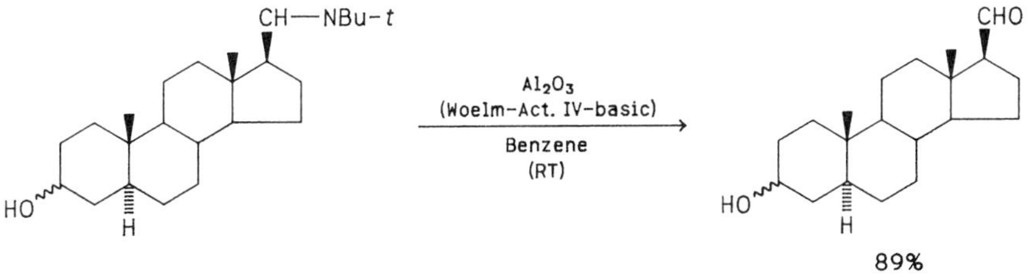

70%

89%

$$\tag{14}$$

Potassium fluoride on alumina is used to hydrolyze aromatic and aliphatic nitriles into the corresponding amides in 95–98% and 65–77% yields, respectively (Rao, 1982). Apparently the potassium fluoride accentuates hydrogen bonding of water on the alumina surface to the nitrogen atom of the nitrile thereby facilitating nucleophilic addition of water to the nitrile triple bond.

3. Carbon–Carbon Double Bonds

Aumina catalyzes Michael addition of secondary amines to the β-carbon atom of exocyclic α-β-ethylenic ketones [Eq. (15) (Pelletier *et al.*, 1980)]; the yields are excellent; the reaction conditions are mild; and amine and ether groups in the enone molecule are inert.

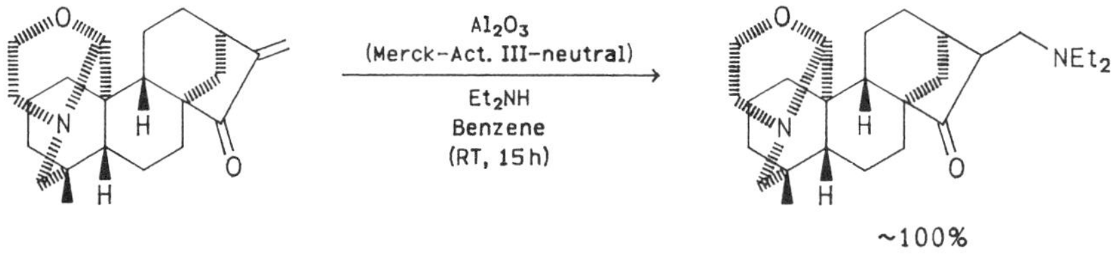

$$\tag{15}$$

Catalytic hydrogenation of alkenes is an important industrial process. Often effective homogeneous catalysts are adsorbed or attached covalently to solid supports such as alumina. These immobilized catalysts then can be separated easily from the reaction mixtures, and they can be reused. Without being exhaustive in coverage, we site here a few example of several hydrogenation catalysts immobilized on alumina: a paired chromium catalyst (Iwasawa *et al.*, 1981), a zero-valent iron catalyst (Kazusaka *et al.*, 1983), a ruthenium carbonyl cluster catalyst (Lausarot *et al.*, 1984), and several organoactinide catalysts (He *et al.*, 1985). These references should be consulted for more detailed discussions of these industrially significant catalytic systems.

4. Cycloadditions.

Only a couple of examples of alumina-promoted cycloadditions have been reported. Diels–Alder [2+4] cycloaddition of 1,3-cyclopentadiene and methyl acrylate occurs on alumina to give almost exclusively ($>30:1$) the *endo* adduct shown in Eq. (16) (Parlar and Baumann, 1981). Under comparable homogeneous conditions, the *endo : exo* product ratio is about $3:1$. This dramatic effect of alumina on the stereochemistry of this cycloaddition was attributed to adsorption of the reactants to the alumina surface and to enhanced secondary orbital interactions.* Photochemical

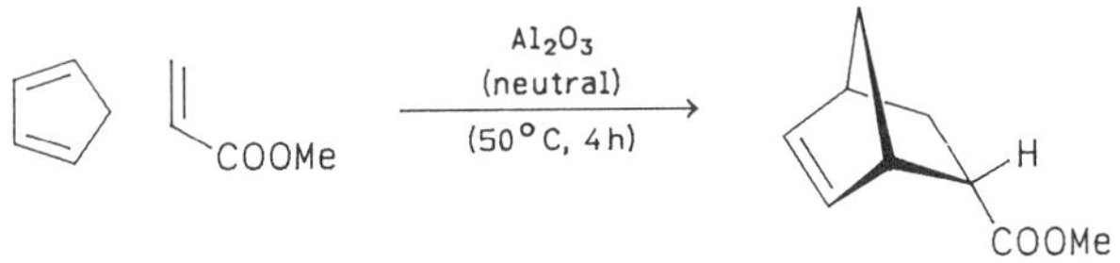

$$\tag{16}$$

---

* These results are highly reminiscent of those in aqueous solution, whose use was pioneered by R. Breslow. Similar results have been achieved on the surface of modified montmorillonites (see Chapter 24).—ED.

[2 + 2] cycloadditions also are facilitated by alumina, as exemplified by Eq. (17) (Farwaha *et al.*, 1985).

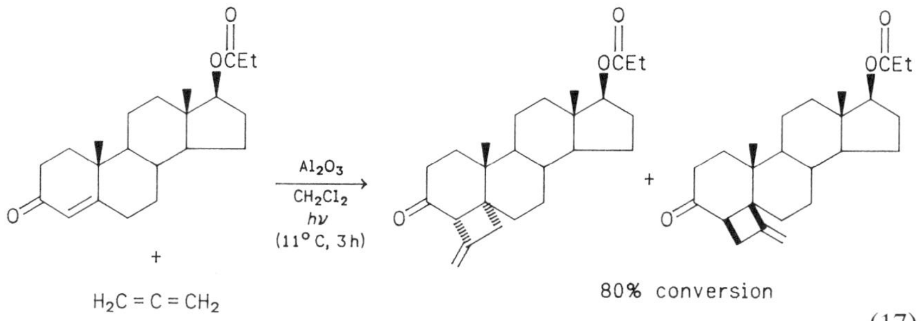

$$\tag{17}$$

III. REDOX REACTIONS

A. Reductions

1. Aldehydes

Homogeneous Meerwein–Ponndorf–Verley reductions of aldehydes by aluminum triisopropoxide can be achieved effectively also under heterogeneous conditions by adsorbing isopropyl alcohol on alumina. Selective hydride transfer to (i.e., reduction of) aldehydes in the presence of many other functional groups (even ketones) is accomplished successfully under mild conditions (Posner *et al.*, 1977). Reports indicate that *sec*-butyl alcohol on alumina can be used effectively in place of isopropyl alcohol on alumina (Su *et al.*, 1982) and that hydrous zirconium oxide can be used effectively in place of alumina (Matsushita *et al.*, 1985). Sodium borohydride supported on alumina is a commercially availble reagent (Alfa, Fluka) capable of reducing aldehydes rapidly at room temperature in nonpolar organic solvents (Hodosan and Serban, 1969; Santaniello *et al.*, 1979). Even hydrolytically sensitive acid chlorides are chemoselectively reduced to alcohols [Eq. (18) (Santaniello *et al.*, 1979)], and alkali-sensitive groups such as dienyl esters survive exposure to sodium borohydride on alumina.

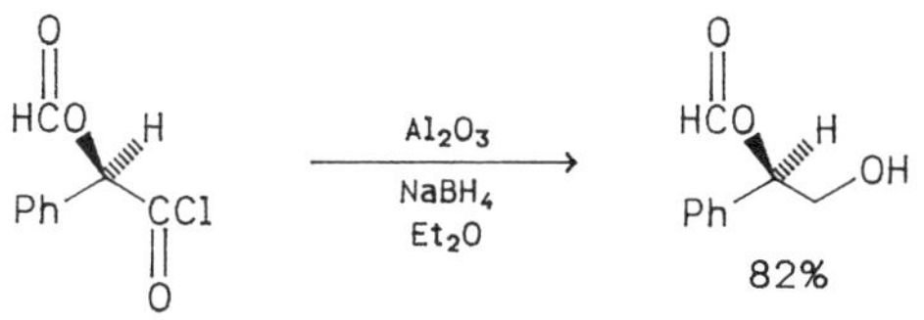

$$\tag{18}$$

Aldehydes can be reduced chemoselectively also by sodium sulfide on alumina [Eq. (19) (Kamitori *et al.*, 1985a)]; note that alumina reverses the inherent reactivity of sodium sulfide so that an aromatic aldehyde can be reduced by Na_2S/Al_2O_3 in preference to an aromatic nitro group! Ketones and esters appear to be stable under these conditions. The exact nature of the active reducing species is not clear.

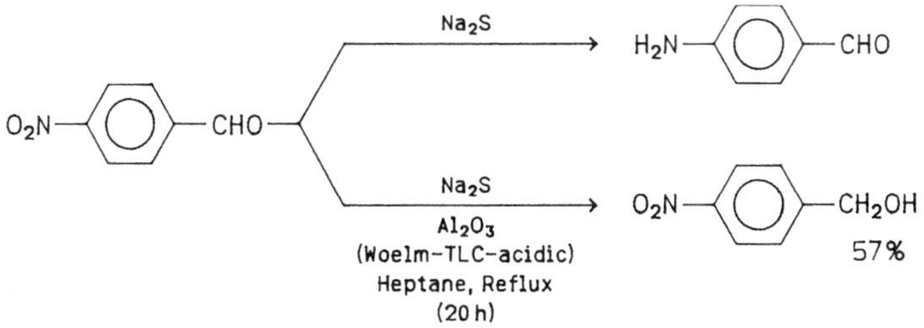

$$O_2N-\text{⟨○⟩}-CHO \begin{cases} \xrightarrow{Na_2S} H_2N-\text{⟨○⟩}-CHO \\ \\ \xrightarrow[\substack{Al_2O_3 \\ (Woelm-TLC-acidic) \\ Heptane, Reflux \\ (20\,h)}]{Na_2S} O_2N-\text{⟨○⟩}-CH_2OH \quad 57\% \end{cases} \tag{19}$$

2. Cyanoalkanes

Although many alumina-supported metals have been used as hydrogenation catalysts, potassium on alumina effects reductive decyanation of cyano groups attached to primary, secondary, and tertiary carbon centers even in the presence of acetal and olefinic functionalities. The reductive decyanation shown in Eq. (20) (Savoia *et al.*, 1980) is a key step in a short total synthesis of an insect sex pheromone.

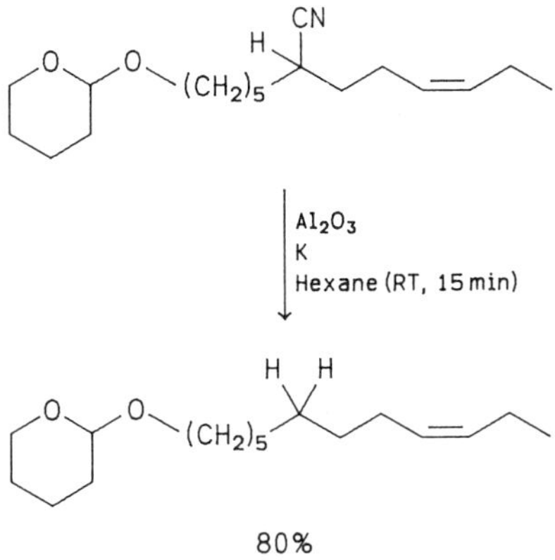

$$\tag{20}$$

80%

3. Haloalkanes

Haloalkanes react with tin hydrides covalently bound to alumina to replace the halogen atom by a hydrogen atom [Eq. (21) (Schumann and

Pachaly, 1981)]:

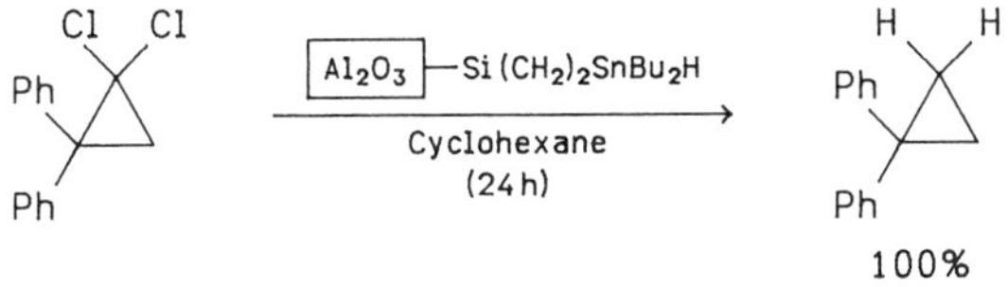

$$\text{(21)}$$

100%

This is an example of a homogeneous reagent (Bu_3S_nH) immobilized by covalent attachment to a solid support.

B. Oxidations

1. Alcohol

Pyridinium chlorochromate (PCC) supported on alumina (Cheng et al., 1980) can be used to oxidize alcohols into carbonyl compounds chemoselectively even in the presence of such acid-sensitive functional groups as tetrahydropyranyl acetals [Eq. (22) (Savoia et al., 1982)]. The alumina effectively neutralizes the acidic character of the pyridinium chlorochromate reagent. Alumina particles dispersed on glassy carbon electrode surfaces adsorb catechols and catalyze their oxidation (Zak and Kuwana, 1982). Glycols (1,2-diols) are oxidatively cleaved by periodic acid on alumina (Villemin and Ricard, 1982).

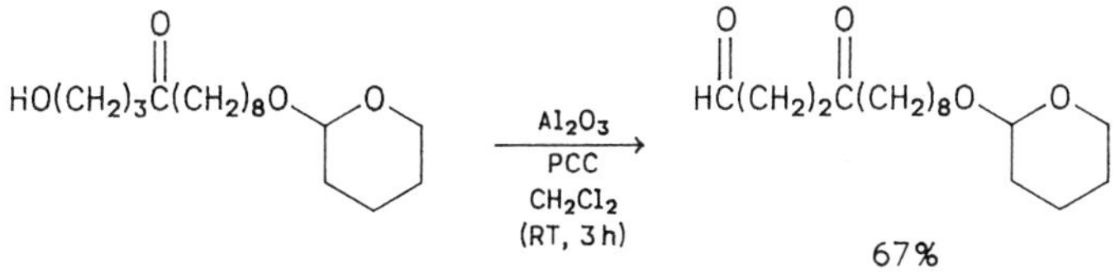

$$\text{(22)}$$

67%

2. Alkenes

Aryl-substituted olefins are oxidatively cleaved into the corresponding aromatic aldehydes without over-oxidation by the stable, nontoxic, and easily handled reagent prepared by placing periodic acid and a catalytic amount of potassium permanganate on alumina [Eq. (23) (Villemin and Ricard, 1984)]. When triphenylsilyl hydroperoxide is placed on alumina, olefins (including aryl-substituted olefins) can be epoxidized easily and mildly [Eq. (24) (Rebek and McCready, 1979)]. Manganese dioxide on alumina is a commercially available (Strem) oxidation catalyst.

$$PhCH{=}CHPh \xrightarrow[\substack{H_5IO_6 \\ KMnO_4 (catalytic) \\ CCl_4 (RT, 24\,h)}]{\substack{Al_2O_3 \\ (Woelm-Act.\ I-neutral)}} 2\ PhCH{\overset{O}{\parallel}} \qquad (23)$$

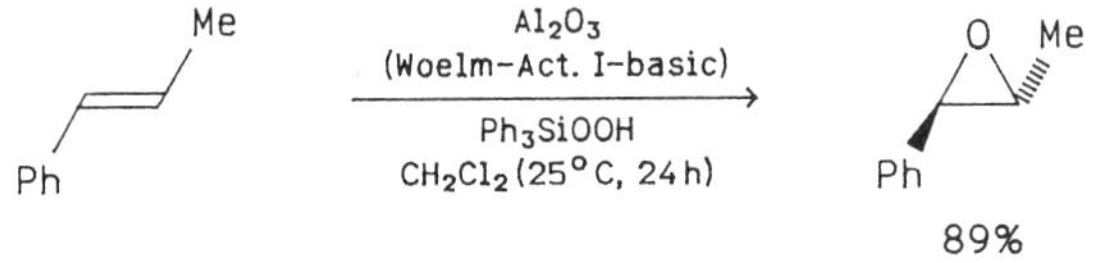

$$(24)$$

89%

3. Sulfides

Selective oxidation of various sulfides into the corresponding sulfoxides can be achieved generally in greater than 85% yield by sodium metaperiodate on alumina. No over-oxidation to sulfones is observed, and carbon–carbon double bonds are inert to this procedure [Eq. (25) Liu and Tong, 1978)].

$$(CH_2{=}CHCH)_2S \xrightarrow[\substack{NaIO_4 \\ 95\%\ EtOH \\ (RT,\ 3\,h)}]{\substack{Al_2O_3 \\ (Merck-acidic)}} (CH_2{=}CHCH)_2S{=}O \qquad (25)$$

87%

IV. SUBSTITUTION REACTIONS

A. Alkyl Electrophiles

1. Halides

Substitution of bromide in 1-bromalkanes by iodide is achieved effectively using aqueous potassium iodide and several onium-salt phase-transfer catalysts immobilized on alumina (Tundo *et al.*, 1982). Substitution of bromide in 1-bromoalkanes by acetate also is facilitated by alumina (Ando *et al.*, 1982) and by attaching polyethyleneglycol to alumina as an immobilized phase-transfer catalyst for potassium acetate (Sawicki, 1982). Substitution of chloride in benzyl chloride by an alkoxy group is effected by alumina [Eq. (26) (Onaka *et al.*, 1983)]; even better results are reported with alkali cation–exchanged Y zeolites.

$$PhCH_2Cl \xrightarrow[\substack{n-C_{10}H_{21}OH \\ Hexane,\ Reflux \\ (2\,h)}]{\substack{Al_2O_3 \\ (Woelm-200-basic)}} PhCH_2OC_{10}H_{21}{-}n \qquad (26)$$

48%

Two special cases have been reported in which anchimeric assistance is likely. A β-bromoethylamine (i.e., a nitrogen mustard) was hydrolyzed into the corresponding β-hydroxyethylamine by alumina [Eq. (27) (Otomasu *et al.*, 1982)], and a homoallylic primary alkyl bromide reacted with silver nitrate on alumina to produce a cyclo-propylcarbinyl nitrate [Eq. (28) (Hrubiec and Smith, 1984)].

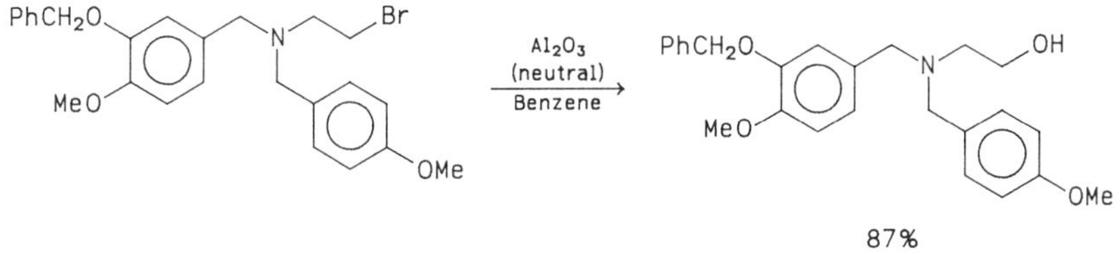

(27)

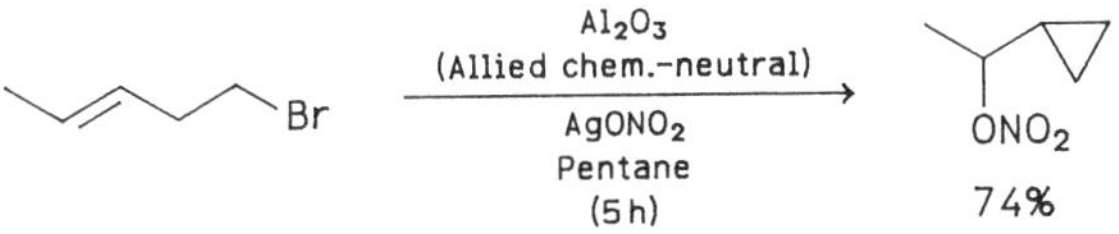

(28)

Substitution of bromide in various primary bromoalkanes by the nitrogen atom of various *N*-unsubstituted amides on alumina carrying potassium hydroxide leads very selectively to the corresponding *N*-monoalkylated amides [Eq. (29) (Sukata, 1985)].

$$\mathrm{R\overset{O}{\overset{\|}{C}}NH_2} + \mathrm{R'CH_2Br} \xrightarrow[\substack{\text{KOH} \\ \text{Dioxane} \\ (40\text{–}60^\circ\text{C, 8 h})}]{\substack{Al_2O_3 \\ \text{(Merck-basic)}}} \underset{56\text{–}83\%}{\mathrm{R\overset{O}{\overset{\|}{C}}NHCH_2R'}} \qquad (29)$$

Replacement of the halide ion in various alkyl (Ando *et al.*, 1984 a,b) and benzylic halides (Clark and Duke, 1985) by the cyano group is achieved by potassium cyanide on alumina.

2. Sulfonate Esters

Ditosylates react with diols (Yamawaki and Ando, 1980) and with diamines [Eq. (30) (Pietraszkiewicz and Jurczak, 1983)] on potassium fluoride-doped aluimina to effect double substitution reactions leading to crown and diazacrown ethers. No high-dilution techniques are necessary in these heterogeneous macrocyclizations.

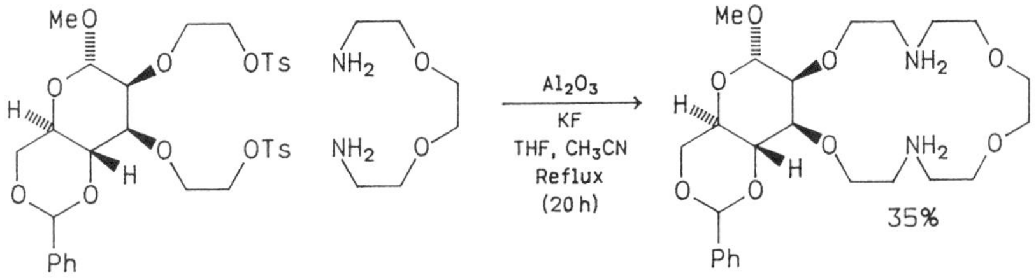

(30)

3. Ethers

Tetrahydrofuran is cleaved in low yield by trimethylsilyl chloride on alumina [Eq. (31) (Kawakami and Yamashita, 1982)].

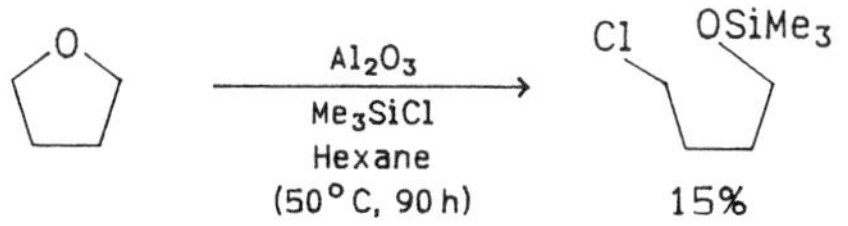

$$
\underset{\text{(50°C, 90 h)}}{\overset{\displaystyle \text{Al}_2\text{O}_3}{\underset{\displaystyle \text{Me}_3\text{SiCl}}{\text{Hexane}}}}
$$

(31)

15%

B. Acyl Electrophiles

1. Carboxylate Ester

Replacement of the —OR group in an R'COOR ester by an —OH group represents hydrolysis of the ester. Alumina impregnated with potassium hydroxide is useful for such high-yield substitution reactions at ambient temperature [Eq. (32) (Regen and Mehrotra, 1981)].

$$
\text{PhCOBu-}t \xrightarrow[\substack{\text{KOH} \\ \text{Et}_2\text{O} \\ \text{(RT, 21 h)}}]{\substack{\text{Al}_2\text{O}_3 \\ \text{(Bio-Rad-neutral)}}} \text{PhCOH} + \text{HOBu-}t
$$

$$\qquad\qquad 100\% \qquad 80\%$$

(32)

Alumina is also able to facilitate mild, neutral transesterifications. When ethyl acetate is used as solvent and at the same time as acetylating reagent, primary alcohols are acetylated on a gram scale effectively and conveniently [Eq. (33) (Posner et al., 1981c)]; even some base-sensitive (e.g., chlorohydrin) and acid-sensitive (e.g., ethylenic and pyridyl) functional groups are tolerated. Primary carbohydrate alcohols are also effectively acetylated. Silica is not able to achieve similar transesterifications. Chemoselective acetylation of a primary alcohol in the presence of a secondary alcohol is also easily achieved even with carbohydrate and steroidal diols [Eq. (34) (Posner and Oda, 1981)]. Aryl amines also are acetylated by ethyl acetate on alumina (Posner and Oda, 1981).

$$
\text{RCH}_2\text{—OH} + \text{EtOCCH}_3 \xrightarrow[\text{(25–30°C, 1 h)}]{\substack{\text{Al}_2\text{O}_3 \\ \text{(Woelm-200-neutral)}}} \text{RCH}_2\text{—OCCH}_3
$$

$$\qquad\qquad\qquad 64\text{–}99\%$$

(33)

$$
\underset{\displaystyle +\ \text{EtOCCH}_3}{\overset{\displaystyle \text{OH}}{\text{CH}\!\sim\!\!\sim\!\!\sim\!\text{CH}_2\text{OH}}} \xrightarrow[\text{(75–80°C)}]{\substack{\text{Al}_2\text{O}_3 \\ \text{(Woelm-200-neutral)}}} \overset{\displaystyle \text{OH}}{\text{CH}\!\sim\!\!\sim\!\!\sim\!\text{CH}_2\text{OCCH}_3}
$$

$$\qquad\qquad\qquad 45\text{–}71\%$$

(34)

2. Acid Fluorides

Alumina itself (without potassium hydroxide) facilitates hydrolysis of acid fluorides. Compared to the corresponding homogeneous hydrolysis, heterogeneous alumina-promoted hydrolysis of acyl fluorides occurs roughly 250–350 times faster and alumina-promoted hydrolysis of phosphoryl fluorides occurs 1800–3000 times faster [Eq. (35) (Posner *et al.*, 1981a)]. Presumably the very strong affinity of aluminum ions for fluoride ions combined with the basic properties of alumina provide a favorable synergistic pull–push mechanism for these unusually fast, mild, and neutral hydrolyses.

$$(i-PrO)_2\overset{\overset{\text{O}}{\|}}{P}F \quad \xrightarrow[\substack{Et_2O \\ (25°C, \, 0.5\,min)}]{\substack{Al_2O_3 \\ (Woelm-200-neutral)}} \quad (i-PrO)_2\overset{\overset{\text{O}}{\|}}{P}OH \qquad (35)$$

$$99\%$$

3. Acylmetallics

Alcoholysis of an acylcobalt intermediate on alumina leads to an ester [Eq. (36) (Sawicki, 1982)].

$$\begin{array}{c} PhCH_2Br \\ + \\ CO \\ + \\ Co(CO)_4^- \end{array} \xrightarrow{MeOH} [PhCH_2\overset{\overset{\text{O}}{\|}}{C}Co(CO)_4] \xrightarrow[KOMe]{Al_2O_3} PhCH_2\overset{\overset{\text{O}}{\|}}{C}OMe \qquad (36)$$

$$70\%$$

C. Carbon–Hydrogen Bonds

1. Acidic C—H Units

Replacement of the hydrogen atom in an acidic C—H bond (e.g., phenylacetylene, cyclohexanone) by a deuterium atom (Gaetano *et al.*, 1985) and by a tritium atom (Peng and Buchman, 1985) is achieved using deuterium-exchanged (D_2O) and tritium-adsorbed (T_2) alumina, respectively. The tritium exchange tolerates such normally hydrogenated groups as alkene and nitro.

2. Aromatic C—H Units

Friedel–Crafts alkylation of aromatic compounds is an electrophilic aromatic substitution process in which an aromatic C—H unit is replaced by a new C—C linkage. Boron trifluoride on alumina [Eq. (37) (Baek *et al.*, 1985)] and potassium cyanide on alumina [Eq. (38) (Ando *et al.*,

1984a)] are useful for such C—H→C—C transformations. These heterogeneous conditions avoid further reactions [e.g., cyclization to a tetrahydro cannabinol in Eq. (37) or dialkylation in Eq. (38)].

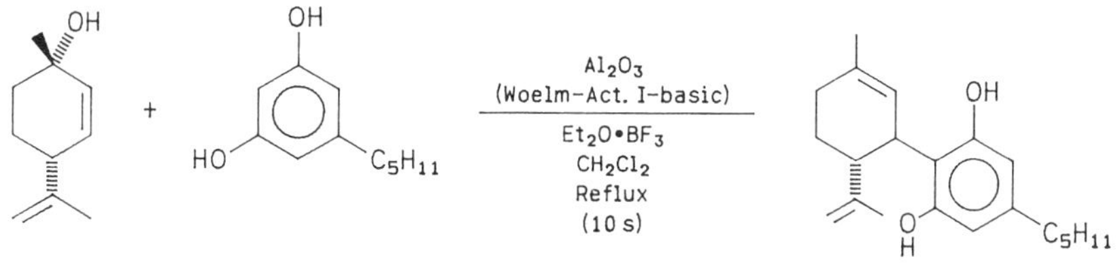

$$\tag{37}$$

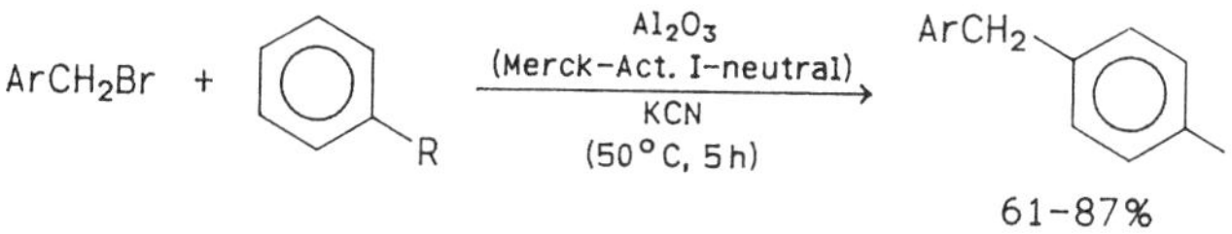

$$\tag{38}$$

61–87%

Photochemically stimulated monochlorination of some polycyclic aromatic hydrocarbons is achieved by irradiating the hydrocarbons adsorbed to alumina in the presence of excess ferric chloride [Eq. (39) (Hasebe *et al.*, 1981)].

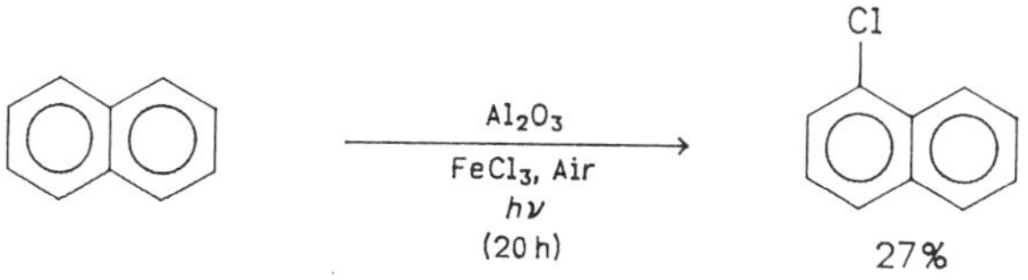

$$\tag{39}$$

27%

V. ELIMINATION REACTIONS

A. Halides

1,2-Dehydrohalogenation of unactivated secondary and tertiary halides is promoted by alumina, as exemplified by a total synthesis of pleasant-smelling (+)-nootkatone [Eq. (40) (Yanami *et al.*, 1979; Inokuchi *et al.*, 1982)] and by a synthesis of some polycyclic hydrocarbons [Eq. (41) (Park *et al.*, 1979)]. Note that the alumina-mediated double β-elimination without skeletal rearrangement shown in Eq. (41) suggests that these are concerted rather then stepwise dehydrohalogenations (Posner and Gurria, 1976).

Potassium fluoride on alumina also is a very effective means of facilitating 1,2-dehydrobromination to form alkenes and alkynes (Yamawaki *et al.*, 1983).

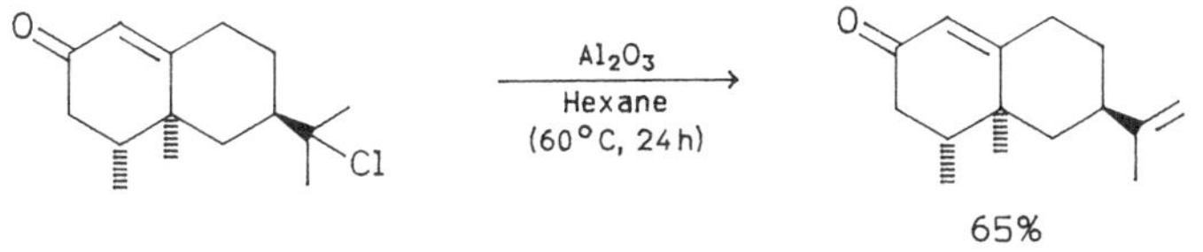

$$(40)$$

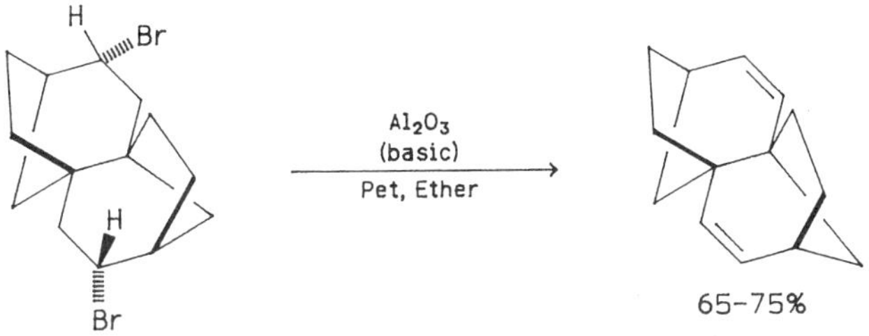

65%

$$(41)$$

65–75%

1,2-Desilicohalogenation of α,β-dihaloethylsilanes occurs easily on alumina to form vinylic halides [Eq. (42) (Miller and McGarvey, 1977)]; the stereoselectivity of this process for overall replacement of trimethylsilyl by halide is high starting with *trans*-vinylsilanes and somewhat lower starting with *cis*-vinylsilanes.

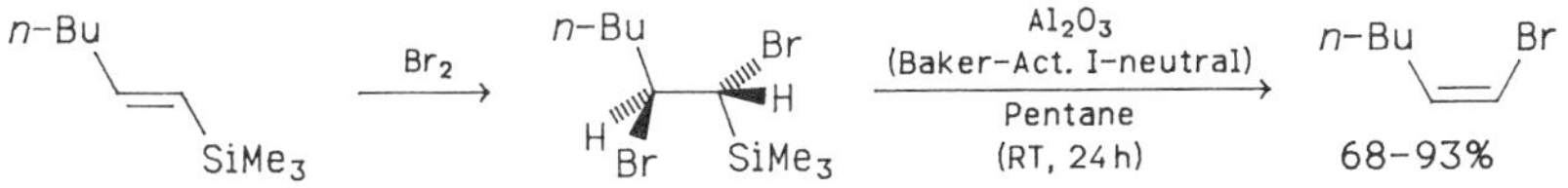

$$(42)$$

Full details of alumina-promoted 1,3-dehydrochlorinations of some γ-chloroketones to form cyclopropylketones have been reported (Ruppert and White, 1981).

B. Oxygen Derivatives

1. Alcohols

Pines has studied extensively and over many years alumina-promoted dehydration of alcohols at elevated temperatures (Pines and Stalick, 1977). Generally most alcohols are stable to alumina at room temperature. Benzylic-type hydroxyl derivatives of several polycyclic hydrocarbons are dehydrated by alumina at 80°C without formation of isomeric or polymeric side products which are formed even under mild Brønsted acid-promoted dehydration [Eq. (43) (Sangaiah *et al.*, 1983)]. Similar results are reported using zeolites for room-temperature dehydration of some allylic and benzylic alcohols (Markgraf *et al.*, 1984).

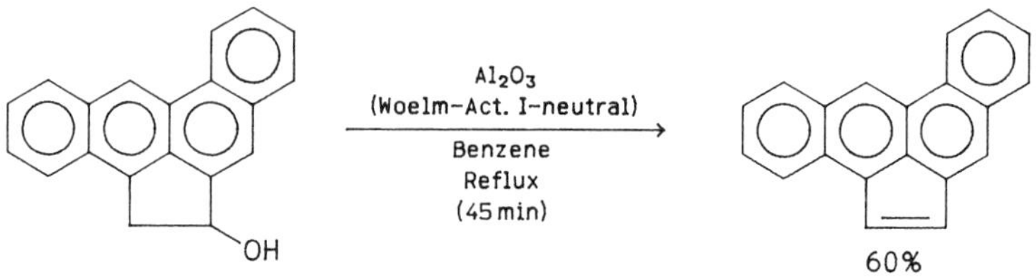

$$\text{(43)}$$

$$
\begin{array}{c}
\xrightarrow[\substack{\text{Benzene} \\ \text{Reflux} \\ \text{(45 min)}}]{\substack{\text{Al}_2\text{O}_3 \\ \text{(Woelm-Act. I-neutral)}}}
\end{array}
$$

60%

2. Acetates

Several β-acetoxyketones undergo facile alumina-mediated β-elimination of acetic acid to produce conjugated enones (Sugihara *et al.*, 1977). This procedure works well even for exclusive preparation in gram quantities of 1-acetyl-cycloalkenes to the virtually complete exclusion of β,γ-isomers which are formed under most other basic conditions [Eq. (44) (Hudlicky and Srnak, 1981; see also Mekhciev *et al.*, 1977)]. A 210–250°C alumina-promoted β-elimination of some enamino ethyl esters has been reported (Rigo *et al.*, 1985).

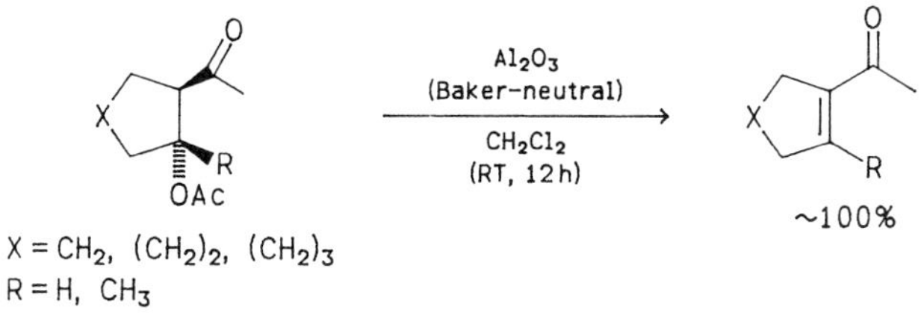

$$
\begin{array}{c}
\xrightarrow[\substack{\text{CH}_2\text{Cl}_2 \\ \text{(RT, 12 h)}}]{\substack{\text{Al}_2\text{O}_3 \\ \text{(Baker-neutral)}}}
\end{array}
\qquad \text{(44)}
$$

~100%

$$X = CH_2,\ (CH_2)_2,\ (CH_2)_3$$
$$R = H,\ CH_3$$

3. Sulfonate Esters

We have shown that alumina very effectively promotes 1,2-elimination of sulfonate esters in a concerted (i.e., not stepwise via carbonium ions) process that has much preparative value in organic synthesis (Posner and Gurria, 1976; Posner, 1978). A polyfunctional organometallic compound carrying a mesylate ester is converted by alumina on 0.5-gram scale into the corresponding olefin which was elaborated into a member of the tricothecene family of biologically active sesquiterpenes [Eq. (45) (Pearson and Ong, 1981).

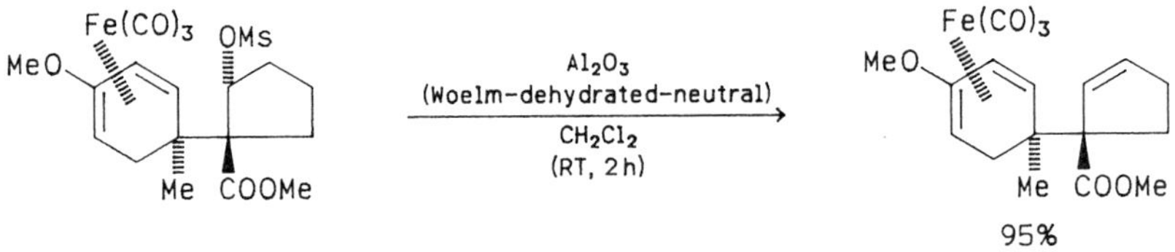

$$
\begin{array}{c}
\xrightarrow[\substack{\text{CH}_2\text{Cl}_2 \\ \text{(RT, 2 h)}}]{\substack{\text{Al}_2\text{O}_3 \\ \text{(Woelm-dehydrated-neutral)}}}
\end{array}
$$

95%

$$\text{(45)}$$

4. Carbamate Esters

β-Carbamoyl ketones undergo facile elimination reactions on alumina to form intermediate amines and vinyl ketones which combine via a Michael addition [Eq. (15)] to form β-aminoketones [Eq. (46) (Reitz *et al.*, 1982)]. Without alumina, the reaction shown in Eq. (46) does not proceed even in refluxing chloroform for 2 days.

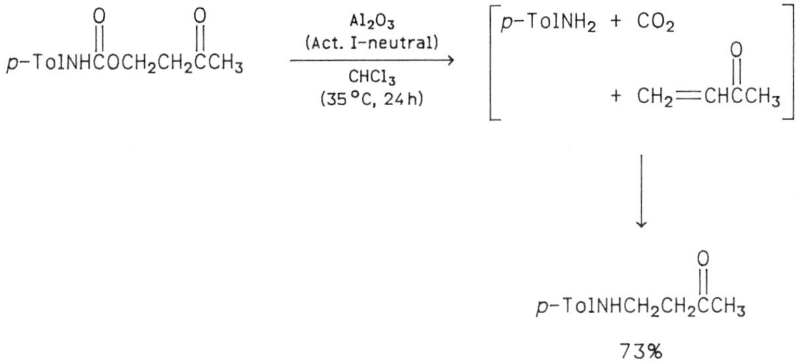

$$p\text{-TolNHCOCH}_2\text{CH}_2\text{CCH}_3 \quad \xrightarrow[\substack{\text{CHCl}_3 \\ (35\,°\text{C, 24 h})}]{\substack{\text{Al}_2\text{O}_3 \\ \text{(Act. I-neutral)}}} \quad \left[\begin{array}{l} p\text{-TolNH}_2 + \text{CO}_2 \\ \\ + \ \text{CH}_2{=}\text{CHCCH}_3 \end{array} \right]$$

$$\downarrow \qquad (46)$$

$$p\text{-TolNHCH}_2\text{CH}_2\text{CCH}_3$$

$$73\%$$

5. Ethers

In certain structures, alumina successfully accelerates elimination of ethereal oxygen substitutents. For example, alumina promotes cleavage of a 3,4-epoxyketone to produce a 4-hydroxy-2-enone [Eq. (47) (Suzuki *et al.*, 1979)], a valuable intermediate in prostaglandin synthesis. Also, alumina catalyzes the mutarotation of α-D-glucose which involves a 1,2-elimination of an ethereal oxygen group from a hemiacetal [Eq. (48) (Dunstan and Pincock, 1985)].

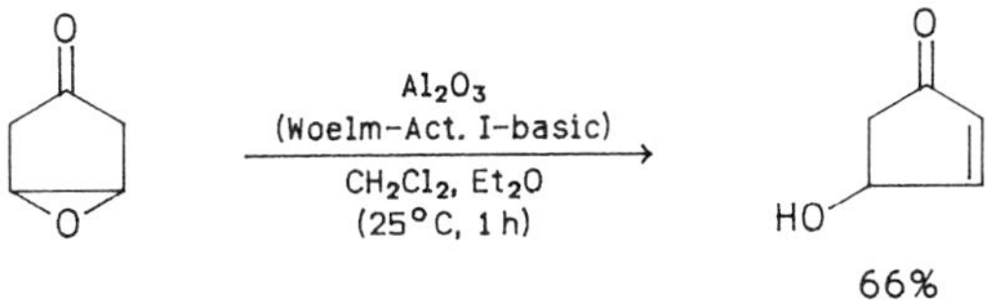

$$\xrightarrow[\substack{\text{CH}_2\text{Cl}_2, \text{ Et}_2\text{O} \\ (25\,°\text{C, 1 h})}]{\substack{\text{Al}_2\text{O}_3 \\ \text{(Woelm–Act. I-basic)}}} \qquad (47)$$

$$66\%$$

$$\xrightleftharpoons[\substack{\text{DMSO} \\ (25\,°\text{C})}]{\substack{\text{Al}_2\text{O}_3 \\ \text{(Woelm–dehydrated–neutral)}}} \qquad (48)$$

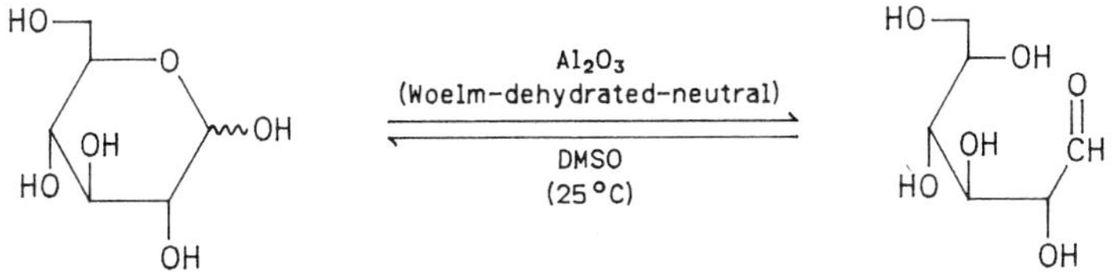

C. Fragmentation Reactions

Retro-aldolization of β-hydroxyketones into the parent ketones is mediated by alumina [Eq. (49) (Campelo *et al.*, 1984; and McMurry and Isser, 1972)]. Even some β-ethylenic ketones can be hydrated to β-hydroxyketones and then fragmented *in situ* [Eq. (50) (Endo and Hikino, 1979)].

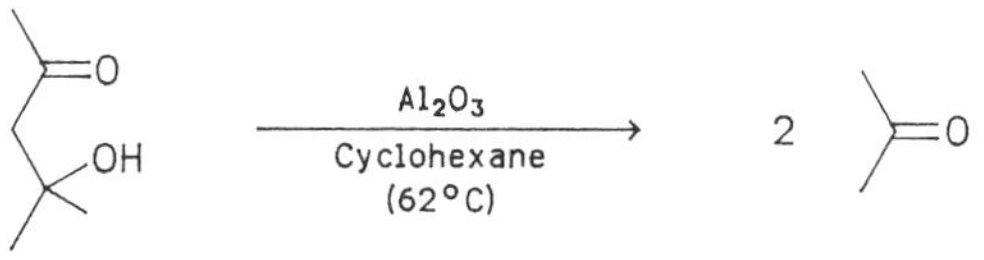

$$ (49) $$

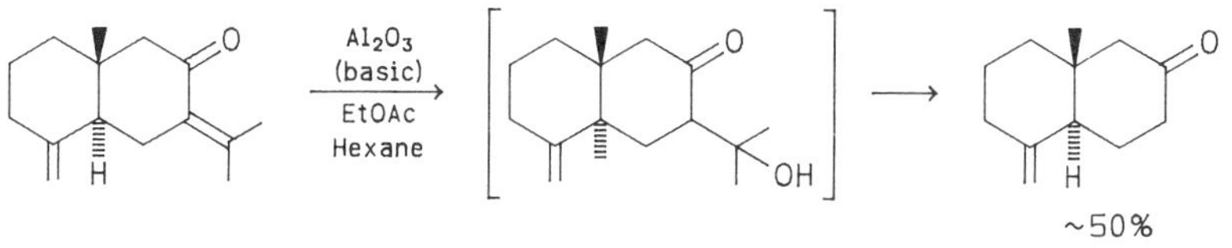

$$\sim 50\%$$

$$ (50) $$

Alumina promotes decarbonylation of unsymmetrical diarylcyclopropenones into the corresponding unsymmetrical diarylacetylenes in a convenient and versatile process which avoids dimer formation [Eq. (51) (Wadsworth and Donatelli, 1981)].

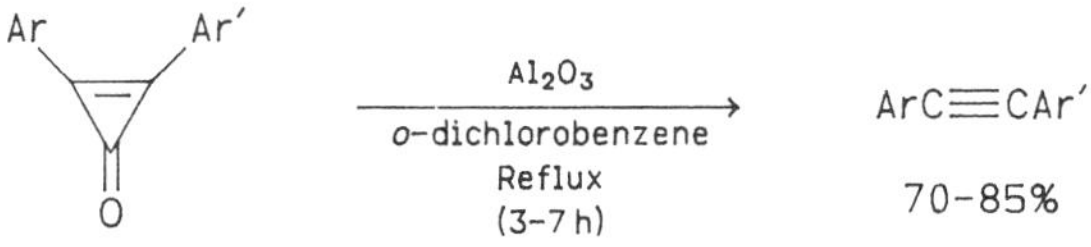

$$ArC \equiv CAr'$$

$$70-85\%$$

$$ (51) $$

VI. REARRANGEMENT REACTIONS

A. Multiple Bonds

1. Alkenes

Alumina itself is used often industrially for *cis*–trans and positional isomerization and for skeletal rearrangement of alkenes. Fluoride ions on alumina are reported to provide improved catalytic ability (Irvine *et al.*, 1980). Homogeneous isomerization catalysts can be attached to alumina via capping groups; a phosphorus-capped rhodium cluster, $H_2Ru_3(CO)_9$-$[\mu\text{-PCH}_2Si(OEt_3)]$, is attached to alumina to provide a tethered isomerization catalyst of improved stability (Cook and Evans, 1983).

2. β-γ-Unsaturated Carbonyl Compounds

β,γ-Ethylenic carbonyl compounds are isomerized on alumina into carbonyl-conjugated α,β-ethylenic carbonyl compounds [Eq. (52) (McMurry and Donovan, 1977)]. The same is true for β,γ-acetylenic carbonyl compounds [Eq. (53) (Gras and Guérin, 1985)]:

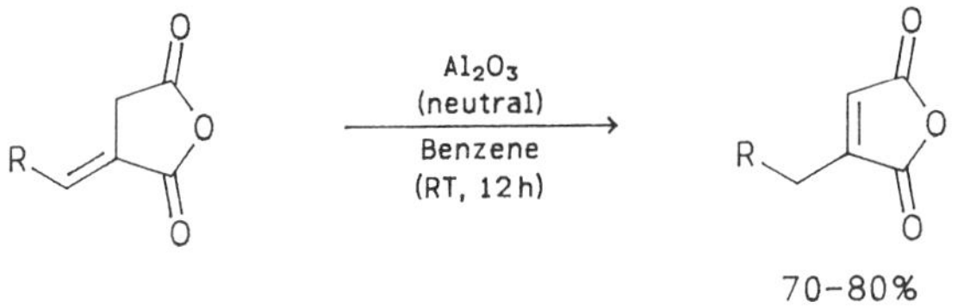

$$(52)$$

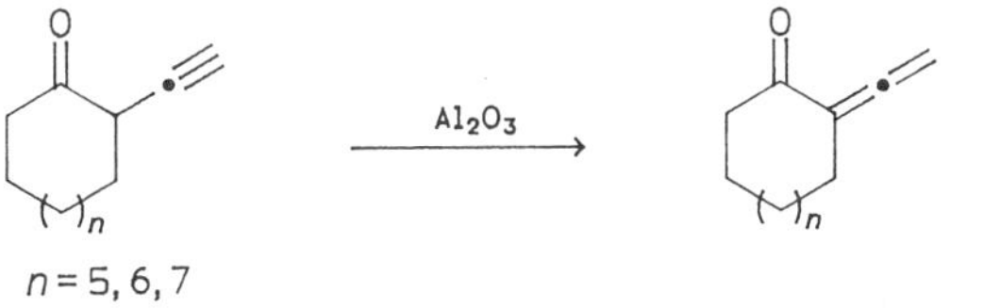

70–80%

$$(53)$$

n = 5, 6, 7

β-Allenic esters rearrange on alumina in aprotic solvents into *trans,cis*-2,4-dienoic esters in good yields and with 91–100% stereoselectivity (Tsuboi *et al.*, 1982a,b, 1983a,b; see also Zakharova *et al.*, 1978). This double-bond isomerization on alumina provides a convenient route to synthesis of several biologically active natural products which are easily elaborated from *trans,cis*-2,4-dienoic esters; examples include flavor components and insect pheromones (Tsuboi *et al.*, 1982b, 1983b) and a leukotriene important in asthma and other respiratory diseases [Eq. (54) (Tsuboi *et al.*, 1983a)].

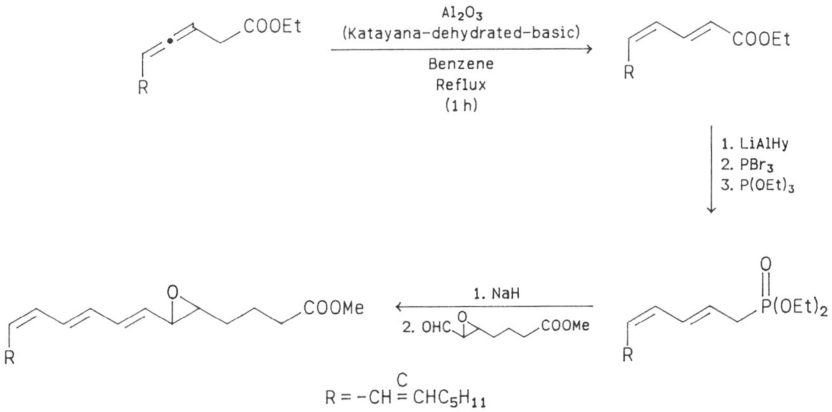

$$R = -CH \overset{C}{=} CHC_5H_{11}$$

$$(54)$$

Isomerization of α,β-acetylenic carbonyl compounds into the corresponding carbonyl-conjugated allenes occurs when electrophilic acetylenes are complexed with a cyclopentadienyl manganese species on alumina. Oxidative cleavage of the manganese organometallic complex with ceric ammonium nitrate (CAN) provides a gram-scale, preparatively useful, and novel method for synthesis of carbonyl-conjugated allenes [Eq. (55) (Franck-Neumann and Brion, 1979)]:

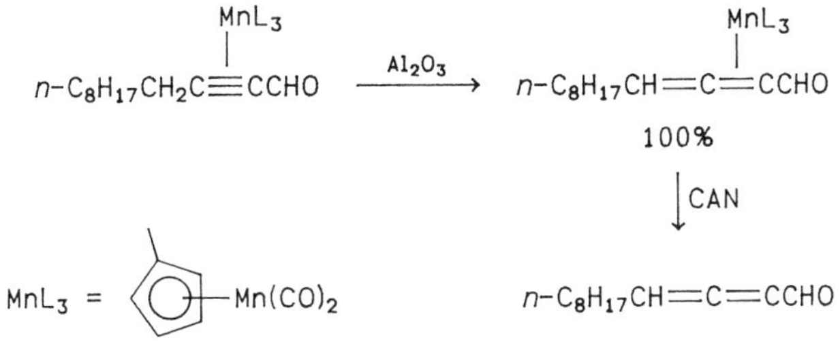

$$n\text{-}C_8H_{17}CH_2C\equiv CCHO \xrightarrow{\quad Al_2O_3 \quad} n\text{-}C_8H_{17}CH=C=CCHO \qquad (55)$$

with MnL$_3$ substituents; 100%; CAN gives $n\text{-}C_8H_{17}CH=C=CCHO$

$$MnL_3 = \text{cyclopentadienyl}-Mn(CO)_2$$

B. Strained Rings

1. Epoxides

Chromatography of epoxides on alumina often leads to skeletal reorganizations (Dev, 1972; Bang and Ourisson, 1973; Taran and Delmond, 1982). For example, alumina promotes skeletal rearrangement of six-membered ring α-pinene oxide mainly into a five-membered ring aldehyde [Eq. (56) (Arata and Tanabe, 1979)]. An unusual result involves chromatographic alumina-induced rearrangement of a three-membered ring epoxide into a four-membered ring oxetane among other products [Eq. (57) (Morelli *et al.*, 1979)].

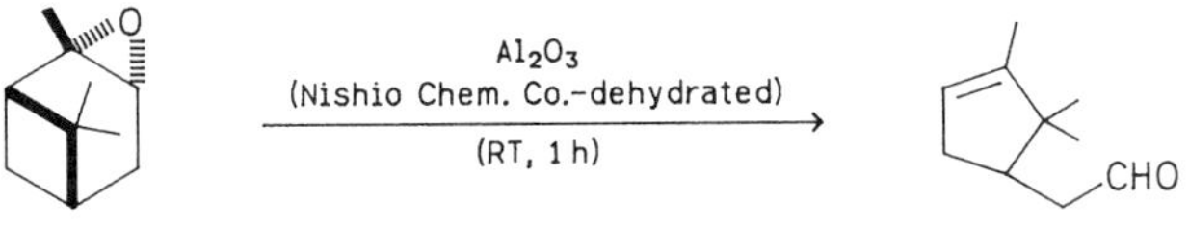

$$\xrightarrow[\text{(RT, 1 h)}]{\substack{Al_2O_3 \\ \text{(Nishio Chem. Co.-dehydrated)}}} \qquad (56)$$

$$\xrightarrow[\text{Pet. Ether}]{\substack{Al_2O_3 \\ \text{(Act. II–III–neutral)}}}$$

$\sim 20\%$

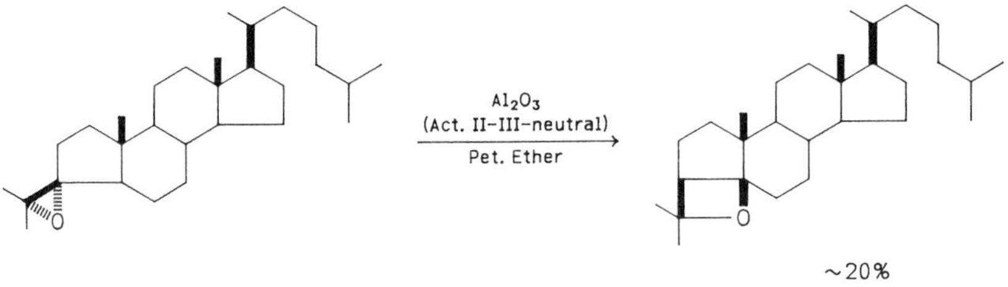

$$(57)$$

2. Cyclopropanes

Rearrangement of the three-membered ring of a cyclopropyl ketone into the five-membered ring of a dihydrofuran occurs on alumina with complete retention of cyclopropane stereochemistry [Eq. (58) (Alonso and Morales, 1980)]. The stereospecificity of this rearrangement, analogous to a vinylcyclopropane → cyclopentene rearrangement, implies a concerted rather than a stepwise process. We have previously emphasized that several alumina-promoted reactions (e.g., β-elimination of sulfonate esters and nucleophilic opening of epoxides) also appear to be concerted processes which do not involve free ionic intermediates (Posner, 1978).

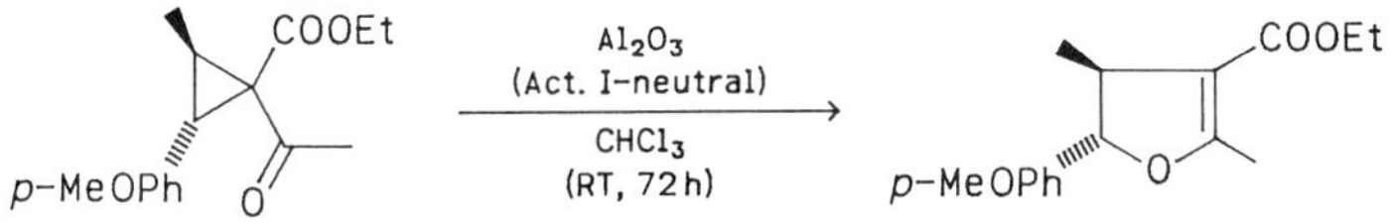

$$\tag{58}$$

A skeletal reorganization involving a proposed cyclopropylcarbinyl carbonium ion intermediate is reported for reaction of a homoallylic neopentylic tosylate on alumina [Eq. (59) (Yeats, 1983)].

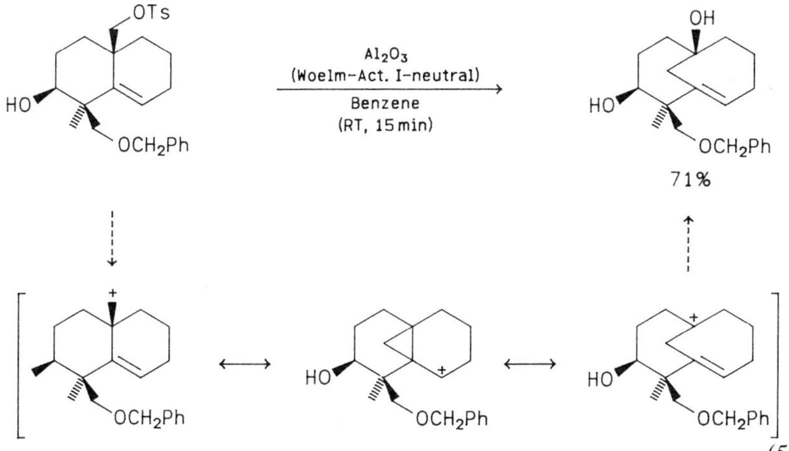

$$\tag{59}$$

3. Quadricyclanes

Various quadricyclanes undergo cycloreversions on alumina to form the corresponding norbornadienes (bicyclo[2.2.1]heptadienes) cleanly and at 0°C [Eq. (60) (Butler and Gupta, 1982)]: functional groups such as a cyclopropyl bromide, alkene, and allene survive these mild rearrangement conditions.

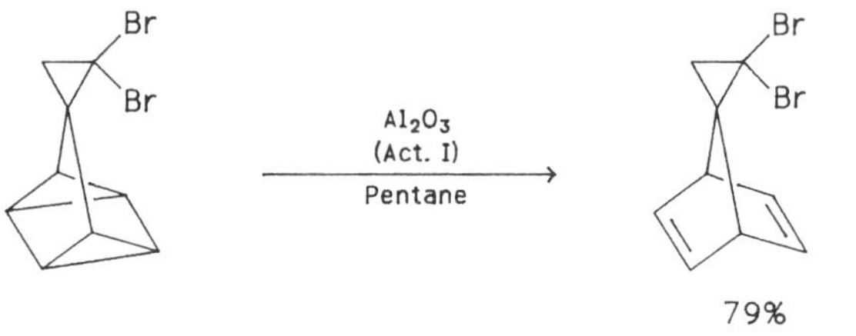

$$(60)$$

79%

C. Nitrogen Compounds

Oxime tosylates undergo Beckmann rearrangements on alumina to produce the corresponding amides [Eq. (61) (Glass *et al.*, 1978; Eq. 62, Greene *et al.*, 1966)].

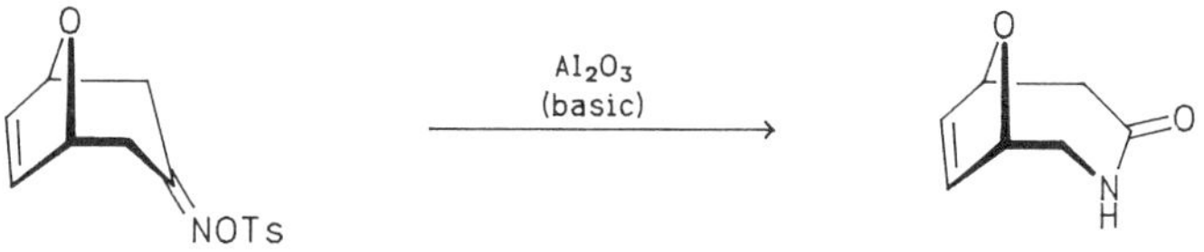

$$(61)$$

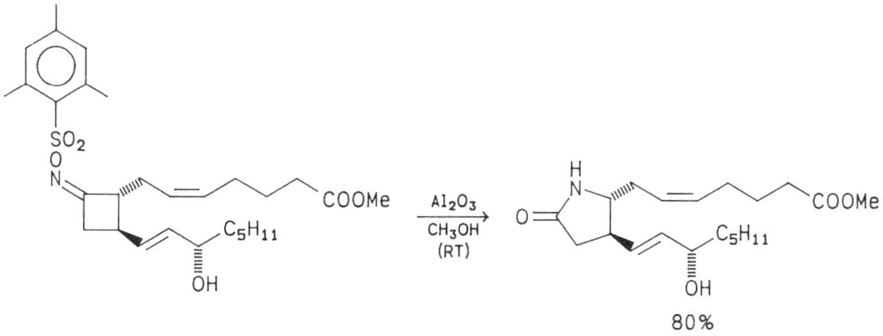

80%

$$(62)$$

Several strained bicyclic *N*-chloroamines undergo mild, alumina-promoted rearrangement to C-chloroamines in which the nitrogen atom is at a bridgehead position; because chlorine surprisingly is retained in the rearranged products, this process is proposed to be a dyotropic (Reetz, 1973) rearrangement involving concerted, simultaneous, intramolecular migration of two sigma bonds [Eqs. (63) and (64) (Davies *et al.*, 1985)].

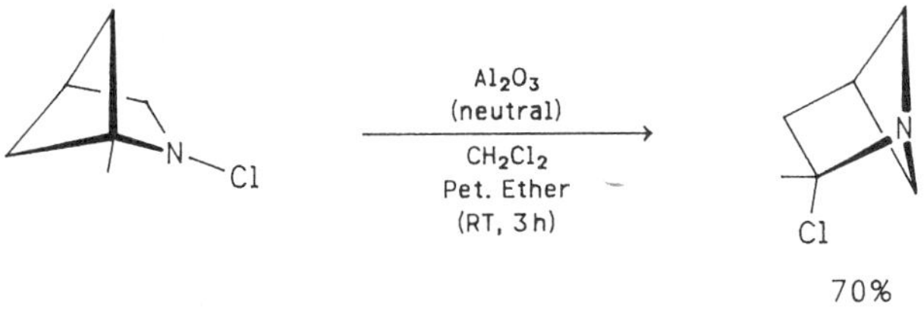

$$(63)$$

70%

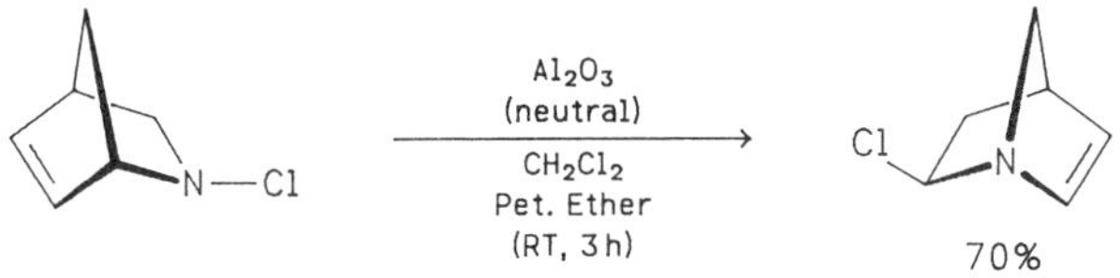

$$(64)$$

70%

D. Sigmatropic

Rearrangement on alumina of a 4-hydroxy-2-cyclopentenone into its isomeric 4-hydroxy-2-cyclopentenone is proposed, based on deuterium labeling experiments and structural studies, to occur via an enolate-induced, concerted, suprafacial, [1,5]-sigmatropic shift of an hydroxyl group rather than by a stepwise elimination–addition pathway [Eq. (65) (Scettri *et al.*, 1979; Novak *et al.*, 1985)].

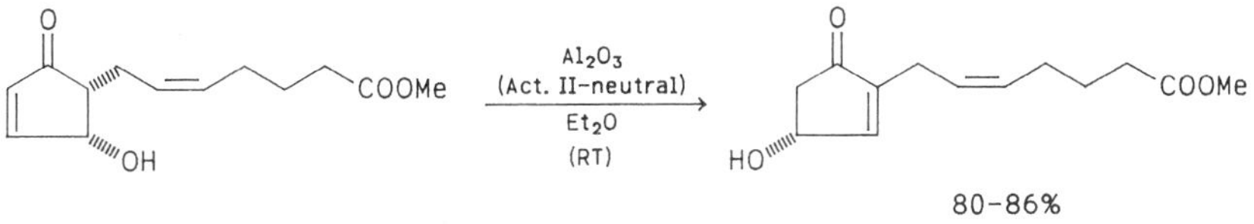

80–86%

$$(65)$$

E. Organometallics

Alkyl derivatives of some transition metal (e.g., Mn and Fe) carbonyls undergo alumina-facilitated alkyl migration to form acylmetal carbonyl species doubly bound to the alumina surface as shown by Fourier-transform infrared spectroscopy (FT-IR) [Eq. (66) (Correa *et al.*, 1980)].

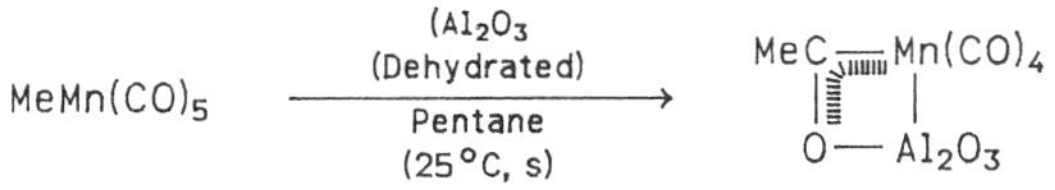

$$(66)$$

VII. CONCLUSIONS

Although some of the examples provided in this chapter are based on inadvertent exposure of organic compounds to chromatographic alumina, the vast majority of examples involve deliberate and planned use of alumina to achieve mild, selective, and convenient organic transformations of significant preparative value. More and more, the general chemical community is becoming aware of the usefulness of alumina not only for chromatographic purifications but also for mediation of many important organic reactions. This trend undoubtedly will continue at an accelerated rate, helped in large part by publications such as this book.

Some areas of research in which alumina might be useful in the future can be suggested. First, asymmetric catalysis is an important subject not only for industrial-scale manufacture of bulk chemicals, fine chemicals, and pharmaceuticals but also for development of new and/or improved catalyst systems. Alumina-supported optically active organic reagents represent challenging and potentially rewarding areas of research. The coorperative pull–push effect of the acidic and basic sites on alumina should make study of alumina-supported reagents at least as interesting and useful as the study of the corresponding silica-supported species.

Second, selective adsorption to alumina of one isomer of an isomeric pair should remove that isomer from solution and thus should diminish or stop its reaction with any reagent in solution. Likewise, a bifunctional (or indeed polyfunctional) molecule could be adsorbed (or covalently bound) to alumina by only one of its functional groups thereby allowing the remaining functional group(s) to react with a reagent in solution (cf. Ogawa *et al.*, 1985; Ricci *et al.*, 1985). For example, unsymmetrical primary–primary diols should be associated to alumina preferentially at the less hindered hydroxyl group thereby leaving the more hindered hydroxyl group available for reaction (e.g., with oxidizing, esterifying, or halogenating reagents) in solution [Eq. (67)]. This process would represent using alumina as a temporary protecting group for the normally more reactive functionality of a bifunctional molecule. The advantage of such a process over a classical two-step protection–deprotection strategy is that the alumina protection–deprotection sequence can be done conveniently *in situ* without isolation and purification of any intermediate monoprotected diol and without a separate deprotection step.

$$
HO \diagdown \diagup\diagdown\diagup\diagdown OH \xrightarrow[\text{Reagent}]{Al_2O_3} Y \diagdown \diagup\diagdown\diagup\diagdown OH \tag{67}
$$

$$Y = CHO, \ CH_2O\overset{\overset{\displaystyle O}{\|}}{C}R, \ CH_2Cl$$

REFERENCES

Alonso, M. E., and Morales, A. (1980). *J. Org. Chem.* **45**, 4530–4532.
Ando, T., Kawate, T., Yamawaki, J., and Hanafusa, T. (1982). *Chem. Lett.* 935–938.
Ando, T., Sumi, S., Kawate, T., Ichihara, J., and Hanafusa, T. (1984a). *J.C.S. Chem. Commun.* 439–440.
Ando, T., Kawate, T., Ichihara, J., and Hanafusa, T. (1984b). *Chem. Lett.* 725–728.
Arata, K., and Tanabe, K. (1979). *Chem. Lett.* 1017–1018.

Baek, S., Srebnik, M., and Mechoulam, R. (1985). *Tetrahedron Lett.* **26**, 1083–1086.

Bang, L., and Ourisson, G. (1973). *Tetrahedron* **29**, 2097.

Boeckman, R. K., Jr., Bruza, K. J., and Heinrich, G. R. (1978). *J. Am. Chem. Soc.* **100**, 7101–7103.

Butler, D. N., and Gupta, I. (1982). *Can. J. Chem.* **60**, 415–418.

Caine, P., and Smith, T. L., Jr. (1980). *J. Am. Chem. Soc.* **102**, 7568–7570.

Campelo, J. M., Garcia, A., Luna, D., and Marinas, J. M. (1984). *Can. J. Chem.* **62**, 638–643.

Cerveny, L., and Ruzicka, V. (1975). *Collect. Czech. Chem. Commun.* **40**, 2622.

Cheng, Y. S., Lin, W. L., and Chen, S. (1980). *Synthesis* p. 223.

Chou, S.-S. P., and Chu, C.-W. (1984). *J. Chin. Chem. Soc.* (*Taipei*) **31**, 351–356.

Clark, J. H., and Duke, C. V. A. (1985). *J. Org. Chem.* **50**, 1330–1332.

Cook, S. L., and Evans, J. (1983). *J.C.S. Chem. Commun.* 713–715.

Cope, A. C., Moon, S., and Peterson, P. E. (1959). *J. Am. Chem. Soc.* **81**, 1650.

Correa, F., Nakamura, R., Stimson, R. E., Burwell, R. L., Jr., and Shriver D. F. (1980). *J. Am. Chem. Soc.* **102**, 5114–5115.

Coutts, I. G. C., Edwards M., Musto, D., and Richards, D. J. (1980). *Tetrahedron Lett.* **21**, 5055–5056.

Danishefsky, S. J., and Pearson, W. H. (1983). *J. Org. Chem.* **48**, 3865–3866.

Davies, J. W., Malpass, J. R., and Walker, M. P. (1985). *J.C.S. Chem. Commun.* 686–687.

Dawson, W. H., Kaiser, S. W., Ellis, P. D., and Inners, R. R. (1981). *J. Am. Chem. Soc.* **103**, 6780–6781.

Dawson, W. H., Kaiser, S. W., Ellis, P. D., and Inners, R. R. (1982). *J. Phys. Chem.* **86**, 867.

Dev, S. (1972). *J. Sci. Ind. Res.* **31**, 60.

Dunstan, T. D. J., and Pincock, R. E. (1985). *J. Org. Chem.* **50**, 863–866.

Endo, K., and Hikino, H. (1979). *Bull. Chem. Soc. J.* **52**, 2439–2440.

Evans, H. E., and Weinberg, W. H. (1980). *J. Am. Chem. Soc.* **102**, 2548–2553.

Evans, J., and Gracey, B. P. (1980). *J.C.S. Chem. Commun.* 852–853.

Farwaha, R., de Mayo, P., Schauble, J. H., and Toong, Y. C. (1985). *J. Org. Chem.* **50**, 245–250.

Franck-Neumann, M., and Brion, F. (1979). *Angew Chem., Int. Ed. Engl.* **81**, 688–690.

Gaetano, K., Pagni, R. M., Kabalka, G. W., Bridwell, P., Walsh, E., True, J., and Underwood, M. (1985). *J. Org. Chem.* **50**, 499–502.

Glass, R. S., Deardoff, D. R., and Gains, L. H. (1978). *Tetrahedron Lett.* 2965–2968.

Gras, J.-L., and Guérin, A. (1985). *Tetrahedron Lett.* **26**, 1781–1784.

Greene, A. E., Depres, J. P., Nagano, H., and Crabbé, P. (1966). *Tetrahedron Lett.* 2365–2366.

Hasebe, M., Lazare, C., de Mayo, P., and Weedon, A. C. (1981). *Tetrahedron Lett.* **22**, 5149–5152.

He, M.-Y., Xiong, G., Toscano, P. J., Burwell, R. L., Jr., and Marks, T. J. (1985). *J. Am. Chem. Soc.* **107**, 641–652, and references cited therein.

Hodosan, F., and Serban, N. (1969). *Rev. Roum. Chim.* **14**, 121.

Hrubiec, R. T., and Smith, M. B. (1984). *J.C.S. Perkin I* 107–110.

Hudlicky, T., and Srnak, T. (1981). *Tetrahedron Lett.* **22**, 3351–3354.

Inokuchi, T., Asanuma, G., and Torii, S. (1982). *J. Org. Chem.* **47**, 4622–4626.

Irvine, E. A., John, C. S., Kemball, C., Pearman, A. J., Day, M. A., and Sampson, R. J. (1980). *J. Catal.* **61**, 326–325.

Iwasawa, Y., Sasaki, Y., and Ogasawara, S. (1981). *J.C.S. Chem. Commun.* 140–142.

Kamitori, Y., Hojo, M., Masuda, R., and Yamamoto, M. (1985a). *Chem. Lett.* 253–254.

Kamitori, Y., Hojo, M., Masuda, R., and Yoshida, T. (1985b). *Tetrahedron Lett.* **26**, 4767–4770.

Kan-Fan, G., and Husson, H.-P. (1980). *Tetrahedron Lett.* 1463–1466.

Kawakami, Y., and Yamashita, Y. (1982). *Synth. Commun.* **12**, 253–256.

Kazusaka, A., Suzuki, H., and Toyoshima, I. (1983). *J.C.S. Chem. Commun.* 150–151.

Kropf, H., and Torkler, A. (1985a). *J. Chem. Res. Synop.* p. 304.

Kropf, H., and Torkler, A. (1985b). *J. Chem. Res., Miniprint* 2948–2962.

Lausarot, P. M., Vaglio, G. A., and Valle, M. (1984). *J. Organomet. Chem.* **275**, 233–237.

Lavalley, J.-C., Calliod, J., and Travert, J. (1980). *J. Phys. Chem.* **84**, 2084–2085.

Liu, K.-T., and Tong, Y.-C. (1978). *J. Org. Chem.* **43**, 2717–2718.

McKillop, A., and Young, D. W. (1979). *Synthesis* 401–422, 481–500.

McMurry, J. E., and Donovan, S. F. (1977). *Tetrahedron Lett.* 2869–2872.

McMurry, J. E., and Isser, S. J. (1972). *J. Am. Chem. Soc.* **94**, 7132–7137.

Markgraf, J. H., Burns, D. B., Greeno, E. W., Leonard, K. J., and Miller, M. D. (1984). *Synth. Commun.* **14**, 647–653.

Matsushita, H., Ishiguro, S., Ichinose, H., Izumi, A., and Mizusaki, S. (1985). *Chem. Lett.* 731–734.

Mekhtiev, S. D., Alimardanov, K. M., Musaev, M. R., and Suleimanova, E. T. (1977). *Azerb. Khim. Zh.* 54-7 [*C.A.* **89**, 425245 (1978)].

Miller, C. J., and Majda, M. (1985). *J. Am. Chem. Soc.* **107**, 1419–1420.

Miller, R. B., and McGarvey, G. (1977). *Synth. Commun.* **7**, 475–482.

Misono, M., and Okuhara, T. (1980). *Yuki Gosei Kagaku Kyokaishi* **38**, 47–60 [*C. A.* **93**, 70604 (1980)].

Morelli, I., Catalano, S., Scartoni,, V., Ferretti, M., and Marsili, A. (1979). *J.C.S. Perkin I* 1665–1668.

Niwa, M., Iguchi, M., and Yamanura, S. (1979). *Tetrahedron Lett.* 4291–4294.

Novak, L., Szantay, C., Meisel, T., Aszodi, J., Szabo, E., and Fekete, J. (1985). *Tetrahedron* **41**, 435–450.

Ogawa, H., Chihara, T., and Taya, K. (1985). *J. Am. Chem. Soc.* **107**, 1365–1369.

Onaka, M., Kawai, M., and Izumi, Y. (1983). *Chem. Lett.* 1101–1104.

Onaka, M., Kawai, M., and Izumi, Y. (1985). *Chem. Lett.* 779–782.

Otomasu, H., Higashiyama, K., Honda, T., and Kametani, T. (1982). *J.C.S. Perkin I* 2399–2401.

Paradisi, M. P., and Zecchini, C. P. (1982). *Tetrahedron* **38**, 1827–1829.

Park, H., King, P. F., and Paquette, L. A. (1979). *J. Am. Chem. Soc.* **101**, 4773.

Parlar, H., and Baumann, R. (1981). *Angew. Chem., Int. Ed. Engl.* **20**, 1014.

Pearson, A. J., and Ong, C. W. (1981). *J. Am. Chem. Soc.* **103**, 6686–6690.

Pelletier, S. W., Venkov, A. R., Finer-Moore, J., and Mody, N. V. (1980). *Tetrahedron Lett.* **21**, 809–812.

Peng, C. T., and Buchman, O. (1985). *Tetrahedron Lett.* **26**, 1375–1378.

Pietraszkiewicz, M., and Jurczak, J. (1983). *J.C.S. Chem. Commun.* 132–133.

Pines, H., and Stalick, W. M. (1977). "Base-Catalyzed Reactions of Hydrocarbons and Related Compounds." Academic Press, New York.

Posner, G. H. (1978). *Angew. Chem., Int. Ed. Engl.* **17**, 487–495.

Posner, G. H., and Gurria, G. M. (1976). *J. Org. Chem.* **41**, 578–580.

Posner, G. H., and Hulce, M. (1980). *J. Catal.* **64**, 497–498.

Posner, G. H., and Oda, M. (1981). *Tetrahedron Lett.* **22**, 5003–5006.

Posner, G. H., and Rogers, D. Z. (1977). *J. Am. Chem. Soc.* **99**, 8208–8213.

Posner, G. H., Runquist, A. W., and Chapdelaine, M. J. (1977). *J. Org. Chem.* **42**, 1202–1208.

Posner, G. H., Ellis, J. W., and Ponton, J. (1981a). *J. Fluorine Chem.* **19**, 191–198.
Posner, G. H., Hulce, M., adn Rose, R. K. (1981b). *Synth. Commun.* **11**, 737–741.
Posner, G. H., Okada, S., Babiak, K. A., Miura, K., and Rose, R. K. (1981). *Synthesis* 789.
Rao, C. G. (1982). *Synth. Commun.* **12**, 177–181.
Rebek, J., and McCready, R. (1979). *Tetrahedron Lett.* 4337–4338.
Reetz, M. T. (1973). *Tetrahedron* 2189–2195.
Regen, S. L., and Mehrotra, A. K. (1981). *Synth. Commun.* **11**, 413–417.
Reitz, A., Verlander, M., and Goodman, M. (1982). *Tetrahedron Lett.* **23**, 751–752.
Ricci, A., Roelens, S., and Vannuchi, A. (1985). *J.C.S. Chem. Commun.* 1457–1458.
Rigo, B., Jabre, S., Maliar, F., and Courturier, D. (1985). *Synth. Commun.* **15**, 473–478.
Ripmeester, J. A. (1983). *J. Am. Chem. Soc.* **105**, 2925–2927.
Riseman, S. M., Massoth, F. E., Dhar, G. M., and Eyring, E. M. (1982). *J. Phys. Chem.* **86**, 1760.
Ruppert, J. F., and White, J. D. (1981). *J. Am. Chem. Soc.* **103**, 1808–1813.
Sangaiah, R., Gold, A., and Toney, G. E. (1983). *J. Org. Chem.* **48**, 1632–1638.
Santaniello, C., Farachi, C., and Manzocchi, C. (1979). *Synthesis* p. 912.
Savoia, D., Tagliavini, E., Trombini, C., and Umani-Ronchi, A. (1980). *J. Org. Chem.* **45**, 3227–3229.
Savoia, D., Trombini, C., and Umani-Ronchi, A. (1982). *J. Org. Chem.* **47**, 564–566.
Sawicki, R. A. (1982). *Tetrahedron Lett.* **23**, 2249–2252.
Schumann, H., and Pachaly, B. (1981). *Angew. Chem. Int. Ed. Engl.* **20**, 1043.
Scettri, A., Piancatelli, G., D'Auria, M., and David, G. (1979). *Tetrahedron* **35**, 135–138.
Stair, P. C. (1982). *J. Am. Chem. Soc.* **104**, 4044–4052.
Su, J., Zheng, W., and Deng, Y. (1982). *Huaxe Tongbar* 391–393 [*C.A.* **97**, 181848. (1982)].
Sugihara, Y., Morokoshi, N., and Murata, I. (1977). *Tetrahedron Lett.* 3887–3880.
Sukata, K. (1985). *Bull. Chem. Soc. J.* **58**, 838–843.
Suzuki, M., Oda, Y., and Noyori, R. (1979). *J. Am. Chem. Soc.* **101**, 1623–1625.
Takahashi, K., Nishizuka, T., and Iida, H. (1981). *Tetrahedron Lett.* **25**, 2389.
Tanis, S. P., and Herrinton, P. M. (1983). *J. Org. Chem.* **48**, 4572–4580.
Taran, M., and Delmond, B. (1982). *Tetrahedron Lett.* **23**, 5535–5538.
Texier-Boullet, F. (1985). *Synthesis* 679–681.
Tsuboi, S., Masuda, T., Makino, H., and Takeda, A. (1982a). *Tetrahedron Lett.* **23**, 209–212.
Tsuboi, S., Masuda, T., and Takeda, A. (1982b). *J. Org. Chem.* **47**, 4478–4482.
Tsuboi, S., Masuda, T., and Takeda, A. (1983a). *Chem. Lett.* 1829–1832.
Tsuboi, S., Masuda, T., and Takeda, A. (1983b). *Bull. Chem. Soc. Jpn.* **56**, 3521–3522.
Tundo, P., Venturello, P., and Angeletti, E. (1982). *J. Am. Chem. Soc.* **104**, 6551–6555.
Villemin, D., and Ricard, M. (1982). *Nouv. J. Chim.* **6**, 605.
Villemin, D., and Ricard, M. (1984). *Nouv. J. Chim.* **8**, 185–189.
Vinczer, P., Novak, L., and Szantay, C. (1984). *Synth. Commun.* **14**, 281–288.
Wadsworth, D. H., and Donatelli, B. A. (1981). *Synthesis* 285–286.
Yamamura, S., Niwa, M., Itu, M., and Saito, Y. (1982). *Chem. Lett.* 1681–1684.
Yamawaki, J., and Ando, T. (1980). *Chem. Lett.* 533–536.
Yamawaki, J., Kawate, T., Ando. T., and Hanafusa, T. (1983). *Bull. Chem. Soc. Jpn.* **56**, 1885–1886.
Yanami, T., Miyashita, M., and Yoshikoshi, A. (1979). *J.C.S. Chem. Commun.* 525–527.
Yeats, R. B. (1983). *Tetrahedron Lett.* **14**, 3423–3426.
Zak, J., and Kuwana, T. (1982). *J. Am. Chem. Soc.* **104**, 3314–3315.
Zakharova, N. I., Filippova, T. M., Bekker, A. R., Miropol'skaya, M. A., and Samokhvolovl, G. I. (1978). *Zh. Org. Khim.* **14**, 1413 [*C.A.* **89**, 146369 (1978)].

17 ANIONIC ACTIVATION, REACTIONS IN DRY MEDIA

André Foucaud

Groupe de Chimie Structurale
Université de Rennes
F-35042 Rennes, France

**Georges Bram
and André Loupy**

Laboratoire Réactions Sélectives
 sur Supports
Université Paris–Sud
F-91405 Orsay, France

I. INTRODUCTION

Various types of organic reactions can be promoted under very mild conditions by the acidic and basic sites on the surface of alumina (Posner, 1978; MacKillop and Young, 1979). The emphasis in this chapter is on more recent reactions promoted by anionic activation with alumina and with alumina-supported reagents.

The activation can be performed according to two processes:

(1) The reagents are alumina-supported nucleophiles. The alumina is impregnated with a solution of a stable nucleophilic salt in water or in a hydroxylic organic solvent; then the solvent is evaporated. This impregnated reagent reacts with the electrophilic reagent in the presence or in the absence ("dry-media" conditions) of an organic solvent. Sometimes the salt is directly mixed with alumina, without impregnation (dispersed reagent). In other cases, the nucleophilic species is generated *in situ* by adsorption of the conjugated acids on alumina.

(2) The nucleophilic species are generated from the conjugated acids by alumina made basic by impregnation with a solution of a basic reagent (such as alkali metal hydroxide, alkali metal alkoxide, or alkali metal fluoride) followed by solvent evaporation. The nucleophilic species A^- are generated *in situ* when the conjugated acids AH are reacted with this

317

PREPARATIVE CHEMISTRY
USING SUPPORTED REAGENTS

"basic alumina" (see Chapter 3). The formation of the anions and their subsequent reactions can be performed in the presence or in the absence of an organic solvent (dry media).

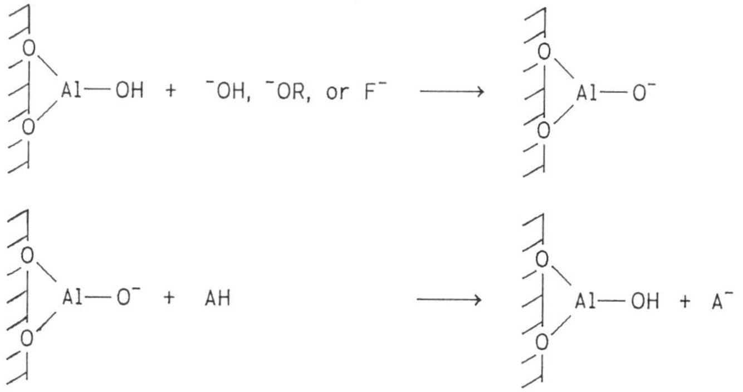

II. ANIONIC ACTIVATION ON ALUMINA

A. Deuteration

Kabalka *et al.* (1981) have studied the deuteration of active methylene compounds with D_2O-treated alumina. The deuterated alumina is prepared by treating dehydrated Al_2O_3 with D_2O in THF:

$$(EtOCO)_2CH_2 \xrightarrow[\text{(5 min)}]{D_2O-Al_2O_3} (EtOCO)_2CD_2$$
$$96\%$$

An efficient method for the preparation of deuterated alumina has been reported by Gaetano *et al.* (1985).

B. Alkylation

1. C-Alkylation

Alkylcyanides have been efficiently prepared by reaction of alumina-impregnated cyanide anion with organic halides in the presence of toluene (Regen *et al.*, 1979a) or in "dry-media" conditions (Bram *et al.*, 1980a). Quici and Regen (1979) and Regen *et al.* (1981) have described the reaction of solid KCN with 1-bromooctane in toluene in the presence of alumina and have proposed that alumina could react as a triphase catalyst. In contrast, Ando *et al.* (1984b) found that the reaction proceeded only very slowly when dry alumina was used but was facilitated in the presence of alumina and a minute amount of water; Sukata (1985a) confirms the inertness of "dispersed" KCN on alumina in the absence of traces of water.

Ando *et al.* (1984b) have also shown that the combined use of alumina, a minute amount of water, and ultrasonic irradiation greatly accelerates nucleophilic substitution by cyanide anion. They suggested that the role of water is to penetrate the crystal lattice of KCN and to produce an active ion or ion aggregate on the alumina surface for promoting an interfacial reaction. It appeared that the use of ultrasonic irradiation not only activates the reaction but can also dramatically change the course of an alumina-promoted reaction. Ando *et al.* (1984a) have found the following spectacular dichotomy:

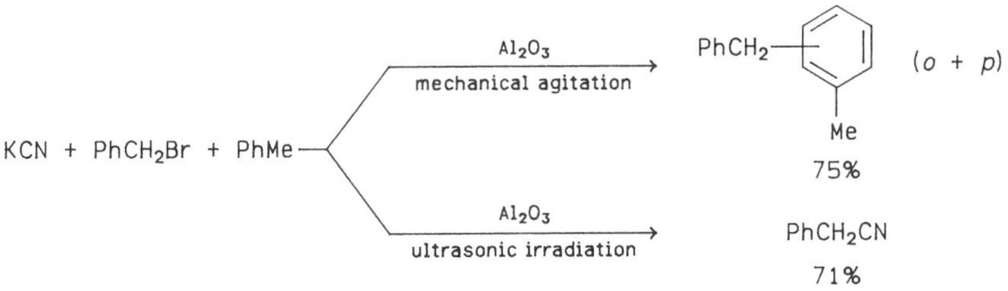

Unexpectedly, alumina under mechanical agitation did not promoted cyanide substitution but instead induced the Friedel–Crafts reaction of benzyl bromide with toluene. The role of sonication could be to enhance contact of KCN with alumina, nucleophilic attack by the cyanide ion on the alumina surface, and to decrease the catalytic ability of Lewis-acid sites of alumina toward the aromatic electrophilic substitution.

Alumina-assisted aryl cyanations have been performed by Dalton and Regen (1979):

$$NaCN/Al_2O_3 + Ar-X \xrightarrow[\text{catalyst}]{\text{toluene}} Ar-CN$$
$$62-95\%$$

The catalyst used was tetrakis(triphenylphosphine) palladium(0). Sodium cyanide impregnated on alumina was recommended for cyanation of aryl iodides, and solid sodium cyanide crushed together with alumina was recommended for reaction with aryl bromides.

2. O-Alkylation

When impregnated acetate and benzoate anions are alkylated on alumina without solvent ("dry media"), excellent yields (85–97%) of alkyl acetates and lower but still useful yields (47–65%) of alkyl benzoates are obtained (Bram *et al.*, 1980a):

$$CH_3CO_2^- + RX \longrightarrow CH_3CO_2R + X^-$$
$$50-97\%$$

$$PhCO_2^- + RX \longrightarrow PhCO_2R + X^-$$
$$47-65\%$$

Quici and Regen (1979) and Regen *et al.* (1981) have described the alkylation of acetate anion in toluene in the presence of alumina acting as a triphase catalyst. This result is quite in contrast with the finding of Ando *et al.* (1982a) who reported that, in the absence of a small quantity of water, the reaction proceeded very slowly, as confirmed by Sukata (1985a).

Chihara (1980) and Ogawa *et al.* (1985) have performed, a very interesting selective esterification of dicarboxylic acids by use of monocarboxylate chemisorption on alumina. This reaction is very difficult by conventional methods.

Thus, terephthalic, isophthalic, *cis*- and *trans*-1,4-cyclohexanecarboxylic acids, and aliphatic dicarboxylic acids $HO_2C(CH_2)_nCO_2H$ ($n = 3-8$), adsorbed on alumina gave the corresponding monomethyl ester quantitatively by reaction with diazomethane. It was suggested that dicarboxylic acids are adsorbed on alumina as monocarboxylate anions; the unadsorbed carboxyl group held on a position remote from the alumina surface is thus selectively esterified. Selective monomethyl esterification of phthalic acid on alumina is not successful, probably as a consequence of the close proximity of the two carboxyl groups and of the forced orientation of the second group when one is adsorbed. Some other methylating reagents also lead to the selective monomethylation of terephthalic acid, but diazomethane is the best. It has the highest reactivity and nitrogen is the only by-product.

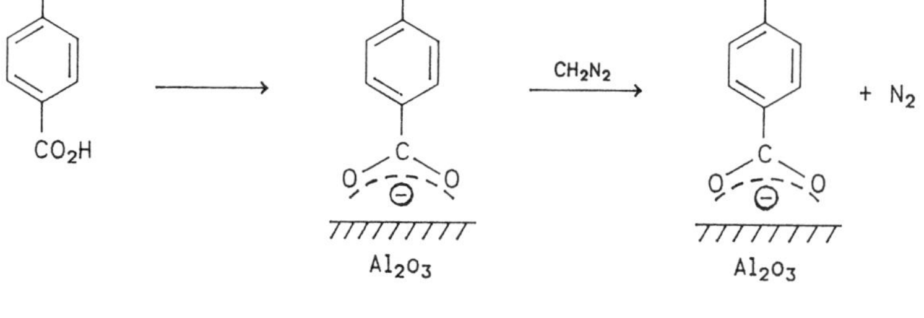

3. *S*-Alkylation

Symmetrical organic sulfides have been conveniently prepared by alkylation of impregnated reagents on alumina (Czech *et al.*, 1980). Thus, n-$C_8H_{17}Br$ reacting in toluene with Na_2S on alumina can afford dioctyl sulfide in good yield:

$$Na_2S + n\text{-}C_8H_{17}Br \xrightarrow[90°C]{Al_2O_3} n\text{-}C_8H_{17}\text{—}S\text{—}n\text{-}C_8H_{17}$$

$$97\%$$

Unsymmetrical sulfides can also be obtained by reaction of organic halides with thiols adsorbed on alumina; but significantly higher yields are obtained when NaOH impregnated on Al_2O_3 is used:

$$PhSH + n\text{-}C_8H_{17}Br \xrightarrow[90°C]{Al_2O_3} Ph\text{—}S\text{—}n\text{-}C_8H_{17}$$

$$55\%$$

4. Halide Exchange

Chloride and iodide displacements on 1-bromooctane on alumina have been described by Quici and Regen (1979).

$$Y^- + R\text{—}Br \longrightarrow R\text{—}Y + Br^-$$
$$Y = I, \ Cl$$

An interesting leaving group effect was observed. 1-Bromooctane exhibits significantly greater reactivity toward nucleophilic displacement by chloride as compared to 1-iodooctane. This unusual reactivity has also been observed during the reaction of cyanide anion and acetate anion with 1-bromooctane and 1-iodooctane.

Tundo *et al.* (1983) have performed halide exchange by passing a mixture of two different gas-phase alkyl halides through a column filled with alumina:

$$R\text{—}X + R'\text{—}Y \longrightarrow R\text{—}Y + R'\text{—}X$$

Addition of phosphonium salts increases the exchange rate by several times.

C. Addition and Condensation Reactions

Various addition reactions, eventually followed by dehydration, have been catalyzed by alumina. Muzart (1982) has shown that the self-condensation of ketones is catalyzed by basic aluminium oxide:

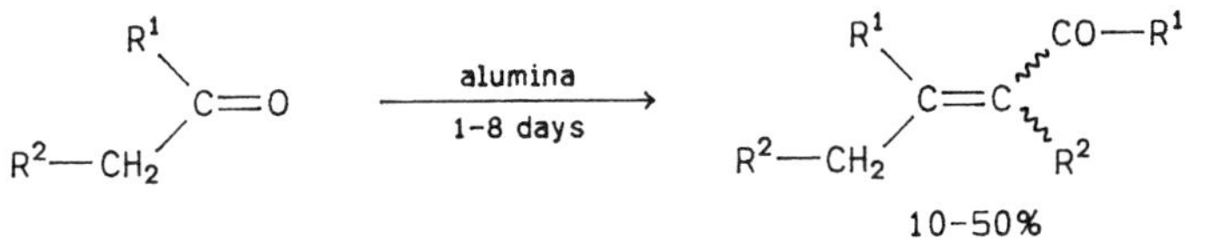

$$10\text{-}50\%$$

The aldol dimerization of ketones catalyzed by basic alumina has also been described by Barot *et al.* (1984). These authors claimed that ultrasonic irradiation improves the yield of the reaction. However, it was shown that an efficient aldol dimerization of ketones can be carried out just by adsorbing the neat ketone on alumina (Muzart, 1985). The condensation of benzaldehyde with enolizable ketones on alumina in dry media has been reported by Varma *et al.* (1985). For example

$$\text{PhCHO} + \text{MeCOPh} \xrightarrow{\text{Al}_2\text{O}_3} \text{Ph—CH=CH—CO—Ph}$$
$$83\%$$

Alumina is a very good catalyst for the Knoevenagel reaction of aldehydes with acidic methylene active reagents such as malonodinitrile or methylcyanoacetate (Texier-Boullet and Foucaud, 1982a) and for the synthesis of imines from aldehydes and amines (Texier-Boullet, 1985):

$$\text{R}^1\text{—CHO} + \text{NC—CH}_2\text{—X} \xrightarrow{\text{alumina}} \begin{array}{c} \text{R}^1 \qquad \text{CN} \\ \diagdown \text{C=C} \diagup \\ \text{H} \diagup \qquad \diagdown \text{X} \end{array} \quad \text{X=CN, CO}_2\text{R}$$

$$\text{R}^1\text{—CHO} + \text{R}^2\text{—NH}_2 \xrightarrow{\text{alumina}} \text{R}^1\text{—CH=N—R}^2$$

The Michael addition of secondary amines to *exo* cyclic α,β-unsaturated ketones has been performed in good yields in the presence of alumina (Pelletier *et al.*, 1980). An interesting application of alumina as a catalyst for the synthesis of 4*H*-chromenes from ortho hydroxybenzaldehydes and malononitrile or methylcyanoacetate has been reported (Roudier and Foucaud 1984).

Rosini *et al.* (1983a,b) have shown that the nitroaldol condensation (Henry reaction) is catalyzed by alumina, at room temperature:

$$\text{R}^1\text{—CH}_2\text{NO}_2 + \text{R}^2\text{—CHO} \xrightarrow{\text{alumina}} \begin{array}{c} \text{R}^1\text{—CH—CH—R}^3 \\ | \quad\; | \\ \text{NO}_2 \;\; \text{OH} \end{array}$$
$$70\text{–}85\%$$

Hanessian and Kloss (1985) have reported a stereocontrolled nitroaldol reaction catalyzed by alumina:

$$\text{O}_2\text{N—(CH}_2)_2\text{—CO}_2\text{Me} \quad + \quad \begin{array}{c} \text{OCH}_2\text{Ph} \\ | \\ \text{Me} \quad \text{CHO} \end{array} \xrightarrow{\text{alumina}} \begin{array}{c} \text{PhCH}_2\text{O} \quad \text{NO}_2 \\ | \qquad\quad | \\ \text{Me} \qquad \text{CH}_2\text{CO}_2\text{Me} \\ \vdots \\ \text{OH} \end{array}$$

The synthesis of nitroalkenes from furaldehydes and nitroalkanes has been achieved in the presence of alumina, without solvent, at room temperature (Rosini *et al.*, 1985):

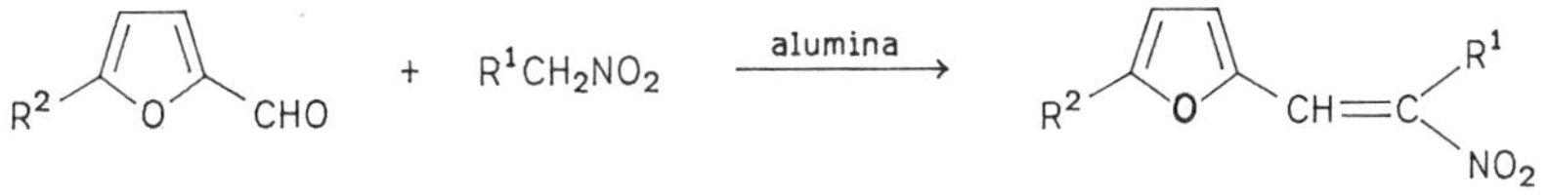

Alumina-promoted addition of dialkylphosphites to aldehydes leads to hydroxyphosphonates (Texier-Boullet and Foucaud, 1982b):

$$R^1—CHO + HPO(OR^2)_2 \xrightarrow{\text{alumina}} R^1—CHOH—PO(OR^2)_2$$

Alumina and potassium cation-exchanged Y-type zeolites have been compared for the catalysis of the *N*-alkylation of aniline derivatives. The zeolite is a better catalyst in most cases (Onaka *et al.*, 1985).

D. Rearrangement Reactions

The prototropic rearrangement of various 1,2-diarylpropynes into corresponding allenes can be accomplished by adsorption on basic alumina (Jacobs and Singer, 1952; Jacobs and Dankner, 1957; Jacobs *et al.*, 1964). The use of $KOH–Al_2O_3$ led to an improvement in the yield of the reaction.

$$Ph—C≡C—CH_2—Ph \longrightarrow Ph—CH=C=CH–Ph$$

The rearrangement on alumina of β-allenic esters into *trans*, *cis*-2,4-dienoic esters has been reported:

The first step of the reaction may be the coordination of the carbonyl oxygen of the ester group to the alumina surface followed by a proton abstraction to produce an enolate intermediate (Tsuboi *et al.*, 1982a,b).

III. ALUMINA-SUPPORTED ALKALI METAL HYDROXIDES AND ALKOXIDES

A. Alkylation

1. *C*-Alkylation

Malonic ester anions have been generated on "basic" alumina prepared by impregnation of the support with NaOMe. Using $Br(CH_2)_5Br$ as the alkylating agent and working in "dry-media" conditions, intramolecular or intermolecular alkylations have been obtained selectively in high yield depending on the amount of base used (Bram and Fillebeen-Khan, 1979):

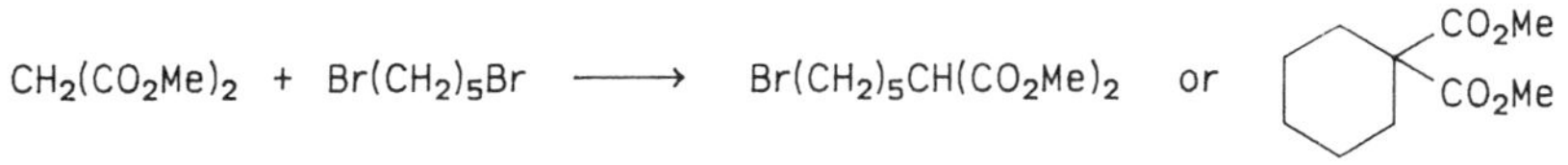

$$CH_2(CO_2Me)_2 \; + \; Br(CH_2)_5Br \; \longrightarrow \; Br(CH_2)_5CH(CO_2Me)_2 \quad or$$

Sukata (1983) has shown that the use of NaOH or KOH impregnated on alumina allows selective α-monoalkylation of phenylacetonitrile. With regard to the selectivity for α-monoalkylation, $NaOH/Al_2O_3$ was better than KOH/Al_2O_3. In contrast to KOH/Al_2O_3, the reaction with alkyl iodides was slower than that with alkyl bromides when $NaOH/Al_2O_3$ was used:

$$PhCH_2CN + RX \xrightarrow[(2-7\,h,40°C)]{Al_2O_3/KOH} PhCHRCN$$
$$69\text{--}94\%$$

2. O-Alkylation

Potassium hydroxide impregnated on alumina was found to be equally or even a little more effective than KF impregnated on alumina for the methylation of 1-octanol and diethylene glycol. However, the former reagent was less effective than the latter for the cyclization of a polyethylene glycol ditosylate with polyethylene glycol into a crown ether (Yamawaki and Ando, 1980).

The anions of phenol and 2-naphtol have been generated on alumina impregnated with NaOMe, and alkylating agents were deposed on the "basic" support without solvent. Yields are modest; in the case of Me_2SO_4, pure O-alkylated material is obtained whereas alkyl halides give small quantities of C-alkylated products in addition (Bram et al., 1980a).

3. Ambident C-Alkylation versus O-Alkylation

The alkylation of acetoacetate anion presents two types of selectivity:

(1) Ambident reactivity, C-alkylation versus O-alkylation, depending on orbital control versus charge control, and related to the distribution of the electronic density in the mesomeric anion.

(2) Mono-C versus di-C selectivity, depending on the position of the equilibrium involving acetoacetate and mono-C-alkylated acetoacetate anions.

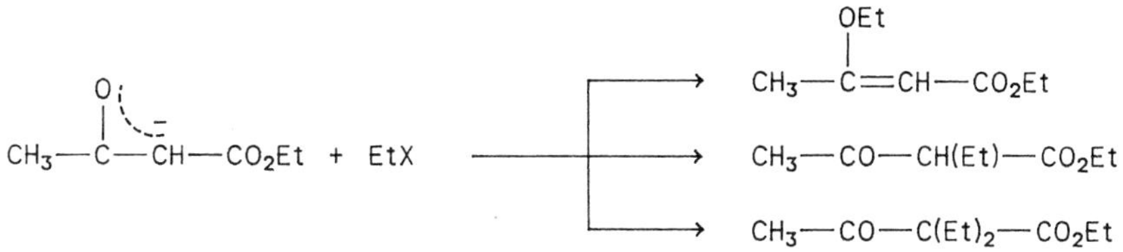

$$CH_3-C-CH-CO_2Et + EtX \longrightarrow$$

$$CH_3-\underset{\underset{OEt}{|}}{C}=CH-CO_2Et$$

$$CH_3-CO-CH(Et)-CO_2Et$$

$$CH_3-CO-C(Et)_2-CO_2Et$$

The alkylation of acetoacetate anion with alkylating agents has been performed in "dry-media" conditions on alumina impregnated with NaOMe: with EtI, EtBr, and even the Et_2SO_4 mono-C-alkylated product is formed selectively at room temperature and only traces of di-C-alkylated compound and O-alkylated enol ether are formed (Bram *et al.*, 1980a,b).

Diphenylethylsulfonium tetrafluoroborate $Ph_2S^+EtBF_4^-$ reacted much more rapidly under the same conditions, but a sizeable proportion of di-C-alkylated compound was formed. However, with a lesser amount of NaOMe on alumina (1 meq/g instead of 5 meq/g), very selective mono-C-alkylation was observed (Badet *et al.*, 1984).

A high proportion of C-alkylated products is obtained when the anion of β-naphtol, generated on NaOMe-impregnated alumina, reacts with MeI or PhCH$_2$Cl in "dry media". However, by contrast with acetoacetate anion under the same conditions, selective alkylations are not observed and mixtures of the different C-alkylated compounds are obtained (Bram *et al.*, 1980b).

The amount of adsorbed water appears very important to control the regioselectivity of alkylation of the ambident enolate and phenoxide anions. Thus, Benson and Geraghty (1983) have performed the alkylation of the anion of 1-methyl-2-napthol on NaOMe-impregnated alumina, with MeI in "dry media." They have shown that the C/O ratio increases with the temperature of alumina activation, i.e., with the weight loss of the adsorbed water. Thus the C/O ratio increases from 0.66 (temperature of activation = 30°C) to 3.80 (temperature of activation = 308°C). An interpretation based on change in the hydrogen bonding ability of alumina and modification of the overall orientation of the adsorbed molecules, has been proposed.

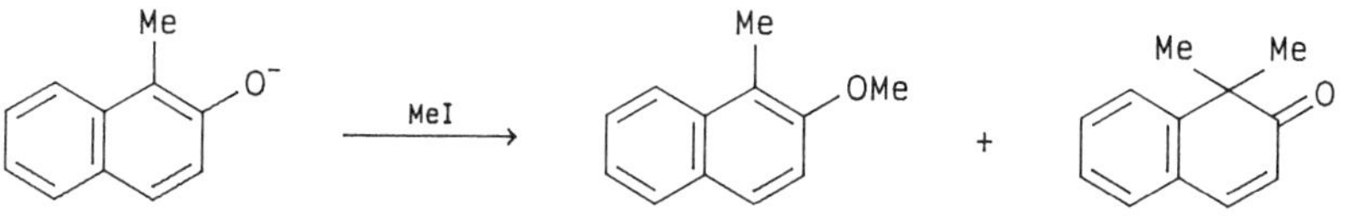

These ambident anion alkylations are evidence for the ability of base-impregnated alumina to promote a high degree of C-substitution when alkylations of ambident enolate or phenoxide anions are concerned.

4. N-Alkylation

Efficient and highly selective N-alkylations of amides have been performed by Sukata (1985b) using KOH-impregnated alumina or KOH dispersed on alumina, in the presence of dioxane.

$$R-\overset{\overset{\displaystyle O}{\|}}{C}-NH_2 + R'-X \longrightarrow R-\overset{\overset{\displaystyle O}{\|}}{C}-NHR'$$

In this case, for both a good yield and selectivity for N-monoalkylation, solid KOH dispersed on alumina is superior to KOH impregnated on the support.

Indole anion formation has been performed with KOH-impregnated alumina or KOH dispersed on alumina; the alumina-dispersed base is more effective. Efficient alkylation with Et_2SO_4 has been performed in the absence of solvent. The yields of alkylation with Et_2SO_4 and $n\text{-}C_8H_{17}Br$ increased when catalytic amounts of a quaternary ammonium salt (1%) were added. In fact, the presence of the quaternary ammonium salt allowed the reaction to be run in the absence of alumina, i.e., by solid–liquid phase transfer catalysis without added solvent (Barry *et al.*, 1983).

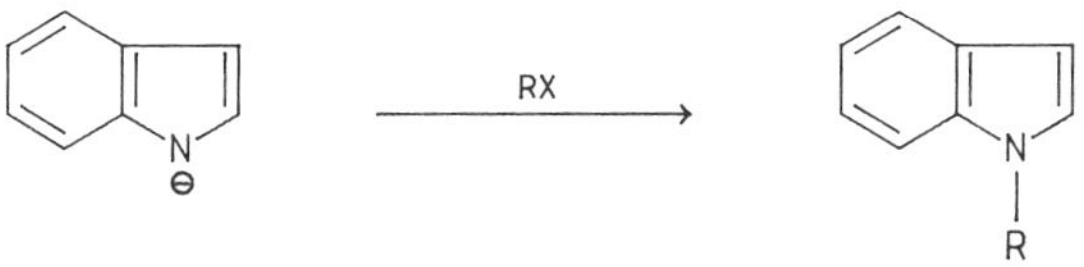

N-Alkylation on various inorganic solid supports has been successfully applied to 2-oxazolidone. Good yields are obtained when KOH dispersed on alumina is used as the basic reagent.

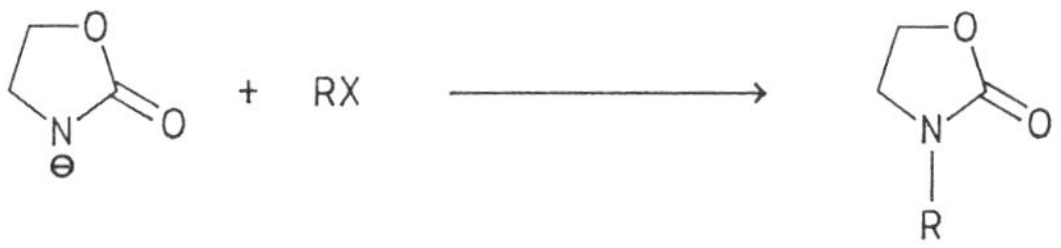

In this case, the presence of a quaternary ammonium salt clearly increases the yield of the reaction (Nguyen Hien Duc *et al.*, 1985).

The usual synthesis of N-propargyl derivatives of heterocycles **I** by alkylation in basic media is impeded by the formation of resonance stabilized allenes **II** and ynamines **III**:

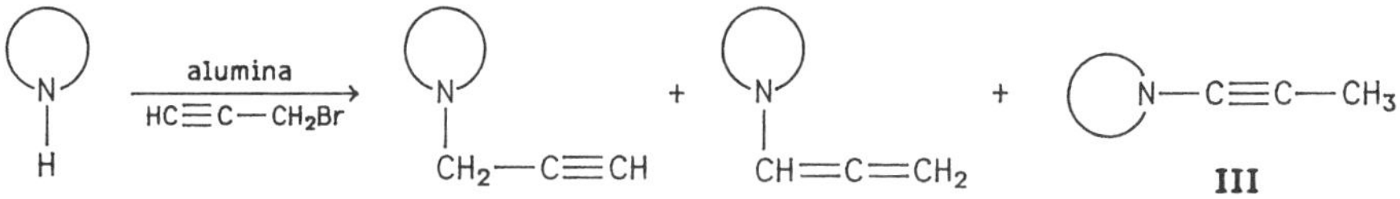

Galons *et al.* (1985) have shown that selective *N*-propargylation of heterocycles can be achieved by solid–liquid phase-transfer catalysis without solvent together with use of solid inorganic supports, especially alumina. On the other hand, alkylation of heterocycles on alumina impregnated with NaOMe in the absence of phase-transfer catalysts always gives selectively the *N*-propargyl heterocycle **I** but appears to be less efficient than when the phase-transfer catalyst is used.

Uracil has been made to react with Me_2SO_4 on alumina impregnated with NaOMe, giving rise to the dimethylated derivative. A better yield was obtained when solid–liquid phase-transfer catalysis without solvent was used; no alkylation took place with MeI (Bram *et al.*, 1985).

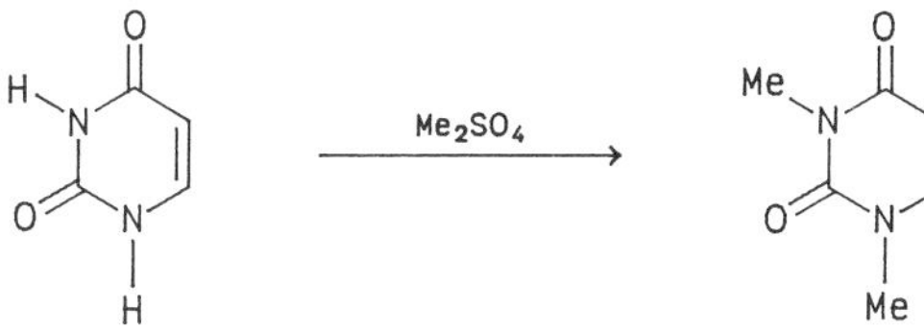

5. *S*-Alkylation

Czech *et al.* (1980) have reported the preparation of unsymmetrical sulfides by reaction of alkyl halides with thiols on alumina impregnated with NaOH. Use of NaOH either dispersed on alumina or powdered or reaction on alumina alone gave significantly lower yields:

$$R_1-X + R_2-SH \xrightarrow{Al_2O_3/NaOH} R_1-S-R_2$$

B. Other Reactions

There are examples of other reactions performed on alumina impregnated with hydroxide or alkoxide. Castells and Fletcher (1956) have reported that alumina impregnated with a solution of KOH is an effective reagent for the hydrolysis of 3,5-dinitrobenzoic acid esters. The use of this "basic alumina" for the preparation of dihalo carbenes in modest yield has been proposed by Serratosa (1964).

Regen and Mehrotra (1981) have shown that solid KOH crushed with neutral alumina is an effective reagent for carrying out ester hydrolysis. This "dispersed reagent" effects a more rapid hydrolysis than KOH impregnated on alumina, even after drying:

$$RCO_2R^1 \xrightarrow[2.\ H_3O^+]{1.\ KOH/Al_2O_3} RCO_2H + R^1OH$$
$$60{-}100\%$$

Under the same conditions, amides are not hydrolysed.

O-ethyl thioesters have been reacted with aromatic aldehydes on alumina impregnated with NaOMe in "dry media" (Davrinche *et al.*, 1984) and 2,3-unsaturated thioesters $(Z + E)$ have been obtained; good yields are obtained (except when $R_2 = CH_3$). Compared to the reference reaction in homogeneous solution, the dry-media reaction on alumina is the only one to give the Z stereoisomer in the case of $R^1 = H$:

$$p\text{-}R^1\text{---}C_6H_4\text{---}CHO + R^2CH_2\text{---}CSOEt \xrightarrow[\text{dry media}]{Al_2O_3/NaOMe} p\text{-}R^1\text{---}C_6H_4\text{---}CH\text{==}C(R^2)CSOEt$$

$$(Z + E)$$

IV. ALUMINA-SUPPORTED ALKALI METAL FLUORIDES

Alkali metal fluorides are strong bases which have found wide application in organic synthesis. The subject has been reviewed by Clark (1980) and Yakobson and Akhmetova (1983). The nucleophilicity of protic compounds is greatly enhanced by the formation of strong hydrogen bonds with the fluoride anion. Hence, nucleophilic substitution proceeds with hydrogen-bond assistance (Clark and Emsley, 1973, 1975; Clark and Miller, 1977; Clark *et al.*, 1983a; Miller and Clark, 1982):

$$M^+F^- + H\text{---}Y\text{---}R^1 \longrightarrow [F\cdots H \cdots Y\text{---}R^1]^-M^+ \xrightarrow{R^2\text{---}X} R^1\text{---}Y\text{---}R^2 + [FHX]^-M^+$$

Moreover, the high affinity of the fluoride ion toward silicon atoms has been used to generate anions by desilylation of the silyl enol ether. This procedure has been used to effect aldol reactions between silyl enol ethers and carbonyl compounds (Nakamura *et al.*, 1983) or alkylation of silyl enol ethers (Kuwajima and Nakamura, 1975):

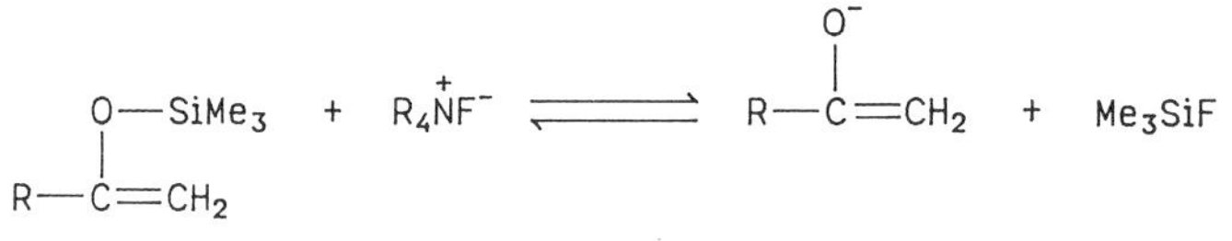

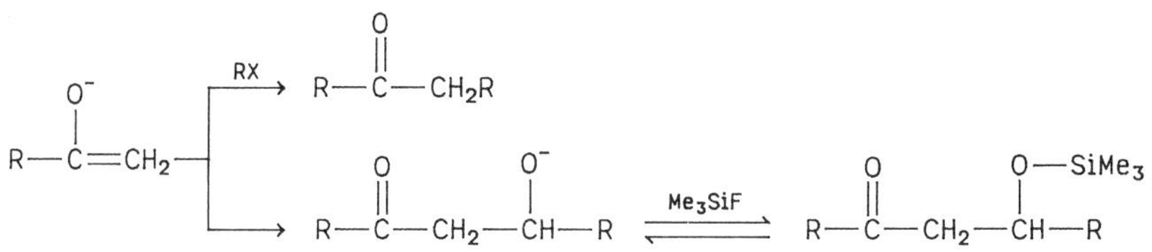

The silyl enol ether has been generated from the ketone and $Si(OR)_4$ in the presence of the fluoride anion. Hence, CsF in the presence of $Si(OR)_4$

is very efficient for carrying out Michael additions (Boyer *et al.*, 1983):

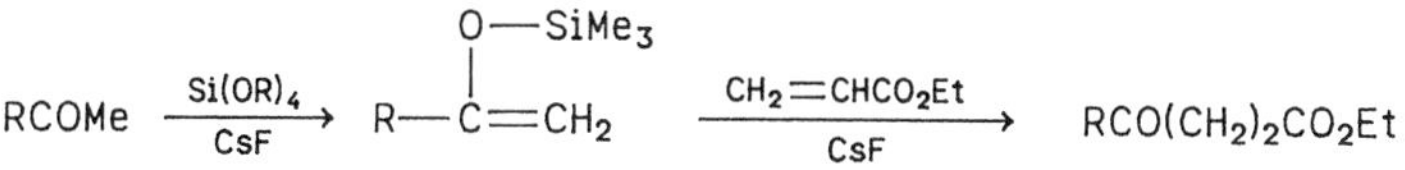

$$RCOMe \xrightarrow[CsF]{Si(OR)_4} R-\overset{\overset{\displaystyle O-SiMe_3}{|}}{C}=CH_2 \xrightarrow[CsF]{CH_2=CHCO_2Et} RCO(CH_2)_2CO_2Et$$

The activity of alkali metal fluorides is dependent on the cation. Cesium fluoride has the largest cation diameter and thus, the anion is nearly a naked fluoride with high reactivity. The low solubilities of alkali fluorides in aprotic organic solvents restrict their application in organic synthesis. Dry tetraalkylammonium fluorides are also very strong bases, more soluble than alkali metal fluorides. However, these reagents are hygroscopic and of low stability (Sharma and Fry, 1983) and the reproducibility of the reactions can be poor. The use of alumina as a support for reagents generating fluoride ion provides several advantages. These advantages may arise from the adsorption of reactants on the surface, the large effective reagent surface area due to adsorption, and the presence of pores (Yakobson and Akhmetova, 1983; Ando *et al.*, 1982b).

A. Alkylation

1. *C*-Alkylation

Yamawaki and Ando (1979) have shown that the reaction of 2,4-pentanedione with methyl iodide, in the presence of KF–alumina gives the dialkylderivative **IV** only and in the presence of KF–celite, the mono-*C*-alkylated product **V** is isolated.

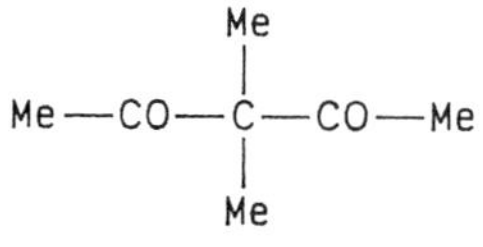

$$Me-CO-\overset{\overset{\displaystyle Me}{|}}{\underset{\underset{\displaystyle Me}{|}}{C}}-CO-Me \qquad\qquad Me-CO-\overset{\overset{\displaystyle }{|}}{\underset{\underset{\displaystyle Me}{|}}{CH}}-CO-Me$$

$$\textbf{IV} \qquad\qquad\qquad\qquad \textbf{V}$$

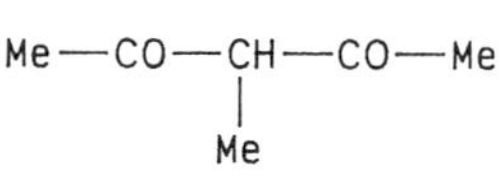

2. *O*-Alkylation

The *O*-methylation of phenol in the presence of alkali metal fluorides and tetrabutylammonium fluoride impregnated alumina has been studied (Ando *et al.*, 1982b). The activity of alumina supported fluorides was found to increase in the sequence: $Bu_4NF > CsF > KF \gg NaF > LiF$.

This is the same as the reactivity order of fluorides without a support (Yakobson and Akhmetova, 1983; Clark and Miller, 1977). The alumina has been compared with various other supports. The effectiveness of the supports varies in the order (Yamawaki and Ando, 1979; Ando and Yamawaki, 1979) alumina > molecular sieves > montmorillonite > celite.

The alumina-supported metal fluorides are more effective than metal fluorides without a support. Bu_4NF–silica gel was more stable than Bu_4NF supported on alumina (Clark, 1978). The effect of the reaction solvent was investigated. It was shown that the reaction became faster as the polarity of the solvent was increased. Ando *et al.* (1982b) have used KF–alumina for O-alkylation of substituted phenols and alcohols in good yields under mild conditions.

$$R^1OH + R^2X \xrightarrow{\text{KF-alumina}} R^1{-}O{-}R^2$$

Yamawaki and Ando (1980) have used alumina coated with potassium fluoride for the synthesis of crown ethers in moderate to good yields:

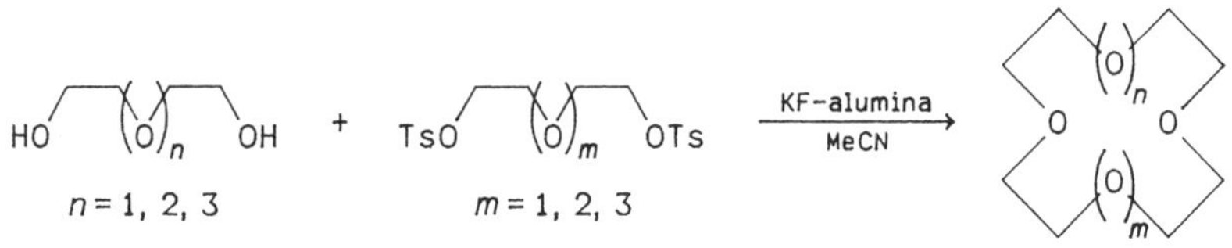

3. *N*-Alkylation

N-alkylation of amides and lactams have been performed at room temperature in the presence of KF–alumina by Yamawaki *et al.* (1981):

$$MeCONHPh + MeI \xrightarrow[\text{1,2-dimethoxyethane}]{\text{KF-alumina}} Me{-}CO{-}\underset{\underset{Me}{|}}{N}{-}Ph$$

93%

KF–alumina has been found also to be effective for *N*-methylation of uracil and xanthine derivatives.

Sebti (1985) has shown that the amide **VI** treated with KF–alumina gives the epoxide **VII**. In the presence of benzyltriethylammonium chloride, a mixture of the epoxide **VII** and of the β-lactame **VIII** (in the ratio 2:3) was obtained. Alumina alone or KF alone gives no reaction.

Potassium fluoride dihydrate, more reactive than dry KF (Carpino and Sau, 1979) or cesium fluoride, gave the β-lactam **VIII** (Sebti and Foucaud, 1983). Adsorption of KF on alumina leads to a partial solvation of potassium ion which enhances the basicity of the fluoride ion. Moreover, the OH groups present on alumina are activated, enough for nucleophilic attack on the acetoxy group of **VI**, converted into the epoxide **VII**. KF–alumina is less selective than CsF alone.

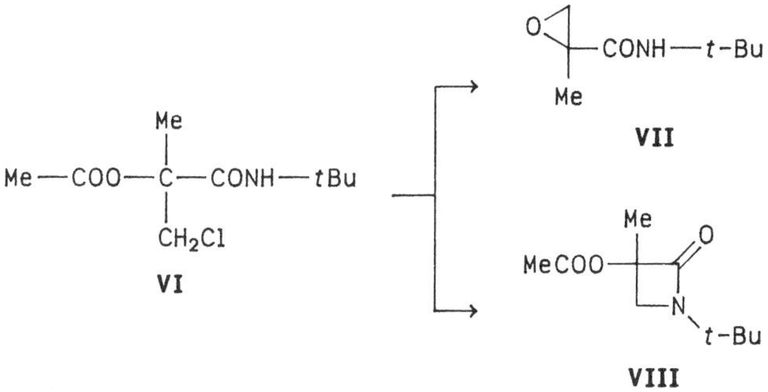

B. Eliminations

1. β-Elimination

Alumina-supported potassium fluoride has been used for olefin and acetylene forming elimination (Yamawaki *et al.*, 1983):

$$\text{PhCH}_2\text{CH}_2\text{Br} \xrightarrow[\text{MeCN}]{\text{KF-alumina}} \text{Ph}-\text{CH}{=}\text{CH}_2$$
$$86\%$$

$$\text{Ph}-\text{CH}{=}\text{CHBr} \xrightarrow[\text{diglyme}/100°C]{24\ h} \text{PhC}{\equiv}\text{CH}$$
$$70\%$$

The elimination did not proceed when KF or alumina was used separately. The double elimination of vicinal bromides has been reported (Yamawaki *et al.*, 1983):

$$\underset{meso}{\text{Ph}-\text{CHBr}-\text{CHBr}-\text{Ph}} \xrightarrow[\text{DME(48 h)}]{\text{KF–alumina}} \underset{89\%}{\text{PhC}{\equiv}\text{C}-\text{Ph}}$$

2. α-Elimination

KF-Alumina promotes the generation of dichlorocarbene, which is trapped by cyclohexene to give dichloronorcarane (Yamawaki *et al.*, 1983):

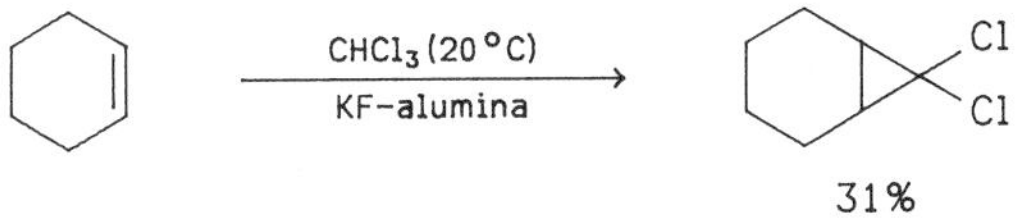

A similar yield was obtained when tetraethylammonium fluoride was used to produce the α-elimination of hydrogen halide from haloform (Hayami *et al.*, 1968). However, these methods are not as efficient as the method using phase transfer catalysts.

C. Additions

1. Hydration

The selective hydrations of nitriles to yield amides have been performed with alumina-supported potassium fluoride (Rao, 1982):

$$\text{Ph—CN} \xrightarrow[\text{t-BuOH, Δ(3h)}]{\text{KF–alumina}} \underset{98\%}{\text{Ph—CONH}_2}$$

Alumina without KF or KF without alumina is ineffective.

2. Michael Reaction

Yamawaki *et al.* (1983) have reported a Michael reaction catalyzed by alumina-supported KF:

$$\text{Ph—CH=CH—CO—Ph} + \text{CH}_3\text{NO}_2 \xrightarrow[\text{MeCN}]{\text{KF–alumina}} \begin{array}{c} \text{Ph—CH—CH}_2\text{COPh} \\ | \\ \text{CH}_2\text{NO}_2 \\ 74\% \end{array}$$

Clark *et al.* (1983b) have shown that KF or CsF supported on alumina were efficient catalysts for the Michael reaction. The catalyst, loaded with 2.5 molecules of fluoride/nm^2 of surface, was dried at 200°C under vacuum.

The Michael addition of Meldrum's acid with chalcone without solvent has been catalyzed by KF–alumina (Villemin, 1983):

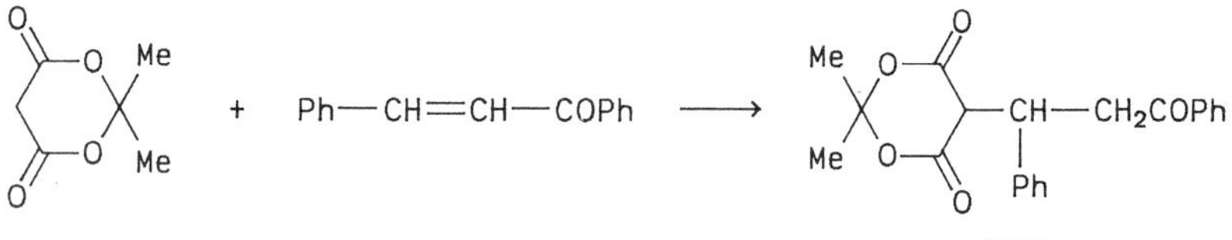

78%

The order of reactivities of various catalysts, based on the times required for total reaction of nitro ethane with methyl vinyl ketone:

$$\text{MeCH}_2\text{NO}_2 + \text{MeCO—CH=CH}_2 \longrightarrow \begin{array}{c} \text{MeCO—CH}_2\text{—CH}_2\text{—CH—Me} \\ | \\ \text{NO}_2 \end{array}$$

is CsF/–alumina > KF/–alumina > CsF in ethanol > wet CsF > dry CsF > dry alumina > KF–18-crown-6 > KF (Clark, *et al*, 1983b).

Base-catalyzed nitroaldol condensation (Henry reaction) has been achieved with alumina-supported potassium fluoride. The reaction times are shorter than with alumina alone, and this procedure can be extended to the reaction of nitroalkanes with aromatic aldehydes to give the nitroalcohols (Melot *et al.*, 1986).

Laszlo and Pennetreau (1985) have used alumina-supported potassium fluoride as a catalyst for the addition of the carbanion of dimedone on methyl vinyl ketone in tetrahydrofuran at room temperature (100% yield after three days). Villemin (1985) has catalyzed by KF–alumina the condensation of poorly acidic carbon compounds ($pK_a = 13$–30) with benzaldehydes without solvent:

$$Ph—C≡CH + PhCHO \xrightarrow{(20°C, 18 h)} Ph—C≡C—CHOH—Ph$$

$$58\%$$

$$Me—SO_2—Me + PhCHO \xrightarrow{(20°C, 15 h)} Ph—CHOH—CH_2—SO_2Me$$

$$52\%$$

Joucla and Le Brun (1985) have reported formation of cyclopropanes in the presence of KF-alumina, from nucleophiles and Michael acceptors:

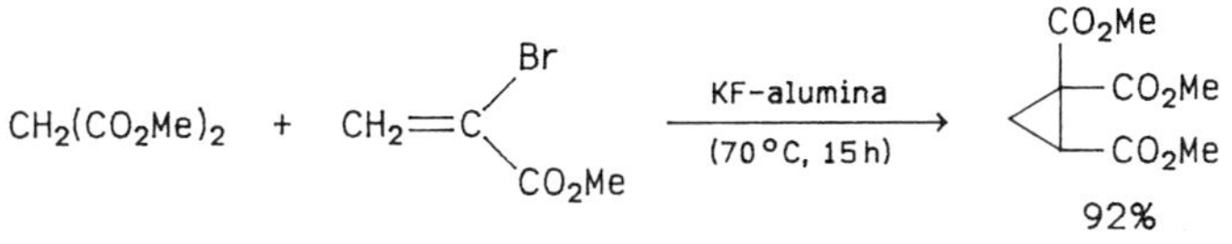

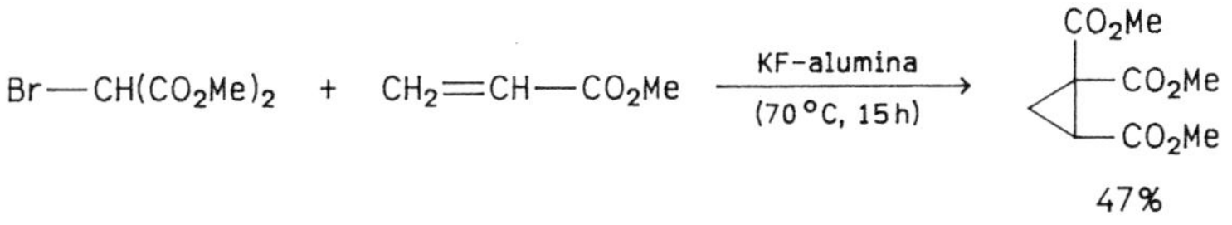

KF without alumina was ineffective; cesium fluoride and KF–alumina gave similar yields.

3. Darzens Reactions

Yamawaki *et al.* (1983) have shown that the Darzens reaction is catalyzed by KF–alumina:

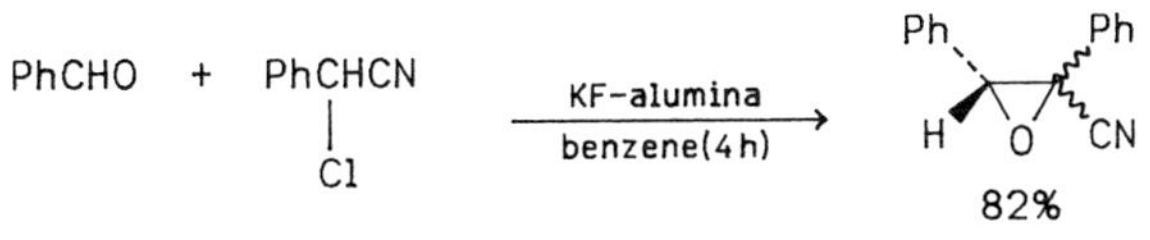

The *cis*–trans isomer ratio is dependent on the solvent. With benzene as solvent, the ratio is 56 : 44 ; with 1,2-dimethoxyethane as solvent, the ratio is 28 : 72; and with addition of 1.0 equivalent of water to KF in benzene, the ratio switches to 71 : 29. These results suggest a reaction in the inorganic solid–organic liquid interfacial region.

An analogous effect has been reported for the same reaction in a two-phase system (aq. NaOH, benzene) with or without a phase-transfer catalyst. When the halohydrin anion is located at the phase boundary, the interactions of O^-, Cl^-, and CN^- substituants with Na^+ located at the surface of the aqueous phase favors a transition state for the cyclization into the cis isomer (Jończyk *et al.*, 1977).

4. Knoevenagel Reactions

The Knoevenagel condensation has been catalyzed by KF–alumina (Yamawaki *et al.*, 1983):

$$Ph—CHO + CH_2(CN)_2 \xrightarrow[\text{(3 h)}]{\text{MeCN}} Ph—CH{=}C(CN)_2$$
$$86\%$$

The Knoevenagel condensation of aldehydes with malononitrile and ethylcyanoacetate has also been carried out on an $AlPO_4$–Al_2O_3 catalyst in the absence of solvent (Cabello *et al.*, 1984). However, alumina alone is a better catalyst than the alumina supported KF or $AlPO_4$ for the Knoevenagel reaction of aldehydes with acidic methylene active reagents (Texier-Boullet and Foucaud, 1982a).

Aldol condensation of weakly acidic species such as phenyl acetylene or fluorene with piperonal catalyzed by KF–alumina, has been reported (Villemin, 1983).

5. Wittig and Wittig–Horner Reactions

Ester-stabilized or nitrile-stabilized ylides can be generated from phosphonium salts with KF (Kossmehl and Nuck, 1979) or alumina. However, it is necessary to use KF–alumina to generate aryl stabilized ylides which are trapped in the presence of aldehydes to give a mixture of Z and E alkenes:

$$PhCHO + Ph_3\overset{+}{P}—CH_2Ph, Cl^- \xrightarrow[\text{RT}]{\text{KF–alumina}} Ph—CH{=}CH—Ph$$

The reaction was favored by the addition of a small amount of water (Texier-Boullet *et al.*, 1985).

The reactions of trimethylsulfonium iodide or trimethylsulfoxonium iodide with benzaldehyde, in the presence of KF–alumina give epoxystyrene:

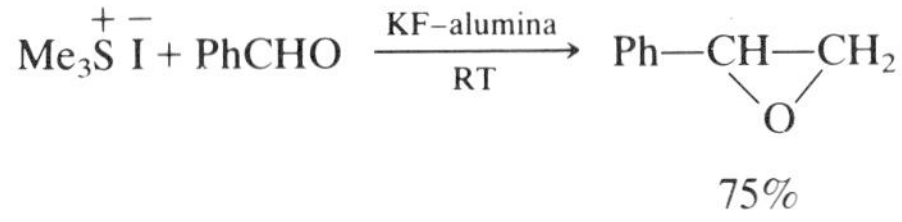

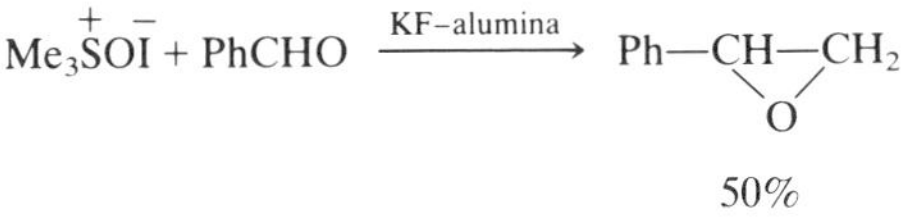

$$Me_3\overset{+}{S}\overset{-}{I} + PhCHO \xrightarrow[RT]{KF-alumina} Ph-\underset{\underset{O}{\diagdown\diagup}}{CH-CH_2}$$

75%

$$Me_3\overset{+}{S}O\overset{-}{I} + PhCHO \xrightarrow{KF-alumina} Ph-\underset{\underset{O}{\diagdown\diagup}}{CH-CH_2}$$

50%

The trimethylsulfoxonium iodide reacts at room temperature (RT) with the chalcone in the presence of KF–alumina, generating the 1-benzoyl-2-phenylcyclopropane (Texier-Boullet et al., 1985):

$$Me_3\overset{+}{S}O^- + PhCH=CH-COPh \longrightarrow Ph-\underset{\underset{CH_2}{\diagdown\diagup}}{CH-CH}-COPh$$

60%

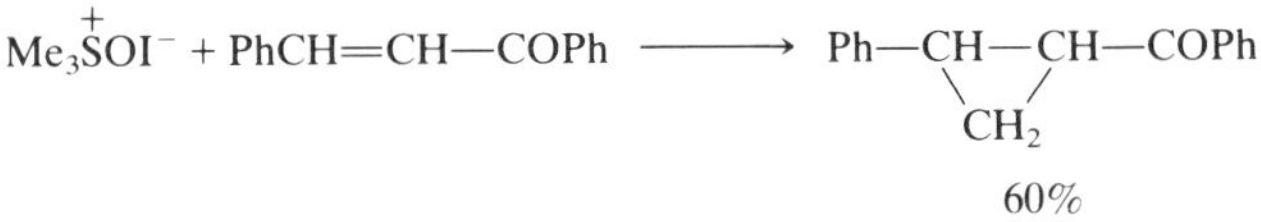

The condensation of various phosphonates **IX** with aldehydes and ketones may be carried out in the presence of KF–alumina and with a small amount of water, at room temperature (Texier-Boullet et al., 1985):

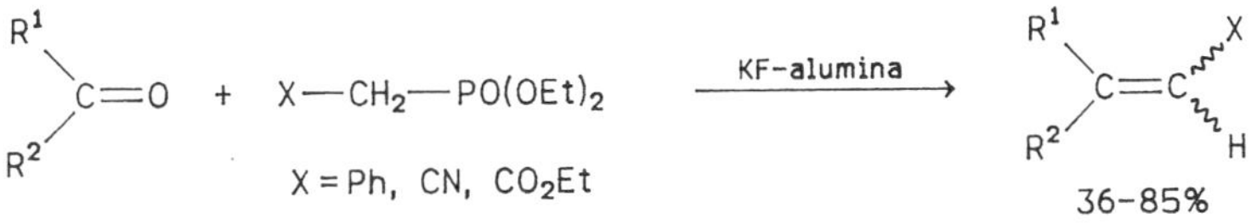

$$\underset{R^2}{\overset{R^1}{\diagdown}}C=O \quad + \quad X-CH_2-PO(OEt)_2 \xrightarrow{KF-alumina} \underset{R^2}{\overset{R^1}{\diagdown}}C=C\underset{H}{\overset{X}{\diagup}}$$

X = Ph, CN, CO₂Et

IX

36–85%

D. Enolization

The deprotonation of methyl dithiopropanoate **X** gives a mixture of dithioenolates which are trapped with benzylchloride to give **XI** and **XII**. With t-BuOLi in tetrahydrofuran or with t-BuOLi on alumina, the selectivity is low (the ratio **XI** : **XII** is 68 : 32). An improved stereoselectivity is obtained when KF–alumina is used as the base (the ratio **XI** : **XII** becomes 85 : 15)(Villemin, 1985):

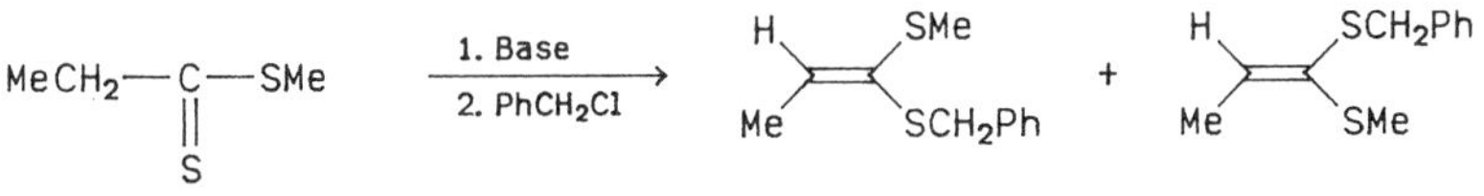

$$MeCH_2-\underset{\underset{S}{\parallel}}{C}-SMe \xrightarrow[2.\ PhCH_2Cl]{1.\ Base} \underset{Me}{\overset{H}{\diagdown}}C=C\underset{SCH_2Ph}{\overset{SMe}{\diagup}} + \underset{Me}{\overset{H}{\diagdown}}C=C\underset{SMe}{\overset{SCH_2Ph}{\diagup}}$$

X   **XI**   **XII**

E. Fluoride-Induced Desilylation Reactions

Tetraalkylammonium fluorides are the most frequently used reagents for desilylation, and though it has been shown that KF alone is inefficient to perform this reaction (Majetich et al., 1983) there are few examples of the use of alumina-supported metal fluorides for desilylation.

The formation of enolate anion by desilylation of silylnitronates **XIII** with tetrabutylammonium fluoride in a tetrahydrofuran solution or adsorbed on alumina has been used to effect the nitroaldol addition:

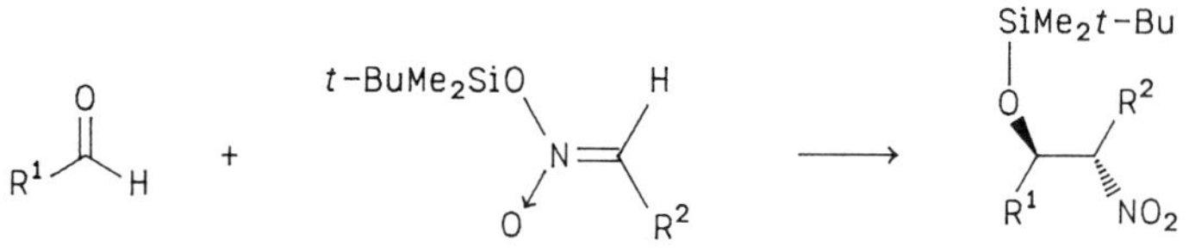

but less *erythro*-selectivity is observed with KF–alumina as catalyst (Seebach et al., 1982).

The synthesis of cyclopropane **XV** with α-chloroacrylonitrile and diethylmalonate with (MeO)$_4$Si and KF on alumina as catalyst has been described. KF or alumina alone gives no reaction (Joucla and Le Brun, 1985):

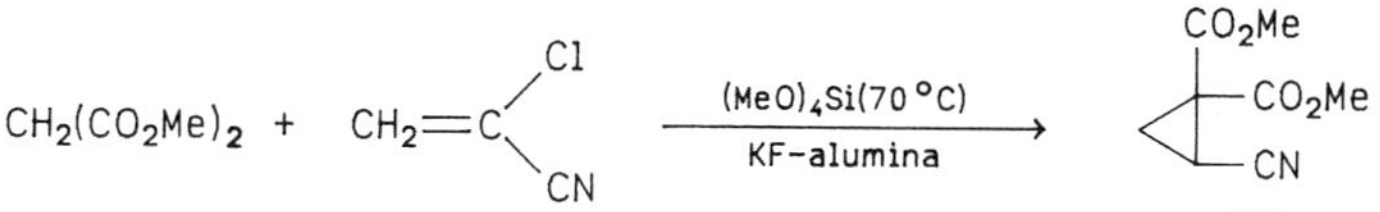

V. ALUMINA-SUPPORTED POTASSIUM

This reagent is prepared by melting potassium over neutral alumina at 150°C in an inert atmosphere (argon); it has been shown to act as a metallating agent toward enolizable compounds and the organopotassium anions have been alkylated with primary alkyl bromides. Tetrahydrofuran is the solvent of choice for the reaction of nitriles, aldehyde N,N'-dimethylhydrazones, and N-cyclohexyl ketimines, whereas hexane must be used for ketone alkylation (Savoia et al., 1981). Thus, the behavior of potassium on alumina appears analogous to that of graphite intercalated potassium (Savoia et al., 1978) and potassium dispersed on charcoal (Hart et al., 1977). On the other hand, potassium dispersed on silica (Levy et al., 1981) did not seem to have been used as a metallating agent toward enolizable compounds.

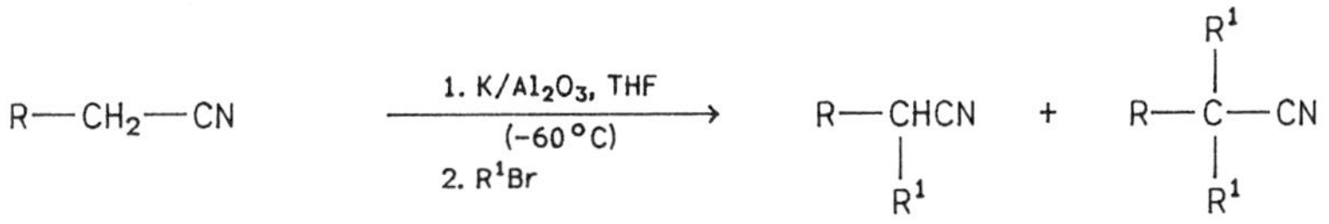

$$R—CH_2—CN \xrightarrow[\text{2. } R^1Br]{\substack{\text{1. } K/Al_2O_3, \text{ THF} \\ (-60\,^\circ C)}} R—\underset{R^1}{CHCN} \quad + \quad R—\underset{\underset{R^1}{|}}{\overset{\overset{R^1}{|}}{C}}—CN$$

R = H	$R^1 = C_8H_{17}$	54%	30%
R = C_2H_5	$R^1 = C_8H_{17}$	55%	3%

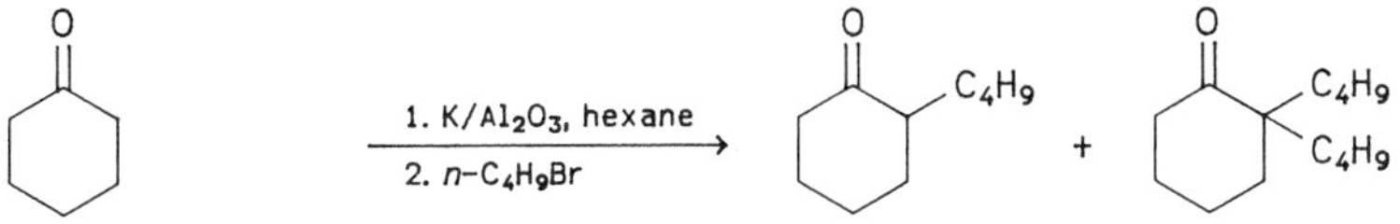

73%   9%

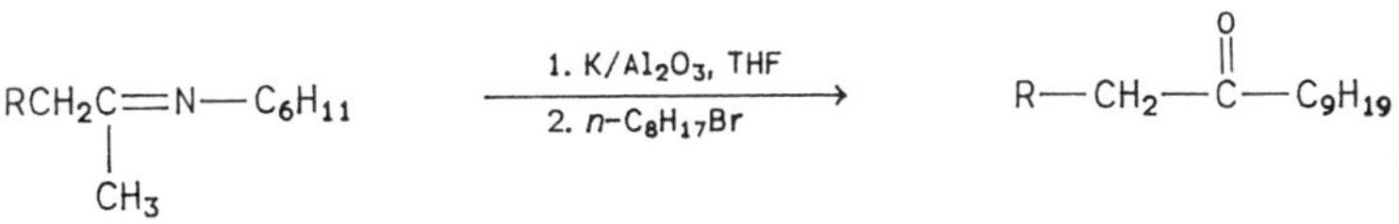

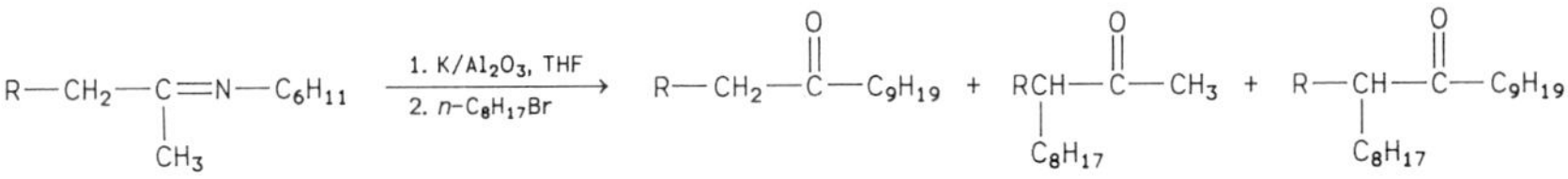

$$RCH_2C\!=\!N—C_6H_{11} \xrightarrow[\text{2. } n-C_8H_{17}Br]{\text{1. } K/Al_2O_3, \text{ THF}} R—CH_2—\overset{O}{\overset{||}{C}}—C_9H_{19}$$

with CH_3 substituent

$$R—CH_2—\underset{CH_3}{C}\!=\!N—C_6H_{11} \xrightarrow[\text{2. } n-C_8H_{17}Br]{\text{1. } K/Al_2O_3, \text{ THF}} R—CH_2—\overset{O}{\overset{||}{C}}—C_9H_{19} \; + \; R\underset{C_8H_{17}}{CH}—\overset{O}{\overset{||}{C}}—CH_3 \; + \; R—\underset{C_8H_{17}}{CH}—\overset{O}{\overset{||}{C}}—C_9H_{19}$$

R = H	41%		20%
R = CH_3	63%	7%	

VI. ONIUM SALTS IMMOBILIZED ON ALUMINA

Onium salts immobilized on alumina have been prepared by the reaction of 3-bromopropyltriethoxysilane with alumina and subsequent quaternization (Tundo *et al.*, 1982):

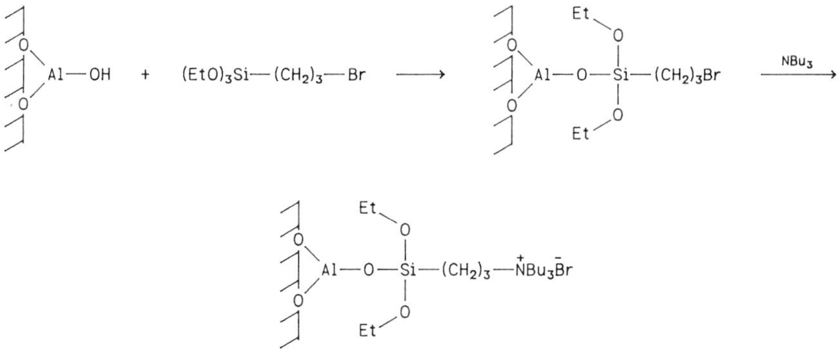

$$Al—OH \; + \; (EtO)_3Si—(CH_2)_3—Br \longrightarrow Al—O—Si(Et,O)_2—(CH_2)_3Br \xrightarrow{NBu_3}$$

$$Al—O—Si(Et,O)_2—(CH_2)_3—\overset{+}{N}Bu_3\bar{B}r$$

These salts immobilized on alumina are effective phase-transfer catalysts in nucleophilic substitution in comparison to analogous soluble catalysts adsorbed on alumina.

VII. CONCLUSIONS

The effectiveness of various inorganic supports, especially alumina and silica, for anionic activation has been studied by different groups. Alumina has generally emerged as the most efficient for alkylation reactions (Regen *et al.*, 1979a,b; Yamawaki and Ando, 1979; Bram and Fillebeen-Khan, 1979; Bram *et al.*, 1980a). Nevertheless, if KCN impregnated on silica is quite inert (Regen *et al.*, 1979a; Bram *et al.*, 1980a), CH_3CO_2K and malonate anions impregnate on the same support can be alkylated, but less efficiently than on alumina (Bram and Fillebeen-Khan, 1979; Bram and Decodts, 1980). The same behavior has been observed for the selective monoesterification of dicarboxylic acids on alumina or silica (Ogawa *et al.*, 1985).

The Knoevenagel (Texier-Boullet and Foucaud, 1982a) and Michael reactions are also efficiently catalyzed by alumina. In comparison to the known procedures, anionic activation on alumina has undoubtedly great advantages: experimental simplicity, high yields, and sometimes high selectivity.

If an overall interpretation of the precise role of the support has not yet been proposed, an interpretation based upon the superficial charge of the solid surfaces has been proposed (Bram and Decodts, 1980). Clark and Duke (1985) have shown by IR studies that impregnation of KCN on silica results in the formation of Si—CN groups on the surface, with a large shift of 200 cm^{-1} in the v (C≡N) band (see Chapter 18). Thus, we can understand that the nucleophilicity of such chemisorbed cyanide species will be quite low. In contrast, KCN impregnated on alumina gave a broadened but only slightly shifted v (C≡N) band.

More spectroscopic work of this type (see Chapter 7), concerning the structure of adsorbed species and performed in conditions where the reactivity of these species is known, is needed to understand the precise role of the inorganic supports.

REFERENCES

Ando. T., and Yamawaki, J. (1979). *Chem. Lett.* 45–46.
Ando, T., Kawate, T., Yamawaki, J., and Hanafusa, T. (1982a). *Chem. Lett.* p. 935.
Ando, T., Yamawaki, J., Kawate, T., Sumi, S., and Hanafusa, T. (1982b). *Bull. Chem. Soc. Jpn.* **55**, 2504–2507.

Ando, T., Sumi, S., Kawate, T., Ichihara-Yamawaki, J., and Hanafusa, T. (1984a). *J.C.S. Chem. Commun.* 439–440.

Ando, T., Kawate, T., Ichihara-Yamawaki, J., and Hanafusa, T. (1984b). *Chem. Lett.* 725–728.

Badet, B., Julia, M., and Lefeuvre, C. (1984). *Bull. Soc. Chim. Fr. II* 431–434.

Barot, B. C., Sullins, D. W., and Eisenbraun, E. J. (1984). *Synth. Commun.* **14**, 397–400.

Barry, J., Bram, G., Decodts, G., Loupy, A., Pigeon, P., and Sansoulet, J. (1983). *Tetrahedron* **39**, 2669–2672.

Benson, R. G., and Geraghty, N. W. A. (1983). *J. Chem. Res., Synop.* 290–291.

Boyer, J., Corriu, R. J. P., Perz, R., and Reye, C. (1983). *Tetrahedron* **39**, 117–122.

Bram, G., and Decodts, G. (1980). *Tetrahedron Lett.* 5011–5014.

Bram, G., and Fillebeen-Khan, T. (1979). *J.C.S. Chem. Commun.* 522–523.

Bram, G., Fillebeen-Khan, T., and Geraghty, N. (1980a). *Synth. Commun.* **10**, 279–289.

Bram, G., Geraghty, N., Nee, G., and Seyden-Penne, J. (1980b). *J.C.S. Chem. Commun.* 325–326.

Bram, G., Decodts, G., Bensaid, Y., Combet-Farnous, C., Galons, H., and Miocque, M. (1985). *Synthesis* 543–545.

Cabello, J. A., Campelo, J. M., Garcia, A., Luna, D., and Marinas, J. M. (1984). *J. Org. Chem.* **49**, 5195–5197.

Carpino, L. A., and Sau, A. C. (1979). *J.C.S. Chem. Commun.* 514–515.

Castells, J. J., and Fletcher, G. A. (1956). *J. Chem. Soc.* p. 3245.

Chihara, T. (1980). *J.C.S. Chem. Commun.* 1215–1216.

Clark, J. H. (1978). *J.C.S. Chem. Commun.* 789–791.

Clark, J. H. (1980). *Chem. Rev.* **80**, 429–452.

Clark, J. H., and Duke, V. A. (1985). *J. Org. Chem.* **50**, 1330–1332.

Clark, J. H., and Emsley, J. (1973). *J.C.S. Dalton* 2154–2159.

Clark, J. H., and Emsley, J. (1975). *J.C.S. Dalton* 2129–2134.

Clark, J. H., and Miller, J. M. (1977). *J. Am. Chem. Soc.* **99**, 498-504.

Clark, J. H., Goodall, D. M., and White, M. S. (1983a). *Tetrahedron Lett.* **24**, 1097–1100.

Clark, J. H., Cork, D. G., and Gibbs, H. W. (1983b). *J.C.S. Perkin I* 2253–2258.

Czech, B., Quici, S., and Regen, S. L. (1980). *Synthesis* p. 113.

Dalton, J. R., and Regen, S. L. (1979). *J. Org. Chem.* **44**, 4443–4444.

Davrinche, C., Brion, J. D., and Reynaud, P. (1984). *Synth. Commun.* **14**, 1181–1190.

Gaetano, K., Pagni, R. M., Kabalka, G. W., Bridwell, P., Walsh, E., True, J., and Underwood, M. (1985). *J. Org. Chem.* **50**, 499–502.

Galons, H., Bergerat, I., Comdet-Farnoux, C., Miocque, M., Decodts, G., and Bram, G. (1985). *J.C.S. Chem. Commun.*, p. 1730–1731.

Hanessian, S., and Kloss, J. (1985). *Tetrahedron Lett.* **26**, 1261–1264.

Hart, H., Chen, B. L., and Peng, C. T. (1977). *Tetrahedron Lett.* 3121–3124.

Hayami, J. I., Uno, N., and Kaji, A. (1968). *Tetrahedron Lett.* 1385–1386.

Jacobs, T. L., and Danker, D. (1957). *J. Org. Chem.* **22**, 1424–1427.

Jacobs, T. L., and Singer, S. (1952). *J. Org. Chem.* **17**, 475–478.

Jacobs, T. L., Dankner, D., and Singer, S. (1964). *Tetrahedron* **20**, 2177–2180.

Jończyk, A., Kwast, A., and Makosza, M. (1977), *J.C.S. Chem. Commun.* p. 902.

Joucla, M., and Le Brun, J. (1985). *Tetrahedron Lett.* **26**, 3001–3004.

Kabalka, G. W., Pagni, R. M., Bridwell, P., Walsh, E., and Hassaneen, H. M. (1981). *J. Org. Chem.* **46**, 1513–1514.

Kossmehl, G., and Nuck, R. (1979). *Chem. Ber.* **112**, 2342–2346.

Kuwajima, I., and Nakamura, E. (1975). *J. Am. Chem. Soc.* **97**, 3257–3258.

Laszlo, P., and Pennetreau, P. (1985). *Tetrahedron Lett.* **26**, 2645–2648.

Levy, J., Tamarkin, D., Selig, H., and Rabinovitz, H. (1981). *Angew. Chem., Int. Ed. Engl.* **20**, 1033.

MacKillop, A., and Young D. R. (1979). *Synthesis* 481–500.

Majetich, G.,Casares, A. M. Chapman, D., and Behnke, M. (1983). *Tetrahedron Lett.* **24**, 1909–1912.

Melot, J. M., Texier-Boullet, F., and Foucaud, A. (1986). *Tetrahedron Lett.* **27**, 493–496.

Miller, J. M., and Clark, J. H. (1982). *J.C.S. Chem. Commun.* 1318–1319.

Muzart, J. (1982). *Synthesis* 60–61.

Muzart, J. (1985). *Synth. Commun.* 285–289.

Nakamura, E., Shimizu, M., Kuwajima, I., Sakata, J., Yokohama, K., and Noyori, R. J. (1983). *J. Org. Chem.* **48**, 932–945.

Nguyen Hieu Duc, Nguyen Ba Hiep, Le Thi Nhut Hoa, and Chu Pham Ngoc Son (1985). *C. R. Seances Acad. Sci., Ser. 2* **300**, 799–802.

Ogawa, H., Chihara, T., and Taya, K. (1985). *J. Am. Chem. Soc.* **107**, 1365–1369.

Onaka, M., Umezono, A., Kawai, M., and Izumi, Y. (1985). *J.C.S. Chem. Commun.* 1202–1203.

Pelletier, S. W., Venkov, A. P., Finer-Moore, J., and Mody, N. V. (1980). *Tetrahedron Lett.* **21**, 809–812.

Posner, G. H. (1978). *Angew. Chem., Int. Ed. Engl.* **17**, 487–496.

Quici, S., and Regen, S. L. (1979). *J. Org. Chem.* **44**, 3436-3437.

Rao, C. G. (1982). *Synth. Commun.* **12**, 177–181.

Regen, S. L., and Mehrotra, A. K. (1981). *Synth. Commun.* **11**, 413–417.

Regen, S. L., Quici, S., and Liaw, S. J. (1979a). *J. Org. Chem.* **44**, 2029–2030.

Regen, S. L., Quici, S., and Ryan, M. D. (1979b), *J. Am. Chem. Soc.* **101**, 7629–7630.

Regen, S. L., Czech, B., and Quici, S. (1981). *Pol. J. Chem.* **55**, 843–847.

Rosini, G., Ballini, R., and Sorrenti, P. (1983a). *Synthesis* 1014–1016.

Rosini, G., Ballini, R., and Sorrenti, P. (1983b). *Tetrahedron* **39**, 4127–4132.

Rosini, G., Ballini, R., Petrini, M., and Sorrenti, P. (1985). *Synthesis* 515–517.

Roudier, J. F., and Foucaud, A. (1984). *Synthesis* 159–160.

Savoia, D., Brombini, C., and Umani-Ronchi, A. (1978). *J. Org. Chem.* **43**, 2907–2910.

Savoia, D., Tagliavini, E., Trombini, C., and Umani-Ronchi, A. (1981). *J. Organomet. Chem.* **204**, 281–286.

Sebti, S. (1985). Thesis, Univ. Rennes, Rennes, France.

Sebti, S., and Foucaud, A. (1983). *Synthesis* 546–549.

Seebach, D., Beck, A. K., Mukhopadyay, T., and Thomas, E. (1982). *Helv. Chim. Acta* **65**, 1101–1133.

Serratosa, F. (1964). *J. Chem. Educ.* **41**, 564.

Sharma, R. K., and Fry, J. L. (1983). *J. Org. Chem.* **48**, 2112–2114.

Sukata, K. (1983). *Bull. Chem. Soc. Jpn.* **56**, 3306–3307.

Sukata, K. (1985a). *J. Org. Chem.* **50**, 4388–4390.

Sukata, K. (1985b). *Bull. Chem. Soc. Jpn.* **58**, 838–843.

Texier-Boullet, F. (1985). *Synthesis* 679–681.

Texier-Boullet, F., and Foucaud, A. (1982a). *Tetrahedron Lett.* 4927–4928.

Texier-Boullet, F., and Foucaud, A. (1982b). *Synthesis* p. 916.

Texier-Boullet, F., Villemin, D., Ricard, M., Moison, H., and Foucaud, A. (1985). *Tetrahedron* **41**, 1259–1266.

Tsuboi, S. Masuda, T., Makino, H., and Takeda, A. (1982a). *Tetrahedron Lett.* **23**, 209–212.

Tsuboi, S., Masuda, T., and Takeda, A. (1982b). *J. Org. Chem.* **47**, 4478–4482.

Tundo, P., Venturello, P., and Angeletti, E. (1982). *J. Am. Chem. Soc.* **104**, 6551–6555.

Tundo, P., Venturello, P., and Angeletti, E. (1983), *J.C.S. Perkin II* 485–491.

Varma, R. S., Kabalka, G. W., Evans, L. T., and Pagni, R. M. (1985). *Synth. Commun.* **15**, 279–284.

Villemin, D. (1983). *J.C.S. Chem. Commun.* 1092–1093.

Villemin, D. (1985). *J.C.S. Chem. Commun.* p. 870.

Villemin, D., and Ricard, M. (1984). *Tetrahedron Lett.* **25**, 1059–1060.

Yakobson, G. G., and Akhmetova, N. E. (1983). *Synthesis* 169–184.

Yamawaki, J., and Ando, T. (1979). *Chem. Lett.* 755–758.

Yamawaki, J., and Ando, T. (1980). *Chem. Lett.* 533–536.

Yamawaki, J., Ando, T., and Hanafusa, T. (1981). *Chem. Lett.* 1143–1146.

Yamawaki, J., Kawate, T., Ando, T., and Hanafusa, T. (1983). *Bull. Chem. Soc. Jpn.* **56**, 1885–1886.

Part **VI**

Silica-Supported and Silica-Gel-Supported Reagents

18

REDUCTIONS, OXIDATIONS, AND ANIONIC ACTIVATIONS: CATALYTIC REDUCTIONS AND OXIDATIONS

Enzo Santaniello

Dipartimento di Chimica e Biochimica Medica
Università degli Studi di Milano
20133 Milano, Italy

I. INTRODUCTION

Immobilization of a reagent onto an insoluble support provides several practical advantages such as simplified workup procedures, since in most cases spent reagents can be generally removed by a simple filtration (see Chapter 6). The procedure offers milder conditions than the corresponding homogenous reactions, so that the selectivity becomes higher. Hence, many classes of reactions have been explored using a wise variety of inorganic supports. Silica gel has been used whenever a highly hydroxylated support or a solid Lewis catalyst was needed.

This chapter is devoted to oxidations, reductions, and anionic activations following chemical or physical adsorption onto silica gel. Experimental procedures in solution are privileged. Reactions in dry media are covered in Chapter 20. Special attention is called to the reactions in which silica gel serves as a catalyst.

II. OXIDATIONS AND CATALYTIC OXIDATIONS

A. General Considerations

Adsorption of oxidation reagents onto inorganic supports is convenient, since an excess of reagent does not cause separation problems, and better

345

PREPARATIVE CHEMISTRY
USING SUPPORTED REAGENTS

yields can be achieved. Additionally, intermediates often prove difficult or impossible to isolate under homogeneous conditions. On the contrary, in several heterogeneous reactions the intermediates can be isolated, in line with the high selectivity offered by supported redox reactions. A few representative oxidizing reagents are considered. Reactions are described in which silica gel is either the inorganic support or the catalyst.

B. Chromium(VI) Reagents

In 1977, San Filippo and Chern reported that chromyl chloride CrO_2Cl_2, a vigorous oxidant that generally produces complex mixtures of products (Stairs et $al.$, 1963), can be advantageously used for oxidation of primary and secondary alcohols when chemisorbed onto silica–alumina:

$$R^1—CHOH—R^2 \xrightarrow{CrO_2Cl_2/SiO_2-Al_2O_3} R^1—CO—R^2 \qquad (1)$$

The adsorbed reagent was inert toward halocarbons, esters, lactones, nitriles, and ethers whereas alkenes and alkynes were oxidized. Preparation of the supported reagent required addition of a solution of CrO_2Cl_2 in methylene chloride to a slurry of $SiO_2-Al_2O_3$ in the same solvent. It can be stored indefinitely if protected from moisture. For a typical oxidation, a solution of the alcohol (1.10 mmol) in methylene chloride (5 mL) is added to 2 g of the supported reagent [1.10 mmol of Cr(VI)] suspended in methylene chloride (20 mL). Depending on the substrate, reaction times of 5–24 h at room temperature gave good yields (77–100%).

Chromic acid (H_2CrO_4) absorbed on silica gel (Santaniello et $al.$, 1978a) is another useful reagent for the oxidation of alcohols to aldehydes and ketones:

$$R^1—CHOH—R^2 \xrightarrow{H_2CrO_4/SiO_2} R^1—CO—R^2 \qquad (2)$$

Addition of silica gel (20 g) to a solution of chromic anhydride (2 g) in water (50 mL), evaporation in $vacuo$ and drying of the yellow solid (100°C, overnight) afforded the supported reagent which should be used within a week in order to achieve the best yields. Reactions were very rapid in diethyl ether. Products can be also eluted from a column of the reagent where the substrate is absorbed for a few hours (yields 60–70%). On the contrary, adsorption on alumina gave an inactive reagent. Yields were usually in the range of 68–90%, using 3 g of the reagent for 1 mmol of the substrate. Solutions of CrO_3 in water and sulfuric acid could also be deposited on silica gel affording a reagent displaying the same kind of oxidations (Santaniello et al., 1978a, and unpublished observations).

Another preparation of chromic acid absorbed onto silica gel has been described (Singh et al., 1979). For oxidation of $(-)$-menthol, 0.8 mole equivalent of chromic acid per mole of substrate is needed for complete conversion in a reasonable time (1–2 h). Reducing the silica gel/$Na_2Cr_2O_7 \cdot 2H_2O \cdot H_2SO_4$ ratio to an optimal level of 5, yields of menthone were 96% in 2 h:

$$R^1—CHOH—R^2 \xrightarrow[\text{Pyridinium chromate/SiO}_2]{\text{Na}_2\text{Cr}_2\text{O}_7 \cdot 2\,\text{H}_2\text{O} \cdot \text{H}_2\text{SO}_4/\text{SiO}_2} R^1—CO—R^2 \qquad (3)$$

In the same paper, the authors reported that pyridinium chromate–silica gel is a convenient off-the-shelf reagent for efficient oxidation of a variety of secondary and primary alcohols containing acid-labile functions. The reagent is prepared first by adsorption of CrO_3 in water onto silica gel. After water removal, the material is covered with light petroleum, and keeping the temperature between 20 and 24°C pyridine is slowly introduced, yielding after filtration a dark-brown powder stored in a brown bottle. The reagent keeps intact at room temperature for over a year. Oxidations of alcohols are easily carried out by stirring a methylene chloride solution of substrate with 2–4 molar equivalents of supported reagent and 1–2 molar equivalents of acetic acid at room temperature for 4–12 h (56–92%). The reaction can be speeded up (1–3 h) in refluxing benzene (62–87%).

Some chromium(VI) species such as chromates or dichromates oxidize alcohols (Lee and Spitz, 1970; Hutchins et al., 1977; Santaniello and Ferraboschi, 1980a; Santaniello et al., 1980) and halides (Cardillo et al., 1976; Landini and Rolla, 1979). The interesting feature is that the oxidations are carried out virtually in a neutral medium and that they are selective with benzylic and allylic alcohols. $K_2Cr_2O_7$ adsorbed on SiO_2, Al_2O_3, Florisil, and MgO selectively oxidizes benzylic alcohols and halides (Santaniello and Ferraboschi, 1980b):

$$\underset{\text{I}}{R^1—CHOH—R^2} \xrightarrow{\text{K}_2\text{Cr}_2\text{O}_7/\text{SiO}_2} \underset{\text{II}}{R^1—CO—R^2}, \qquad R^1\text{=Aryl, } R^2\text{=H, Alkyl, Aryl} \quad (4)$$

$K_2Cr_2O_7/SiO_2$ is prepared by adding to 100 g of silica gel 42.8 g of salt dissolved in 400 mL of water and using 1.5 g of supported reagent per mmol of substrate in refluxing toluene or benzene (1–4 h, yields 58–90%). With 1-phenyl-1,3-propanediol (I; $R^1 = C_6H_5$, $R^2 = CH_2CH_2OH$) as a substrate having both benzylic and nonbenzylic hydroxy groups, $K_2Cr_2O_7/SiO_2$ was the reagent of choice (90% yield, 80% selectivity).

An example of dichromate coated on silica gel has been reported by Fischer and Henderson (1985), who described a mild and efficient method of preparation of quinones from hydroquinones and catechols employing either cerium(IV)/SiO$_2$ and chromium(VI)/SiO$_2$ reagents. Ammonium dichromate coated on silica was obtained as a free-flowing orange powder and only a 20% reagent was used (1 mmol of chromium(VI) contained in 0.79 g of reagent). Hydroquinone (1–5 mmol) in dichloromethane was added to chromium(VI)/SiO$_2$ (2.07 g/mmol substrate) and after 5 min at room temperature, 98% of pure benzoquinone was obtained:

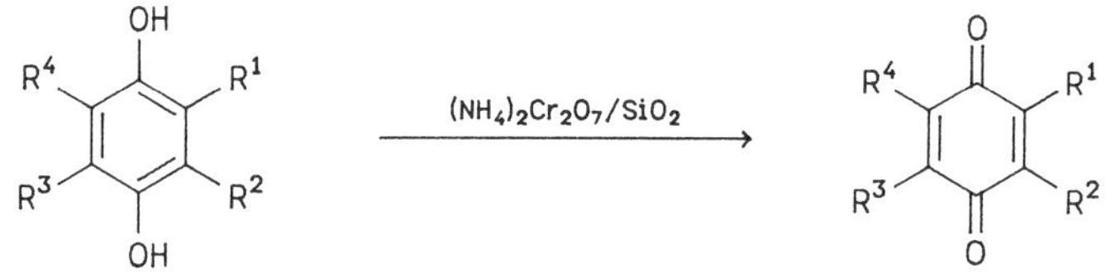

$$\text{(5)}$$

C. Permanganate

Potassium permanganate is activated by impregnation on solid supports such as Linde molecular sieves, silica gel, and certain clays (Regen and Koteel, 1977). Oxidations of alcohols in benzene thus become possible, whereas permanganate alone in benzene is unreactive due to reagent insolubility. Optimization was later performed, and the best $KMnO_4$: alcohol ratio was established (Al Jazzaa *et al.*, 1982). Solid $KMnO_4$ shows a broad infrared adsorption band at 900 cm^{-1} due to degenerate $\nu_3(F_2)$ vibration of the tetrahedral MnO_4^-. Loading 0.51 mmol $KMnO_4$/g silica gel and pumping at temperatures around 100°C leaves a highly unsymmetrical triplet centered at ~920 cm^{-1}. This loading corresponds to an optimum monolayer coverage of MnO_4: surface OH ratio 1:5 with a surface area effectively covered by permanganate moiety of ~1 nm^2. By comparison, the surface area of 150-Å average pore diameter silica gel is 320 m^2/g and for a normal silica gel (4–5 surface OH/nm^2) corresponds to 1.4×10^{21} hydroxyl groups/g. Using the supported reagent with loading characteristics so finely defined, the optimum $KMnO_4$: alcohol ratio for quantitative conversion of benzyl alcohol to benzaldehyde (benzene, 70°C,

2 h) turns out to be 2.8:

$$\text{(structure)}-CH_2OH \xrightarrow{\ \ KMnO_4/SiO_2\ \ } \text{(structure)}-CHO \qquad (6)$$

The optimized reagent is also very efficient for the transformation of nitro groups to carbonyl groups (Clark and Cork, 1983):

$$R-CHNO_2-(CH_2)_2-COCH_3 \xrightarrow{\ \ KMnO_4/SiO_2\ \ } R-CO-(CH_2)_2-COCH_3 \qquad (7)$$

Previous applications of permanganates to the Nef reaction have been severely hindered by the conditions of temperature and pH to be used. The supported reagent afforded 1,4-diketones from nitroketones in refluxing benzene (3–10 h), but product recovery required multiple washings with diethyl ether, presumably due to surface hydroxyl–carbonyl hydrogen bonding (yields 4–55%).

An amazing gain of selectivity was observed by adsorption of zinc permanganate on silica gel (Wolfe and Ingold, 1983). This salt reacts instantly with common laboratory solvents, sometimes with fires. Its use for oxidation in water solutions is inconvenient, as well as for two-phase water–methylene chloride oxidations. Use of inorganic supports such as celite, Florisil, and Linde 13 × molecular sieves did not help, whereas adsorption on silica gel provided an oxidizing agent that could be handled safely and employed under mild conditions. Interestingly, the zinc salt was more reactive than magnesium, and the potassium salt was virtually inactive. Alkynes were oxidized to diketones, ethers to lactones, cyclic ketals to ketones, ciclic ketones to diacid, acylated amines to acylimides, and sulfides to sulfones (Fig. 1).

D. Periodate

For oxidation of sulfides to sulfoxides adsorption of sodium meta-periodate was systematically studied on some 10 inorganic supports, including alumina, celite, charcoal, florisil, montmorillonite clays, and silica gel (Liu and Tong, 1978). The acidic clays and Girdler Catalyst K-10 and KO, were by far the most effective.

For oxidation of other substrates, periodates supported on silica gel are efficient (Gupta et al., 1981). Sodium metaperiodate supported on silica gel was used for various oxidations in methylene chloride, benzene, or ether. The supported reagent was prepared as a free-flowing solid containing 0.36 mmol of sodium periodate per gram by evaporating to dryness a solution of sodium metaperiodate (1 part by weight) in water in the presence of chromatographic-grade silica gel (10 parts). Laser Raman and

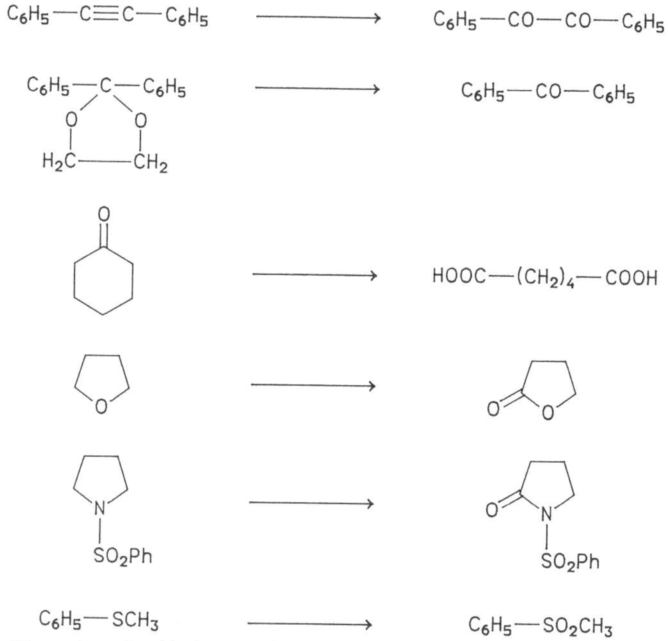
C6H5—C≡C—C6H5
C6H5—CO—CO—C6H5
C6H5—C—C6H5
O
O
H2C
CH2
C6H5—CO—C6H5
O
HOOC—(CH2)4—COOH
O
O
O
N
SO2Ph
O
N
SO2Ph
C6H5—SCH3
C6H5—SO2CH3

Fig. 1. Examples of oxidations performed by $ZnMnO_4/SiO_2$. [From Wolfe and Ingold (1983).]

x-ray spectroscopy of the supported reagent show that it consists mainly of a monomolecular layer of $NaIO_4$ bound to silica gel by surface hydroxy groups along with minor amounts of crystalline anhydrous salt. At 20°C 2,5-di-*t*-butylhydroquinone (DBHQ) (**III**; $R^1{=}R^3{=}t$-butyl, $R^2{=}R^4{=}H$), a typical hydroquinone, reacted smoothly with a 10% molar excess of the supported reagent in methylene chloride or benzene in 3 h (98 and 55%, respectively). No reaction occurs with ether as solvent. The sodium iodate formed by reaction of the sodium metaperiodate with DBHQ could itself oxidize DBHQ. 2-(*p*-Tosylsulfenyl)hydroquinone reacts with a 10% molar excess of the reagent to give the corresponding quinone, thus suggesting that hydroquinones are more readily oxidized than sulfides. In fact, oxidation of dibenzyl sulfide to the corresponding sulfoxide proceeds in good yield at 20°C only with a substantial excess of the reagent and after 4 d. *Trans*-cyclohexane-1,2-diol and hydrazobenzene react smoothly with the supported reagent in methylene chloride or ether to give hexane-1,6-dial and azobenzene, respectively (Fig. 2).

Villemin and Ricard (1982) have studied dimesoperiodate $K_4I_2O_9$ on different supports. They found that fixation on alumina was as effective as on silica gel. However, although handling of the reagent supported on

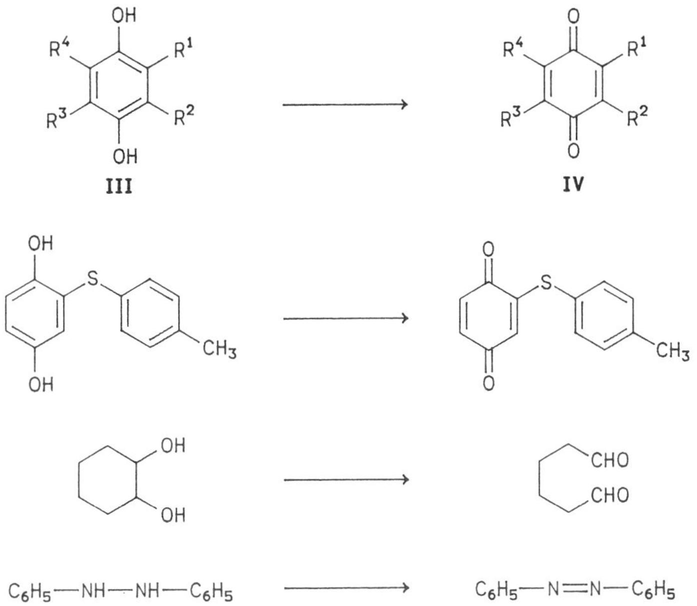

$$C_6H_5\!-\!NH\!-\!NH\!-\!C_6H_5 \longrightarrow C_6H_5\!-\!N\!\!=\!\!N\!-\!C_6H_5$$

Fig. 2. Examples of oxidations realized using $NaIO_4/SiO_2$. [From Gupta *et al.* (1981).]

silica gel can pose problems, due to its volume, the oxidation of *meso*-1,2-diphenyl-1,2-ethanediol proceeded with the same yields with $K_4I_2O_9$ supported on alumina or silica gel (55 and 50%, respectively):

$$C_6H_5\!-\!CHOH\!-\!CHOH\!-\!C_6H_5 \xrightarrow[\text{K}_4\text{I}_2\text{O}_9/\text{Al}_2\text{O}_3]{\text{K}_4\text{I}_2\text{O}_9/\text{SiO}_2} 2\ C_6H_5\!-\!CHO \quad (8)$$

Paraperiodic acid, H_5IO_6, was another oxidant tried; it is unable to convert stilbene into benzaldehyde. On the other hand, it is well known that in hydro-organic medium H_5IO_6 is able to oxidize alkenes to carboxylic acids if $KMnO_4$ is present in catalytic amount (Lemieux and Von Rudloff, 1955). In carbon tetrachloride, H_5IO_6 is unable to oxidize stilbene to benzaldehyde, even in the presence of manganese(III) acetate. When alumina is present or when periodic acid is supported on alumina, the same oxidation proceeds almost quantitatively:

$$C_6H_5\!-\!CH\!\!=\!\!CH\!-\!C_6H_5 \xrightarrow{\text{H}_5\text{IO}_6/\text{SiO}_2 \text{ or Al}_2\text{O}_3} 2\ C_6H_5\!-\!CHO \quad (9)$$

A similar result is obtained with H_5IO_6 in the presence of or supported on alumina or silica gel with catalytic amounts of potassium permanganate.

E. Cerium(IV) Salts

Cerium(IV) salts have been coated on inorganic supports and used advantageously for oxidations. Fischer and Henderson (1985) have described a mild and efficient preparation of quinones from hydroquinones and catechols with a reagent prepared by coating silica with ceric ammonium nitrate. The reaction is the same as described for ammonium dichromate on SiO_2 (Fischer and Henderson, 1985) (Section II.A):

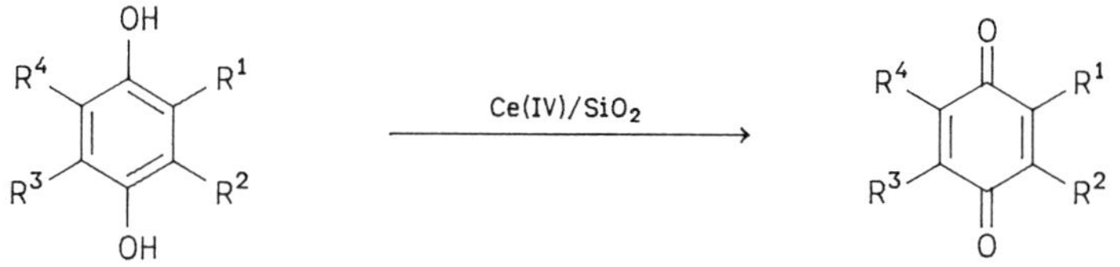

$$ (10) $$

This cerium(IV)/SiO_2 reagent was obtained as a free-flowing yellow powder. Preliminary investigation with hydroquinones showed no significant difference between the 10 and 20% reagents. Drying of the supported reagent did not improve the results. No change in the rate and extent of oxidation was apparent between cerium(IV) and ammonium dichromate adsorbed on silica gel.

The same reagent effects selective oxidative nitrations of aromatic compounds (Chawla and Mittal, 1985). Treatment without solvent of polynuclear arenes and hydroxynaphthalenes with cerium(IV) ammonium nitrate (CAN) absorbed on silica gel gave mononitro derivatives:

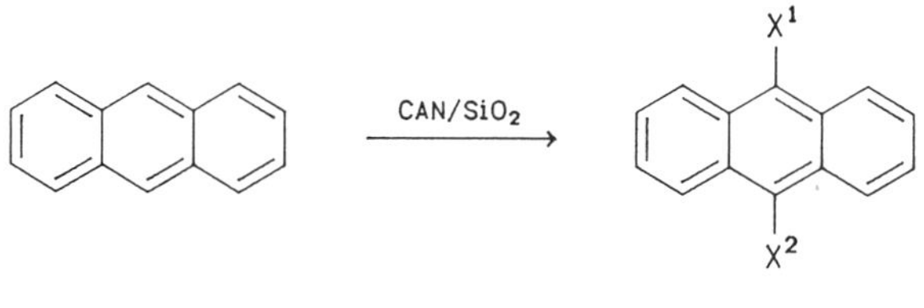

$$ X^1, X^2 = NO_2 \quad (11) $$

Reaction of the same substrates with cerium(IV) ammonium nitrate *in solution* gives a considerable percentage of dinitro derivatives (Chawla and Mittal, 1983) or the corresponding quinones (Hatenaka *et al.*, 1983). Acetoxylated products observed in the functionalization of anisole with cerium ammonium nitrate under homogenous conditions are not formed in the alternative procedure. Furthermore, yields of substitution products with substituents in positions otherwise unfavorable for direct electrophilic attack are improved in reactions with silica gel-supported cerium(IV) ammonium nitrate.

F. Sulfuryl Chloride

Reactions of sulfuryl chloride with methyl phenyl sulfide affords chloromethyl phenyl sulfide in high yields (Bordwell and Pitt, 1955), but in the presence of a small amount of wet silica gel, methyl phenyl sulfoxide is produced in almost quantitative yield, without any formation of chloromethyl phenyl sulfide or sulfones (Hojo and Masuda, 1976a):

$$R^1\!-\!S\!-\!R^2 \xrightarrow{\ SO_2Cl_2,\ SiO_2\cdot H_2O\ } R^1\!-\!SO\!-\!R^2 \qquad (12)$$
$$\mathbf{V} \qquad\qquad\qquad\qquad \mathbf{VI}$$

In a typical experiment a solution of sulfuryl chloride (8.0 mmole) in methylene chloride (6 mL) is added dropwise at room temperature to a stirred mixture of wet silica gel (0.6 g of silicic acid, 100 mesh, and 0.6 g of water) and thioanisole (**V**; $R^1\!=\!C_6H_5$, $R^2\!=\!CH_3$) (7.6 mmole) in methylene chloride (6 mL). After 2 h, methyl phenyl sulfoxide (**VI**; $R^1\!=\!C_6H_5$, $R^2\!=\!CH_3$) is obtained (95% yield). Methyl aryl sulfoxides and diaryl sulfoxides are obtained in high yields. For dialkyl and benzylic sulfides ice cold temperature is used to avoid α-chlorination and subsequent C—S bond cleavage. Allylic sulfoxides were also synthesized conveniently and very easily by this method (at 0°C) without any allylic chlorination or any addition to the double bond. With commercial silica gel, only chloromethyl phenyl sulfide is produced.

The same oxidation applied to thioacetals caused oxidative cleavage of the thioacetal groups cleanly at room temperature, and the original carbonyl compounds were regenerated in almost quantitative yields (Hojo and Masuda, 1976b):

$$R^1\!-\!C(SR^3)_2\!-\!R^2 \xrightarrow{\ SO_2Cl_2/SiO_2\cdot H_2O\ } R^1\!-\!CO\!-\!R^2 \qquad (13)$$
$$\mathbf{VII}$$

This reaction proceeds probably via the monosulfoxides. It can be applied efficiently to thioacetals derived from α,β-unsaturated aldehydes without any chlorination at the allylic position or any addition reaction to the double bond. Thioacetals bearing amino or sulfide groups are also converted to the corresponding parent carbonyl compounds in high yields. Also in the case of the diphenyldithioacetal of p-methylthioacetophenone (**VII**, $R^1\!=\!H_3CS\!-\!C_6H_4$, $R^2\!=\!CH_3$) dethioacetalization occurred in preference to chlorination or oxidation, giving p-methylthioacetophenone in quantitative yield.

This led to a new synthesis of ^{18}O-labeled sulfoxides and carbonyl compounds, using ^{18}O-enriched water (Hojo et al., 1978). About two-thirds of ^{18}O enrichment was found with the labeled sulfoxide, whereas with the aldehyde the ^{18}O enrichment was about one-half that of the added water. Incomplete isotopic incorporation is probably due to partial scrambling of ^{18}O between the added water and water loosely bound on the surface of the silica gel.

G. Selenium Dioxide

In medium-ring sesquiterpenes, selenium dioxide oxidation of an allylic methyl group selectively to the corresponding primary allylic alcohol was unsuccessful (Chhabra et al., 1981). Epoxidation of the double bonds occurred instead. Although some success was achieved using selenium dioxide together with t-butylhydroperoxide (Umbreit and Sharpless, 1977), yields were poor and the reactions were accompanied by undesired epoxidations of double bonds and formation of t-butyl ethers together with a complex mixture of oxidation products. When selenium dioxide supported on silica gel was used with t-butylhydroperoxide in hexane or dichloromethane as the solvent, only oxidations were observed at allylic methyls to afford the corresponding allylic primary alcohols and α,β-unsaturated aldehydes. The supported reagent was prepared by adding to selenium dioxide (5 g) the minimum amount of distilled water and the volume was raised to 150 mL by the addition of methanol to get a clear solution. Addition of silica gel to this solution yielded a fine slurry and after evaporation of the methanol–water mixture under reduced pressure, a free-flowing powder was obtained. For the general oxidation procedure, to selenium dioxide supported on silica gel (5%, 0.5 g) in dichloromethane or hexane (5 mL) t-butylhydroperoxide (70%, 3 eq) is added, and the mixture is stirred for 15 min at room temperature under argon. A solution of 6,7-diacetoxybicyclohumulene (**VIII**, 200 mg) in the same solvent (1 mL) is added, and after 3 h at room temperature, the reaction mixture is filtered and the residue is washed with dichloromethane. Evaporation of solvent under reduced pressure gives 6,10-diacetoxybicyclohumulen-12-ol (**IX**, 80%) and unchanged substrate (20%):

VIII		IX

$$\text{VIII} \xrightarrow{\text{SeO}_2/\text{SiO}_2} \text{IX}$$

(14)

Other examples of oxidations of terpenes have been presented by the authors and all proceeded smoothly and were almost complete in a period of 2–4 h with the formation of alcohols sometimes accompanied by α,β-unsaturated aldehydes. The course of the reaction and yields were independent of the solvent used. A few terpenes remained unchanged under these reaction conditions possibly because, by the process of adsorption, selenium dioxide may be able to attack only the exposed methyl groups, the other allylic positions in the medium ring compounds being untouched.

H. Vacillating Reactions

In 1981, the term *wolf and lamb reactions* was introduced (Cohen *et al.*, 1981) to describe two-stage reactions in which a starting material is modified successively by two polymeric transfer reagents. The analogous soluble reagents would react with each other rapidly in solution, but they could be rendered mutually inactive on their attachment to the respective polymeric phases and therefore coexist in the same reaction vessel "like the proverbial wolf and lamb" (Isaiah 11:6). An application of this concept has been reported for the cleavage of a diol and reduction of the formed aldehyde by a 1:1 mixture of periodate and borohydride supported on the same resin (Bessodes and Antonakis, 1985). This concept has not been extended to reagents adsorbed on inorganic supports, although the method appears very attractive and promising. However, basically on the same principle, Kim and Regen (1983) have realized a "vacillating reaction," using a couple of redox reagents separately adsorbed onto inorganic supports. An alcohol was subjected to a mixed suspension of suitably supported oxidizing and reducing reagents in order to undergo a continuous oxidation–reduction in one pot. Addition of 1-dodecanol to a suspension of H_2CrO_4/SiO_2 plus $NaBD_4/Al_2O_3$ in ether at room temperature led to the formation of 1-dodecanol-*1-1-d_2*:

$$CH_3-(CH_2)_{10}-CH_2OH \xrightarrow[\text{NaBD}_4/\text{Al}_2\text{O}_3]{\text{H}_2\text{CrO}_4/\text{SiO}_2} CH_3-(CH_2)_{10}-CD_2OH \qquad (15)$$

The alcohol was subsequently monitored for deuterium content by GC–MS and the overall formation of 1-dodecanol-*1-1-d_2* clearly showed that the soluble alcohol underwent continuous oxidation and reduction although at relatively slow rates of conversion.

III. REDUCTIONS AND CATALYTIC REDUCTIONS

A. General Remarks

Reductions are well suited for use of silica gel either as inorganic support or catalyst, since here also reagent excess can be used without complicating final workup. Problems of overreductions can be sometimes minimized or eliminated, and in these cases the selectivity becomes astonishingly higher. Silica gel may offer many distinct advantages over alumina, when a solid Lewis catalyst is necessary, as indicated by the beneficial effect of the inorganic support on a few reducing reagents.

B. Sodium Borohydride

Sodium borohydride has been the reagent of choice for selective reductions of organic compounds in the presence of other functional groups and has been a choice target for preliminary studies of supported reducing reagents. In fact, adsorption of sodium borohydride on neutral alumina (Hodosan and Serban, 1969; Santaniello et al., 1978b) or silica gel (Hodosan and Ciurdaru, 1971; Ciurdaru and Hodosan, 1977) has expanded the scope of its applications. A distinct advantage of supported borohydride in nonpolar solvents is that no boron contamination was detected in the reaction medium by atomic adsorption (Snyder, 1981). The reagent has been prepared by Hodosan and Ciurdaru (1971) and is now commercially available. The degree of hydration of the silica gel is extremely important (Bram et al., 1982). In fact, preparation of the reagent from an aqueous solution of $NaBH_4$ and silica gel has resulted in a completely inactive supported reagent in the hands of the French authors. On the contrary, a solution of tetrabutylammonium borohydride in tetrahydrofuran absorbed on the same type of silica gel afforded a stable and efficient reagent (Bram et al., 1982).

In any event, the traditional reduction of carbonyl compounds can be realized in nonpolar solvents using $NaBH_4/SiO_2$ (Hodosan and Ciurdaru, 1971; Ciurdaru and Hodosan, 1977). Remarkably, silica gel affects the reduction capability of $NaBH_4$. It renders possible the reduction of oximes which are not reduced by $NaBH_4$ under normal conditions. In a benzene solution at room temperature, 5α-cholestan-3-one oxime was reduced to a mixture of two epimeric 3α- and 3β-hydroxylamino(N-B) borane compounds. Reduction of 5α-androstan-17-one oxime proceeded in analogous fashion, and in this case it was also possible to isolate the initial product of reduction, an unstable oxime–boron adduct. From these, findings, the reductions effected by $NaBH_4/SiO_2$ appear to proceed according to the scheme outlined in Fig. 3.

The active reducing species on the silica surface should indeed be able to transfer simultaneously a hydride ion and a proton to the acceptor, in this case the oxime group.

Reducing properties of sodium borohydride could also be modified profitably by silica gel in a mixture of polar solvents, as shown by reduction

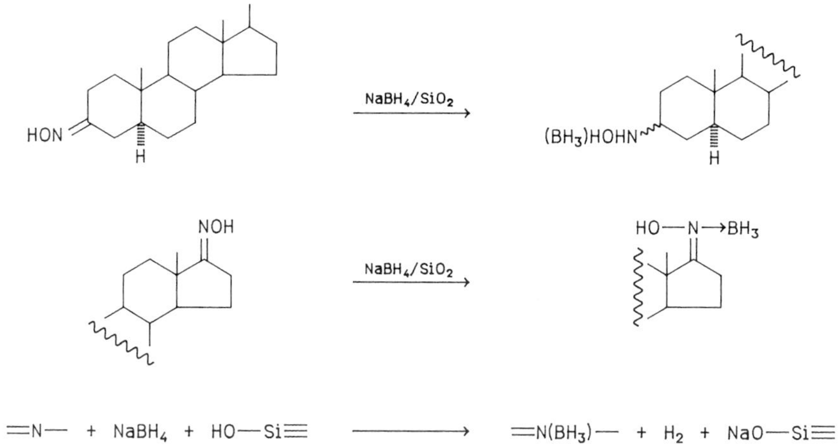

$$\equiv N\!\!-\!\! + NaBH_4 + HO\!\!-\!\!Si\!\!\equiv \quad\longrightarrow\quad \equiv N(BH_3)\!\!-\!\! + H_2 + NaO\!\!-\!\!Si\!\!\equiv$$

Fig. 3. Reduction of steroidal oximes by $NaBH_4/SiO_2$ and proposed mechanism. [From Hodosan and Ciurdaru (1971).]

of nitrostyrene (Sinhababu and Borchardt, 1983). These compounds can be reduced by $NaBH_4$ to the corresponding nitroalkanes, together with a dimer:

$$R^1\!\!-\!\!CH\!\!=\!\!CR^2NO_2 \xrightarrow{\ NaBH_4\ } R^1\!\!-\!\!CH_2CHR^2NO_2 + (R^1\!\!-\!\!CH_2CHR^2NO_2)_2 \quad (16)$$
$$\mathbf{X} \qquad\qquad\qquad\qquad \mathbf{XI}$$

Formation of these dimers can be prevented by acidic conditions, except in the case of vinyl unsubstituted nitrostyrenes (Meyers and Sircar, 1967). Sinhababu and Borchardt (1983) successfully minimized the amount of dimer in the unfavorable case of nitrostyrene ($\mathbf{X}$; $R^1 = C_6H_5$, $R^2 = H$) by carrying out the reduction in the presence of silica gel in a combination of aprotic and protic solvents, a mixture of chloroform and 2-propanol. Yields of the expected reduction products ranged between 92 and 98%. The amount of silica gel required to suppress dimer formation was in the range of 1–3 g/mmol of nitrostyrene ($\mathbf{XI}$; $R^1 = C_6H_5$, $R^2 = H$). The dimeric product is formed by Michael addition of the resonance stabilized α-carbanion to the starting nitroalkene. Formation of the dimer in acidic conditions may be due to the stability of the carbanion, the high reactivity of nitrostyrenes toward nucleophiles, and the sensitivity of $NaBH_4$ towards acids. Silica gel can be considered as an insoluble acid protonating the carbanion formed at the silica gel surface before the Michael addition can become effective.

C. Lithium Aluminum Hydride

Lithium aluminium hydride is well known to organic chemists for reducing carbonyl compounds and carboxy esters to alcohols and many other functional groups as well. This eclecticism of $LiAlH_4$ also constitutes its limit, since its selectivity is not sufficient to discriminate between two reducible functions present in the same molecule. Hojo and co-workers have introduced silica gel either as catalyst or as inorganic support for this reactive hydride. The $LiAlH_4/SiO_2$ reagent can be prepared by mixing dried silica gel and $LiAlH_4$ in dry ethyl ether. It decomposes slowly when stored in a dessiccator (Kamitori *et al.*, 1983a). Reduction of carbonyl compounds and acyl chlorides can be carried out in nonpolar solvents such as *n*-hexane or benzene at room temperature:

$$R^1—CO—R^2 \xrightarrow{\text{LiAlH}_4/\text{SiO}_2} R^1—CHOH—R^2 \qquad R^2 = \text{Alkyl, Aryl, Cl, OR}^3 \quad (17)$$

Reduction of esters proceeded more slowly and required refluxing *n*-hexane for 3 h. Camphor was readily reduced to isoborneol. This contrasts with $NaBH_4/Alox$ which reacted only to a minor extent with the same carbonyl compound (Santaniello *et al.*, 1978b).

The presence of a small amount of dry silica gel changed the reactivity of $LiAlH_4$ (Kamitori *et al.*, 1982). When reduction of ketoesters with $LiAlH_4$ was carried out under classical conditions invariably hydroxyesters were accompanied by diols. A small amount of added silica gel effectively increased selectivity of lithium aluminum hydride and various ketoesters **XII** were converted into corresponding hydroxyesters **XIII** in fair yields [Eq. (18)]. The reaction proceeded quite cleanly at room temperature, and the raw products were practically pure hydroxyesters, no diol being detected by ^{1}H-NMR:

$$R^1—CO—(CH_2)_n—COOR^2 \xrightarrow{\text{LiAlH}_4, \text{SiO}_2} R^1—CHOH—(CH_2)_n—COOR^2 \quad (18)$$
$$\mathbf{XII} \qquad\qquad\qquad\qquad\qquad\qquad \mathbf{XIII}$$

In a typical reaction, commercial-grade silica gel for column chromatography was dried. To a mixture of silica gel (750 mg) thus prepared, $LiAlH_4$ (1.4 mmol) and 5 mL of dry ether were added under an atmosphere of nitrogen. Spontaneous evolution of some hydrogen occurred and after stirring (2 h), 1 mmol of ketoester **XII** was added to the mixture and stirring continued for additional 3 h. The whole mixture was then heated to reflux temperature for 15 min. After workup good yields of hydroxyesters **XIII** were obtained. When silica gel was present in the reaction mixture, selective reduction by $LiAlH_4$ of ketones and carboxyesters could be realized also in the presence of nitro and cyano groups (Kamitori *et al.*, 1983b):

$$X—\bigcirc—CO—R^1 \xrightarrow{\text{LiAlH}_4,\ \text{SiO}_2} X—\bigcirc—CHOH—R^1, \quad X = CN,\ NO_2$$

(19)

In order to realize the same selectivity by $LiAlH_4$, carefully controlled conditions should be used involving lower temperatures, inverse additions, and dilution methods (Felkin *et al.*, 1952; Avison and Morrison, 1950).

Similar conditions succeed in the reduction of the sulfinyl group of methyl *p*-nitrophenyl sulfoxide:

$$O_2N—\bigcirc—SOCH_3 \xrightarrow{\text{LiAlH}_4,\ \text{SiO}_2} O_2N—\bigcirc—SCH_3 \quad (20)$$

D. Pyridine Borane

The air stability, ease of preparation, and excellent solubility characteristics of pyridine borane should qualify it as an attractive reducing agent, but its sluggish reactivity in the absence of acid can severely limit its applications (Lane, 1973). For example, only under vigorous conditions (refluxing toluene or benzene) does pyridine borane reduce aldehydes, ketones, and acid chlorides to the corresponding alcohols (Barnes *et al.*, 1958). Under such conditions, yields are not always satisfactory. Although pyridine borane does function as a more potent reducing agent in acidic media such as trifluoroacetic acid, these conditions are too harsh for use with acid-sensitive substrates (Kikugawa and Ogawa, 1979).

The reactivity of pyridine borane could be modified by adsorption on silica gel or alumina (Babler and Sarussi, 1983). Treatment of a solution of benzaldehyde in cyclohexane with $\frac{1}{3}$ molar equivalent of pyridine borane in the presence of silica gel resulted in >95% reduction of the aldehyde after 45 min at room temperature. When a similar reaction was conducted with 4-phenyl-2-butanone **XV** the reduction was slower and 4-phenyl-2-butanol **XVI** comprised >40% of the product mixture after a reaction time of 2 h. Use of Florisil in place of silica gel slowed the reaction slightly (*i.e.*, 25% versus 40% reduction after 2 h). A more dramatic effect was seen when ethyl ether was used as the solvent in lieu of cyclohexane. Under such conditions (pyridine borane, silica gel or Florisil, ethyl ether, 20°C, 2 h) reduction of 4-phenyl-2-butanone occurred only to a slight extent (<5%). Conversely, under similar conditions reduction of benzaldehyde was nearly quantitative (95%):

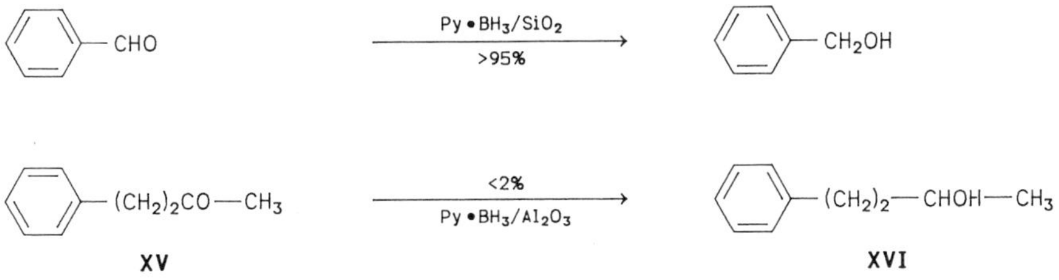

$$\text{C}_6\text{H}_5\text{—CHO} \xrightarrow[>95\%]{\text{Py} \cdot \text{BH}_3/\text{SiO}_2} \text{C}_6\text{H}_5\text{—CH}_2\text{OH}$$

$$\text{C}_6\text{H}_5\text{—(CH}_2)_2\text{CO—CH}_3 \xrightarrow[\text{Py} \cdot \text{BH}_3/\text{Al}_2\text{O}_3]{<2\%} \text{C}_6\text{H}_5\text{—(CH}_2)_2\text{—CHOH—CH}_3$$

XV   XVI

(21)

Selective reduction of an aldehyde rather than a ketone was finally achieved with pyridine borane in the presence of activated alumina. In this case, only trace ($<2\%$) reduction of the ketone resulted after 2 h at room temperature, whereas benzaldehyde was reduced almost quantitatively (95%). Lauric acid in 1 : 1 (v/v) cyclohexane/benzene in the presence of either silica gel or alumina gave no reduction products after 20 h at room temperature and reduction of a 4-ketocarboxylic acid afforded in quantitative yield a 1 : 1 mixture of starting ketone and the expected hydroxyacid.

E. Tributyltin Hydride

Tributyltin hydride is one of the more readily available and least reactive organotin hydrides (Kupchik, 1971). It will readily reduce only strongly electrophilic species such as carbonyl groups bearing powerful electron-withdrawing functions, i.e., alkyl iodides and bromides. Alkyl chlorides, aryl halides, esters, ketones, and other functional groups are reduced only at high temperatures in the presence of a radical initiator. In the presence of a cyclohexane slurry of dried silica gel, tributyltin hydride cleanly reduced aldehydes and ketones to give high yields of the corresponding alcohols (Fung *et al.*, 1978):

$$\text{R}^1\text{—CO—R}^2 \xrightarrow{\text{Bu}_3\text{SnH/SiO}_2} \text{R}^1\text{—CHOH—R}^2 \qquad (22)$$

Other functional groups such as sulfoxides, nitro groups, esters, arylinitriles, and alkyl, aryl, and benzylic chlorides were not effectively reduced. The rate of reduction of carbonyl groups varies in the order: aldehydes > dialkyl ketones > aralkyl ketones < diaryl ketones. As a consequence, when equimolar mixtures of aldehydes and ketones were treated with 1 eq. of tributylin hydride in the presence of silica gel, selective aldehyde reduction was achieved (69–94% yields at room temperature in 0.5–6 h):

No reduction was observed in the presence of methanol or in the absence of silica gel. Reduction occurred inefficiently in the presence of dried basic alumina, proceeded well in the presence of zinc chloride. From these observations, silica gel acts as a mild acid catalyst, polarizing the carbonyl group of the ketone sufficiently to favor reduction by the weakly nucleophilic hydride. If silica gel is not correctly dried, no reduction occurs presumably because of hydrolysis of the tin hydride. Tributyltin hydride does not undergo exchange with the solid support to produce a "reducing silica gel." Treatment of dried silica gel with a cyclohexane solution of tributyltin, filtration and treatment of silica gel with a cyclohexane solution of a ketone gave no reduction. On the other hand, the filtrate was able to reduce norcamphor when fresh, newly dried silica gel was added. After reduction, no tin-containing compounds were isolated when the filtrate and washings were evaporated, indicating that the latter were strongly bound to the silica gel. In fact, only elution with an ethanol–acetic acid mixture allowed the isolation of tributyltin acetate, identical with a prepared authentic sample. These observations suggest (as one possibility) a six-center reduction mechanism to give stannylated silica gel and the alcohol.

F. Hantzsch Ester

Hantzsch ester (HEH, 1,4-dihydro-3,5-diethoxycarbonyl-2,6-dimenthylpyridine) is easily synthetized (Singer and McElvain, 1943) and is a reagent stable at light and air, often used as an NAD (P)H model. Japanese authors (Nakamura *et al.*, 1985) reported that Hantzsch ester (HEH) reduces nitro olefins in good to excellent yields in the presence of silica gel in benzene:

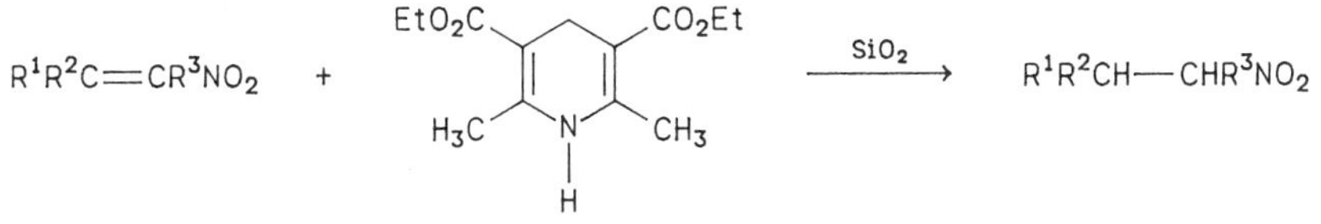

$$R^1R^2C{=}CR^3NO_2 \quad + \qquad \xrightarrow{\;SiO_2\;} \qquad R^1R^2CH{-}CHR^3NO_2$$

XXIII

(23)

The method can be compared with the $NaBH_4/SiO_2$ reported reduction of the same class of organic compounds (Sinhababu and Borchardt, 1983) (Section III.B).

Aliphatic and aromatic nitro alkenes were reduced smoothly, with neither reduction of nitro group nor dimerization of olefins. Other functional groups such as esters were inert under the reaction conditions.

Only a carbon–carbon double bond conjugated with a nitro group was reduced, while other carbon–carbon double bonds remained practically inert. For instance, an equimolar mixture of β-nitrostyrene **XVII** and *trans*-4-phenyl-3-butene-2-one **XVIII** was subjected to reduction with 1.1 equivalent moles of HEH. After 18 h, the nitrostyrene **XVII** was reduced to 2-phenyl-1-nitroethane **XIX** in 94% yield, while the enone was reduced only in 2% yield to compound **XX** and 96% of the unreacted enone was recovered:

$$Ph—CH{=}CH—NO_2 + PhCH{=}CH—COCH_3 \xrightarrow{\text{HEH, SiO}_2}$$

XVII   **XVIII**

$$PhCH_2CH_2\,NO_2 + PhCH_2CH_2COCH_3 \quad (24)$$

XIX   **XX**

94%   2%

The presence of silica gel also altered the reactivity of HEH itself, since it is known that carbonyl groups are usually sensitive to reduction with NAD(P)H or its models. HEH in the presence of silica gel, on the contrary, reduces only the nitroalkene moiety of the molecule, leaving untouched the carbonyl functions. Furthermore, HEH alone in the absence of silica gel is unable to reduce nitroalkenes, and therefore silica gel is essential for the higher selectivity of HEH, although its role is not completely clear at the time being. In a typical experiment, a mixture of β-nitrostyrene **XVII** (149 mg, 1 mmol), HEH (1.1 mmol), and silica gel (100 mg) in 4 ml of benzene was heated at 60°C. After 20 h, 2-phenyl-1-nitroethane **XIX** (146 mg) was obtained in 96% yield.

IV. ANIONIC ACTIVATION

A. Introduction

Most studies of anionic activation by silica gel for carbon–carbon bond formation have been carried out in "dry media." In most cases, electrostatic forces favor adsorption of anions on alumina but not on silica or clays (Bram and Decots, 1980). The reason is that at the solid–liquid interface with an aqueous solution and at the pHs at which adsorption of anionic reagents is performed, alumina is positively charged (Dobias, 1978) whereas silica (Unger, 1979) and clays (Theng, 1974) are negatively charged.

B. Carbon–Carbon Bond Formation

A few examples of activation of carbon–hydrogen bond cleavage on silica gel have been reported in "dry media." Villemin and Ricard (1984) have shown that condensation of poorly acidic carbon compounds with pK_a ranging between 18 and 30 with benzaldehyde can be achieved in the absence of solvents on KF adsorbed onto alumina:

$$C_6H_5-C{\equiv}CH \xrightarrow[C_6H_5-CHO]{KF/Al_2O_3} C_6H_5-C{=}C-CHOH-C_6H_5 \qquad (25)$$

This reagent, KF/Al_2O_3, appears to be a stronger base than KF itself. However, KF adsorbed on Fontainebleau sand, a poorly hydroxylated silica, is inactive.

Malonic ester synthesis is possible on silica gel in the absence of solvent (Bram and Fillebeen Khan, 1979):

$$CH_2(COOCH_3)_2 + Br-(CH_2)_3\text{-}Br \xrightarrow{\text{"basic" } SiO_2} Br-(CH_2)_3CH(COOCH_3)_2 \qquad (26)$$

Malonic ester anions were generated on "basic" inorganic solid supports (alumina and silica gels) where they undergo either intramolecular or intermolecular alkylations selectively in good yields. The "basic" silica gel used in this study was prepared by treating silica gel with a methanolic solution of sodium methoxide, followed by evaporation to dryness and heating at 400°C for several hours (Keinan and Mazur, 1977).

C. Nucleophilic Displacements

Studying the function of the support in the reactions of alkylation of anions on solid supports carried out in "dry media," Bram observed that reaction of acetate anion with 1-bromooctane proceeded smoothly on alumina, but that using silica gel yields were significantly lower (Bram and Decots, 1980):

$$R-Br \xrightarrow{AcO^-/Al_2O_3} R-OAc \qquad (27)$$

However, impregnation of silica with quaternary ammonium salts in catalytic amounts improved yields of n-octyl acetate, the best yields were obtained with a cationic surfactant, cetyltrimethylammonium bromide. Using this method, the yields were nearly the same as on alumina (96%). These results agreed with the previously suggested hypothesis that displacement reactions by an anion should be faster on a positively charged surface. Use of a chemically modified silica gel with a quaternary ammonium group covalently bound to the silica backbone "Spherosil QMA" led to quantitative formation of n-octyi acetate:

$$CH_3—(CH_2)_6—CH_2Br \xrightarrow{AcO^-/SiO_2, R_4N^+Br^-} CH_3—(CH_2)_6—CH_2OAc \quad (28)$$

Activation of potassium cyanide onto silica gel for synthesis of nitriles in dry media was not as effective as on alumina (Bram *et al.*, 1980). Reaction in toluene of alkyl bromide with sodium cyanide adsorbed on silica gel afforded only traces of the corresponding nitrile (Regen *et al.*, 1979):

$$R—X + KCN \quad \begin{array}{c} \xrightarrow{Al_2O_3} \\ \\ \xrightarrow{SiO_2} \end{array} \quad R—CN + KX \quad (29)$$

This negative result has been explained by use of infrared spectroscopy and x-ray diffraction. The most striking change observed in the infrared spectrum was the large shift in the $v(C≡N)$ band from 2076 cm^{-1} in KCN to 2300 cm^{-1} in KCN silica. Such a shift derives from chemisorption resulting in the formation of Si—CN groups on the silica surface. A 200 cm^{-1} shift cannot be attributable to hydrogen bonding or to another weak interaction with the surface but most probably to a $(—O)_3$ SiCN species. To further support this conclusion, KCN could not be removed from KCN–silica reagents with cyanide loadings of less than 5 molecules/nm^2. In contrast to the KCN–silica reagent, KCN–alumina gave a broadened but only slightly shifted $v(C≡N)$ band. The reaction of 2.5 mol equiv of optimized KCN–alumina and KCN–F–alumina reagents with benzyl chloride in a variety of solvents gave rapid and quantitative conversion of the substrate to benzyl cyanide (Clark and Duke, 1985). On the contrary, no reaction was observed in systems employing KCN–silica, which is consistent with the presence of a chemically bonded and inert CN group:

$$C_6H_5—CH_2Cl \quad \begin{array}{c} \xrightarrow{KCN/Al_2O_3} \\ \\ \xrightarrow{KCN/SiO_2} \end{array} \quad C_6H_5—CH_2CN \quad (30)$$

Differently from cyanides, thiocyanates can be conveniently adsorbed onto silica, gel and the supported reagent can be advantageously used for nucleophilic displacements. Epoxides can be converted into thiiranes using silica gel either as a support or catalyst for potassium thiocyanate (Brimeyer *et al.*, 1980):

$$R{=}CH—CH_2 \underset{O}{\diagdown \diagup} \xrightarrow{KSCN/SiO_2} R—CH—CH_2 \underset{S}{\diagdown \diagup} \quad (31)$$

Finely ground KSCN suspended in toluene at 90°C was unable to convert 1-decene oxide to 1-decene sulfide, due to a very low solubility of the salt in the solvent (8×10^{-4} M at 90°C, by atomic emission). In contrast, when KSCN was mixed or coated on silica gel the same reaction proceeded quantitatively. The supported reagent as well as the reaction catalyzed by silica gel required a small but finite amount of water for high activity.

Alkyl thiocyanates can also be conveniently synthetized with silica gel-supported potassium thiocyanate (Kodomari *et al.*, 1983). The reaction of metal thiocyanates with organic halides is one of the most used thiocyanate syntheses; dipolar aprotic solvents such as dimethylformamide are the most suitable (Guy, 1977). Thiocyanates are prepared by the reaction of polymer-supported thiocyanate with organic halides in benzene or toluene under reflux (Harrison and Hodge, 1980), but dodecyl thiocyanate was obtained in only 10% yield. The synthesis of alkyl thiocyanates using $KSCN/SiO_2$ has been described (Kodomari *et al.*, 1983) mainly in the absence of any organic solvent:

$$R-X \xrightarrow{\text{KSCN/SiO}_2} R-SCN \tag{32}$$

Solvents are compatible with the supported reaction, but they slow the reactions somewhat. Silica gel was the most efficient among the inorganic supports tested. Molecular sieves or Kieselgur were much less effective for activating potassium thiocyanate.

D. Tetrabutylammonium Fluoride

As indicated in Section IV.B, potassium fluoride on silica gel is unable to carry out condensations effected instead by KF/Al_2O_3 (Villemin and Ricard, 1984). On the contrary, tetraalklammonium fluorides supported on silica gel are a convenient source of nonhygroscopic fluoride ions for a variety of reactions. [The extreme hygroscopy of tetrabutylammonium fluorides (TBAF) had restricted the widescale use of these valuable reagents for organic synthesis.] The value of these salts depends on the initial formation of a hydrogen bond to a reactant molecule (Clark and Miller, 1977a,b, 1978) and any water present should consequently strongly reduce their effectiveness. Clark (1978) has shown that deposition of the salts on silica gel circumvents this problem, since silica gel presents a surface having many hydroxy groups and is capable of strong hydrogen bonding to H-bond electron acceptors and donors. The resulting "driflour" reagent is a stable, nonhygroscopic catalyst that shows all of the properties of "naked" fluoride and is therefore able to promote a variety of typically base-assisted reactions. Thus, Michael and thiol additions to α,β-unsaturated ketones, phenacyl ester synthesis, enolisation, sulphenylation,

$$CH_3NO_2 \; + \; Ph—CH=CH—CO—Ph \longrightarrow Ph—CH(CH_2NO_2)—CH_2COPh$$

$$PhSH \; + \; CH_3CO—CH=CH_2 \longrightarrow CH_3COCH_2CH_2SPh$$

$$PhCOCH_2Br \; + \; AcOH \longrightarrow PhCOCH_2OAc$$

$$PhCO—CH_2COPh \; + \; EtI \longrightarrow PhCO—CHEt—COPh$$

$$CH_3COCH_2CH_2COCH_3 \longrightarrow CH_3COCH=CHOH—CH_3$$

$$PhSH \longrightarrow PhSSPh$$

Fig. 4. Reactions of tetrabutylammonium fluoride supported on silica gel (TBAF/SiO$_2$). [From Clark (1978).]

and C-alkylation of β-diketones, oxidations of fluorene to fluorenone and thiophenol to diphenyl disulfide all proceed smoothly (1–24 h, 20–60°C) in excellent yields (30–100%) (Fig. 4).

An interesting use of TBAF on silica gel has been envisaged for organosilanes. Fluoride ion has a marked nucleophilic affinity for silicon in organosilanes, due to the high Si—F bond energy and therefore has been used for the cleavage of X—Si bonds (Colvin, 1981). In these F$^-$-induced reactions, a suitable organosilicon compound containing the C—SiR$_3$ moiety serves as a masked carbanion which can be generated under mild conditions. Silica- gel- supported TBAF can transfer benzyl, 4-picolyl, and phenylallyl groups onto weak electrophiles such as benzaldehyde or benzyl bromide in THF at room temperature in 12 h (Ricci et al., 1982a):

$$ArCH_2SiMe_3 \xrightarrow[PhCHO]{TBAF/SiO_2} ArCH_2CHOHPh,$$

$$Ar=o\text{-}NO_2—C_6H_5; \; 3,5\text{-}CH_3—C_6H_4; \; 4\text{-picolyl}; \; PhCH=CH_2 \quad (33)$$

Yields vary between 40 and 80% and it appears that organosilanes corresponding to carbon acids RH with pK_a <37 can be effectively used as an agent for transferring the organic group R to an electrophilic center.

Another application has been proposed by Ricci and co-workers (1982b). In this case, benzyl and heteroalyl carbanions generated from the corresponding organosilanes can be extended to other electrophilic centers. Using various temperatures and catalytic amounts of TBAF/SiO$_2$ in THF such transfers could be realized on saturated lactones or α,β-enones. Yields of reactions of organosilanes with δ- and ε-lactones to afford the corresponding ketoalcohols were in the range of 45–65%, whereas γ-lactones gave only modest conversion and β-propiolactone saw only ring opening and polymerization. With cyclohexen-2-one yields of the Michael reaction were lower than with benzaldehyde. Moreover, the regioselectivity of the reaction depends on the catalyst, since addition of allylsilane on cyclohexene-2-one proceeds in two different ways according to the use of CsF or TBAF/SiO$_2$ as catalyst. Double allylation was the main process when TBAF was used and with CsF only the expected product of conjugate addition was formed (Fig. 5).

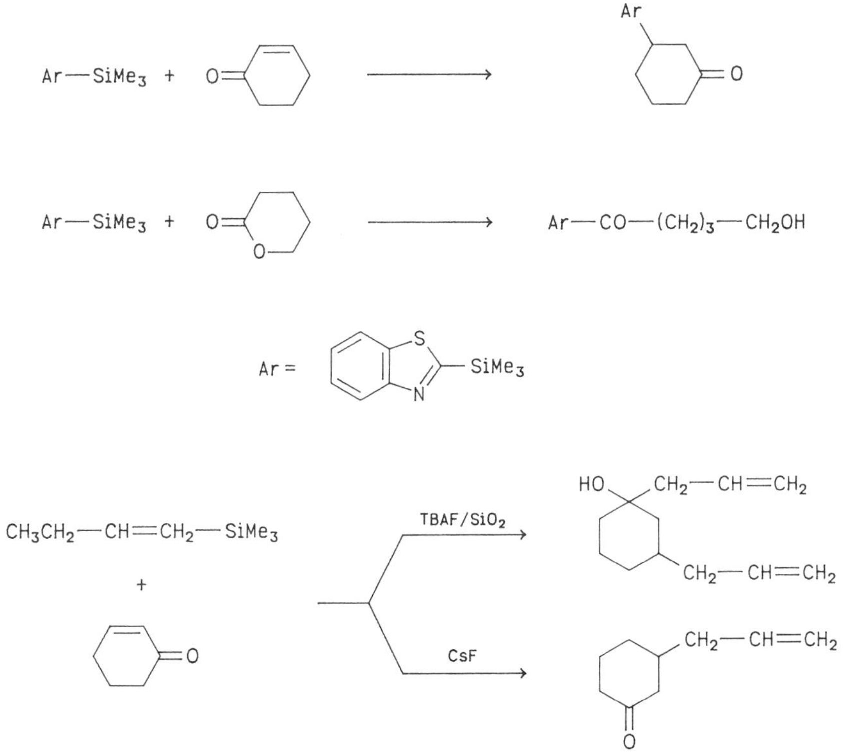

Fig. 5. Reactions of aryl- and alkylsilanes in the presence of supported fluorides (TBAF/SiO$_2$) and under homogeneous conditions (CsF). [From Ricci *et al.* (1982b).]

V. CONCLUSIONS

The number of redox reagents that can be adsorbed on silica gel appears unlimited. Additional work on this area is thus desirable. A better knowledge of the chemico-physical properties of the adsorption process and of the chosen support appears as the best guideline to the optimum reagent/support ratio. The reactivity of reactants, due to their chemical interaction with the support, is significantly altered. Thus, new reactions of old reagents can be discovered, and marked improvements of selectivity can be observed.

An interesting area of applications seems to be the exploration of the coexistence of different reagents coated on the same support or reagents separately absorbed on differ supports for "wolf and lamb"-type reactions.

REFERENCES

Al Jazzaa, A., Clark, J. H., and Robertson, M. (1982). *Chem. Lett.* p. 405.
Avison, A. W. D., and Morrison, A. L. (1950). *J. Chem. Soc.* p. 1474.
Babler, J. H., and Sarussi, S. J. (1983). *J. Org. Chem.* **48**, 4416.
Barnes, R. P., Graham, J. H., and Taylor, M. D. (1958). *J. Org. Chem.* **23**, 1561.
Bessodes, M., and Antonakis, K. (1985). *Tetrahedron Lett.* **26**, 1305.
Bordwell, F. G., and Pitt, B. M. (1955). *J. Am. Chem. Soc.* **77**, 572.
Bram, G., and Decots, G. (1980). *Tetrahedron Lett.* **21**, 5011.
Bram, G., and Fillebeen Khan, T. (1979). *J.C.S. Chem. Commun.* p. 522.
Bram, G., Fillebeen Khan, T., and Geraghty, N. (1980). *Synth. Commun.* **10**, 279.
Bram, G., D'Incan, E., and Loupy, A. (1982). *Nouv. J. Chim.* **6**, 573.
Brimeyer, M. O., Mehrota, A., Quici, S., Nigam, A., and Regen, S. L. (1980). *J. Org. Chem.* **45**, 4254.
Cardillo, G., Orena, M., AND Sandri, S. (1976). *J.C.S. Chem. Commun.* p. 190.
Chawla, H. M., and Mittal, R. S. (1983). *Indian J. Chem.* **22B**, 753.
Chawla, H. M., and Mittal, R. S. (1985). *Synthesis* p. 70.
Chhabra, B. R., Hayano, K., Ohtsuka, T., Shirahama, H., and Matsumoto, T. (1981). *Chem. Lett.* p. 1703.
Ciurdaru, V., and Hodosan, F. (1977). *Rev. Roum. Chim.* **22**, 1027.
Clark, J. H. (1978). *J.C.S. Chem. Commun.* p. 789.
Clark, J. H., and Cork, D. G. (1983). *J.C.S. Perkin I* p. 2253.
Clark, J. H., and Duke, C. V. A. (1985). *J. Org. Chem.* **50**, 1330.
Clark, J. H., and Miller, J. M. (1977a). *J.C.S. Perkin I* p. 1743.
Clark, J. H., and Miller, J. M. (1977b). *J. Am. Chem. Soc.* **99**, 498.
Clark, J. H., and Miller, J. M. (1978). *Can. J. Chem.* **56**, 141.
Cohen, B. J., Kraus, M. A., and Patchornik, A. (1981). J. Am. Chem. Soc. **103**, 7620.
Colvin, E. (1981). "Silicon in Organic Synthesis," Chaps. 7–9. Butterworth, London.
Dobias, B. (1978). *Colloid Polym. Sci.* **256**, 465.
Felkin, I. E., Felkin, H., and Wervart, Z. (1952). *Hebd. Seances Acad. Sci.* **234**, 1789.

Fischer, A., and Henderson, G. N. (1985). *Synthesis* p. 641.
Fung, N. Y. M., De Mayo, P., Schauble, J. H., and Weedon, A. C. (1978). *J. Org. Chem.* **43**, 3977.
Gupta, D. N., Hodge, P., and Davies, J. E. (1981). *J.C.S. Perkin I* p. 2970.
Guy, R. G. (1977). *In* "The Chemistry of Cyanates and Their Thio Derivatives" (S. Patai, ed.), p. 819. Wiley, New York.
Harrison, C. R., and Hodge, P. (1980). *Synthesis* p. 299.
Hatenaka, Y., Imamota, T., and Yokoyamma, M. (1983). *Tetrahedron Lett.* **24**, 2399.
Hodosan, F., and Ciurdaru, V. (1971). *Tetrahedron Lett.* p. 1997.
Hodosan, F., and Serban, N. (1969). *Rev. Roum. Chim.* **14**, 121.
Hojo, M., and Masuda, R. (1976a). *Tetrahedron Lett.* p. 613.
Hojo, M., and Masuda, R. (1976b). *Synthesis* p. 678.
Hojo, M. Masuda, R., and Hakotani, K. (1978). *Tetrahedron Lett.* p. 1121.
Hutchins, R. O., Natale, N. R., Cook, W. J., and Ohrr, J. (1977). *Tetrahedron Lett.* p. 4176.
Kamitori, Y., Hojo, M., Masuda, R., Inoue, T., and Izumi, T. (1982). *Tetrahedron Lett.* **23**, 4585.
Kamitori, Y., Hojo, M., Masuda, R., Izumi, T., and Inoue, T. (1983a). *Synthesis* p. 387.
Kamitori, Y., Hojo, M., Masuda, R., Inoue, T., and Izumi, T. (1983b). *Tetrahedron Lett.* **24**, 2575.
Keinan, E., and Mazur, Y. (1977). *J. Am. Chem. Soc.* **99**, 3861.
Kikugawa, Y., and Ogawa, Y. (1979). *Chem. Pharm. Bull. Jp.* **27**, 2405.
Kim, B., and Regen S. L. (1983). *Tetrahedron Lett.* **24**, 689.
Kodomari, M., Kuzuoka, T., and Yoshitomi, S. (1983). *Synthesis* p. 141.
Kupchik, E. J. (1971) *In* "Organotin Compounds" (A. K. Sawyer, ed.), Vol. 1, Chap. 2. Dekker, New York.
Landini, D., and Rolla, F. (1979). *Chem. Ind. (London)* p. 1213.
Lane, C. F. (1973). *Aldrichim. Acta* **6**, 51.
Lee, D. G., and Spitz, U. A. (1970). *J. Org. Chem.* **35**, 3589.
Lemieux, R. U., and Von Rudloff, E. (1955). *Can. J. Chem.* **33**, 1701.
Liu, K. T., and Tong, Y. C. (1978). *J. Org. Chem.* **43**, 2717.
Meyers, A. I., and Sircar, J. C. (1967). *J. Org. Chem.* **32**, 4134.
Nakamura, K., Fujii, M., Oka, S., and Ohno, A. (1985). *Chem. Lett.* 523.
Regen, S. L., and Koteel, C. (1977). *J. Am. Chem. Soc.* **99**, 3837.
Regen, S. L., Quici, S., and Liaw, S. J. (1979). *J. Org. Chem.* **44**, 2029.
Ricci, A., Degl'Innocenti, A., Fiorenza, M., Taddei, M., Spartera, M. A., and Walton, D. R. M. (1982a). *Tetrahedron Lett.* **23**, 577.
Ricci, A., Fiorenza, M., Grifagni, M. A., Bartolini, G., and Seconi, G. (1982b). *Tetrahedron Lett.* **23**, 5079.
San Filippo, J., and Chern, C. I. (1977). *J. Org. Chem.* **42**, 2182.
Santaniello, E., and Ferraboschi, P. (1980a). *Synth. Commun.* **10**, 75.
Santaniello, E., and Ferraboschi, P. (1980b). *Nouv. J. Chim.* **4**, 279.
Santaniello, E., Ponti, F., and Manzocchi, A. (1978a). *Synthesis* p. 534.
Santaniello, E., Ponti, F., and Manzocchi, A. (1978b). *Synthesis* p. 891.
Santaniello, E., Ferraboschi, P., and Sozzani, P. (1980). *Synthesis* p. 646.
Singer, A., and McElvain, S. M. (1943). *Org. Synth., Collect.* **2**, 214.
Singh, R. P., Subbarao, H. N., and Dev, S. (1979). *Tetrahedron* **35**, 1789.
Sinhababu, A. K., and Borchardt, R. T. (1983). *Tetrahedron Lett.* **24**, 227.
Snyder, S. L. (1981). Ventron Corp., unpublished results; *J. Org. Chem.* **46**, 1A.
Stairs, R. A., Diaper, D. G. M., and Gatzke, A. L. (1963). *Can. J. Chem.* **41**, 1059.

Theng, B. K. G. (1974). "The Chemistry of Clay-Organic Reactions," p. 211. Adam Hilger,
    Bristol, England.
Umbreit, M. A., and Sharpless, K. B. (1977). *J. Am. Chem. Soc.* **99**, 5526.
Unger, K. K. (1979). "Porous Silica," p. 138. Elsevier, Amsterdam.
Villemin, D., and Ricard, M. (1982). *Nouv. J. Chim.* **6**, 605.
Villemin, D., and Ricard, M. (1984). *Tetrahedron Lett.* **25**, 1059.
Wolfe, S., and Ingold, C. F. (1983). *J. Am. Chem. Soc.* **105**, 7755.

19 SILICA-SUPPORTED REAGENTS: POLYMERIZATIONS

Peter W. Lednor

Koninklijke/Shell-Laboratorium, Amsterdam
Shell Research B.V.
1003 AA Amsterdam, The Netherlands

I. INTRODUCTION

The literature on silica-based catalysts or initiators for polymerization reactions is dominated by work related to two commercial processes for the production of polyethylene: the Phillips process, first commercialized in 1956, based on chromium oxides, and the Union Carbide process based on bis(cyclopentadienyl)Cr. The present chapter provides a guide to this area but deals in more depth with other work, perhaps of greater significance to the preparative chemist. Coverage is from 1967 to mid-1985; it is hoped that, outside the polyethylene and polypropylene literature, all work of interest to the readers of this chapter has been included. However, two areas that have been excluded are those of $CO + H_2$ conversion to hydrocarbons (Fischer–Tropsch chemistry) and coke formation with SiO_2-supported heterogeneous catalysts.

Silica is most commonly encountered in catalysis in the form of granules or spheres produced by dehydration of aqueous gels or fine powders (e.g., aerosil) produced by high-temperature processes. Both types are amorphous; crystalline forms of silica such as quartz or silicalite (a pure SiO_2 analog of the zeolite ZSM-5) have hardly, if at all, been used in polymerization reactions. Important properties of silica for catalysis are texture (surface area, pore volume, and pore distribution) and the concentration of surface hydroxyl groups. For example, in the Phillips catalyst the surface area influences activity, and pore size affects both

371

PREPARATIVE CHEMISTRY
USING SUPPORTED REAGENTS

activity and molecular weight of the polymer (Hogan, 1983). In this catalyst fracturing of the silica during polymerization, which exposes fresh surface, is essential (McDaniel, 1981). The effect of silica (and other oxide) surfaces on catalysts derived from organometallic complexes has been reviewed by Yermakov *et al.* (1981; Yermakov, 1983), and the chemistry of silica is covered in an extensive monograph (Iler, 1979).

A major theme in the work reviewed here is the attachment of species to the silica surface by reactions of metal complexes with surface hydroxyl groups:

$$Si—OH + LMY \longrightarrow Si—O—ML + YH$$

Examples are wide ranging: Y can be a hydrocarbon group (e.g., alkyl, allyl, benzyl, or cyclopentadienyl), halide (as in $TiCl_4$), BH_4^- (liberating $BH_3 + H_2$), alkoxide (as in R_3SiOEt), or oxide (as in reactions of CrO_3 with SiO_2 to form surface chromate and water).

Rather notable by their absence in this chapter are examples of condensation polymerizations, silica-supported anionic initiators (radical and cationic ones are known) and application of solid-state NMR spectroscopy to catalyst characterization. It will also be observed that use of main group elements on SiO_2 is a relatively little developed area.

II. POLYMERIZATION REACTIONS INVOLVING SiO$_2$

A. Mechanochemical Polymerizations

Mechanochemistry deals with the unusually high reactivity of mechanically worked solids and has been studied in connection with catalysis, boundary lubrication, and corrosion (Tamai and Mori, 1981). Two complementary studies have appeared on polymerizations induced by vibromilled SiO_2, one concentrating on the mechanism of this effect and the other on polymer properties. In the former (Tamai and Mori, 1981) it was found that vibromilled SiO_2 was still active for polymerization of methyl methacrylate after exposure of the milled solid to air for 2 h. The polymerization was then carried out by contacting the milled powder (surface area = 0.24 m^2/g) with monomer in air at 20°C for 2 h. The weight increase of the SiO_2 was up to five times that calculated for monolayer coverage. Monomer reactivity followed the sequence: acrylonitrile > methyl methacrylate > styrene, based on weight increases (all greater than a monolayer). In contrast, vinyl acetate or benzene gave less than monolayer adsorption. This order of monomer reactivity is consistent with an anionic polymerization. Evidence was obtained that exoelectron formation was a more probable cause than lattice defects or long-lived radicals. This paper also includes results with Al or Fe powders.

In a separate study (Murakami *et al.*, 1984a) ethylene was exposed to SiO_2 undergoing milling. Extraction with hot heptane gave a polyethylene with a molecular weight [by gel-permeation chromatography (GPC)] in the range 10^2–10^4, and shown by ^{13}C NMR to be highly branched. Based on infrared measurements, some 20% of the polymer was extracted, the rest being bonded to the surface. In an extension of this work to acetylene (primarily over Al_2O_3) a significant degree of reaction was also observed over SiO_2, but product characterization was limited to the observation of a purplish color, and a broad, strong ESR signal ($g = 2.0034$) superimposed on a sharp, weak signal ($g = 2.0015$) originating from the SiO_2 (Murakami *et al.*, 1984b).

Chemisorption of ethene, propene, isobutene, 1,3-butadiene, formaldehyde, and benzene on the surface of ground SiO_2 or GeO_2 has been studied by ESR. Structures were established, and the role of graft hydrocarbon radicals in association, isomerization, and dissociation reactions was investigated (Radtsig, 1983).

B. "Reactive" SiO_2

Low *et al.* have investigated the preparation and properties of an especially reactive SiO_2 surface, formed by methoxylating the (Cabosil) surface [$(CH_3O)_3CH$ was the most efficient reagent] followed by pyrolysis and degassing at high temperature, e.g., 850°C. This procedure is believed to create pairs of silicon-centered radicals, $\equiv Si^{\cdot}$, on the surface. These surfaces show high reactivity in chemisorbing a variety of gases and can function as free-radical initiators for the polymerization of acetylenes (Low and Mark, 1980), ethylene (Low and Mark, 1977a), dienes and trienes (Low and Mark, 1977b), and ethylene oxide (Sotani and Low, 1980; Sotani, 1982). With ethylene oxide, the hydroxyl groups on the untreated surface reacted to give $\equiv SiCH_2CH_2OH$, whereas both the silicon-centered radicals, and the Si—O$^{\cdot}$ species derived from these by reaction with O_2 at 25°C, gave grafted $+CH_2CH_2O+_n$ polymers. These investigations were based on IR spectroscopy, and further characterization of the products was not carried out.

C. Effect of SiO_2 on Radical Polymerizations

Several studies have been concerned with the effect of a SiO_2 surface on a polymerization reaction, a topic of practical importance in the context of using SiO_2 as a filler.

Up to 10 wt. % of colloidal SiO_2 was found to accelerate the polymerization of methyl methacrylate when using (1) azo-bis(isobutyronitrile) (AIBN) + UV irradiation, (2) AIBN + heat, or (3) γ-radiation as the initiator. The time taken to reach the maximum of the polymerization exotherm was used to monitor the effect of the SiO_2: this ranged from 180 min (no SiO_2) to 80 min for initiating system (1) at 40°C with 10% SiO_2. Extraction experiments (refluxing chloroform or acetone for 100 h) showed that very little grafting had occurred. SiO_2 was found to increase the viscosity of the monomer; this increases the polymerization rate by reducing the rate constant for termination reactions. However, this effect was not large enough to explain the results, and a mechanism was favored in which radical chain ends were stabilized by interaction with the SiO_2 surface, this again leading to a reduction in the rate of termination (Manley and Murray, 1972).

In a related study, the effect of aerosil (175 m^2/g) on the thermally induced, AIBN-initiated polymerization of butyl methacrylate has been investigated (Eliseeva *et al.*, 1983). The mechanism was still found to be a radical one in the presence of 7 wt.% SiO_2, (above this value the SiO_2 was no longer well dispersed in the monomer), based on the reaction order in initiator being the same (0.5) and an inhibiting effect of diphenylpicrylhydrazyl (DPPH) or O_2. The presence of aerosil did not affect the decomposition of AIBN (in benzene) nor the activation energy of the overall polymerization reaction but did reduce the induction period and raise the rate of initiation (by a factor of 1.46 for 4% aerosil and 1.72 for 7% aerosil polymerizations at 60°C). AIBN, monomer, and polymer were all found by IR spectroscopy to adsorb on the aerosil surface, and it was concluded that the accelerating effect of aerosil on the polymerization was related to these adsorptions, by reducing (1) recombination reactions of the radicals derived from AIBN and (2) the termination reactions of the growing polymer chains. This investigation also includes data on the effect of poly(vinyl chloride) powder instead of aerosil and on the properties of the filled polymers obtained.

In contrast, Tsvetkov and Koval'skii (1982) have found that aerosil ($\leqslant$6 wt. %) inhibits the polymerization of styrene at 65–80°C, using polymeric sebacic acid peroxide as initiator. This was due to a nonradical decomposition of the peroxide via reaction with surface silanol groups. Decomposition of the less labile benzoyl peroxide was not affected by aerosil. Related to this work is a kinetic study of styrene polymerization initiated by peroxide groups grafted to aerosil, at 373–393 K in PhCl (Ivanchev *et al.*, 1981). The ratio of grafted polymer to homopolymer increased with increasing surface concentration of peroxide groups and degree of conversion. Aerosil containing no peroxide inhibited polymerization at <393 K.

D. Condensation Reactions at the SiO₂ Surface

The polycondensation of $Ph_2Si(OH)_2$ in the presence of aerosil gave a non-cross-linked structure that melted at up to 30% SiO_2 concentration. In contrast, highly cross-linked structures were formed in the presence of Al_2O_3 (Bryk *et al.*, 1974).

III. POLYMERIZATION REACTIONS INVOLVING SiO₂-SUPPORTED MAIN GROUPS SPECIES

The literature is classified according to the main group element. Monomers oligomerized or polymerized with these systems include 1-decene (B), isobutene (C,S), styrene (C), alkenes (P), vinyl monomers (C,S,Zn), and acetaldehyde (Ba).

A. Boron

The oligomerization of 1-decene over a $SiO_2-BF_3-H_2O$ catalyst has been investigated by Madgavkar and Swift (1983). The products after hydrogenation were synthetic lubricants particularly suitable for use at both very high and very low temperatures. Using a fixed bed of SiO_2 (248 m²/g and a feed of 3.1 wt. % BF_3 and 30 ppm H_2O in 1-decene, a high (89%) stable (at least three weeks) conversion was obtained at 250 psig, 27°C, and a liquid hourly space velocity of 4. The BF_3 could be recovered as such and recycled. In the absence of added water the SiO_2-BF_3 catalyst deactivated rapidly due to formation of polymers. It was thought that water maintained a hydroxylated surface and displaced oligomers from the surface, thereby preventing further growth.

B. Carbon

A cationic polymerization of isobutene at a modified SiO_2 surface has been carried out by Vidal *et al.* (1980) leading to an SiO_2-grafted poly(isobutene). The compound $ClSiMe_2CH_2CH_2C_6H_4CH_2Cl$ (**I**) was reacted with the surface hydroxyl group of 150 m²/g SiO_2, giving 1.04 molecules of **I** per 100 Å² of surface, attached by SiOSi linkages. Reaction with isobutene in the presence of Et_2AlCl gave poly(isobutene) attached to the benzylic carbon of SiO_2-grafted **I**. A maximum grafting efficiency (grams grafted PIB/grams converted isobutene) of 50% was obtained after a reaction time of 0.5 h at $-50°C$ using an Al/Cl ratio of 20. The composite product, with a solid hydrophilic core attached to flexible hydrophobic branches, might have interesting properties. In a subsequent paper (Vidal *et al.*, 1982) the method was extended to isobutene/isoprene copolymers.

A radical graft polymerization of styrene to an SiO_2 surface has been carried out by thermal decomposition of a phenyldiazothioether group, $—C_6H_4NNSC_{10}H_7$, attached to the silica surface by a Si—C bond, in the presence of styrene. The surface concentration of diazo groups was 0.03–$0.1/100$ Å^2. The effect of reaction time, temperature, and concentration of monomer and diazo groups on the amount of grafted copolymer was investigated. The structure of the grafted polymer could be described as mutually penetrating ellipsoidally deformed random coils attached to the surface (Laible and Hamann, 1975).

Andrianov *et al.* (1983, 1984, 1985) have studied the photoinitiated radical graft polymerization of vinyl monomers on SiO_2 surfaces. For example (Andrianov *et al.*, 1984, 1985), the SiO_2 surface was modified with (dichloromethylsilyl)propyl trichloroacetate and the resulting supported $—CCl_3$ groups were grafted with methyl methacrylate in a reaction initiated by $Mn_2(CO)_{10}$ in the presence of visible light. The polymerization is presumably initiated by abstraction of Cl$^.$ from $—CCl_3$ by $(OC)_5Mn^.$ The same monomers were also grafted to a silica surface modified with $Me_2N(CH_2)_3Si(OEt)_2Me$ or $Et_2NCH_2Si(OEt)_3$ using fluorene, anthroquinone, benzoin methyl ether, or benzophenone and visible light (Andrianov *et al.*, 1983).

C. Aluminum

Catalysts for the polymerization of cyclohexene oxide, derived from $Al(acac)_3$ and $RR'Si(OH)_2$ (R = Ph, R' = Ph or Me) were more active in the presence of certain SiO_2 gels (Hayase *et al.*, 1981). In the most pronounced case, conversion was $\sim2\%$ with $Al(acac)_3$–$MePhSi(OH)_2$ and 25% for the same combination in the presence of SiO_2 after 60 h at 40°C. Based on GPC and ^1H NMR data it was concluded that the effect of SiO_2 was to increase the acidity of the silane diol and to stabilize the latter against condensation reactions. Results with zeolites were also reported.

Treatment of dry SiO_2 gel with Et_3Al, Et_2AlCl, $EtAlCl_2$, or Me_2-AlCl led to formation of surface $—O—Al\ R_{n-1}\ Cl_{3-n}$ groups. Exposure of this modified SiO_2 gel to an air or Ar plasma discharge provided an initiating system for vinyl chloride polymerization (Papoyan *et al.*, 1984).

D. Phosphorus

Phosphoric acid on SiO_2 (kieselguhr or quartz chips) has been used as a strong acid catalyst in the petrochemical industry since the 1930s. Applications include alkene oligomerization at 200°C, at which temperature the catalyst is a viscous liquid. An acid strength of 100–110% is used (achieved by adding P_2O_5 to H_3PO_4); beyond 110% polymer formation is promoted. Further information is contained in a review by Villadsen and Livbjerg (1978) on supported liquid-phase catalysts. Phosphoric acid on aerosil catalyzes butene isomerization whereas diphosphate causes polymerization, a seemingly attractive example of selectivity determined by site nuclearity (Walkov *et al.*, 1978).

E. Sulfur

Reaction of γ-mercaptopropyltriethoxysilane with SiO_2 (370 m^2/g) gave a $\equiv Si\sim\!\!\sim\!\! SH$ species (9.3% S) which, in combination with CCl_4 at 75°C, initiated the radical polymerization of acrylic acid to a water-soluble polymer. This initiating system allowed a water-soluble polymer to be obtained at acrylic acid concentrations in water of up to 67% (compared to 25% with other initiators) and was effective at an acrylic acid/—SH molar ratio of 10^5. Propyl mercaptan did not cause any reaction. Carrying out the polymerization in the presence of divinyl benzene as a cross-linking agent and sodium hydroxide gave a super-absorbent sodium polyacrylate capable of absorbing 300 g/g of water (Huang *et al.*, 1983).

With the aim of providing a thermostable analog of sulfonated styrene–divinylbenzene copolymers, trimethoxybenzylsilane was grafted to SiO_2 by reaction at 100°C, followed by sulfonation of the benzene ring (Saus and Schmidl, 1985). The resulting material had the following properties: 0.68 mmol H$^+$/g, 3.0 mmol OH$^-$/g, Hammett acidity $-6.2 > H_0 > -6.6$, pore volume 0.93 ml/g, surface area 540 m^2/g. It was tested for the oligomerization of isobutene as well as for the reverse reaction. No loss in acid capacity occurred on heating the material for 240 h at 300°C under N_2. Maximum conversion of isobutene into oligomers over this modified SiO_2 was 97%, obtained at 130°C and 1 bar, giving a weight ratio of di : tri : tetraisobutylene of 20 : 15 : 1. Initial activity at this temperature was 0.86 mol/g h, declining by 50% over 120 h, and thereafter with a half life of 350 h (activity 0.14 mol/g h after 31d). At temperatures above 130°C conversion decreased due to scission reactions of the oligomers.

Absorption of methyl methacrylate on SiO_2 (or zeolites) followed by treatment with gaseous SO_2 led to formation of a polymeric coating which could not be extruded by hot benzene (Itabashi *et al.*, 1983).

F. Zinc

The copolymerization of epoxides with CO_2 to give aliphatic polycarbonates has been chiefly studied with combinations of diethyl zinc and hydroxy compounds as catalysts (Rokicki and Kuran, 1981). Aqueous impregnation of $ZnBr_2$ on various oxides has also resulted in active catalysts. In the case of SiO_2 a propene oxide–CO_2 copolymer yield of 48 g/g Zn was obtained after 40 h at 60°C (Soga, 1982), this yield being relatively high. Novel microstructures have been observed for this copolymer, resulting from aselective attack on the propene oxide ring by the growing chain (Lednor and Rol, 1985).

Zinc-modified aluminosilicates, silica gels, and Al_2O_3 form, in combination with potassium persulfate, effective redox systems for the graft polymerization of vinyl compounds on the oxide surfaces (Tutaeva *et al.*, 1978).

G. Barium

Acetaldehyde has been polymerized at −78°C, in aqueous solution, over SiO_2 containing 1.5, 2, or 5% BaO. The polymer was soluble in acetone; molecular weight ranged between 4.5×10^5 and 3.6×10^6 (Kasatkina and Kirpicheva, 1967).

H. Lead

The effect of PbO–SiO_2 glasses on the aqueous polymerization of methyl methacrylate at 40°C, initiated by sodium bisulfite, has been studied by Moustafa and Badran (1979). In air, both conversion and rate were about the same with either PbO or SiO_2 and higher than with mixed oxide compositions (50–80 mol. % PbO). Under nitrogen, however, the activity order was: $SiO_2 \approx$ crystalline PbO $\cdot$ $SiO_2 >$ amorphous PbO $\cdot$ $SiO_2 >$ PbO. Molecular weights of the polymers obtained showed some dependence on glass composition but were all of the order of 10^5.

IV. REACTIONS INVOLVING SiO_2-SUPPORTED TRANSITION METALS

As indicated in Section I, of the various commercial processes for producing polyethylene, two use a SiO_2-supported transition metals: the Phillips process based on Cr oxides and the Union Carbide process based on chromocene, $[(C_5H_5)_2Cr]$. In the case of propylene, attempts to support Ziegler–Natta catalysts (e.g., $TiCl_3$ + aluminium alkyls) on inorganic oxides containing surface hydroxyl groups were not sufficiently successful, and attention subsequently switched to the highly active and stereospecific compositions based on magnesium chlorides.

These commercial developments have led to an extensive literature over the past 30 years. In the first section a guide is provided to the recent literature on ethylene and propylene polymerization in the form of reviews, papers on catalysts, and of a brief survey of characterization techniques. In subsequent sections copolymerizations (all based on ethylene) and polymerizations with monomers other than ethylene or propylene are reviewed.

A. Ethylene and Propylene

1. Reviews

Choi and Ray (1985a) cover more recent developments in ethylene and propylene (Choi and Ray, 1985b) polymerization, including comparison of commercial processes and references to the patent literature. An extensive review of catalytic olefin polymerization has appeared, with close to 500 references (Pasquon and Giannini, 1984). Karol (1984) has discussed high-activity catalysts for olefin polymerization, and Hsieh (1984) has covered some of the historical developments. Hogan (1983) and McDaniel (1985) have reviewed the Phillips polyethylene process. A book by Yermakov et al. (1981), has chapters on olefin polymerization by supported organometallics and by supported transition metal halides, with details of much work originally published in Russian; this material has been updated (Yermakov 1983; Zakharov, 1983). Papers presented at a symposium, held in 1981, with the title *Transition Metal Catalysed Polymerizations: Alkenes and Dienes* have been published with transcripts of the discussions (Quirk, 1983).

2. Catalysts

The most recent work (1983–1985) on SiO_2-based catalysts for ethylene polymerization is that concerning (1) SiO_2–$TiCl_4$–alkyl aluminium catalysts (Munoz-Escalona, 1983), (2) the effect of Ti on Cr/SiO_2 (Pullukat et al., 1983), (3) catalysts derived from SiO_2–R_2Mg–$TiCl_4$–$R_3'Al$ (Shida et al., 1983), (4) highly active Cr/SiO_2 catalysts (Rebenstorf, 1984), (5) the heterogenization of Ni complexes as catalysts for ethylene oligomerization (Peuckert and Keim, 1984), and (6) the photoreduction of supported Ni(II) ions to give catalytically active Ni(I) (Kazanskii et al., 1983). For propene, the only recent paper is that on oligomerization over Ni oxides, including comparison with Cp_2Ni/SiO_2 (Kvisle et al., 1984).

3. Catalyst Characterization

In this section a brief survey is given of more recent techniques used to characterize SiO_2-supported polymerization catalysts. Much work has been done on the characterization of the Phillips catalyst and for further literature on this topic reference should be made to papers by McDaniel (1983), Merryfield *et al.* (1982), Myers and Lunsford (1985), Rebenstorf (1984, 1985a,b), and Ellison (1984a,b).

Groeneveld *et al.* (1983) have made *in-situ* IR measurements on Cr/SiO_2; Fourier-transform IR/photoacoustic spectroscopy has been applied to chromocene on SiO_2 (McKenna *et al.*, 1984) and to $TiCl_4$ or $AlMe_3$ on SiO_2 (Kinney and Staley, 1983). Ellison has characterized Cr/SiO_2 with ESR spectroscopy (Ellison, 1984a) and fast atom bombardment–secondary ion mass spectrometry (Ellison, 1984b). ESR has also recently been used to study the interaction of ethylene with Ti^{3+} sites derived from $Ti(CH_2Ph)_4$ on SiO_2 (Poluboyarov *et al.*, 1984). X-ray photoelectron spectroscopy (McDaniel, 1983; Merryfield *et al.*, 1982) has been used to determine the oxidation state of the Phillips catalyst; this has also been characterized by chemiluminescence measurements (Myers and Lunsford, 1985) and UV–VIS diffuse reflectance spectroscopy (Rebenstorf, 1985a,b). ^{1}H-NMR spectroscopy has only been applied in one case: that of ethylene adsorption on $Zr(BH_4)_4$ or $Zr(CH_2Ph)_4$ on SiO_2 (Nesterov *et al.*, 1982). A study of the adsorption of chromocene, from solution, on silica has been made by Union Carbide workers (Karol *et al.*, 1979).

The single most important issue in the characterization of SiO_2-supported polymerization catalysts is probably that of the oxidation state of Cr in the Phillips catalyst, a debate that is still vigorous some 30 years after commercialization of the process. After catalyst preparation and activation (Hsieh, 1984) Cr is in the +6 state. Reduction to Cr(II) takes place on exposure to ethylene or to CO, and the view taken by Phillips scientists is that this is probably the active site (Hsieh, 1984; Hogan, 1983; McDaniel, 1983; Merryfield *et al.*, 1982). However, Cr(III) may also play a role: Rebenstorf (1984) reported much higher turnover numbers for Cr(III) compared to Cr(II) at 20°C, although Myers and Lunsford have found that the activity of Cr(II), while still less than that of Cr(III), increases sharply above ~40°C (paper submitted for publication).

It should of course be remembered that the formal oxidation state is only a partial description of an active site and that coordination number and nature of the ligating atoms are also important. For further discussion of these and related aspects reference should be made to a review article entitled "Supported Chromium Catalysts for Ethylene Polymerization" (McDaniel, 1985).

B. Ethylene-Based Copolymers

Soga *et al.* (1982) have described an active catalyst for the copolymerization of ethylene and propylene to moderately random copolymers. The catalyst was prepared by adding $TiCl_3(pyr)_3$, $MgCl_2(THF)_n$, and SiO_2 to *n*-heptane containing $AlEt_2Cl$. The Ti^{3+} ion was then found to be stable to further reduction, which is important since Ti^{2+} was active only for ethylene polymerization. The $\bar{M}_n$ value for a copolymer with $C_2/C_3 = 1$ was 29,000, $q = 6.8$. An activity of 176 kg/(g Ti) h was obtained using 0.13 mmol C_2 and 0.54 mmol C_3 at 65°C.

This work was extended (Soga *et al.*, 1983) to catalysts prepared by reacting $TiCl_4$ with SiO_2, followed by treatment with $MgCl_2(THF)_2$ and $AlEt_2Cl$; monomer reactivity ratios were very similar to those in the work with $TiCl_3$. The distribution of monomers in these polymers has been determined by ^{13}C NMR (Doi *et al.*, 1983).

Copolymerization of ethylene with 1-decene in hexane has been carried out over Cr_2O_3 on silica gel (Babaeva *et al.*, 1984).

C. Monomers Other than Ethylene or Propylene

1. Monoolefins

Cr(II) on SiO_2 is known to polymerize propene or butene to a limited extent at high temperature and pressure. It has been found that by carrying out the reactions at $\leq -20°C$ and 1 bar for 20 h, linear 1-alkenes, C_nH_{2n} ($n = 3, 4, 5, 6, 7, 8$, and 12) can be polymerized in high yield (37, 69, 99, 80, 70, 99, and 99%, respectively). Molecular weights were of the order of 10^4, with a broad distribution, and increased with lower reaction temperature. Catalyst activity was maintained after extraction of the polymer (Weiss and Krauss, 1984).

Catalysts derived from the reaction of $TiCl_4$ with an SiO_2 surface, or the hydrolyzed products from this reaction, gave predominantly dimers of α-methylstyrene. After 18 h at 30°C a maximum of 25% higher oligomers was found at a monomer conversion of 98%. Styrene could be polymerized to 38% conversion after 0.5 h at 30°C (Kawakami *et al.*, 1981).

Cr(II) on SiO_2 polymerized 2,3-dihydrofuran via the double bond, no ring opening occurring. The yield increased steadily over 16 days, at 22°C, to 20%. The 2,5 isomer, furan, and tetrahydrofuran were detected in the reaction mixture, and these monomers also gave a low yield of poly(2,3-dihydrofuran). Clearly, the catalyst also possesses activity for hydrogenation, dehydrogenation, and isomerization. A mechanism involving Cr hydrides as initiator was suggested by Hueholt *et al.* (1982).

2. Dienes

Yermakov *et al.* (1981) have reviewed polymerization of dienes by supported organometallic complexes in a chapter which contains details of much work originally published in Russian; the main points are summarized here. The catalytic systems studied were principally allyl complexes of Ni, Cr, Zr, Mo, Ti, or Co on SiO_2, Al_2O_3 or SiO_2–Al_2O_3. The activity of the supported complexes was often much higher than that of the complexes in solution: $[Ni(C_3H_5)I]_2$, one of the most active homogeneous catalysts, gave a rate of 38 mol butadiene/(g atom Ni) h, whereas the corresponding figure for $Ni(C_3H_5)_2/SiO_2$ was 25,000, the data being obtained at 45°C with a monomer concentration in benzene of 2.5 mmol/L. This difference was shown to be in the rate constant for the propagation reaction. Stereospecificity can vary widely with these systems, as illustrated by the following examples of isomeric poly(butadienes): $Ni(C_3H_5)_2$, no activity; $Ni(C_3H_5)_2/SiO_2$, 95% 1,4-*cis*; $Cr(C_3H_5)_3$, 80% 1,2; $Cr(C_3H_5)_3/SiO_2$, 99% 1,4-*trans*; $Mo(C_3H_5)_4$, no activity; $Mo(C_3H_5)_4/SiO_2$, 90% 1,2. Further parameters of importance are metal loading, the activity of $Ni(C_3H_5)_2/SiO_2$ passing through a maximum at 0.6 wt. % Ni, and the temperature of support pretreatment, which can have a major effect on activity and in some cases on stereospecificity.

Two reports have appeared since on diene polymerization with SiO_2-supported catalysts. Skupinski *et al.* (1980) investigated $Ni(\pi\text{-allyl})_2$ supported on SiO_2, Al_2O_3, and a range of SiO_2–Al_2O_3 mixed oxides, as catalysts for butadiene polymerization. 97% 1,4-*Trans*-poly(butadiene) was formed over SiO_2 or Al_2O_3; the maximum deviation from this value was found over 82% SiO_2, 18% Al_2O_3, which gave 71% 1,4 *cis*, 11% 1,4-*trans*, and 18% 1,2, the reactions being carried out for 2 h at 18°C in benzene. It should be noted that the selectivity reported here for $Ni(C_3H_5)_2/SiO_2$ is completely different to that described in Yermakov's review (*vide supra*).

Highly active catalysts for isoprene polymerization have been prepared from the reaction of $CoCl_2$ (pyridine)$_2$ with SiO_2 (300 m^2/g, average pore diameter $= 21$ nm) in heptane at room temperature. Diethyl aluminium chloride was used as a cocatalyst. A maximum activity of 220 kg polymer/(g Co) h was obtained (0.25 mmol isoprene in 10 mL *n*-heptane at 65°C for 30 min); under these conditions the polymer was 66% *cis*-1,4, 2% *trans*-1,4, and 32% 3,4 with a molecular weight of 5.3×10^4 (Soga *et al.*, 1981).

3. Miscellaneous

Vinyl monomers (acrylamide, methyl methacrylate, and 4-vinylpyridine) have been graft polymerized onto the Co^{2+} forms of silica gel,

Al_2O_3, or montmorillonite in the presence of $K_2S_2O_8$. After extraction of homopolymer, weight gains due to grafting ranged from 1–54% (Tutaeva *et al.*, 1980).

Manganese on SiO_2 has been used as a catalyst for preparing poly(ethylene terephthalate) from dimethyl terephthalate and ethylene glycol. The catalysts were prepared from $MnCl_2$ or $Mn(OAc)_2$ via adsorption or impregnation and gave a product of better color than when using amorphous MnO_2 (Paryjczak *et al.*, 1980).

REFERENCES

Andrianov, A. K., Olenin, A. V., Zubov, V. P., Kashutina, E. A., Zhdanov, A. A., and Kabanov, V. A. (1983). *Vysokomol. Soedin.*, *Ser. A* **25**, 1987–1992 [*C.A.* **100**, 7316z].

Andrianov, A. K., Olenin, A. V., Saprygin, O. N., Garina, E. S., Zubov, V. P., Kashutina, E. A., and Zhdanov, A. A. (1984). *Vysokomol. Soedin.*, *Ser. A* **26**, 2599–2603 [*C.A.* **102**, 132515f].

Andrianov, A. K., Mislavskii, B. V., Olenin, A. V., Garina, E. S., Zubov, V. P., and Kabanov, V. A. (1985). *Dokl. Akad. Nauk SSSR* **280**, 900–904 [*C.A.* **102**, 204361c].

Babaeva, M. A., Buniyat-Zade, A. A., and Bulatnikova, E. L. (1984). *Azerb. Khim. Zh.* 110–115 [*C.A.* **102**, 46346v].

Bryk, M. T., Pavlova, I. A., and Kurilenko, O. D. (1974). *Ukr. Khim. Zh.* **40**, 154–160 [*C.A.* **81**, 26036p].

Choi, K.-Y., and Ray, W. H. (1985a). *J. Macromol. Sci.*, *Rev. Macromol. Chem. Phys.* **C25**, 1–55.

Choi, K.-Y., and Ray, W. H. (1985b). *J. Macromol. Sci.*, *Rev. Macromol. Chem. Phys.* **C25**, 57–97.

Doi, Y., Ohnishi, R., and Soga, K. (1983). *Makromol. Chem.*, *Rapid Commun.* **4**, 169–74.

Eliseeva, V. I., Morozova, E. M., and Aslamasova, T. R. (1983). *Acta Polym.* **34**, 197–204.

Ellison, A. (1984a). *J.C.S. Faraday I* **80**, 2581–2597.

Ellison, A. (1984b). *J.C.S. Faraday I* **80**, 2567–2579.

Groeneveld, C., Wittgen, P. P. M. M., Swinnen, H. P. M., Wernsen, A., and Schuit, G. C. A. (1983). *J. Catal.* **83**, 346–361.

Hayase, S., Ito, T., Suzuki, S., and Wada, M. (1981). *J. Polym. Sci.*, *Polym. Chem. Ed.* **19**, 2541–2550.

Hogan, J. P. (1983). *In* "Applied Industrial Catalysts" (B. E. Leach, ed.), Vol. 1, pp. 149–176. Academic Press, New York.

Hsieh, H. L. (1984). *Catal. Rev.—Sci. Eng.* **26**, 631–651.

Huang, M. Y., Wu, R., and Jiang, L. R. (1983). *Polym. Bull.* **9**, 5–10.

Hueholt, S., Meyer, G., Moeseler, R., Schriefer, A., and Woehrle, D. (1982). *Polym. Bull.* **6**, 315–320.

Iler, R. K. (1979). "The Chemistry of Silica." Wiley (Interscience), New York.

Itabashi, O., Yokoyama, T., and Yamazaki, H. (1983). *Tohoku Kogyo Gijutsu Shikensho Hokoku* **16**, 1–6 [*C.A.* **98**, 216118r].

Ivanchev, S. S., Enikolopyan, N. S., Polozov, B. V., Dmitrenko, A. V., Demidova, V. A., Krupnik, A. M., and Litkovets, A. K. (1981). *Vysokomol. Soedin.*, *Ser. A* **23**, 2064–2070 [*C.A.* **96**, 7494e].

Karol, F. J. (1984). *Catal. Rev.—Sci. Eng.* **26**, 557–595.

Karol, F. J., Wu, C., Reichle, W. T., and Maraschin, N. J. (1979). *J. Catal.* **60**, 68–76.

Kasatkina, N. G., and Kirpicheva, M. V. (1967). *Vestn. Leningr. Univ., Fiz. Khim.* **4.** 153–155 [*C.A.* **68**, 50156e].

Kawakami, Y., Sato, W., Miki, Y., and Yamashita, Y. (1981). *Polymer* **22**, 859–862.

Kazanskii, V. B., Elev, I. V., and Shelimov, B. N. (1983). *J. Mol. Catal.* **21**, 265–274.

Kinney, J. B., and Staley, R. H. (1983). *J. Phys. Chem.* **87**, 3735–3740.

Kvisle, S., Blindheim, U., and Ellestad, O. H. (1984). *J. Mol. Catal.* **26**, 341–354.

Laible, R., and Hamann, K. (1975). *Angew. Makromol. Chem.* **48**, 97–133.

Lednor, P. W., and Rol, N. C. (1985). *J.C.S. Chem. Commun.* 598–599.

Low, M. J. D., and Mark, H. (1977a). *J. Catal.* **48**, 104–110.

Low, M. J. D., and Mark, H. (1977b). *J. Catal.* **50**, 373–378.

Low, M. J. D., and Mark, H. (1980). *J. Res. Inst. Catal., Hokkaido Univ.* **27**, 129–144.

McDaniel, M. P. (1981). *J. Polym. Sci., Polym. Chem. Ed.* **19**, 1967–1976.

McDaniel, M. P. (1983). *In* "Transition Metal Catalysed Polymerizations: Alkenes and Dienes" (R. P. Quirk, ed.), Part B, MMI Press Symp. Ser. No. 4, pp. 713–735. Harwood Academic, New York.

McDaniel, M. P. (1985). *In* "Advances in Catalysis" (D. D. Eley, H. Pines, and P. B. Weisz, eds.), Vol. 33. Academic Press, Orlando, Florida.

McKenna, W. P., Bandyopadhyay, S., and Eyring, E. M. (1984). *Appl. Spectrosc.* **38**, 834–837.

Madgavkar, A. M., and Swift, H. E. (1983). *Ind. Eng. Chem., Prod. Res. Dev.* **22**, 675–680.

Manley, T. R., and Murray, B. (1972). *Eur. Polym. J.* **8**, 1145–1150.

Merryfield, R., McDaniel, M., and Parks, G. (1982). *J. Catal.* **77**, 348–359.

Moustafa, A. B., and Badran, A. S. (1979). *J. Polym. Sci., Polym. Chem. Ed.* **17**, 603–612.

Munoz-Escalona, A. (1983). *In* "Transition Metal Catalysed Polymerizations: Alkenes and Dienes" (R. P. Quirk, ed.), Part A, MMI Press Symp. Ser. No. 4, pp. 323–340. Harwood Academic, New York.

Murakami, S., Tabata, M., and Sohma, J. (1984a). *J. Appl. Polym. Sci.* **29**, 291–298.

Murakami, S., Tabata, M., Sohma, J., and Hatano, M. (1984b). *J. Appl. Polym. Sci.* **29**, 3445–3455.

Myers, D. L., and Lunsford, J. H. (1985). *J. Catal.* **92**, 260–271.

Nesterov, G. A., Mastikhin, V. M., Lapina, O. B., and Zakharov, V. A. (1982). *React. Kinet. Catal. Lett.* **19**, 175–179.

Papoyan, A. T., Matkovskii, P. E., Shestakov, A. F., and Enikolopyan, N. S. (1984). *Dokl. Akad. Nauk SSSR* **275**, 97–101 [*C.A.* **101**, 55568a].

Paryjczak, T., Kazmierczak, A., Leo, J., Balczewska, H., Bielawska, Z., and Czajkowski, J. (1980). *Polymery* **25**, 58–61 [*C.A.* **93**, 72397c].

Pasquon, I., and Giannini, U. (1984). *In* "Catalysis (Science and Technology)" (J. R. Anderson and M. Boudart, eds.), Vol. 6, pp. 65–159. Springer-Verlag, Berlin and New York.

Peuckert, M., and Keim, W. (1984). *J. Mol. Catal.* **22**, 289–295.

Poluboyarov, V. A., Nesterov, G. A., Zakharov, V. A., and Anufrienko, V. P. (1984). *React. Kinet. Catal. Lett.* **25**, 33–38.

Pullukat, T. J., Shida, M., and Huff, R. E. (1983). *In* "Transition Metal Catalysed Polymerizations: Alkenes and Dienes" (R. P. Quirk, ed.), Part B, MMI Press Symp. Ser. No. 4, pp. 697–712. Harwood Academic, New York.

Quirk, R. P., ed. (1983). "Transition Metal Catalysed Polymerizations: Alkenes and Dienes," MMI Press Symp. Ser. No. 4. Harwood Academic, New York.

Radtsig, V. A. (1983). *Kinet. Katal.* **24**, 173–180 [*C.A.* **98**, 222547w].

Rebenstorf, B. (1984). *Z. Anorg. Allg. Chem.* **513**, 103–113.

Rebenstorf, B. (1985a). *Acta Chem. Scand., Ser. A* **39**, 133–141.

Rebenstorf, B. (1985b). *Acta Chem. Scand.*, *Ser. A* 39, 370–372.

Rokicki, A., and Kuran, W. (1981). *J. Macromol. Sci.*, *Rev. Macromol. Chem.* 21, 135–186.

Saus, A., and Schmidl, E. (1985). *J. Catal.* 94, 187–194.

Shida, M., Pullukat, T. J., and Hoff, R. E. (1983). *Polym. Prepr.*, *Am. Chem. Soc.*, *Div. Polym. Chem.* 24, 110–111.

Skupinski, W., Zawartke, M., and Malinowski, S. (1980). *React. Kinet. Catal. Lett.* 13, 319–322.

Soga, K. (1982). *Nippon Kagaku Kaishi* 295–300 [*C.A.* 96, 123412a].

Soga, K., Yamamoto, K., and Chen, S.-I. (1981). *Polym. Bull.* 5, 1–4.

Soga, K., Ohnishi, R., and Sano, T. (1982). *Polym. Bull.* 7, 547–552.

Soga, K., Ohnishi, R., and Doi, Y. (1983). *Polym. Bull.* 9, 299–304.

Sotani, N. (1982). *Bull. Chem. Soc. Jpn.* 55, 1992–1998.

Sotani, N., and Low, M. J. D. (1980). *React. Kinet. Catal. Lett.* 13, 339–344.

Tamai, Y., and Mori, S. (1981). *Z. Anorg. Allg. Chem.* 476, 221–228.

Tsvetkov, N. S., and Koval'skii, Y. P. (1982). *React. Kinet. Catal. Lett.* 21, 335–340.

Tutaeva, N. L., Komarov, V. S., and Belyakova, M. D. (1978). *Vestsi Akad. Navuk BSSR*, *Ser. Khim. Navuk* 19–23 [*C.A.* 89, 75703c].

Tutaeva, N. L., Komarov, V. S., and Belyakova, M. D. (1980). *Zh. Prikl. Khim.* 53, 933–934 [*C.A.* 94, 47949x].

Vidal, A., Guyot, A., and Kennedy, J. P. (1980). *Polym. Bull.* 2, 315–320.

Vidal, A., Guyot, A., and Kennedy, J. P. (1982). *Polym. Bull.* 6, 401–407.

Villadsen, J., and Livbjerg, H. (1978). *Catal. Rev.—Sci. Eng.* 17, 203–272.

Walkov, W., Hanke, W., Jerschkewitz, H. G., and Voelter, J. (1978). *Geterog. Katal.*, *Tr. Mezhdunar*, *Simp. 3rd*, *Varna*, *Bulg. 1975*, 196–202 [*C.A.* 89, 221528t].

Weiss, K., and Krauss, H. L. (1984). *J. Catal.* 88, 424–430.

Yermakov, Y. I. (1983). *J. Mol. Catal.* 21, 35–55.

Yermakov, Y. I., Kuznetsov, B. N., and Zakharov, V. A. (1981). "Catalysis by Supported Complexes." Elsevier, Amsterdam.

Zakharov, V. A. (1983). *Indian J. Technol.* 21, 353–357.

20 SILICA-SUPPORTED REAGENTS: REACTIONS IN DRY MEDIA

Georges Bram and André Loupy

Laboratoire des Reactions Sélectives sur Supports
Université Paris–Sud
F-91405 Orsay, France

I. WHY DRY-MEDIA REACTIONS?

Reactions involving reagents supported on inorganic supports are usually run in the presence of an organic solvent, as are heterogeneous solid–liquid process. However, many examples show that organic syntheses can also be efficiently performed with these supported reagents in the absence of any organic solvent in so-called dry-media conditions.

This methodology of dry-media reactions has been advocated by the Mazur group (Mazur, 1975; Mazur *et al.*, 1975; Cohen *et al.* 1975) in their work on "dry ozonation." Ozonation of saturated hydrocarbons inserts an oxygen atom into the C—H bonds, resulting in alcools and ketones. This useful reaction was impeded by the low solubility of ozone in organic solvents and also by its reactivity toward most of the organic solvents, which limited its practical use for hydroxylations. Since silica gel adsorbs ozone efficiently it served as the reaction matrix, thus overcoming the drawbacks of solution ozonations.

Furthermore, even in the absence of competing reactions with the solvent, there may be other advantages to performing reaction in dry media. These include faster reactions than in the presence of a solvent, different selectivities observed, and more economical reactions due to the absence of solvent.

387

PREPARATIVE CHEMISTRY
USING SUPPORTED REAGENTS

Some of the most important reactions performed in dry media on silica gel are discussed here: reactions with simple inorganic reagents supported on silica and then other transformations. Some comparisons with dry-media reactions using other inorganic supports are also presented.

II. OXIDATION REACTIONS

A. Ozone

Ozonation in dry media of substrates adsorbed on silica gel is a convenient method for inserting an oxygen atom into unactivated C—H bonds (Mazur, 1975; Cohen et al., 1975). The synthesis of 1-adamantanol from adamantane is illustrative (Cohen et al., 1975, 1980):

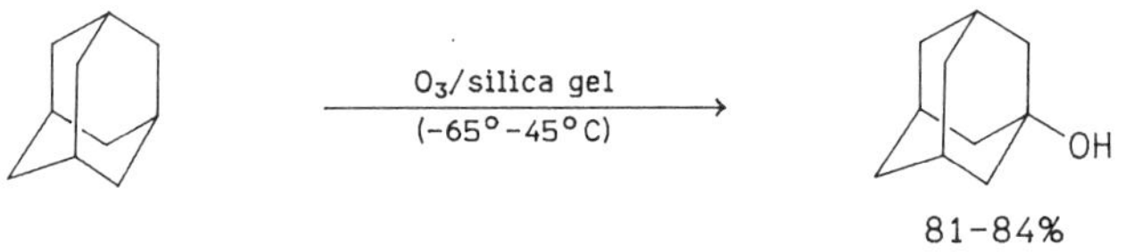

$$O_3/\text{silica gel} \quad (-65° - 45° C)$$

81–84%

Steroids (Cohen and Mazur, 1979), natural products (Trifilieff et al., 1977, 1978), and other cyclic substrates have been hydroxylated. This method is less favorable for the oxidation of secondary C—H bonds (Beckwith and Thach Duong, 1978), but a quantitative conversion of adamantan-2-ol to adamantanone has been observed (Cohen et al., 1975). Methylene groups adjacent to a cyclopropyl group have been oxidized efficiently to ketones (Proksch and de Meijere, 1976a,b):

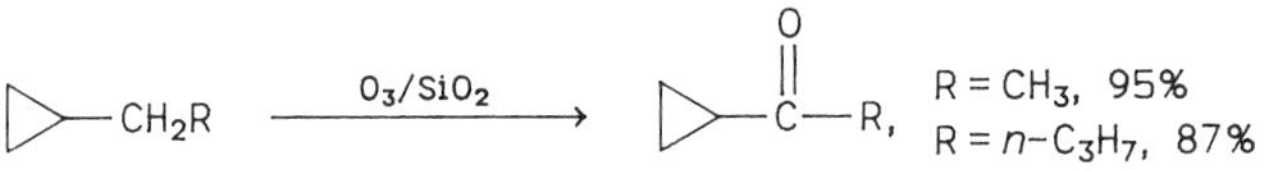

$$O_3/SiO_2$$

R = CH₃, 95%
R = n-C₃H₇, 87%

Dry ozonation on silica gel of cyclododecyl acetate, dodecyl acetate, and acetate derived from cyclopentadiene trimer affords ketoacetates by oxidation of methylene groups remote from functional groups; a similarity was found between the results of dry ozonation and biological oxidation of the substrates (Beckwith and Thach Duong, 1978).

When aliphatic hydrocarbons were reacted with ozone on silica, the yields of tertiary alcools were considerably lower than those obtained on the ozonation of cyclic hydrocarbons. Not only C—H bonds but also C—CH₃ bonds have been cleaved by direct insertion of ozone into C—C bonds (Cohen et al., 1975; Beckwith et al., 1977a,b; Tal et al., 1979).

Reaction in dry media with ozone on silica gel is also useful for the oxidation of aliphatic and aromatic amines into the corresponding nitro compounds. Unlike the reaction with saturated hydrocarbons this reaction is relatively fast, and the absence of solvent prevents the formation of undesired side products from interaction of the partially oxidized amine with the solvent (Keinan and Mazur, 1977a) Ozonation of aromatic amines gives rise to low yields of nitro derivatives as pure crystalline materials:

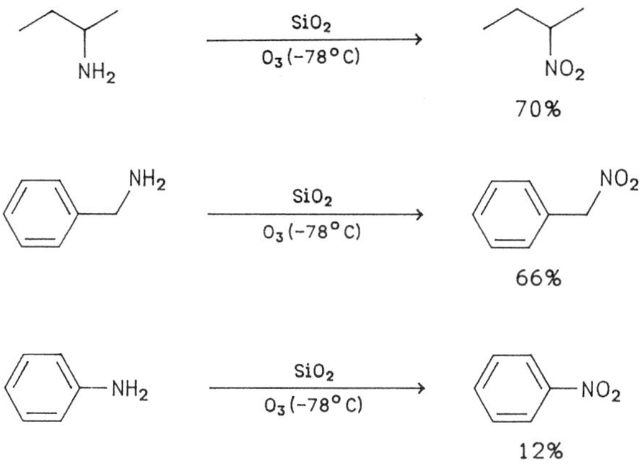

70%

66%

12%

Ozonation of olefins (Den Besten and Kinstle, 1980; Aronovitch *et al.*, 1982) on silica gel in dry media has been studied. In fact, in ozonolysis of aliphatic mono-, di- and trisubstituted olefins, a mixture of ozonides (or polymeric peroxides) and of carbonyl compounds was formed, similar to that obtained in an aprotic solvent (Aronovitch *et al.*, 1982). Tetramethylene, which by ozonolysis in aprotic solvents gives polyperoxides, produces acetone when adsorbed on silica gel. Ozonolysis of phenylethylethylenes in dry media on silica gel takes a different course than the ozonolysis of other olefins, yielding almost exclusively aromatic aldehydes and ketones, certainly with the participation of adsorbed water when untreated silica is used. When acetylenes are ozonized in dry media on silica gel, a mixture of acid and carbonyl compounds forms, diphenylacetylene giving only one product, benzil (Aronovitch *et al.*, 1982). Dry ozonation on silica has been reviewed (Keinan and Varkony, 1983).

B. Oxygen Atoms

Oxygen atoms generated by microwave discharges in O_2 or in CO_2 are efficient oxidants of organic compounds (Zadok *et al.*, 1980; Zadok and Mazur, 1982). Dry-media conditions are interesting because of the high reactivity of the oxygen atom toward conventional solvents. Thus, after having performed synthetic work on neat liquid or solid substrates, Mazur and his group used silica gel as a reaction medium (Zadok *et al.*, 1982,

1983). With discharges in CO_2 or O_2 at temperatures of 0°C or higher, olefins adsorbed on silica gel gave mainly epoxide and carbonyl compounds, as in neat liquid reactions. Below $-60°C$ ozonolysis products were formed; microwave discharge in O_2 at low temperatures is thus an alternative to the ozonation of organic compounds: the advantage of this method lies in the relatively slow formation *in situ* of ozone which may give more selective ozonations.

C. Dioxygen under Light Irradiation

Substituted phenylethylenes adsorbed on silica gel or other inorganic supports (alumina, Florisil) have been oxidatively cleaved to ketones or aldehydes by irradiation with 350-nm light in the presence of molecular oxygen (Aronovitch and Mazur, 1985). The intermediacy of cation radicals, whose formation is initiated by contact charge-transfer interactions between the olefins on the inorganic support and oxygen molecules, has been proposed.

D. Ferric Chloride as Oxidant and as Dehydrating Agent

Ferric chloride hexahydrate, $FeCl_3 \cdot 6H_2O$, dissolved in a polar volatile solvent can be impregnated on silica gel. After evaporation under high vacuum (60°C, 0.1 torr, 3 h), a dry yellowish-brown powder is obtained. This powder is an effective reagent for dehydration, in dry media, of allylic, tertiary, and sterically hindered secondary alcohols (Keinan and Mazur, 1978). The results obtained show a selectivity of this reagent for dehydration of polyhydroxy compounds, a selectivity normally difficult to achieve in solution without special protection of the additional hydroxy functions in the molecule or without recourse to specially designed reagents:

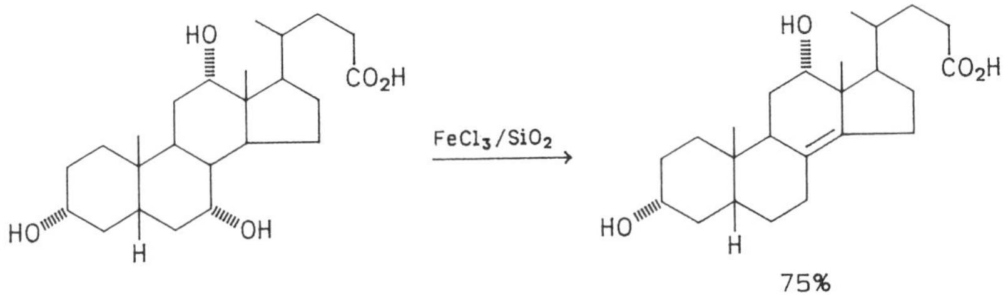

Addition of 2% water to the dry $FeCl_3$–SiO_2 reagent results in a bright yellow powder which epimerizes tertiary alcohols and converts epoxides into 1,2-diols.

The cholestane–diacholestene rearrangement has been achieved by heating 5-cholestene or hydroxy- and halogen-substituted cholestane derivatives at 100°C with silica-bound $FeCl_3$; a 1:1 mixture of 20-epimeric diacholestenes had been obtained (Tal *et al.*, 1981).

The backbone rearrangement of cholestane-5α, 6β-diol occurred under milder conditions and gave rise to diacholesten-6β-ol with the natural R configuration at C_{20}. It is of interest to note that the free OH at C_6 survived the reaction conditions:

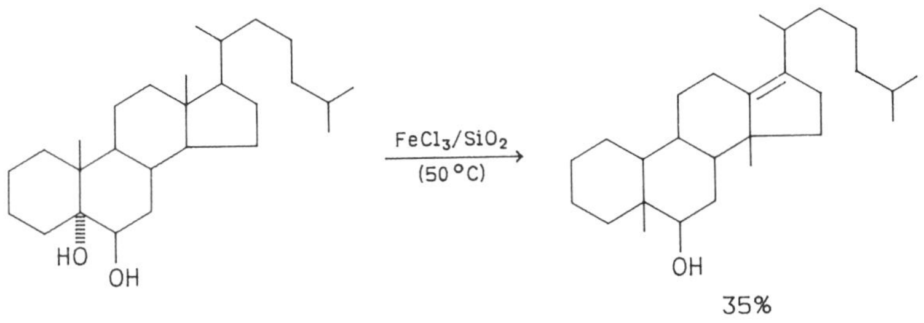

$$\xrightarrow[\text{(50°C)}]{\text{FeCl}_3/\text{SiO}_2}$$

35%

Jempty *et al.* (1981) have shown that $FeCl_3/SiO_2$ reacts as an oxidant or as a Lewis acid with phenol ethers. Thus, dimethoxy aromatics have been inter- or intramoleculary cleaved to phenols. These reactions are very fast *once the solvent is completely removed*: these are therefore truly heterogeneous reactions in dry media and not reactions in which $FeCl_3$ desorbs and reacts in solution:

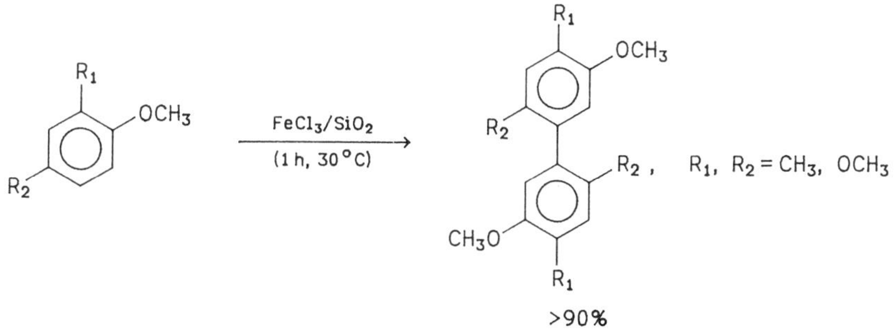

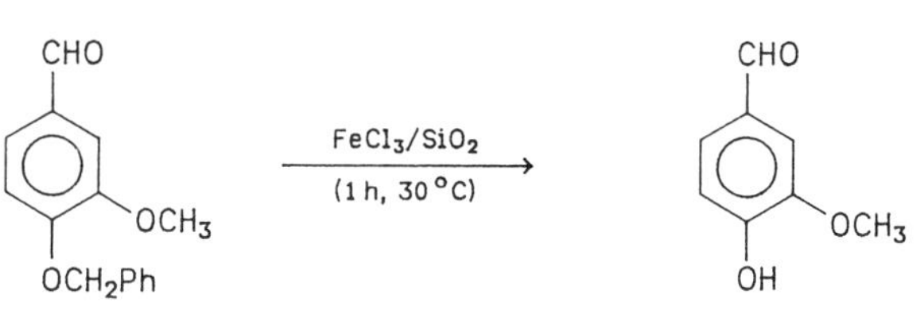

$$\xrightarrow[\text{(1 h, 30°C)}]{\text{FeCl}_3/\text{SiO}_2}$$

R_1, $R_2 = CH_3$, OCH_3

>90%

$$\xrightarrow[\text{(1 h, 30°C)}]{\text{FeCl}_3/\text{SiO}_2}$$

slurry in CH_2Cl_2 6%
dry media 99%

Control experiments have shown that a minimal amount of debenzylation occurred when the substrates were reacted on either finely ground $FeCl_3 \cdot 6\ H_2O$ or on chromatographic-grade silica gel.

Stirring anhydrous $FeCl_3$ with silica gel resulted in a pale yellowish-green powder which is, like Mazur's reagent, effective for dehydration of tertiary alcohols (Fadel and Salaün, 1985a,b).

The same authors also found that this reagent induced, in dry media, $C_4 \rightarrow C_5$ and $C_5 \rightarrow C_6$ ring expansion of tertiary cyclobutanols, cyclization of olefinic alcohols, and cleavage of tetrahydropyranyl ethers:

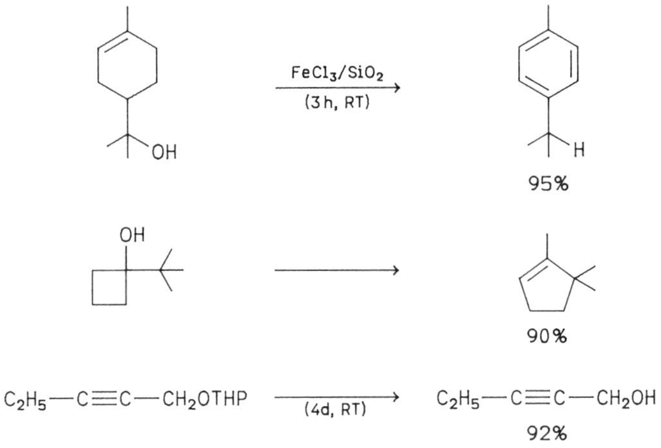

OH
FeCl3/SiO2
(3h, RT)
H
95%
OH
90%
C2H5—C≡C—CH2OTHP
(4d, RT)
C2H5—C≡C—CH2OH
92%

E. The Nef Reaction

Keinan and Mazur (1977b) have reported the use of silica gel in dry-media conditions both as a reaction medium and as a reagent to convert nitro compounds to ketones and aldehydes. This conversion, previously accomplished by acid catalyzed hydrolysis of nitronate salts (the Nef reaction), has been performed with a very simple, one-step procedure. "Basic" silica gel is readily prepared by mixing silica gel with a methanolic solution of sodium methoxide, followed by evaporation to dryness and heating at 400°C. The nitro compound is then embedded into the activated basic silica gel, left at room temperature (RT) for 48 h or heated for 2 h at 80°C, and the resulting carbonyl compound is then eluted with ether:

NO2
MeONa/SiO2
(48h, RT)
O
99%

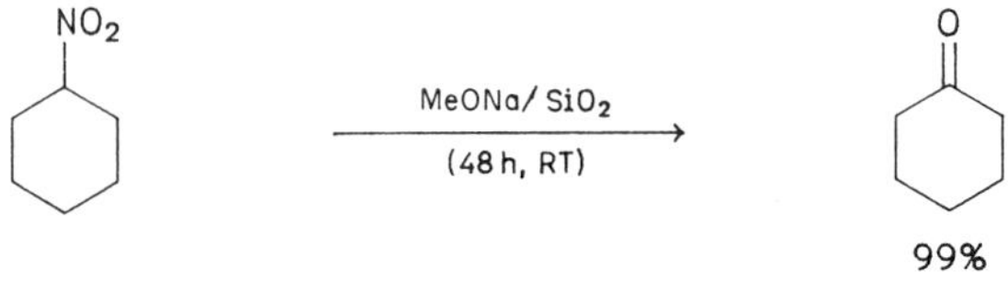

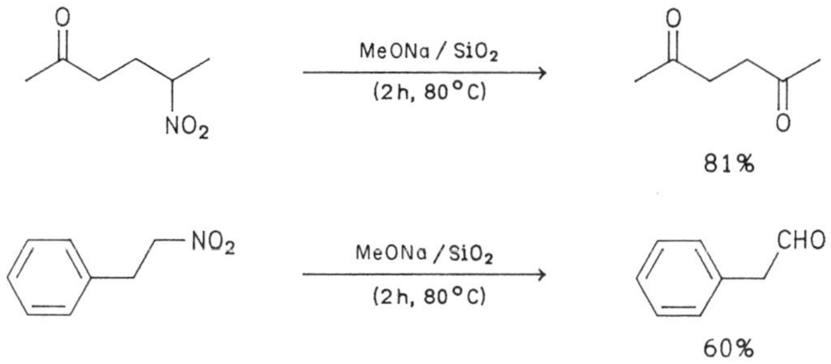

A plausible mechanism for these conversions involves formation of the nitronate anion which reacts with a siloxane function of the silica gel resulting in a mixed anhydride of the nitronic acid and poly(silicic acid). This intermediate is then attacked by a silanoxide anion, resulting in the carbonyl derivative plus nitrogen oxide.

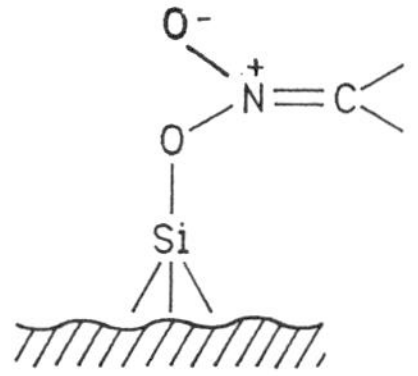

III. REDUCTION REACTIONS

A. Diborane

A combination in sequence of $NaHSO_3$, as a protecting reagent, and of solid inorganic supports effects the selective reduction of 4-acetylbenzaldehyde by B_2H_6 (Chihara et al., 1981b):

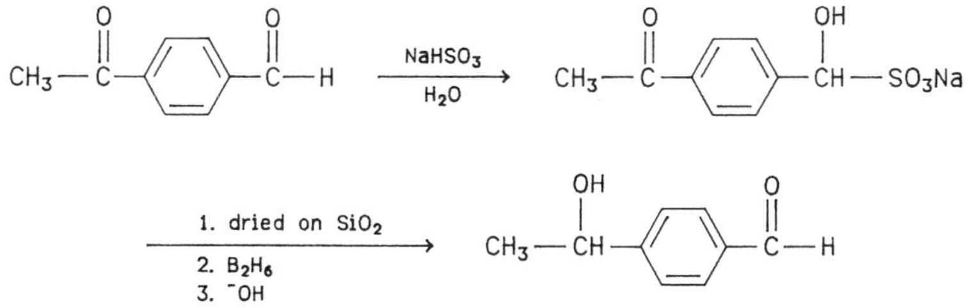

After protection of the formyl group by addition of $NaHSO_3$, the adduct is selectively reduced by diborane on silica to 4-(1-hydroxyethyl) benzaldehyde. The selectivity of reduction at approximately 60% conversion reached 93%. When 4-acetylbenzaldehyde was reduced by B_2H_6 without a protecting group, both in diglyme or on silica gel, practically all of the formyl groups remained intact in both cases. When the adduct was reduced by $NaBH_4$ in aqueous solution, selectivity was low. In place of silica gel, alumina gave reasonably good results.

B. Borohydrides

Fontainebleau sand, a slightly hydroxylated microcrystalline silica, can be used as a support for the $NaBH_4$ reduction of carbonyl compounds in dry media. This reagent is stable and efficient. However, sodium borohydride impregnated on chromatographic silica gel, an amorphous and hydroxylated silica, has no reducing properties. This is due to hydrolysis of $NaBH_4$ catalyzed by the silica gel support producing $NaB(OH)_4$, or it results from formation of borates $NaB(OSi{\equiv})_4$ due to the reaction of BH_4^- with the silanol groups on silica gel (Bram et al., 1981, 1982). This finding contrasts with the results by Hodosan and Ciurdaru (1971) and Ciurdaru and Hodosan (1977) in which $NaBH_4$ impregnated on silica gel effectively reduced ketones and oximes. The discrepancy may stem from differences in the structural properties of the silica gels, especially with respect to the acid properties of silanol groups. On the other hand, NBu_4BH_4 decomposes immediately on Fontainebleau sand, but is stable on silica gel and behaves thus as a very efficient and selective reducing agent. It gives rise mainly to 1-4 reduction derivatives in reductions of α-enones.

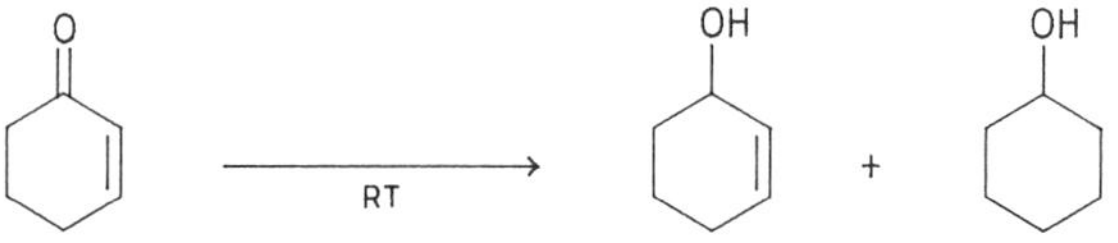

| NaBH₄/Fontainebleau sand | $t_{1/2} < 1$ min | 48% | 52% |
| NB₄BuH₄/silica gel | $t_{1/2} < 1$ min | 6% | 94% |

Alumina is also an effective support for borohydride carbonyl reductions in dry media; but, when stable, borohydrides on silica gel or Fontainebleau sand react faster and with comparable selectivity of reduction. Steroidal ketones are not reduced in dry media; this is related to the small vapor pressures of these compounds (Bram et al., 1982).

IV. ANIONIC CONDENSATIONS

A. Alkylation

The alkylation of some anions such as malonates (Bram and Fillebeen-Khan, 1979), cyanide (Bram et al., 1980), and acetate (Bram and Decodts, 1980) has been performed on silica gel in the dry state. Actually alumina has generally emerged as the most efficient inorganic support for anionic alkylation in dry media conditions (Bram and Fillebeen-Khan, 1979; Bram et al., 1980) or in the presence of an organic solvent (Regen et al., 1979; Yamawaki and Ando, 1980). Very few physico-chemical explanations of the role of the support have so far been offered. In an anionic alkylation, (CH_3CO_2K + n-$C_8H_{17}Br$), silica gel impregnated with a cationic surfactant (cetyltrimethylammonium bromide) appears as effective as alumina (Bram and Decodts, 1980). Furthermore, Spherosil QMA (Rhône–Poulenc), a silica gel bearing ammonium groups covalently bound to the silica, exhibits increased effectiveness. It appears to be due to the change of the negative superficial charge of the silica gel by adsorption or grafting of cationic surfactants moities (Bram and Decodts, 1980).

B. Cycloaddition of β-Dicarbonyl Compounds

Benzofuroxan and 1,3-dicarbonyl compounds react on silica under dry-media conditions giving rise conveniently to 2,3-disubstituted quinoxaline-1,4-bis oxides. The efficacy of the reactions varies considerably with the silica gel used, the more acidic silica gels being less efficient. Attempts to use alumina in place of silica gel led to less satisfactory results, with contamination by side products (Hasegawa and Takabatake, 1985):

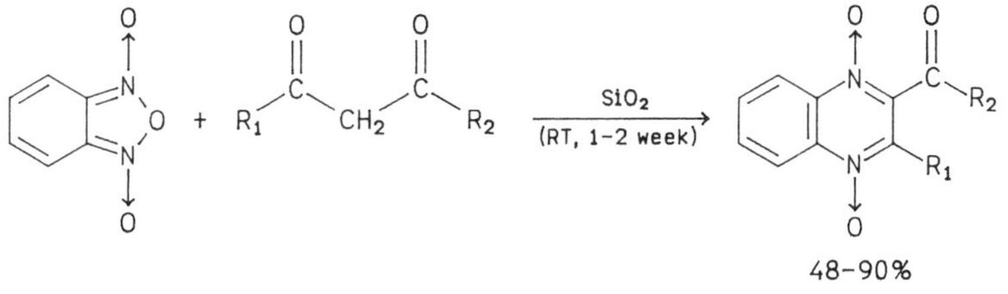

R_1, R_2 = CH_3, OCH_3, C_6H_5, etc.

V. OTHER REACTIONS

A. Acetylation with Ketene

Phenols (Chihara et al., 1981a) and primary, secondary, and tertiary alcohols (Chihara et al., 1982) supported on silica have been acetylated by ketene in good yields at room temperature in dry media:

$$H_2C=C=O + ROH \longrightarrow CH_3CO_2R$$

70–91%

Polyhydroxy compounds such as *trans*-1,4-cyclohexanediol, ethylene glycol, and glycerol were also acetylated. Other inorganic supports (alumina, celite, MgO, ZnO, and TiO_2) are also effective for the acetylation reaction.

B. Halogenation

Only a few halogenation reactions of unsaturated or aromatic systems have been performed on silica under dry-media conditions. Cyclohexene reacting with bromine on silica gel gave only *trans*-1,2-dibromocyclohexane (Stoldt and Turk, 1969). Diethyl fumarate, a deactivated olefin, has been brominated on silica; the reaction is very effective, completely stereospecific, and yields exclusively the trans bromination product, *meso*-diethyl-2,3-dibromosuccinate. A mechanism involving a complex between bromine molecules and silanol O-H groups on the surface as the effective brominating agent has been proposed (Rosen and Eden, 1970).

Toluene, when brominated on silica, gave rise to 37% ortho and 63% para substitution, which is very similar to the result obtained when bromination is performed in 85% CH_3CO_2H (Stoldt and Turk, 1969). Yaroslavsky (1974) has performed a similar reaction on ethylbenzene and cumene. A quantitative yield of aryl-substitued monobromoethylbenzene in which the isomer ratio of $p:o$-bromoethylbenzene was 2.8:1 was obtained. In a similar manner, bromination of cumene yields 88% of monobrominated products; the isomer ratio of $p:o$-bromocumene was 8:1.

Biphenyl,4-bromobiphenyl, and the still less reactive 4-nitrobiphenyl, impregnated on silica, have been efficiently brominated by gaseous Br_2. The $p:o$ isomer ratio for the reaction of 4-nitrobiphenyl is similar to that obtained in solution in presence of $FeCl_3$, regardless of the coverage of the silica surface by the 4-nitrobiphenyl. A control experiment with biphenyl showed no perceptible reaction in carbon tetrachloride in the absence of silica (Rosen and Gandler, 1971):

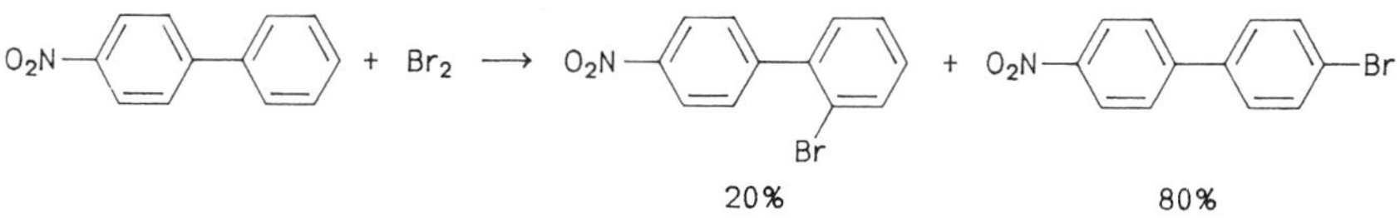

Chlorination of adamantane and other hydrocarbons adsorbed on silica gel with Cl_2 or SO_2Cl_2 has been studied. High yield has been achieved if the chlorination is at a bridgehead carbon. The reaction in dry media is more effective and more selective than in solution in dichloromethane or on silica gel in presence of dichloromethane (Gonzalez et al., 1985).

Chlorination of anisole and chlorobenzene by SO_2Cl_2 is more effective in dry media than in solution. Nitrobenzene could not be chlorinated under these conditions (Hojo and Masuda, 1975).

C. Elimination

In a synthesis of the diterpene cembrene skeleton, two important iso-propenyl intermediates are the sole products from a dehydrochlori-nation performed on silica chromatographic plates:

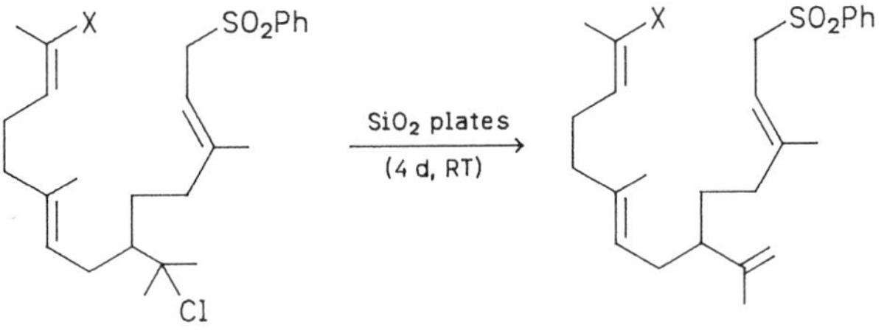

In contrast, dehydrochlorination in DMF at 100°C with $LiBr$–Li_2CO_3 gave a 1:1 mixture of the previous compounds and the isomeric iso-prop-ylidene derivatives (Takayanagi et al., 1978).

D. Preparation of Acylsilanes

Acylsilanes may be prepared in high yield by treating α,α-dibromo-benzylsilanes with silica gel in dry media at 50°C (Degl'Innocenti and Walton, 1980):

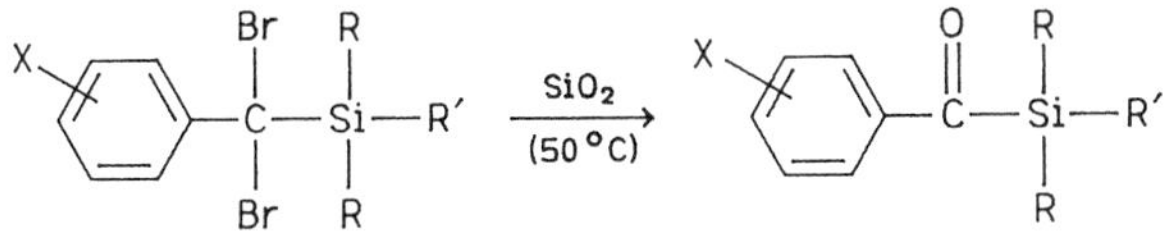

X = H, p-Cl, m-Cl; R, R' = Me, Ph, $mClC_6H_4$

The degree of hydration is important; the rate of product formation decreases markedly if the silica gel is preheated at 200°C for 48 h prior to use. Alumina has also been used as support, but the conversion rate was markedly lower.

E. Self-Condensation of Phenolic Compounds

In order to investigate the reactivity of lignin models on solid supports, Girard *et al.* (1985) have studied the self-condensation of vanillyl alcohol on silica gel under dry media conditions. Dimeric, trimeric, and oligomeric products have been obtained. The intermediacy of quinone methides has been invoked to account for the results, especially for the selectivity of *C*-alkylation on silica as compared to the solution results. This use of silica gel as a support is a valuable new approach into the reactivity of the para hydroxyl group, believed to be of importance both in the formation and the ageing of lignin.

F. Diels–Alder Reactions

A way to catalyze a Diels–Alder reaction is to deposit the reactants into a silica gel surface (Trost *et al.*, 1980). Adsorption of a mixture of 2-methoxy-3-phenylthio-1,3-butadiene and methyl vinyl ketone on silica gel without solvent led to a mixture of the cycloadducts in 42% yield after standing at room temperature for 24 h (Trost *et al.*, 1980):*

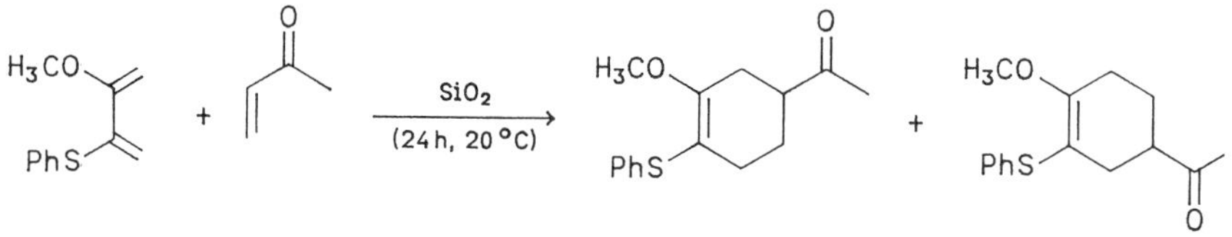

V. CONCLUSION

Different organic reactions of synthetic significance have been efficiently performed under dry-media conditions on silica gel and also on some other supports. Such media offer milder reactions conditions, simpler work up, and often a peculiar or higher selectivity than the same reactions performed in homogeneous solution or on a mineral support in the presence of an organic solvent. Furthermore, the absence of a solvent during the reaction can be of economical interest, especially when simple inexpensive supports such as Fontainebleau sand are used.

An overall interpretation of the precise role of silica gel, in such dry-media reactions has yet to be reached. A difficult and unresolved question is whether a substance adsorbed on an inorganic support is mobile

---

* A fascinating instance of an alumia-promoted Diels–Alder, with very high attendant stereoselectivity, is documented in Chapter 16. —Ed.

and may then react, in the absence of an organic solvent. Adamson and Slawson (1981) have studied the rate of transport of ^{14}C-labeled palmitic acid between silica gel coated chromatographic plates. Rates of transfer of the deposited acid to an adjacent plate were determined, and the temperature dependence was shown to correspond to that for vaporization of palmitic acid. It was then concluded that vapor-phase hopping between particles and within particle pores is fast enough to permit significant molecular mobility and to account for bimolecular reactions taking place on the support. From these results an inference relevant to dry-media reactions can be drawn: if not the only factor of molecular mobility, such evaporative hopping accounts for reactant mobility and for the attendant reactions in dry media.

The fact of the nonreduction of steroidal ketones by NaBH$_4$ on silica in dry media (Bram *et al.*, 1981, 1982) is in agreement with this evaporative hopping concept (Adamson and Slawson, 1981). The ketone should vaporize and then reach the supported NaBH$_4$ provided that the ketone vapor pressure is high enough, which is not the case for steroids.

[This chapter illustrates the strong changes in selectivity—usually, the considerable improvements—when a reaction is run on a solid surface instead of in a homogeneous medium. Our collective task is how to gain an understanding of what goes on in these complex examples of surface chemistry.—ED.]

REFERENCES

Adamson, A. W., and Slawson, V. (1981). *J. Phys. Chem.* **85**, 116–119.
Aronovitch, C., and Mazur, Y. (1985). *J. Org. Chem.* **50**, 149–150.
Aronovitch, C. Tal, D. and Mazur, Y. (1982). *Tetrahedron Lett.* **23**, 3623–3626.
Beckwith, A. L. J., and Thach Duong (1978). *J.C.S. Chem. Commun.* 413–414.
Beckwith, A. L. J., Bodkin, C. L., and Thach Duong (1977a). *Aust. J. Chem.* **30**, 2177–2181.
Beckwith, A. L. J., Bodkin, C. L., and Thach Duong (1977b). *Chem. Lett.* 425–428.
Bram, G., and Decodts, G. (1980). *Tetrahedron Lett.* **21**, 5011–5014.
Bram, G. and Fillebeen-Khan, T. (1979). *J.C.S. Chem. Commun.* 522–523.
Bram, G., Fillebeen-Khan, T., and Geraghty, N. (1980). *Synth. Commun.* **10**, 279–289.
Bram, G., d'Incan, E., and Loupy, A. (1981). *J.C.S. Chem. Commun.* 1066–1067.
Bram, G., d'Incan, E., and Loupy, A. (1982). *Nouv. J. Chim.* **6**, 573–579.
Chihara, T., Teratani, S., and Ogawa, H. (1981a). *J.C.S. Chem. Commun.* 1120.
Chihara, T., Wakabayashi, T., and Taya, K. (1981b). *Chem. Lett.* 1657–1660.
Chihara, T., Takagi, Y., Teratani, S., and Ogawa, H. (1982). *Chem. Lett.* 1451–1452.
Ciurdaru, V., and Hodosan, F. (1977). *Rev. Roum. Chim.* **22**, 1027–1035.
Cohen, Z., and Mazur, Y. (1979). *J. Org. Chem.* **44**, 2318–2320.
Cohen, Z., Keinan, E., Mazur, Y., and Varkony, T. H. (1975). *J. Org. Chem.* **40**, 2141–2142.
Cohen, Z., Varkony, H, Keinan, E., and Mazur, Y. (1980). *Org. Synth.* **59**, 176–182.
Degl'Innocenti, A., and Walton, D. R. M. (1980). *Tetrahedron Lett.* **21**, 3927–3928.

Den Besten, I. E., and Kinstle, T. H. (1980). *J. Am. Chem. Soc.* **102**, 5968–5969.
Fadel, A., and Salaün, J. (1985a). *Tetrahedron* **41**, 413–420.
Fadel, A., and Salaün, J. (1985b). *Tetrahedron* **41**, 1267–1275.
Girard, P., Yianni, P., Desvergne, J. P., Castellan, A., and Bouas-Laurent, H. (1985). *J. Chem. Res. Synop.* 358–359.
Gonzalez, A. G., De La Fuente, G., and Trujillo Vasquez, J. (1985). *Anal. Quim.* **81**, 38–40.
Hasegawa, M., and Takabatake, T. (1985). *Synthesis*, 938.
Hodosan, F., and Ciurdaru, V. (1971). *Tetrahedron Lett.* 1997–1998.
Hojo, M., and Masuda, R. (1975). *Synth. Commun.* **5**, 169–171.
Jempty, T. C., Gogins, K. A. Z., Mazur, Y., and Miller, L. L. (1981). *J. Org. Chem.* **46**, 4545–4551.
Keinan, E., and Mazur, Y. (1977a). *J. Org. Chem.* **42**, 844–847.
Keinan, E., and Mazur, Y. (1977b). *J. Am. Chem. Soc.* **99**, 3861–3862.
Keinan, E., and Mazur, Y. (1978). *J. Org. Chem.* **43**. 1020–1022.
Keinan, E., and Varkony, T. H. (1983). *In* "The Chemistry of Functional Groups, Peroxides" (S. Patai, ed.), pp. 649–683. Wiley, New York.
Mazur, Y. (1975). *Pure Appl. Chem.* **41**, 145–148.
Mazur, Y., Cohen, Z., Keinan, E., and Varkony, T. H. (1975). Isr. Pat. 47,344.
Proksch, E., and de Meijere, A. (1976a). *Angew. Chem. Int. Ed. Engl.* **15**, 761–762.
Proksch, E., and de Meijere, A. (1976b). *Tetrahedron Lett.* 4851–4854.
Regen, S. L., Quici, S., and Liaw, S. J. (1979). *J. Org. Chem.* **44**, 2029–2030.
Rosen, M. J., and Eden, C. (1970). *J. Phys. Chem.* **74**, 2303–2309.
Rosen, M. J., and Gandler, J. (1971). *J. Phys. Chem.* **75**, 887–890.
Stoldt, S. H., and Turk, A. (1969). *J. Org. Chem.* **34**, 2370–2375.
Takayanagi, H., Uyehara, T., and Kato, T. (1978). *J.C.S. Chem. Commun.* 359–360.
Tal, D., Keinan, E., and Mazur, Y. (1979). *J. Am. Chem. Soc.* **101**, 502–503.
Tal, D., Keinan, E., and Mazur, Y. (1981). *Tetrahedron* **37**, 4327–4330.
Trifilieff, E., Bang, L., and Ourisson, G. (1977). *Tetrahedron Lett.* 2991–2994.
Trifilieff, E., Bang, L., Narula, A. S., and Ourisson, G. (1978). *J. Chem. Res.*, pp. 601–637.
Trost, B. M., Vladuchick, W. C., and Bridges, A. J. (1980). *J. Am. Chem. Soc.* **102**, 3554–3572.
Yamawaki, J., and Ando, T. (1980). *Chem. Lett.* 533–536.
Yaroslavsky, C. (1974). *Tetrahedron Lett.* 3395–3396.
Zadok, E., and Mazur, Y. (1982). *J. Org. Chem.* **47**, 2223–2225.
Zadok, E., Amar, D., and Mazur, Y. (1980). *J. Am. Chem. Soc.* **102**, 6369–6370.
Zadok, E., Aronovitch, C., and Mazur, Y. (1982). *Nouv. J. Chem.* **6**, 695–698.
Zadok, E., Rubinraut, S., and Mazur, Y. (1983). *Isr. J. Chem.* **23**, 457–460.

Part VII

Use of Zeolites as Supports

21 USE OF MOLECULAR SIEVES AS SUPPORTS: NOVEL ALUMINOPHOSPHATE-BASED MOLECULAR SIEVES

Brent M. Lok and Stephen T. Wilson

Union Carbide Corporation
Tarrytown, New York 10591

I. INTRODUCTION

Zeolite molecular sieves (Breck, 1974), a class of microporous crystalline aluminosilicate materials, have been extensively studied over the last 30 years (see Csicsery and Laszlo, Chapter 22, this volume). During the first 10 years of development, zeolite molecular sieves with low silica-to-alumina ratios ($SiO_2/Al_2O_3 \leqq 10$) were discovered. The synthesis gels showed very high apparent pH ($\geqq 13$) and were usually crystallized at 100°C or less. Well-known examples are zeolites A, X, Y, L, and synthetic mordenite. This group of materials is characterized by high ion-exchange capacity, extremely hydrophilic surfaces, many acid sites with low to high acid strength, and low to moderate thermal and hydrothermal stability.

Zeolites have been widely used as supports for catalysts containing highly dispersed metals to provide product selectivity and polyfunctional action (Minachev and Isakov, 1976; Jacobs, 1977, 1982; Guisnet and Perot, 1984; Gallezot, 1979; Uytterhoeven, 1978). Among the important commercial applications, hydrocracking and isomerization (selectoform and hysomer processes) are good examples of the use of metal-loaded zeolites

PREPARATIVE CHEMISTRY
USING SUPPORTED REAGENTS

as catalysts (Hansford *et al.*, 1960; Bolton, 1976; Chen *et al.*, 1970; Kouwenhoven and Van Zijll Langhokt, 1971: Simoniak *et al.*, 1973).

In 1967, a second class of zeolites emerged when the first member of the high-silica zeolite ($SiO_2/AlO_3 \cong 20$) family, zeolite β, was synthesized (Wadlinger *et al.*, 1967). Shortly after a number of high-silica zeolites including ZSM-5, 8, and 11, were reported in patent publications (Argauer and Landolt, 1972; Mobil Oil Corp., 1971; Chu, 1973). Additional compositions of this type, commonly referred to as the ZSM-5 family, continue to appear.

In 1977, silicalite, the first member of another family of molecular sieves composed of pure silica, was reported (Grose and Flanigen, 1977; Flanigen *et al.*, 1978). Other important members of the crystalline microporous silica family, silicalite-2 (Bibby *et al.*, 1979), TEA-silicate (Grose and Flanigen, 1978), and holdstite (R. W. Grose and E. M. Flanigen, 1985 unpublished data), have also been discovered. The high silica-to-alumina ratio zeolites and silica polymorphs are somewhat different from the low silica-to-alumina ratio zeolites. They have low-to-nil cation exchange capacity, few acid sites with high acid strength, and high thermal and hydrothermal stability. Some have hydrophobic surfaces.

Substantial synthetic efforts have been devoted to the incorporation of other elements into zeolite frameworks. Among the most promising, iron (Kouwenhoven *et al.*, 1979; Kouwenhoven and Stork, 1980; Rubin *et al.*, 1980; Lok and Messina, 1984; Messina and Lok, 1984), boron (Klotz and Ely, 1981; Hinnenkamp and Walatka, 1981; Marosi *et al.*, 1980a; Tarramasso *et al.*, 1980), germanium (Breck, 1974, p. 321; Barrer, 1982), gallium (Breck, 1974, p. 321; Barrer, 1982), titanium (Young, 1967; Tarramasso *et al.*, 1981; Lok *et al.*, 1985b), and chromium (Klotz, 1981; Marosi; *et al.*, 1980b); have been introduced.

Secondary synthesis is also important for generating new zeolite materials. Both gas-phase treatments and liquid-phase reactions of existing zeolite materials have been used. These processes tend to change the silica-to-alumina ratio of a zeolite and hence to either increase the thermal and hydrothermal stability or alter the acidity. Among the known methods are high-temperature steaming (McDaniel and Maher, 1968a,b), reaction with $SiCl_4$ (Beyer and Belenykaja, 1980) and SiF_4 (Gortsema and Lok 1984), reaction with fluorine gas (Lok and Izod, 1982; Lok *et al.*, 1983), and reaction with $AlCl_3$ (Chang *et al.*, 1984). In the liquid-phase reactions, either hexafluorosilicate (Breck and Skeels, 1985) or hexafluoroaluminate (Chang *et al.*, 1984) is used to insert silicon or aluminum into the zeolite framework accompanied by removal of aluminum or silicon. The secondary synthesis process has provided a practical means of continuously varying the silica-to-alumina ratio.

This chapter reviews the newest framework molecular sieves, specifically the aluminophosphate, silicoaluminophosphate, and metal aluminophosphate families.

II. ALUMINOPHOSPHATE MOLECULAR SIEVES

The aluminophosphate molecular sieves ($AlPO_4$-n) (Wilson *et al.*, 1982a,b, 1983, 1985) represent a new class of microporous inorganic solids. With phosphorus in the 5+ oxidation state, aluminum phosphate possesses many structural similarities to silica: (1) $AlPO_4$ is isoelectronic with Si_2O_4, (2) the average of the ionic radii of Al^{3+} (0.39 Å) and P^{5+} (0.17 Å) is 0.28 A, which is very close to the ionic radius of Si^{4+} (0.26 Å) (Shannon, 1976), and (3) there are isomorphic dense phases of $AlPO_4$ with Al^{3+} alternating with P^{5+} in tetrahedral oxide networks corresponding to several structural forms of SiO_2: α- and β-quartz; α- β-, and γ-tridymite; and α- and β-cristobalite (Dietzel and Poegel, 1953).

In addition, there is an extensive nonmolecular sieve aluminophosphate synthesis literature. Several metastable hydrated networks have been synthesized (D'Yvoire, 1961). For example, two forms of $AlPO_4 \cdot 2H_2O$ are known, metavariscite and variscite (Kniep and Mootz, 1973; Kniep *et al.*, 1977). The partial removal of the waters from metavariscite or variscite is reversible but complete dehydration results in structural collapse. In addition to those hydrated forms of $AlPO_4$, the dense-phase $AlPO_4$ can also be synthesized hydrothermally.

The synthesis of aluminophosphate molecular sieves is carried out hydrothermally from reactive aqueous gels containing alumina, phosphate, and an organic templating agent. The templating agent R is an organic amine or quaternary ammonium hydroxide. The mixture is digested quiescently in a temperature range of 125–250°C. The product compositions are

$$xR \cdot Al_2O_3 \cdot 1.0 \pm 0.2P_2O_5 \cdot yH_2O$$

where the quantities x and y represent the amounts needed to fill the microporous voids within the neutral $AlPO_4$ framework. The voids are freed of both organic and water by calcination at 400–600°C for a period of an hour to a few hours.

The large number of different structure-types, as determined by the x-ray powder diffraction method, made it necessary to assign each new structure-type a number (Table I). Over 20 three-dimensional structures have been observed, of which 13 are microporous. These 13 include structure analogs of two zeolites, erionite ($AlPO_4$-17) and sodalite ($AlPO_4$-20). Some of the novel structures are $AlPO_4$-5,-11,-14,-18,-31, and -33.

TABLE I

Properties of Selected $AlPO_4$ Molecular Sieves[a]

		Adsorption properties[b]			
				Intracrystalline pore volume (cm^3/g)	
	Structure	Pore size (nm)	Ring size[c]	O_2	H_2O
$AlPO_4$-5	Determined, novel	0.8	12	0.18	0.3
$AlPO_4$-11	Determined, novel	0.61	10	0.11	0.16
$AlPO_4$-14	Determined, novel	0.41	8	0.19	0.28
$AlPO_4$-16	Determined, novel	0.3	6	nil	0.3
$AlPO_4$-17	Erionite type	0.46	8	0.20	0.28
$AlPO_4$-18	Novel	0.46	8	0.27	0.35
$AlPO_4$-20	Sodalite type	0.3	6	nil	0.24
$AlPO_4$-31	Novel	0.8	12	0.09	0.17
$AlPO_4$-33	Determined, novel	0.41	8	0.23	0.23

[a] Revised and reprinted with permission from Wilson et al., J. Am. Chem. Soc, **104**, 1146 (1982b). Copyright 1982 American Chemical Society.

[b] Determined by standard McBain–Bakr gravimetric techniques after calcination (500–600°C in air) to remove R; pore size determined from measurements on molecules of varying size (kinetic diameter of Ref. 1); pore volumes near saturation, O_2 at −183°C, H_2O at ambient temperature.

[c] Number of tetrahedral atoms (Al or P) in ring that controls pore size; when structure not known, estimated from adsorption measurements.

The $AlPO_4$-5 structure has been solved by single-crystal methods (Bennett et al., 1983) and contains a unidimensional channel system with channels bounded by 12-membered oxide rings composed of alternating AlO_4 and PO_4 tetrahedra. Among the new $AlPO_4$s six appear to be layered-structure materials.

The three-dimensional $AlPO_4$ molecular sieves show excellent physical properties. Many are thermally stable and resist structure loss at 1000°C. Those studied for hydrothermal stability, including $AlPO_4$-5,-11, and -17, show no structure loss when treated with 16% steam at 600°C. The adsorption properties of selected $AlPO_4$ molecular sieves are summarized in Table I. $AlPO_4$-5 and $AlPO_4$-31 adsorb neopentane and exhibit structures containing twelve-ring pore openings. The ten-ring structure of $AlPO_4$-11 adsorbs cyclohexane but not neopentane. This pore size is similar to that of ZSM-5. Two of the eight-ring structures, $AlPO_4$-17 and $AlPO_4$-18, sorb butane but not isobutane. $AlPO_4$-14 and $AlPO_4$-33 sorb xenon but not butane indicating a somewhat puckered 8-ring structure.

Finally, the six-ring structures of $AlPO_4$-16 and $AlPO_4$-20 adsorb only water.

The aluninophosphate molecular sieves exhibit intracrystalline pore volumes (measured by H_2O at near saturation) from 0.04 to 0.35 cm^3/g and adsorption pore sizes from 0.3 to 0.8 nm, spanning the entire range of pore volumes and pore sizes known in zeolites and silica polymorphs. Nearly all of the materials with pore sizes from 0.4 to 0.8 nm exhibit a characteristic adsorption pore volume for oxygen that is only 50–80% that of water pore volume. This indicates that the framework structures contain some voids accessible only to water.

Compared to the negatively charged framework of zeolite materials, aluminophosphate molecular sieves have neutral frameworks. Therefore, no extra cation is needed to balance charge in the crystal. Cations in zeolites give rise to three major properties: (1) cation exchange capability, (2) hydrophilic surface, and (3) Brønsted acid sites, upon ammonium exchange and calcination. Therefore, aluminophosphate materials exhibit nil cation exchange capability and extremely weak acidity. The neutral aluminophosphate frameworks are moderately hydrophilic, apparently due to the difference in electronegativity between alumium (1.5) and phosphorus (2.1). The absence of the extra framework cation leaves the pore-opening unobscured. In one instance, $AlPO_4$-20, the pore opening appears to be slightly larger than in the zeolite counterpart, sodalite.

The aluminophosphate molecular sieves have been reported to have applications in both adsorption and catalysis areas. Kaiser (1985c) disclosed the catalytic use of $AlPO_4$-17 for the interconversion of propylene to ethylene and butenes, of ethylene to propylene and butenes, and of 1-butene to ethylene and propylene.

Garska and Tellis (1985) showed that the by-product streams from pyrolytic hydrocarbon cracking processes, containing monolefins and diolefins, can be treated to hydrogenate the olefins and aromatize the aliphatics with aluminophosphates as catalysts. The conversion of halogenated hydrocarbon to hydrocarbon was cited by Kaiser (1985b). Rosenfeld and Barthomeuf (1983) demonstrated that ortho -aromatic isomers can be selectively removed from aromatic mixtures by using crystalline aluminophosphates.

III. SILICOALUMINOPHOSPHATE MOLECULAR SIEVES

Silicoaluminophosphate molecular sieves (Lok *et al.*, 1984a,b) are a further extension of the aluminophosphate materials. The main stimulus for adding silica to the aluminophosphate framework is to generate framework charge and catalyst acidity. In addition, silica incorporation into framework sites could provide further structural diversity. Both goals were achieved with great success.

The new family of silicoaluminophosphate materials (SAPO-*n*) exhibits structural diversity with some 13 three-dimensional microporous framework structures reported to date. These include novel structures, SAPO-40, SAPO-41, and SAPO-44; structures topologically related to the zeolites, chabazite (SAPO-34), levynite (SAPO-35), faujasite (SAPO-37), and A (SAPO-42); structures topologically related to the novel $AlPO_4$-*n* structure-types, including SAPO-5, SAPO-11, SAPO-16, and SAPO-31; and topological relatives of structures found in both zeolites and aluminophosphates, erionite (SAPO-17) and sodalite (SAPO-20).

Members of the SAPO molecular sieves are synthesized hydrothermally at 100–200°C from reactive mixtures containing organic amine or quaternary ammonium templates (R). The crystalline materials have a wide compositional range of

$$0\text{–}0.3\,R \cdot (Si_x Al_y P_z)O_2$$

in the anhydrous form; where x, y and z are the mole fractions of silicon, aluminum, and phosphorus and range from 0.01 to 0.98, 0.01 to 0.60, and 0.01 to 0.52, respectively, with $x + y + z = 1$.

Silicoaluminophosphate molecular sieves have tetrahedral oxide frameworks containing silicon, aluminum, and phosphorus. From a structural standpoint we can consider them in terms of silicon substitution into hypothetical aluminophosphate frameworks (Fig. 1). The substitution can occur via (1) silicon substitution for aluminum, (2) silicon substitution for phosphorus, or (3) simultaneous substitution of two silicons for one aluminum

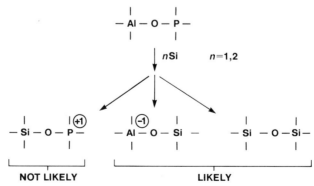

Fig. 1. Effect of silicon incorporation on framework charge of SAPO molecular sieves.

and one phosphorus. The net framework charge per framework silicon atom resulting from each substitution mode would be $+1$, -1, and 0, respectively. Our studies of the SAPO materials indicate that the silicon substitutes via the second and the third mechanisms. Thus, these materials have anionic frameworks like zeolites and should have Brønsted acid sites.

Among the 13 three-dimensional materials (Table II), two have very small pores (six-membered ring), SAPO-16 and SAPO-20, which admit only small molecules such as water and ammonia. The five eight-ring pore opening materials, SAPO-17, SAPO-34, SAPO-35, SAPO-42, and SAPO-44, adsorb n-paraffins but exclude iso-paraffins. SAPO-11, SAPO-31, SAPO-40, and SAPO-41 are intermediate to large in pore size and have pore openings of either ten-rings or puckered twelve-rings. The largest structures include SAPO-5 and SAPO-37, which have known structural analogs with circular twelve-ring pore openings.

The silicoaluminophosphate molecular sieves have intracrystalline pore volumes (measured by water at near saturation) from 0.18 to 0.48 cm^3/g and adsorption pore diameters from 0.3 to 0.8 nm, again spanning the

TABLE II

Adsorptive Properties of Silicoaluminophosphate Molecular Sieves[a]

SAPO species	Pore size[b] (nm)	Ring size[c]	Typical intracrystalline[b] pore volume (cm^3/g)	
			O$_2$	H$_2$O
5	0.8	12	0.23	0.31
11	0.6	10	0.13	0.16
16	0.3	6	(d)	0.3
17	0.43	8	0.25	0.28
20	0.3	6	0	0.24
31	≧0.7	10 or puckered 12	0.13	0.17
34	0.43	8	0.32	0.3
35	0.43	8	0.26	0.3
37	0.8	12	0.35	0.35
40	≧0.7	10 or puckered 12	0.31	0.35
42	0.43	8	(d)	0.3
41	0.6	10 or puckered 12	0.10	0.22
44	0.43	8	0.28	0.34

[a] Partially reprinted with permission from Lok et al., J. Am Chem. Soc. **106**, 6092 (1984b). Copyright 1984 American Chemical Society.

[b] See Table I, footnote b.

[c] See Table I, footnote c.

[d] Insufficient data.

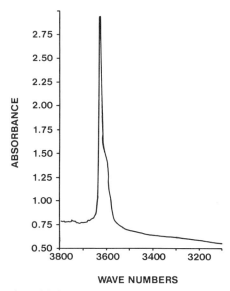

Fig. 2. Hydroxyl region of infrared spectrum of calcined SAPO-34 at 700°C for 1 h.

range of pore volumes and pore sizes known in zeolites and silica polymorphs.

Most three-dimensional silicoaluminophosphates have excellent thermal and hydrothermal stabilities. All remain crystalline after the 400–600°C calcination necessary to remove the organic template and to free the intracrystalline void volume for adsorption or catalysis; some of them even retain their structure up to 1000°C. Hydrothermally, most of the silicoaluminophosphates can withstand a temperature of 600°C and 20% steam without substantial loss of crystallinity. The variable presence of cations and surface hydroxyl groups and the local electronegativity differences between framework Si, Al, and P render the surface moderately to highly hydrophilic.

Hydroxl-region IR studies (R. L. Patton and C. L. Angell, 1985) unpublished data) indicate that some of the calcined SAPO materials have hydroxyl groups resembling those of zeolites. As shown in Figure 2, SAPO-34 has a strong peak around $3610 \, \text{cm}^{-1}$ and a shoulder around $3600 \, \text{cm}^{-1}$. Meanwhile, SAPO-37 (Fig. 3) shows two strong peaks resembling those of Y zeolite. However, the low peak in SAPO-37 occurs at a higher energy than the corresponding peak in Y. Even the $AlPO_4$ analogs such as SAPO-5 and SAPO-11 exhibit hydroxyl groups which are not found in $AlPO_4$ materials (Fig. 4).

The joint presence of a framework with a net negative charge and

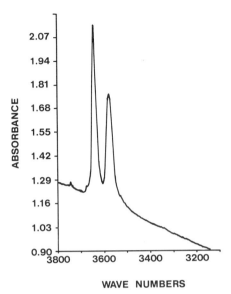

WAVE NUMBERS

Fig. 3. Hydroxyl region of infrared spectrum of calcined SAPO-37 at 700°C for 1 h.

exchangeable cations also provides a means of introducing metal cations into microporous voids. Reduction of these metal cations can introduce highly dispersed metal into SAPO materials, thus, giving them bifunctional (or polyfunctional) catalytic activity in addition to acidity and size selectivity.

The catalytic properties of the silicoaluminophosphate materials as demonstrated by the *n*-butane cracking test (Rastelli *et al.*, 1982) (Table III) are noteworthy. These materials in general are mildly acidic. In some structures, the acidity can be varied substantially by controlling the synthesis conditions. As shown in Table III, the materials are substantially more active than the AlPO$_4$'s but are generally less active than their zeolite analogs.

Although the SAPO materials have been known for only a few years, already many catalytic applications have appeared in the patent literature. Kaiser (1985a) uses SAPO molecular sieves for the production of light olefins from feedstocks comprising methanol, ethanol, dimethyl ether, or mixtures thereof. The conversion of certain hydrocarbons using SAPO catalysts to hydrogenate the olefins and aromatize the aliphatics is cited by Garska and Lok (1985). Small olefin interconversions with SAPO catalysts were discovered by Kaiser (1985c). Long *et al.* (1985a,b) found that the SAPO molecular sieves are not only good catalysts for olefin oligomerization but also can be used as catalysts for conversion of a crude oil feed.

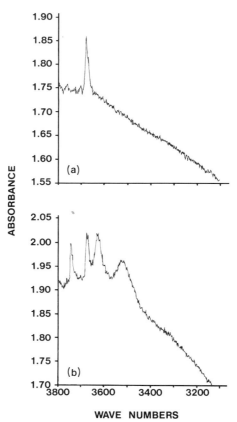

Fig. 4. Comparison of the hydroxyl regions of the infrared spectra of (a) AlPO$_4$-5 at 600°C for 3 h and (b) SAPO-5 at 500° for 1 h.

IV. METAL ALUMINOPHOSPHATE MOLECULAR SIEVES

The third and most recent family of novel molecular sieve materials are the metal aluminophosphates or MeAPO-n (Lok *et al.*, 1985a; Wilson and Flanigen, 1986) where the number n denotes a particular structure-type. The actual values of n are consistent with those used to designate SAPO and AlPO$_4$ molecular sieve structure-types. Thus AlPO$_4$-11, SAPO-11, and MeAPO-11 have the same framework topology. The metal Me can be divalent cobalt (MeAPO species is CoAPO), iron (FAPO), magnesium (MAPO), manganese (MnAPO), zinc (ZAPO), or trivalent iron (FAPO). Metal aluminophosphate molecular sieves have tetrahedral oxide frameworks containing metal, aluminum, and phosphorus and represent the first

TABLE III

n-Butane Cracking Results on
Silicoaluminophosphate
Molecular Sieves[a]

Material tested	k_A[b]
AlPO$_4$-5	~0.05
SAPO-5	0.2–16.1
SAPO-11	0.2–2.0
Erionite[c]	4–5
SAPO-17	0.5
SAPO-31	0.1–0.9
Chabazite[d]	~7
SAPO-34	0.1–3.2
SAPO-35	0.3–1.7
NH$_4$Y[e]	~2
SAPO-37	1.1–1.6
SAPO-40	2.4
SAPO-41	1.3
SAPO-44	1.2–2.4

[a] Reprinted with permission from Lok *et al.*, *J. Am. Chem. Soc.* **106**, 6092 (1984). Copyright 1984 American Chemical Society.

[b] Typical pseudo-first-order rate constant, in (cm^3/min g). Samples were precalcined at 500°C to 600°C for 1–7 h except SAPO-37 and the zeolites which were calcined *in situ*.

[c] Mineral zeolite erionite (Pine Valley, Nevada), NH$_4^+$ exchanged.

[d] Mineral zeolite chabazite (Reese River, Nevada), NH$_4^+$ exchanged.

[e] Synthetic zeolite NaY (SiO$_2$/Al$_2$O$_3$ = 4.8), NH$_4^+$ exchanged.

demonstrated incorporation of these divalent elements into microporous frameworks.

The MeAPO molecular sieves exhibit a wide range of compositions within the general formula,

$$0–0.3 \, R \cdot (Me_x Al_y P_z)O_2$$

where x, y, and z are the mole fractions of Me, Al, and P, respectively, and $x + y + z = 1.00$. The observed range for x is 0.01–0.25, for y, 0.15–0.52, and for z, 0.35–0.60. The R refers to the organic template, an organic amine or quaternary ammonium species. Members of this family are synthesized hydrothermally at 100–200°C from mixtures containing the template and reactive sources of the metal, alumina, and phosphate, such as a metal acetate salt, hydrated alumina, and phosphoric acid. The organic template occluded in the MeAPO as synthesized can be removed thermally at 400–600°C freeing the pore volume for adsorption and catalysis.

The MeAPO structure-types include (1) structures topologically related to zeolites such as MeAPO-34 (chabazite) and MeAPO-35 (levynite), (2) structures topologically related to $AlPO_4$ molecular sieves such as MeAPO-5 and MeAPO-11, and (3) novel structures such as MeAPO-36 (8-Å pore) and MeAPO-39 (4.0-Å pore). In general, any of the five metals Me can be used to synthesize any of the MeAPO structures.

Of the 12 reported structure-types (Table IV) seven have pore openings bounded by eight-rings (effective diameter 4.0–4.5 Å). All of these adsorb oxygen and most of these adsorb butane but all exclude isobutane. Of the remaining structures one (MeAPO-11) has a pore opening consistent with ten-rings and two (MeAPO-5 and MeAPO-36) have large pores (8 Å) controlled by twelve-ring openings. Like their $AlPO_4$ and SAPO counterparts the MeAPO molecular sieves are moderately hydrophilic. The MeAPO molecular sieves have intracrystalline pore volumes of 0.18 to 0.35 cm^3 H_2O/g activated solid, and adsorption pore diameters of 0.3 to

TABLE IV

Selected Properties of MeAPO Molecular Sieves

MeAPO species	Pore size (nm)	Ring size[a]	Typical template
5	0.8	12	N,N-Diethylethanolamine
11	0.61	10 or 12	Di-n-propylamine
14	0.41	8	Isopropylamine
16	0.3	6	Quinuclidine
17	0.46	8	Quinuclidine
20	0.3	6	Me$_4$NOH
34	0.43	8	Et$_4$NOH
35	0.43	8	Quinuclidine
36	0.8	12	Pr$_3$N
39	0.41	8	Di-n-propylamine
44	0.43	8	Cyclohexylamine
47	0.43	8	N,N-Diethylethanolamine

[a] See Table I, footnote b.

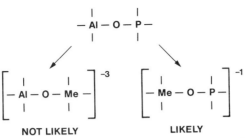

Fig. 5. Effect of divalent metal incorporation on framework charge of MeAPO molecular sieves.

0.8 nm, spanning the range of pore sizes and volumes shown by other framework oxide molecular sieves.

The tetrahedral oxide frameworks of the MeAPO molecular sieves can be envisioned as hypothetical $AlPO_4$ frameworks that have undergone metal substitution. Based on extensive elemental analysis results and other observations, the metal appears to substitute substantially or exclusively for the Al rather than the P (Fig. 5), in contrast to the substitution pattern observed for the SAPOs. This mode of substitution, divalent or trivalent metal for trivalent alumium, results in a net negative or neutral framework charge. A negatively charged framework presents the possibility of Brønsted acid sites and the attendant catalytic activity.

The acid catalytic properties of the MeAPO molecular sieves are demonstrated by their n-butane cracking activity (Table V). The materials

TABLE V

n-Butane Cracking Results on Metal Aluminophosphate
Molecular Sieves

Material	$k_A{}^a$	Material	$k_A{}^a$
AlPO$_4$-5	~0.05	CoAPO-34	5–15
CoAPO-5	0.4	FAPO-34	0.1–0.6
MAPO-5	0.5	MAPO-34	7–29
MnAPO-5	1.2	MnAPO-34	2.5–5.2
SAPO-5	0.2–16	ZAPO-34	13
	11–24	SAPO-34	0.1–7.6
MAPO-36	11	Chabazite	~7
COAPO-36	0.05		
MAPO-39		NH$_4$Y	~2

^{a} Typical pseudo-first-order rate constant in (cm^3/min g).

in general exhibit activities that are both structure and metal dependent. For example, of the two large-pore structures, MeAPO-5 typically has low catalytic activity and MeAPO-36 is typically moderate to high. Among the MeAPO-34s each metal appears to possess its own unique range of activities. The low activity of the MAPO-39 is consistent with the small pore size which excludes n-butane from the intracrystalline voids. This activity, comparable to that of an $AlPO_4$, also indicates that the external surfaces of a MeAPO contribute little to the observed catalytic activity.

The synthesis of the MeAPO molecular sieves is a breakthrough and represents unique crystal chemistry in that (1) divalent cations larger than Al, Si, and P and not normally predisposed to tetrahedral coordination have been incorporated into the lattice and (2) three of these ellements are transition elements of catalytic interest in their own right.

V. SUMMARY

All of the classes of molecular sieves—zeolites, silica polymorphs, $AlPO_4$, SAPO, and MeAPO—have their own interesting and unique properties in catalysis. Zeolites typically exhibit high acidity and are used in the refining industry as cracking, hydrocracking, and isomerization catalysts. The silica polymorphs with essentially no ion-exchange capacity are more hydrophobic than zeolites and offer potential as shape-selective catalyst supports. The $AlPO_4$ molecular sieves also have nil ion-exchange capacity but have hydrophilic surface character and offer a variety of novel structures for use as alternate shape-selective catalyst supports. The low-to-moderate acidity of SAPO materials and the ability to prepare various ion-exchanged forms should provide many opportunities in catalytic reactions where strong acidity is detrimental. Finally, the MeAPO molecular sieves, in addition to offering a wide range of acid strengths and potential ion-exchange capacity, contain catalytically promising metals in framework sites.

ACKNOWLEDGMENT

We acknowledge the major contributions of L. D. Vail, T. R. Cannan, C. A. Messina, R. T. Gajek, and R. L. Patton to the discovery and development of the $AlPO_4$, SAPO, and MeAPO families of materials.

We thank E. M. Flanigen for her leadership in new molecular sieve synthesis, R. L. Patton and C. L. Angell for the hydroxyl-region IR study on SAPO materials, and Union Carbide Corporation for permission to publish this chapter.

REFERENCES

Argauer, R. J., and Landolt, G. R. (1972). U.S. Pat. 3,702,886.

Barrer, R. M. (1982). "Hydrothermal Chemistry of Zeolites," p. 282. Academic Press, New York.

Bennett, J. M., Cohen, J. P., Flanigen, E. M., Pluth, J. J., and Smith, J. V. (1983). "Intrazeolite Chemistry," ACS Symposium Series No. 218, p. 109, Am. Chem. Soc. Washington, D.C.

Beyer, H. K., and Belenykaja, I. (1980). "Catalysis by Zeolites," p. 203. Elsevier, Amsterdam.

Bibby, D. M., Milestone, N. B., and Aldridge, L. P. (1979). *Nature (London)* **280**, 664.

Bolton, A. P. (1976). "Zeolite Chemistry and Catalysis," ACS Monograph No. 171, p. 715. Am. Chem. Soc. Washington, D.C.

Breck, D. W. (1974). Zeolite Molecular Sieves," Wiley, New York.

Breck, D. W., and Skeels, G. W. (1985). Eur. Pat. Appl. 82211; U.S. Pat. 4,503,023.

Chang, C. D., Chu, C. T. W., Miale, J. N., Bridger, R. F., and Calvert, R. B. (1984). *J. Am. Chem. Soc.* **106**, 8143.

Chen, N. Y., Maziuk, J., Schartz, A. B., and Weise, P. B. (1970). *Oil Gas J.* **66**, 154.

Chu, P. (1973). U.S. Pat. 3,709,979.

Dietzel, A., and Poegel, H. J. (1953). *Naturwissenschaften* **40**, 604.

D'Yvoire, F. (1961). *Bull. Soc. Chim. Fr.* p. 1762.

Flanigen, E. M., Bennett, J. M., Grose, R. W., Cohen, J. P., Patton, R. L., Kirchner, R. M., and Smith, J. V. (1978). *Nature (London)*, **271**, 512.

Gallezot, P. (1979). *Catal. Rev.—Sci. Eng.* **20**, 121.

Garska, D. C., and Lok, B. M. (1985). U.S. Pat. 4,499,315.

Garska, D. C., and Tellis, B. (1985). U.S. Pat. 4,499, 316.

Gortsema, F. P., and Lok, B. M. (1984). Eur. Pat. Appl. 100,544.

Grose, R. W., and Flanigen, E. M. (1977). U.S. Pat. 4,061,724.

Grose, R. W., and Flanigen, E. M. (1978). U.S. Pat. 4,104,294.

Guisnet, M., and Perot, C. (1984). "Zeolites: Science and Technology," NATO ASI Science 80, p. 397. Nijhoff, The Hague.

Hansford, R. C., Reeg, C. P., Wood, F. C., and Vaell, R. P. (1960). *Petrol. Eng.* **32**, C7.

Hinnenkamp, J. A., and Walatka, V. V. (1981). U.K. Pat. Appl. 2,062,603.

Jacobs, P. A. (1977). "Carboniogenic Activity of Zeolites," p. 183. Elsevier, Amsterdam.

Jacobs, P. A. (1982). "Metal Microstructures in Zeolites," p. 71. Elsevier, Amsterdam.

Kaiser, S. W. (1985a). U.S. Pat. 4,499,327.

Kaiser, S. W. (1985b). U.S. Pat. 4,524,234.

Kaiser, S. W. (1985c). U.S. Pat. 4,527,001.

Klotz, M. R. (1981). U.S. Pat. 4,299,808.

Klotz, M. R., and Ely, S. R. (1981). U.S. Pat. 4,285,919.

Kniep, R., and Mootz, D. (1973). *Acta Crystallogr.*, *Sect. B* **B29**, 229.

Kniep, R., Mootz, D., and Vegas, A. (1977). *Acta Crystallogr.*, *Sect. B* **B33**, 263.

Kouwenhoven, H. W., and Stork, W. H. J. (1980). U.S. Pat. 4,108,305.

Kouwenhoven, H. W., and Van Zijll Langhokt, W. C. (1971). *Chem. Eng. Prog.* **67**, 65.

Kouwenhoven, H. W., Stork, W. H. J., and Schaper, L. (1979). Br. Pat. Spec. 1,555,928.

Lok, B. M., and Izod, T. P. J. (1982). *Zeolite 2*, 66.

Lok, B. M., and Messina, C. A. (1984). Eur. Pat. Appl. 108271.

Lok, B. M., Gortsema, F. P., Messina, C. A., Rastelli, H., and Izod, T. P. J. (1983). "Intrazeolite Chemistry," ACS Symposium Series No. 218, p. 41. Am. Chem. Soc., Washington, D.C.

Lok, B. M., Messina, C. A., Patton, R. L., Gajek, R. T., Cannan, T. R., and Flanigen, E. M. (1984a). U.S. Pat. 4,440,871.

Lok, B. M., Messina, C. A., Patton, R. L., Gajek, R. T., Cannan, T. R., and Flanigen, E. M. (1984b). *J. Am. Chem. Soc.* **106**, 6092.

Lok, B. M., Messina, C. A., and Flanigen, E. M. (1985a). U.S. Pat. 4,544,143.

Lok, B. M., Marcus, B. K., and Flanigen, E. M. (1985b). U.S. Pat. 4,551,236.

Long, G. N., Pellet, R. J., and Rabo, J. A. (1985a). U.S. Pat. 4,512,875.

Long, G. N., Pellet, R. J., and Rabo, J. A. (1985b). U.S. Pat. 4,528,414.

McDaniel, C. V., and Maher, P. K. (1968a). "Molecular Sieves," p. 186. Soc. Chem. Ind., London.

McDaniel, C. V., and Maher, P. K. (1968b). U.S. Pat. 3,449,070.

Marosi, L., Stabenow, J., and Schwartzmann, M. (1980a). Ger. Pat. 2,909,929.

Marosi, L., Stabenow, J., and Schwartzmann, M. (1980b). Ger. Pat. 2,831,630.

Messina, C. A., and Lok, B. M. (1984). Eur. Pat. Appl. 115,031.

Minachev, K. M., and Isakov, Y. I. (1976). "Zeolite Chemistry and Catalysis," ACS Monograph No. 171, p. 552. Am. Chem. Soc., Washington, D.C.

Mobil Oil Corp. (1971). Neth. Pat. 7,014,807.

Rastelli, H., Lok, B. M., Duisman, J. A., Earls, D. E., and Mullhaupt, J. T. (1982). *Can. J. Chem. Eng.* **60**, 44.

Rosenfeld, D. D., and Barthomeuf, D. M. (1983). U.S. Pat. 4,482,776.

Rubin, M. K., Plank, C. J., and Rosinski, E. J. (1980). Eur. Pat. Appl. 0,013,630.

Shannon, R. D. (1976). *Acta Crystallogr., Sect. A* **A32**, 751.

Simoniak, M. F., Reber, R. A., and Victory, R. M. (1973). *Hydrocarbon Process.* **52**, 101.

Tarramasso, M., Perego, G., and Notari, B. (1980). *Proc. Int. Conf. Zeolites, 5th* p. 40.

Tarramasso, M., Perego, G., and Notari, B. (1981). U.K. Pat. Appl. 2,070,071.

Uytterhoeven, J. B. (1978). *Acta Phys. Chem.* **24**, 53.

Wadlinger, R. L., Kerr, G. T., and Rosinski, E. J. (1967). U.S. Pat. 3,308,069.

Wilson, S. T., and Flanigen, E. M. (1986). U.S. Pat. 4,567,029.

Wilson, S. T., Lok, B. M., and Flanigen, E. M. (1982a). U.S. Pat. 4,440,310.

Wilson, S. T., Lok, B. M., Messina, C. A., Cannan, T. R., and Flanigen, E. M. (1982b). *J. Am. Chem. Soc.* **104**, 1146.

Wilson, S. T., Lok, B. M., Messina, C. A., Cannan, T. R., and Flanigen, E. M. (1983). "Intrazeolite Chemistry," ACS Symposium Series No. 218. p. 79. Am. Chem. Soc., Washington, D.C.

Wilson, S. T., Lok, B. M., Messina, C. A., and Flanigen, E. M. (1985). *Proc. Int. Zeolite Conf., 6th* p. 97.

Wright, L. J., and Milestone, N. B. (1985). Eur. Pat. Appl. 141,662.

Young, D. A. (1967). U.S. Pat. 3,329,480; 3,329,481.

BIBLIOGRAPHY

Bond, G. C., Gelsthorpe, M. R., Sing, K. S. W., and Theocharis, C. R. (1985). *J. Chem. Soc. Chem. Commun.*, pp. 1056–1057.

Derouane, E., and von Ballmoos, R. (1987). U.S. Pats. 4639357, 4639358, and 4647442.

Mitsubishi Chemical Industries Co., Ltd. (1984). Jpn. Pat. Appl. Kokai No. 84217619.

Pyke, D. R., Whitney, P., and Houghton, H. (1985). *Appl. Catal.* **18**, 173–190.

Tapp, N. J., Milestone, N. B., and Wright, L. J. (1985). *J. Chem. Soc. Chem. Commun.*, pp. 1801–1803.

Wright, L. J., and Milestone, N. B. (1985). Eur. Pat. Appl. 141662.

22 SHAPE-SELECTIVE CATALYSIS

Sigmund M. Csicsery

P.O. Box 843
Lafeyette, California 94549

Pierre Laszlo

Institut de Chimie Organique
Université de Liège
and
Ecole Polytechnique
Palaiseau, France

I. INTRODUCTION

Zeolites are porous crystalline aluminosilicates. As they are formed the cavities are filled with ions and water molecules. Both types of inclusions retain enough mobility for ion exchange and reversible dehydration. Zeolites are built from TO_4 modules, corner-linked via the oxygen atoms. In natural zeolites, aluminum and silicon occupy the center of tetrahedra (T-atoms). In synthetic zeolites, other elements such as gallium, germanium, and phosphorus can be incorporated into the framework. About 30 naturally occurring zeolites are known, of which only eight are exploited commercially. They are believed to be detritic minerals, originating in volcanic debris or in deep magma formations subsequently exposed by volcanic extrusion (Haggin, 1982).

The name *zeolite*, i.e., foaming stone, refers to the effervescence displayed by some of these minerals in the presence of water; this was later shown to be due to their molecular sieve properties. Zeolites were discovered a couple of centuries ago in Sweden, but their amazing properties were not used to advantage by chemists until the 1920s. Some natural zeolites include faujasite (isometric) and the two hexagonal structures, gmelinite and mordenite.

Most zeolite catalysts are synthetic because of strict requirements for high purity and controlled composition. They are made by precipitation

419

PREPARATIVE CHEMISTRY
USING SUPPORTED REAGENTS

from supersaturated solutions using appropriate additives to control crystallization. A key aspect in the making of a zeolite is the introduction of a guest ion (organic or inorganic) later exchanged or burnt-off by calcination. Some artificial zeolites have the following compositions: types $A = Na_{12}Al_{12}Si_{12}O_{48} \cdot 27H_2O$; type $X = Na_{86}Al_{86}Si_{106}O_{384} \cdot 264H_2O$; type $Y = Na_{56}Al_{56}Si_{136}O_{384} \cdot 250H_2O$.

Zeolites are molecular sieves. The pores and channels in the zeolite structure can be filled with guest molecules. By contrast with smectites (swelling clays), there is no change in the external dimensions of the crystals. Around 1925, Weigel and Steinhof made the observation that dehydrated zeolite crystals can encapsulate molecules other than water (ethanol, for instance) provided that their size is smaller than the channels and cavities. This idea was enlarged upon by MacBain who introduced, in 1932, the notion of molecular sieve action by zeolites. The work of Barrer, during 1942–1945 made apparent the full potential of zeolites for selective adsorption and separation of gases. Numerous industrial applications were devised during this period, using zeolites with very small pores (3 Å), for separations of gaseous mixtures, the drying of gases and liquids, elimination of sulfur derivatives from petroleum and the separation of air into its constituents (Barthomeuf, 1980).

Zeolites share with clay minerals two properties of interest for heterogeneous catalysis: (1) the presence of exchangable cations, allowing the introduction of catalytic cations and (2) if these are exchanged to H^+ by ammonium ion exchange followed by thermal deammoniation (or, with certain very stable zeolites which can tolerate a low pH, by washing with mineral acid) a very high surface acidity results. In addition, zeolites are precious because of two properties related to pore size: (1) pore diameters less than 10 Å and (2) pores with one or more discreet sizes. The latter two properties are responsible for the molecular sieve properties of zeolites.

Before 1960 silica–alumina and alumina were the most common heterogeneous acid catalysts, as documented elsewhere in this book. The pore distributions in these materials are very broad; pore diameters range from 10 to over 100 Å. With the advent of zeolites as catalysts in 1960 and the attendant much improved control of pore size, shape-selective catalysis became a practical possibility.

Molecular diameters for a number of simple molecules are listed in Table I. Diameters of very small molecules are less than 3 or 4 Å. Normal paraffins are 4.2 Å wide. By contrast, the width of isobutane is 5 Å: the difference between 4.2 and 5 Å is very significant in a number of catalytic applications (see Fig. 1 for the way in which these diameters are measured).

Pore diameters in molecular sieves depend on the number of SiO_4

TABLE I

Kinetic Diameters of Various Molecules from the Lennard–Jones Relationship[a]

Molecule	Kinetic diameter (Å)	Molecule	Kinetic diameter (Å)
He	2.6	C_2H_4	3.9
H_2	2.89	C_3H_8	4.3
O_2	3.46	n-C_4H_{10}	4.3
N_2	3.64	c-C_3H_6	4.23
NO	3.17	i-C_4H_{10}	5.0
CO	3.76	SF_6	5.5
CO_2	3.3	C_5H_{12} (neopentane)	6.2
H_2O	2.65	$(C_4F_9)_3N$	10.2
NH_3	2.6	C_6H_6	5.85
CH_4	3.8	c-C_6H_{12}	6.0
C_2H_2	3.3		

[a] From Breck (1974).

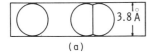

(a)

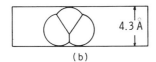

(b)

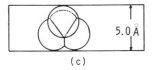

(c)

Fig. 1. Molecular diameters for (a) methane and ethane, (b) propane, and (c) isobutane. [Reproduced from Csicsery (1984), by permission of the publishers, Butterworth & Co. (Publishers) Ltd. ©.]

tetrahedra in a ring (Table II) and on the type of cation present. Small molecules such as oxygen, nitrogen, ammonia, and water can diffuse freely through the channels of practically every type of molecular sieve. Type A sieves have cubic structures whose pores are just about wide enough to let normal paraffins creep through. Cations, however, occupy positions that block part of the pores. Univalent cations (e.g., sodium and potassium) restrict the pore diameter to below ~4 Å. Hence, no organic molecules, with the exception of methane, can enter Na-A or Li-A

TABLE II

Pore Diameters in Zeolites[a]

Number of tetrahedra in ring	Maximum free diameter (Å)	Example
6	2.8	
8	4.3	Erionite (3.6 × 2), phillipsite (4.2 × 4.4), TMA-E(AB) (3.7 × 4.8), chabazite (3.6 × 3.7)
10	6.3	ZSM-5 (5.4 × 5.6), ferrierite (4.3 × 5.5), epistilbite (3.2 × 5.3), stilbite (4.1 × 6.2), diachiardite (3.7 × 6.7)
12	8.0	Mazzite (7.4), Linde type L (7.1), gmelinite (7.0), faujasite (7.4), offretite (6.4), mordenite (6.7 × 7.0)
18	15	Not yet observed

[a] From Meier and Olson (1978); typical hydrocarbon dimensions are benzene = 5.7 × 2.2 Å; n-hexane = 3.5 × 4.2 Å.

zeolites. Bivalent cations, conversely, occupy only every other cationic site, leaving enough space for normal hydrocarbons to pass through. Isobutane is slightly wider than the pores of Ca-A so it cannot enter. However, molecules with nominal dimensions of about 0.5 Å too large can make their way through narrower pores than expected because molecular vibration allows them to wiggle through. In addition, bond cleavage, followed by reformation of the broken bond, could facilitate diffusion of bulkier molecules through narrow channels (Rabo, 1976).

Zeolites are regular, microporous crystals having uniform pore sizes. For instance, in synthetic faujasite, made by combining units called sodalites, the internal BET (Brunauer–Emmett–Teller) surface area is 500–800 m^2/g. There are two kinds of pores, large and small. The entrance ports to the supercages are 12-membered rings about 8 Å in diameter. The smaller pores are about 2.5 Å in diameter (Haggin, 1982).

If most of the catalytic sites occur within this pore structure and if the pores are small, the fate of reactant molecules and the probability of forming product molecules are governed predominantly by molecular dimensions and conformations. Only molecules small enough, with dimensions below a critical size, can enter the pores. Only these molecules will reach internal catalytic sites and react there. Furthermore, only molecules small enough to exit through the available channels will compose the final product. Many reviews have appeared on shape-selective catalysis. (Csi-

Reactant Selectivity

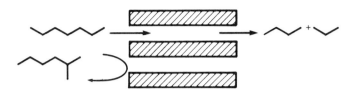

Product Selectivity

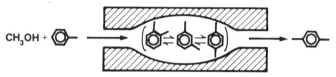

Fig. 2. Reactant selectivity and product selectivity. [Reproduced from Csicsery (1984), by permission of the publishers, Butterworth & Co. (Publishers) Ltd. ©.]

csery, 1976, 1984, 1985; Heinemann, 1981; Whyte and Dalla Betta, 1982; Dwyer and Dyer, 1984). Various types of shape selectivities can occur depending on whether pore size limits the entrance of the reacting molecule, the departure of the product molecule, or the formation of certain transition states.

Reactant selectivity occurs due to sieving by the zeolites of only those reactant molecules small enough to enter the catalyst pores (Fig. 2). Product selectivity arises from the impossibility for too big product molecules to diffuse out of the mineral structure. They are either converted to less bulky molecules, e.g., by equilibration, or eventually deactivate the catalyst by blocking the pores (Fig. 2).

Restricted transition-state selectivity is the name given to the prevention of certain reactions, whenever the attendant transition state would require more space than available in the cavities. Neither reactant nor potential product molecules are prevented from passing through the pores. Reactions with smaller and therefore allowable transition states proceed unhindered (Fig. 3).

The importance of diffusion in shape-selective catalysis cannot be overemphasized. In general (and as a good rule of thumb) one type of molecule will react preferentially and selectively in a shape-selective catalyst if its diffusivity surpasses that of competing molecular species by at least one or two orders of magnitude (Chen et al., 1979; Védrine et al., 1982; Kaeding et al., 1981a,b). Obviously molecules that are too large are unable

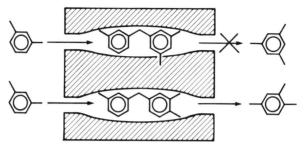

Fig. 3. Restricted transition-state-type selectivity. [Reproduced from Csicsery (1984), by permission of the publishers, Butterworth & Co. (Publishers) Ltd. ©.]

to diffuse through the pores. Even those molecules that react preferentially have much smaller diffusivities in shape-selective catalysts than in large-pore catalysts. For example, normal paraffins have diffusivities at least five orders of magnitude lower in the zeolite KT than in the large-pore Y-type zeolite.

Conformational factors become important when the dimensions of the reacting or diffusing molecule approach those of the pores. Even subtle changes in the dimensions or in the precise local molecular shape of either the reactant or the product can greatly alter the diffusivity (Weisz, 1973). For instance, *trans*-2-butene diffuses in Ca-A zeolite 200 times faster than its configurational isomer *cis*-2-butene (Chen and Weisz, 1967; Olson and Haag, 1984).

Weisz (1980) has pointed out that the selectivity can depend on the diffusivity/acidity ratio. Furthermore, a given hydrocarbon may be endowed with more than one diffusivity in a given zeolite. Fortunately, Yasuda (1982) has evolved a frequency response method for determining diffusion coefficients of small molecules through zeolites. This method resolved the discrepancies that existed between sets of values for diffusivities obtained either by matching experimental sorption curves to solutions of a diffusion equation (Breck, 1974; Eberly, 1976) or by use of the pulsed NMR field gradient technique (Kärger and Karo, 1977; Kärger et al., 1980). The Yasuda method was applied to ethane and propane diffusing through Linde 5A zeolites (Yasuda and Yamamoto, 1985). Whereas atoms of xenon (whose cross section is almost the same as that of *n*-paraffins) have a single diffusion coefficient in the 5A zeolite (Yasuda and Sugasawa, 1984), the diffusion of ethane and propane is dual, and the two diffusivities differ by four orders of magnitude. Based on these results, it would appear that hydrocarbon admolecules are either loosely or tightly bound inside the zeolite cages and show accordingly either a fast or a slow diffusivity (Yasuda and Yamamoto, 1985).

Reactant molecules trapped within the internal network of a zeolite can encounter one another fast in a random walk diffusion, because zeolites have smooth rather than irregular walls. They resemble much more in this respect lamellar inorganic solids such as graphite or clays than macroscopically porous solids such as alumina or silica. This important counterintuitive notion is conveyed by the effective (fractal) dimension D. For a synthetic faujasite, of composition $Na_2O \cdot Al_2O_3 \cdot 2.67SiO_2 \cdot mH_2O$ (Linde Air Products), $D = 2.02$ (Avnir et al., 1984), i.e., very close to a *bona fide* two-dimensional surface. Hence, as explained elsewhere in this book, restriction of the dimensionality of the reaction space to two, instead of three as in a volume, boosts the reaction rates simply because the reactants waste less time encountering one another.

II. DESCRIPTION OF MOLECULAR SIEVES AND OTHER SHAPE-SELECTIVE MATERIALS

Most shape-selective catalysts presently available are zeolites. Some zeolite minerals have been neglected a little in this regard. For instance, because pores of faujasite-type zeolites X and Y are somewhat bigger than the C_4–C_{12} molecules that have attracted most of the interest of the industrial researchers, shape selectivity has been rarely described—even though it may well occur—in these minerals. Mordenite, ZSM-5 (Argauer and Landolt, 1972; Kokotailo et al., 1978a; Rollmann, 1979a), ZSM-11 (Chu, 1973; Kokotailo et al., 1978b), ZSM-21 (Plank et al., 1977a), ZSM-22 (Valyocsik, 1984a), the almost identical Theta-1 (Barri et al., 1984), ZSM-23 (Planck et al., 1978; Valyocsik, 1984b), ZSM-35 (Plank et al., 1977b), ZSM-48 (Chu, 1983), and TMA offretite (Rollmann and Valyocsik, 1983) are shape-selective catalysts. The same is true of many other small- and medium-pore zeolites, whether natural or synthetic (Kibby et al., 1974; Chen et al., 1978a, 1984; Dewing et al., 1985). Of course, the concept has more widespread applicability than to zeolites only. Some of the crystalline aluminophosphates (Wilson et al., 1982, 1983; Flanigan et al., 1985) and silicoaluminophosphates (Lok et al., 1984), some "pillared clays" as well (Kikuchi et al., 1984) can serve as shape-selective catalysts.

Of the synthetic zeolites ZSM-5—a member of the 'pentasil' group of zeolites, so named for the pentagonal elements in its structure (as in that of related materials)—introduced by Mobil, is at present by far the most important because of its stability, catalytic activity, shape selectivity and many useful applications already implemented in industry. The ZSM-5 zeolites can have extremely variable silica/alumina ratios down to almost zero Al content (Dwyer, 1976; Flanigan *et al.*, 1978). For descriptions of the ZSM-5 structure, see Olson *et al.* (1980) and Kerr (1981).

The Si/Al ratio is key to the acid catalytic activity of ZSM-5 as well as of other zeolites. It also governs the acidity of these minerals. The Brønsted acidity stems from the framework Al (see Jacobs, 1982, for a unifying explanation of acidity in zeolites). Accordingly, the number of acid sites is directly proportional to the concentration of framework aluminum. The strength of the acid sites is inversely proportional to the concentration of framework aluminum up to a silica/alumina ratio of about 10 (Derouane and Védrine, 1980; Dejaifve *et al.*, 1980; Ramdas *et al.*, 1981; Nagy *et al.*, 1982; Fyfe *et al.*, 1982; Melchior *et al.*, 1982). Indeed, it is possible to synthesize ZSM-5 with different Si/Al ratios and different Al gradients, resulting in different acid site densities and/or gradients, with a plethora of important consequences for catalytic activity and selectivity (Chen *et al.*, 1978b; von Ballmoos and Meier, 1981; Rollmann, 1978, 1979b). Other examples include dealumination by acid leaching (Dwyer *et al.*, 1982a), by steaming in ammonia (Haag and Lago, 1982, 1983a) with the attendant acidity increase, by steaming without ammonia which also raises the acidity (Dessau, 1984), or with Al halides (Chang *et al.*, 1984a). In ZSM-5 the acid sites are located at channel intersections. A number of other studies have explored various features of the ZSM-5 structure using ion exchange, ammonia desorption, ammonium ion reconstruction, etc. (Haag *et al.*, 1984; Klinowski *et al.*, 1982; Beagley *et al.*, 1984; Jacobs and von Ballmoos, 1982; Topsøe *et al.*, 1981; Mikovsky *et al.*, 1979; Barthomeuf, 1979; Dwyer *et al.*, 1982; Mikovsky and Marshall, 1977).

III. SHAPE SELECTIVITY: REACTANT AND PRODUCT TYPE

Shape selectivity was first described in 1960 (Weisz and Frilette, 1960). P. B. Weisz, N. Y. Chen, V. J. Frilette, and J. N. Miale were not only the pioneers of shape-selective catalysis, but in their subsequent publications they demonstrated its numerous possible applications. Examples include the selective hydrogenation of *n*-alkenes over Ca-A type (Table III) and Pt ZSM-5 (Table IV) molecular sieves (Weisz and Frilette, 1960; Weisz *et al.*, 1962; Dessau, 1982).

In the hydrogenation over calcium-A-type molecular sieve containing platinum (Table III), the linear propene and 1-butene both react. By contrast, isobutene is almost entirely devoid of reactivity.

The ~1% conversion observed with isobutene is most probably due to external sites on the surface of zeolite crystals. Depending on crystallite size, 1–3% of the active sites are located on the outside, available for

TABLE III

Reactant Selectivity over Ca-A Molecular Sieve
Containing 0.31% Pt[a]

Reactant olefin	Hydrogenation conversion[b] (%)
Propene	52
1-Butene	70
Isobutene	<2

[a] From Weisz and Frilette (1960).
[b] at 25°C, 1 atm, H_2/hydrocarbon = 3:1, 0.35 residence time.

TABLE IV

Shape-Selective Hydrogenation Over Pt-ZSM-5[a]

Catalyst	Temperature (°C)	Pt–Al₂O₃	Pt–ZSH-5
Hydrogenated (%)			
Hexene	275	27	90
4,4-Dimethyl-1-hexene	275	35	<1
Styrene	400	57	50
2-Methylstyrene	400	58	<2

[a] From Dessau (1982).

substrate molecules of any size. Furthermore, if 2% of the active sites are superficial, more than 2% of the reaction will originate there; diffusion to such sites is unrestricted. Even small molecules must reach inner sites by slow diffusion, which makes reactions inside the zeolite crystals much slower than on the outside.

Molecular sieve catalysts with incorporated metals are generally made by cation exchange. For example, Pt-HY or Pd-HY are made from solutions of the corresponding tetrammine nitrate complexes and exchange. However, such a direct exchange procedure is not applicable to preparation of the Ca-A molecular sieve; the preceding complexes are much larger than the pores of A zeolites. To bypass this difficulty, Weisz and his co-workers prepared the Ca-A catalyst by first making the Na-A zeolite in a solution of tetrammine platinum nitrate. When the crystals formed, some of the cavities hosted platinum atoms. The initially formed Na form was then calcium-exchanged, with the additional benefit of removing part of the "external" platinum (this is one way to remove 'outside' sites; another way is selective poisoning of the outside sites using very large sulfur-containing molecules that are unable to get inside the crystallites).

Dessau (1982) prepared the Pt-ZSM-5 catalyst by ion-exchanging ZSM-5 with platinum tetrammine chloride, reducing the Pt(II) in H_2, and finally neutralizing all acid with ammonia. With this catalyst, linear olefins were hydrogenated about two orders of magnitude faster than branched ones; a similar selectivity was observed with the styrene/2-methylstyrene system (Table IV).

Somewhat lower selectivity was observed in the oxidation of p-xylene to carbon dioxide over a Cu(II)-exchanged ZSM-5 sieve (Dessau, 1982).

Most applications of shape-selective catalysis involve acid-catalyzed reactions such as isomerization, cracking, or dehydration. The stability and reactivity of carbocationic intermediates is well known. Primary carbon atoms do not form carbocations under ordinary conditions and are thus unreactive. Secondary carbocations are more stable than primary carbocations. Only secondary carbocations can form from normal hydrocarbons. The yet more stable tertiary carbocations are produced from singly branched isoparaffins. Therefore, it was a standard finding that isoparaffins crack and isomerize much faster than normal paraffins. This order is reversed in most shape-selective acid catalysis, with normal paraffins reacting faster than branched ones and being sometimes the only reactive species in mixtures of hydrocarbons (Table VI).

Such a behavior is observed with alcohol dehydration (Table V). Over the small-pore CaA sieve, only the primary alcohol is dehydrated to any significant extent (Weisz and Frilette, 1960; Weisz et al., 1962).

Table VI shows examples of both reactant- and product-type selectivities. Cracking conversions of 3-methylpentane and normal hexane are compared over four catalysts. Virtually no reaction occurs over silica.

TABLE V

Reactant Selectivity in Alcohol Dehydration[a]

Reactant alcohol	Reaction temperature (°C)	Conversion (wt. %)	
		Ca-X (8 Å)	Ca-A (5 Å)
C—C—C—C—OH	260	64	60
C—C—C̣—C (OH on 3rd C)	130	82	0
C—C̣—C—OH (C branch)	260	85	<2

[a] From Weisz et al. (1962).

TABLE VI

Comparison of n-Hexane and 3-Methylpentane Cracking at 500°C[a]

Catalyst	3-Methylpentane cracking conversion (%)	n-Hexane cracking		
		Conversion (%)	iC_4/nC_4	iC_5/nC_5
Silica	<1	1.1		
Amorphous silica-alumina	28	12.2	1.4	10
Linde Na-A	<1	1.4		
Linde Ca-A	<1	9.2	<0.05	<0.05

[a] From Weisz et al. (1962).

However, over the amorphous silica–alumina catalyst, both normal hexane and the branched 3-methylpentane react at significant rates. In Na-A, sodium ions occupy part of the pores, blocking both normal and isoparaffins, and there is no reaction. Furthermore, this sieve has no acid sites. Over Ca-A methylpentane does not react (cannot get in) but normal hexane reacts quite well.

Product selectivity is illustrated here by the iso/normal ratios of the butanes and pentanes formed. Over silica–alumina these ratios for the cracked products are quite high: 1.4 with the butanes and 10 with the pentanes. However, practically no isobutane or isopentane is formed over the shape-selective Ca-A, with its pore opening slightly below 5 Å. Isobutane that forms internally must isomerize first to normal butane before it can diffuse out.

IV. RESTRICTED TRANSITION-STATE-TYPE SELECTIVITY

In restricted transition-state-type selectivity, reactions are prevented because the transition state is too large for the cavities of the molecular sieve, whereas reactants or potential products can diffuse freely through the channels. An interesting example is the acid-catalyzed transalkylation of dialkylbenzenes. This and isomerization are only two of the many possible reactions (Fig. 3) (Csicsery, 1969). In the stepwise isomerization (there are other isomerization pathways) of an ortho dialkyl benzene to a meta and then to a para isomer, a circumambulatory motion of one alkyl group around the ring occurs.

Transalkylation is the intermolecular transfer of an alkyl group from one molecule to another, a bimolecular process involving a diphenylmethane transition state. If m-xylene is the reactant, transalkylation, combined or not with isomerization, can lead *a priori* to 1,2,4- or to 1,3,5-trimethyl-benzene respectively (Fig. 3). Figure 3 clearly shows an akyl group in the transition state for this reaction protruding downward, making this transition state too wide for the pores of mordenite. Thus, whereas the 1,2,4-isomer can form, the 1,3,5-isomer cannot (Csicsery, 1970, 1971).

Mordenite is similar to a bundle of macaroni aligned parallel to each other. It has a one-dimensional pore structure, the diameter of the pores being ~7 Å. Very strongly acidic mordenites can be prepared, and quite large molecules—even symmetrical trialkylbenzenes 7 or 8 Å wide—can pass through the pores. However, if one of them sticks and stops, the whole channel is blocked. Thereby, activities generally decline with time unless hydrogenation activity is present.

When 1-methyl-2-ethylbenzene undergoes acid-catalyzed isomerization and transalkylation, symmetrical trialkylbenzenes (1-ethyl-3,5-dimethyl and 1-methyl-3,5-diethyl) predominate at equilibrium (Csicsery, 1967) (Table VII). Use of an amorphous silica–alumina cracking catalyst and decationized Y-type faujasite leads to qualitatively similar results.

By contrast, Zeolon H-mordenite (with 0.99% Na and a Si/Al ratio of 5) gives product distributions considerably depleted in the symmetrical trialkylbenzenes. Transition-state selectivity is evident from examination of the kinetic course of the reaction. Plots of the amount of symmetrical trialkylbenzene products versus the reaction coordinate, as measured from the extent of methylethylbenzene isomerization, show with alumina–silica and with Y-type faujasite that the former approach equilibrium faster

TABLE VII

Reactions of Trimethylbenzenes at 315°C[a]

	Hemimellitene[b]		Mesitylene[c]
Feed catalyst	H-mordenite	H-Y	H-mordenite
Product composition (mol %)			
C_7	2.3	3.4	2.4
C_8	20.3	25.3	19.4
C_9	56.5	48.4	57.2
C_{10}	20.9	22.8	21.0
Isomer distribution			
1,2,3-Trimethylbenzene	8	8	8
1,2,4-Trimethylbenzene	64	64	60
1,3,5-Trimethylbenzene	28	28	32

[a] From Csicsery (1971). LHSV of 8, H_2/hydrocarbon ratio of 3.

[b] 1,2,3-Trimethylbenzene.

[c] 1,3,5-Trimethylbenzene.

than do methylethylbenzenes. Hence, their concentrations are relatively high even at low (i.e., 11%) methylethylbenzene isomerization levels. Over H-mordenite, methylethylbenzene isomerization is much faster than formation of symmetrical trialkylbenzenes, which are practically absent at low total conversions (Csicsery, 1970, 1971).

Reaction rates of the symmetrical mesitylene and the smaller hemimellitene over mordenite and H-Y are almost identical (Table VII). This shows that symmetrical trialkylbenzenes are themselves not hindered within the pores of H-mordenite. In isomerization the transition state involves only one molecule; so there is enough space to form the transition state in the internal cavity of the sieve.

Another example for transition state selectivity is isobutane isomerization over H-ZSM-5 zeolite. According to Hilaireau et al. (1982), this reaction goes via a bimolecular mechanism over a variety of strongly acidic catalysts including H-mordenite. In every case, isomerization is accompanied by formation of propane and pentanes. In H-ZSM-5, however, there is no room for a bimolecular transition state where it belongs, at the channel intersections where the acidic sites are probably located (Valyon et al., 1979). Indeed the isomerization rate is lower by almost two orders of magnitude on H-ZSM-5 than over H-mordenite, even though the former has stronger acid sites than the latter (Védrine et al., 1979). Pentanes are not formed from isobutane over H-ZSM-5. This, too, suggests that the mechanism with H-ZSM-5 is not bimolecular (Table VIII).

Restricted transition-state-type selectivity can determine the mechanism of cracking in much the same way. Cracking of alkanes over amorphous acid catalysts and large-pore zeolites involves a bimolecular transition state. The rate-determining step consists of a hydride transfer from a neutral molecule to a carbocation. Cracking over H-ZSM-5 can also follow

TABLE VIII

Isobutane Isomeriation with a H_2/Isobutane ratio of 4 at 350°C[a]

	Norton Zeolon 900 H	H-ZSM-5
Si/Al	5	50
Na(wt.%)	0.4	0.01
Rates of formation (10^{-4} mol/h g)		
Propane	41	0.6
Normal butane	24	0.35
Pentanes	11	0

[a] From Hilaireau et al. (1982).

a unimolecular mechanism, according to Haag and Dessau (1984). The transition state for the unimolecular reaction has a smaller steric requirement than that of the bimolecular process. It resembles a pentacoordinate carbonium ion intermediate. Accordingly, the two processes differ markedly with respect to the nature of the resulting products. The predominant low-molecular-weight products from the unimolecular reaction are hydrogen, methane, and ethane. The main end-products of the bimolecular reaction are propene, isobutane, and isobutene.

We can distinguish experimentally between reactant- (and product-) type selectivities and restricted transition state-type selectivity by using particle size effects. In restricted-transition-state-type selectivity, observed rates depend on the intrinsic, uninhibited rate constant. If mass transfer is limiting, though, the rates are determined by the diffusivities of the reactant (or product) molecules and by the catalyst particle size. In other words, whereas reactant and product selectivities are mass-transfer limited and thereby affected by crystallite size, restricted transition-state selectivity is unaffected.

This crystallite-size dependence was used to ascertain the causes of shape selectivity in the cracking of C_6 and C_9 paraffins and olefins over the polyvalent and all important H-ZSM-5 catalyst (Haag *et al.*, 1982a). Crystal diameters were 0.05 and 2.7 μm.

The intrinsic cracking rates of monomethyl hydrocarbons are affected by steric constraints on the transition states (Fig. 4). Cracking of the corresponding olefins does not require such bulky transition states, and

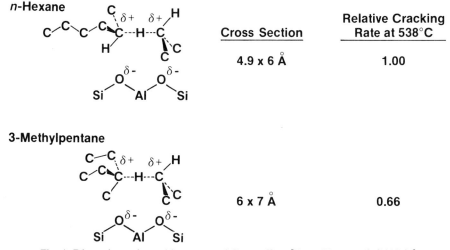

Fig. 4. Dimensions of transition states of C_6 paraffins. [From Haag *et al.* (1982a).]

normal and branched olefins crack indeed with about equal rates. Cracking rates of olefins are two to three orders of magnitude faster than those of the corresponding structurally related alkanes. At these high rates diffusion could be rate limiting even for *n*-hexene above a certain crystallite size. For gem-dimethyl paraffins and olefins, however, diffusivity within the catalyst determines cracking rates; hence, selectivity depends strongly on crystallite size.

Such crystallite size effects allowed Haag *et al.* (1982a) to determine effective diffusivities. This was the first known case in which molecular diffusivities could be obtained in a zeolite under steady-state and actual reaction conditions (Table IX).

Olefins have diffusivities similar to the corresponding paraffins. Diffusivities decrease by four orders of magnitude from normal to *gem*-dimethylalkanes. While branching has a large effect (understandably), lengthening the hydrocarbon chain has a small effect.

One surprising observation is that these diffusivities (Table IX) are about one order of magnitude higher than the calculated Knudsen diffusivities. It may be that once a molecule has entered the narrow zeolite channels, instead of rebounding haphazardly after each collision, it keeps going and maintains part of its forward momentum.

TABLE IX

Diffusivities of Various Olefins and Paraffins in Zeolite ZSM-5 at 538°C[a]

Hydrocarbon	Diffusivity D (cm^2/sec)
C—C—C—C=C	3×10^{-4}
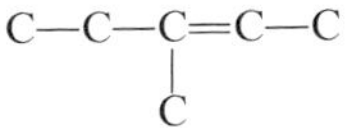 C—C—C=C—C ($\mid$ C)	4×10^{-5}
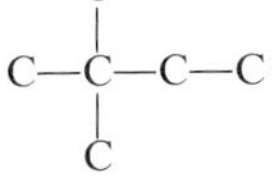 C—C($\mid$ C)—C—C ($\mid$ C)	2×10^{-8}
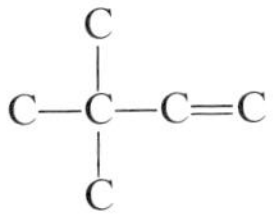 C—C($\mid$ C)—C=C ($\mid$ C)	7×10^{-8}
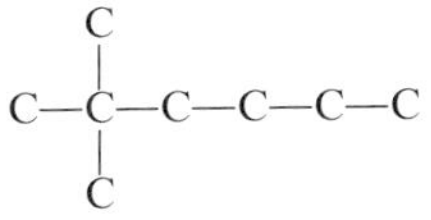 C—C($\mid$ C)—C—C—C—C	3×10^{-8}

[a] From Haag *et al.* (1982a).

An important consequence of the lack of space for the bulky transition state for transalkylation within H-ZSM-5 is that xylene isomerization proceeds without trimethylbenzene formation; this improves xylene yields and increases catalyst life.

The most important example of restricted transition-state selectivity is the near-total absence of coking in ZSM-5-type molecular sieves. This has great significance because certain reactions can occur in the absence of metal hydrogenation components and high hydrogen pressure. Coking is less severe in ZSM-5 than in larger-pore zeolites because the pores lack enough space for the polymerization of coke precursors. Dejaifve and co-workers (1981) have found that coke deposits on the outer surface of the ZSM-5 crystallites; whereas in offretite and mordenite most of the coke forms within the larger pores. Figure 5 shows an oversimplified picture of these phenomena. Activity of the catalyst is barely affected in the former case; it decreases rapidly in the latter. Coke burnoff is also affected.

This point is of great practical import. It gives ZSM-5 the enormous advantage, over competing catalysts, of an increased lifetime (Walsh and Rollmann, 1979; Rollmann and Walsh, 1982; Derouane, 1985).

Low coking is not a feature of hydrocarbon reactions only. Another example is the Beckmann reaction. Thus, when cyclohexanone oxime

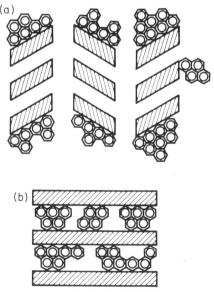

Fig. 5. Coke formation in zeolites: (a) pentasil and (b) mordenite. [From Dejaifve *et al.* (1981).]

rearranges to caprolactam, the ZSM-5 catalyst has much longer lifetime than H-Y (Bell and Chang, 1982). However, the H-Y catalyst shares with ZSM-5 a lower coke buildup and therefore a much slower deactivation than other zeolite catalysts in the *n*-pentane cracking, a process extensively studied using various zeolites as catalysts. The deactivation rate increases with increasing acid-site density, i.e., with increasing Al content (Ione *et al.*, 1984). The cracking of *n*-pentane has also served to establish structure–reactivity and structure–selectivity relationships (Kikuchi *et al.*, 1985a,b). Product selectivity depends strongly on the structure of the zeolite, as does conversion of aromatics to coke (Kikuchi *et al.*, 1985b). Cracking mechanisms differ in medium- and large-pore zeolites, and so do the products (Haag and Dessau, 1984; Kikuchi *et al.*, 1985b).

V. MOLECULAR TRAFFIC CONTROL

Molecular traffic control was an interesting and brilliant idea of Eric Derouane. It is a special type of shape selectivity. It might occur in those zeolites having two or more types of intersecting channel networks. This is the notion that reactant molecules may preferentially enter the catalyst through one of the pore systems while the products exit by the other. This factor would minimize counterdiffusion, and thus it would boost reaction rates. However, this is an hypothesis difficult to prove or to disprove (Derouane and Gabelica, 1980; Derouane, 1980, 1981; Derouane *et al.*, 1981; Pope, 1981).

VI. PARA-SELECTIVE ALKYLATION OF TOLUENE

Product selectivity, restricted transition state-type selectivity, and molecular traffic control may all contribute to several reactions in which *p*-xylene forms above its equilibrium concentrations over ZSM-5 zeolites. Alkylation of toluene by methanol can yield xylenes, and the reaction may be extremely selective in the production of *p*-xylene if ZSM-5 is the catalyst. Selectivities up to 97% have been achieved over ZSM-5 zeolite impregnated with phosphoric acid, boron compounds, or surface-coated with polymers (Chen *et al.*, 1979; Kaeding *et al.*, 1981a). The same systems also catalyze transalkylation to benzene and *p*-xylene molecule (Kaeding *et al.*, 1981b). Selectivity increases with increasing temperature, indicating increasing importance of limitations to diffusion. Deviations from primary distributions are caused (in part) by differences in the diffusivities of three xylenes. In ZSM-5 the diffusivity of *p*-xylene is about 1000 times faster than that of the other two xylenes (Chen and Garwood, 1978a; Chen *et al.*,

TABLE X

Para-Selective Alkylation of Toluene by Methanol over ZSM-5 and Modified
ZSM-5 Zeolite Catalysts[a]

	Catalyst		
	ZSM-5 (3 μm)	ZSM-5 + P	Equilibrium distribution
Temperature (°C)	500	600	
Toluene conversion (%)	39	21	
Xylenes (wt.% of aromatics)	88	90	
p-Xylene	46	97	23
m-Xylene	36	2	51
o-Xylene	18	1	26

[a] From Chen *et al.* (1979).

1979). In unmodified ZSM-5, p-xylene selectivity increases with increasing diffusion limitation. In ZSM-5 crystals smaller than 0.5 μm, the xylenes produced are at equilibrium with each other, producing a mixture containing only 23% p-xylene. When the crystal size reaches 3 μm, the p-xylene content increases to 46%. Impregnation with phosphorus or magnesium or boron compounds apparently increases diffusion limitation yet it greatly reduces the activity (Chen *et al.*, 1979) (Table X). Indeed impregnation of magnesia (Haag and Olson, 1978; Olson and Haag, 1984) or phosphorus (Kaeding and Butter, 1980) on ZSM-5 zeolite increases para-selectivity in xylene production.

Transalkylation into benzene and xylenes is relatively slow within the confines of the pores. Benzene diffuses out of the pores rapidly. The xylenes isomerize rapidly within the pores. *Para*-xylene diffuses out moderately fast. Its ortho and meta isomers diffuse more sluggishly and are converted into more *para*-xylene before their final escape from the channel maze.

The primary product distribution at the catalystic sites is changed by isomerization; selective diffusion favors the species with the highest diffusivity. A mathematical theory explains this enhancement of para-selectivity together with its decrease with increasing total conversion (Wei, 1982); agreement with experimental data is good.

Differences in diffusivity are not the only factor for the high selectivity observed in toluene disproportionation. Restrictions of the xylene-forming transition state may also contribute to the para-selectivity observed (Table X)(Young *et al.*, 1982).

VII. CONTROL OF SHAPE SELECTIVITY

One type of control can be exercised simply by reducing the number of active sites on the outer surface of zeolite crystallites. Surface poisoning by a molecule that is too large to enter the pore inactivates such superficial sites. Both n-hexane and 2,2-dimethylbutane crack over H-ZSM-5. Normal hexane reacts in the pores, 2,2-dimethylbutane reacts mostly on the external surface. If the latter is poisoned with 4-methylquinoline, only the 2,2-dimethylbutane cracking is affected; everything else being equal, conversion drops from 34.5 to 3.6% whereas the n-hexane reaction maintains its high conversion, 95.3% as against 97.6% before poisoning (Anderson et al., 1979).

A second way to control shape selectivity uses cations (Table XI). Over mordenite, xylene transmethylation becomes progressively more selective toward 1,2,4-trimethylbenzene with increasing cation size (Namba et al., 1979). Rapid copper exchange deposits most of the copper near the outside surfaces of the zeolite. Thus, less copper is needed than with slow exchange, which distributes the metal more evenly. Substitution of Na^+ with Cs^+ in X-type zeolite can change its toluene alkylation selectivity. Xylenes are the main products over Na-X; whereas ethylbenzene and styrene form over Cs-X(Freeman and Unland, 1978; Unland, 1978; Sefcik, 1979).

Zinc, a mild hydrogenation component, also increases the selectivity (Chen and Reagan, 1979, 1981). Treatment with antimony also increases

TABLE XI

Cation Control of Pore Dimensions in Mordenite Catalysts and Its Influence on Xylene Transmethylation at 300°C[a]

Cation	Ionic radius (Å)	Conversion (%)	Selectivity[b] (%)
H^+		49	8
Be^{2+}	0.33	29	8
Mg^{2+}	0.62	24	17
Co^{2+}	0.68	23	25
Zn^{2+}	0.74	21	31
Cu^+ (10%; slow exchange)	0.96	41	19
Cu^+ (20–27%; slow exchange)	0.96	26	33
Cu^+ (5.5%; rapid exchange)	0.96	34	39

[a] From Namba et al. (1979).
[b] m-Xylene disproportionation into either toluene + 1,2,4-trimethylbenzene or toluene + mesitylene. 100% 1,2,4-trimethylbenzene = 100% selectivity.

the selectivity (Butter, 1976). As the antimony sesquioxide is reduced during operation it loses its effectiveness; small amounts of hydrogen peroxide prevent such reduction (Chen and Reagan, 1977).

Yet a third group of methods for enhancing selectivity consists of surface treatment of the zeolite catalysts with reactive silanes (Rodewald, 1978a, 1979; Cody, 1984; Niwa et al., 1984). Silane reagent molecules should be larger than the pore entrances so that treatment respects the integrity of the inner active sites and does not reduce interior pore dimensions. The increase in selectivity is due to two factors:

(1) As the silane reacts with surface silanol hydroxyls, it forms a surface layer which extends slightly over part of each pore entrance. As access through this veiled port becomes more restricted, reactant- and product-type selectivites increase. Conversely, restricted transition-state-type selectivity should remain unaffected by such treatment. Note furthermore that the silane reagent is converted to the oxide or the hydroxide after hydration or oxidation. Complete plugging of *some* pores (as opposed to partial plugging of *all* pores) is a better explanation. The effect is similar to that observed in larger crystals: longer diffusion paths, favoring products with higher diffusivities (Olson and Haag, 1984);

(2) Nonselective acid sites on the outside of the zeolite crystallites, such as hydroxyls attached to surface Al atoms, react with the silane compounds and are thus obliterated.

A silane reagent frequently used for this purpose is tetraethyl orthosilicate.

Fourth (but according to a similar principle) decreasing the aluminum content in the last stage of crystallization of ZSM-5 zeolites reduces the number of active sites on the outer surface of crystallites and thus improves shape selectivity (Rollmann, 1979b). Here the outer shell becomes essentially catalytically inactive silica.

In general, the ratio of outer surface catalytic sites to those inside the zeolite mineral is very important to shape selectivity (Chen et al., 1978b; von Ballmoos and Meier, 1981; Rollmann, 1978, 1979b). The catalytic activity of the outer sites becomes significant when the crystallite size falls below a critical size of about 1 μm (Gilson and Derouane, 1984).

VIII. ERIONITE AS AN EXAMPLE OF SELECTIVITY

Offretite and erionite are hexagonal zeolites. The synthetic Linde T consists mostly of offretite, with some erionite intergrowth patterns (Gard and Tait, 1971). Erionite has a hexagonal lattice made of cancrinite cages rotated 60° relative to each other in subsequent layers (ABA formation).

The crystal axes are a and c. Each cancrinite cage adjoins two others through six oxygen atoms, forming hexagonal prisms. The resulting columns form a hexagonal pattern. Each supercage is surrounded by 12 four-membered rings, two planar and three asymmetric six-membered rings, and six eight-membered rings.

No hydrocarbon molecule is able to pass through the four- and six-membered rings. The free diameter of the latter is only about 2.5 Å. This constrains hydrocarbons to diffuse along the c axis of erionite. The slightly asymmetric eight-membered rings line up about 7 Å apart along the a axis. Only straight chain hydrocarbons can pass through these pores.

In offretite the cancrinite cages and the hexagonal rings are all oriented similarly in AAA formation. The supercages open therefore into each other, creating 6.3-Å wide unobstructed channels along the c direction. These channels are wide enough that branched and even hydrocarbons can go through. The pore geometries (Gorring, 1973) of erionite and offretite are compared, along the c axis, in Fig. 6. Erionite—like gmelinite, a member of the same family—can discriminate selectively between normal and isoparaffins. These materials are extremely selective, as shown in their

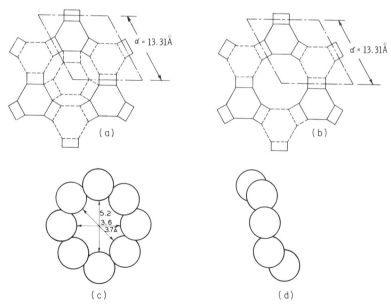

Fig. 6. Projections of (a) erionite and (b) offretite frameworks in a plane parallel to the c-axis, and (c) front and (d) side views of the eight-membered ring of erionite. [From Gorring (1973).]

TABLE XII

Selective Cracking of C_6 Hydrocarbons over H-Gmelinite and H-Erionite
Molecular Sieves[a]

	H-Gmelinite		
	n-Hexane	2-Methylpentane	Methylcyclopentane
Temperature (°C)	370	320–540	510–540
Conversion (%)	47	0–0.7	0.4–1.9

	H-Erionite	
	n-Hexane	2-Methylpentane
Temperature (°C)	320	430
Conversion (%)	52	1

[a] From Miale *et al.* (1966).

reactions with hexane, 2-methylpentane, and methylcyclopentane (Table XII). Methylpentane and methylcyclopentane do not react. Under normal conditions, cracking of n-hexane over nonshape-selective catalysts would be much slower than that of 2-methylpentane or methylcyclopentane. Note the temperatures in Table XII: 320°C for n-hexane and 430°C for 2-methylpentane. The conversion of n-hexane is 52% while only about 1% of the methylpentane reacts, presumably on the outer surface. The same selectivity is obtained with erionite. The otherwise much more reactive 2-methylpentane reacts 50 times slower than normal hexane, although the reaction temperature is lower by 110°C for the latter process!

IX. THE CAGE (OR WINDOW) EFFECT

Reactivities, diffusivities, and most other properties usually change monotonically with the length of the molecule in homologous series. This is no longer true with certain molecules in some molecular sieves. This is referred to as the cage (or window) effect. It has been observed in many zeolites, such as chabazite, levynite, and gmelinite. But the most important studies involved erionite and the closely related KT zeolite (Gorring, 1973; Chen et al., 1969; Chen and Garwood, 1973). The three most significant manifestations of the cage effect in erionite and related zeolites are:

(1) the curve of diffusion coefficient versus chain length shows a minimum at C_8 and a maximum at C_{12} (Fig. 7);

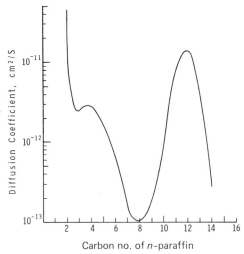

Fig. 7. Diffusion coefficients of normal paraffins in KT at 300°C. [From Gorring (1973).]

(2) hydrocracking rates of *n*-hexane, *n*-decane, and *n*-undecane are much higher than that of *n*-octane (Fig. 8); and

(3) *n*-C$_7$, *n*-C$_8$, and *n*-C$_9$ are missing from the cracked product of longer *n*-paraffins, an example of product selectivity (Fig. 9).

What causes these effects? The paraffin lowest in both diffusion coefficient (Fig. 7) and hydrocracking rate (Fig. 8) is *n*-octane. With a length of 12.82 Å it can fit almost exactly into the 13-Å cavity of erionite. Molecules hosted snugly in such tight-fitting cages have low mobility.

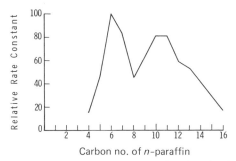

Fig. 8. Hydrocracking rates of normal paraffins over dual-functional erionite. [From Chen and Garwood (1973).]

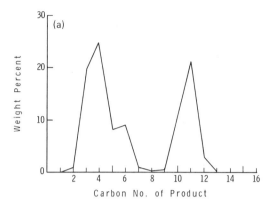

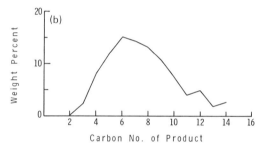

Fig. 9. Product distributions in the cracking of normal docosane: (a) erionite (340°C) and (b) rare earth X fanjasite (316°C). [From Chem *et al.* (1969).]

Molecules with hydrocarbon chains longer than *n*-octane will be unable to fit completely in a single cage, part of each extends into the next cage; their diffusivities are thus enhanced relative to smaller molecules. *n*-Dodecane, 17.85 Å long, extends to the extremes of a unit cell in the *a* direction. It is held by three eight-membered rings, which ensures maximum mobility. Reaction rates of *n*-paraffins in zeolites exhibiting the cage effect vary with chain length in parallel with their diffusivities.

In the third example, shown as Fig. 9, *n*-heptane, *n*-octane, and *n*-nonane are absent from the cracking products of very long paraffins. This product selectivity occurs because, once these products form inside the cavities, they are trapped there until the acid sites present crack them further into butanes. Accordingly, Fig. 9 shows a maximum at butane. The *n*-octane molecule lingers 100 times longer than any other hydrocarbon molecule inside a cavity. Such a residence time is sufficient to crack all of it into smaller fragments. *n*-Dodecane, on the other hand, can diffuse out of the microreactor fast and escapes unscathed.

X. QUANTITATIVE MEASURE OF SHAPE SELECTIVITY

A quantitative measure of shape selectivity—the constraint index—compares the cracking rates of *n*-hexane and 3-methylpentane (Young, 1976; Frilette *et al.*, 1981). Silica–alumina has a constraint index of 0.6. This ratio represents the intrinsic cracking rates of normal alkanes and isoalkanes. Mordenite and rare earth Y are similarly unselective. Erionite has a very high constraint index and so do various ZSM zeolite catalysts. The cyclodecane test (Jacobs and Tielen, 1984) and the H/D exchange test on cyclopentane and methylcyclopentane (Poole and Whan, 1984) are two ways in which to quantify shape selectivity. However, many other methods have been applied or devised for this purpose (Table XIII).

XI. SELECTED EXAMPLES OF SHAPE-SELECTIVE CATALYSIS

We have already touched in this chapter on the isomerization of alkanes (Weitkamp, 1980; Weitkamp *et al.*, 1983). In large-pore zeolites alkane isomerization proceeds through classical hydride and alkyl shifts. In medium-pore zeolites, which do not offer enough space for the transition state necessary for the hydride or alkyl shifts, alkane isomerization may involve protonated intermediates (Weitkamp, 1980). This provides a very important example of restricted transition-state-type selectivity.

TABLE XIII

Some of the Tests Developed to Measure Activities, Selectivities and Other Properties of Zeolite Catalysts

Test	References
n-Pentane cracking	Kikuchi *et al*, (1985a,b)
n-Decane cracking	Martens *et al.* (1984)
Cyclodecane cracking	Jacobs and Tielen (1984)
H/D exchange of cyclopentane and methylcyclopentane	Poole and Whan (1984)
Alkylbenzene transalkylation	Csicsery (1969, 1970, 1971); Jacobs *et al.* (1982); Karge *et al.* (1983)
Sorption	Harrison *et al.* (1983)
Deuterium NMR	Eckman and Vega (1983)
Magic angle spinning NMR	Thomas *et al.* (1982)
Fast atom bombardment MS	Dwyer *et al.* (1982a)
High-resolution electron microscopy	Thomas *et al.* (1982)

We have also described shape selective catalysis for disproportionation of xylene and dialkylbenzenes via transalkylation. Besides work already referred to (Csicsery, 1969, 1970, 1971) the influence of crystallite size has been studied for toluene and xylene disproportionation over H-ZSM-5 zeolites (Beltrame *et al.*, 1985). Propene oligomerization and disproportionation, using ZSM-5 catalysis, has been reported (Haag, 1983).

Alkylation of hydrocarbons is a commercially important process that has also greatly benefitted from zeolite catalysts (Weitkamp, 1985). Specific instances include alkylation of benzene and toluene with ethylene. In the latter case the predominant product, due to shape selectivity, is *p*-ethyltoluene (Keown, 1973; Young, 1976; Lewis and Dwyer, 1977; Dwyer, 1981; Chen, 1977, 1978; Rodewald, 1978b; Kaeding *et al.*, 1982). Benzene has also been alkylated with long chain olefins to form 2-phenylalkanes (Young, 1981a). Likewise, phenol was alkylated with olefins to para-alkylphenols (Young, 1981b,c), and xylene was selectively alkylated with methanol to 1,2,4-trimethylbenzene (Namba *et al.*, 1983).

Hydrogenation is performed with rhodium-loaded ZSM-11 and other zeolites. Hydration of the zeolite has pronounced influence on the activity and on the selectivity of the catalyst, presumably because the intersticial water molecules partly occlude and thus narrow the channels of the sieve (Corbin *et al.*, 1985).

Zeolites with a Si/Al ratio greater than 12 and a constraint index in the 1–12 range catalyze the aromatization of propane, butane, and higher paraffins (British Petroleum's Cyclar-process; Davies and Kolombos, 1979a,b; Dave *et al.*, 1982; Johnson *et al.*, 1984; Haag and Lago, 1983b). This is dehydrocyclodimerization (Csicsery, 1979) in which at least two smaller, i.e., less than six carbon atoms, alkane molecules condense into a aromatic molecule.

Hydrocracking and hydroisomerization of various hydrocarbons are also done with zeolite catalysis (Ernst and Weitkamp, 1985; Giannetto *et al.*, 1985; Martens *et al.*, 1985). Of course zeolite catalysis is not limited to hydrocarbons, even though it has been developed very much by and for the petroleum industry. A few reactions on molecules with heteroatoms have been explored. We have cited already the Beckmann rearrangement (Bell and Chang, 1982). Aniline, 2-picoline, and other aromatic amines can also be prepared by various processes with zeolite catalysis (Chang and Lang, 1983, 1984; Chang and Perkins, 1983).

Considerable promise is found in combining zeolite sieve action with catalysis by transition metal complexes. Inclusion of the latter in the cavities of the former may increase catalytic selectivity, lifetimes, etc. There has already been one pioneering study of the value as an oxidation catalyst of iron phthalocyanin encapsulated in three different zeolite systems (Herron *et al.*, 1984). Finally, selective Fischer–Tropsch synthesis of hydrocarbons has been achieved over ruthenium Y zeolites (Nijs *et al.*, 1980).

XII. APPLICATIONS OF SHAPE-SELECTIVE MOLECULAR SIEVE CATALYSIS

The most important commercial applications of shape-selective catalysis are xylene isomerization to produce *p*-xylene, toluene disproportionation, distillate and lube-oil dewaxing, ethylbenzene production from benzene by alkylation with ethylene, and the methanol-to-gasoline proceess. Other commercially important processes are Selectoforming and M-forming.

The objective of xylene isomerization is to produce *p*-xylene, an intermediate for terephthalic acid, which in turn is used to manufacture polyesters. We have already discussed several aspects of this reaction in this Chapter. Four versions of the process are licensed by Mobil Oil Corporation: the vapor-phase, liquid-phase, high-temperature, and high-severity processes. Combined, they account for about one-half of the free world's *p*-xylene production. The catalyst is most probably HZSM-5 or modified HZSM-5.

Mobil Oil Corporation's distillate and lube-oil dewaxing processes probably also use HZSM-5 or modified HZSM-5 catalysts, whereas the catalyst in the similar British Petroleum process is probably mordernite. These processes allow the cracking of normal paraffins (and some of the monomethyl paraffins) to valuable propane and butane without appreciable cracking of branched and cyclic molecules (Weisz, 1980; Chen *et al.*, 1977; Chen and Garwood, 1972, 1978b; Garwood *et al.*, 1983; Bennett *et al.*, 1975; Donaldson and Pout, 1972; see also Cody, 1983, for an Exxon process). Other Mobil processes in which undesirable impurities were removed by cracking to easily removable molecules are Selectoforming and M-forming. These increase gasoline octane numbers by selectively cracking normal paraffins which have low-octane numbers (Chen *et al.*, 1968; Anonymous, 1970; Bonacci and Patterson, 1981). Both processes were operated in combination with reforming. The catalyst in Selectoforming was erionite, and in M-Forming it was ZSM-5. M-Forming was more sophisticated and complex than Selectoforming. The olefins produced in the cracking of the low-octane *n*-paraffins were made to alkylate the benzene and toluene present in the gasoline *in situ* to alkylbenzenes. M-Forming offers three advantages: (1) a high octane gasoline, (2) a low-benzene, and therefore environmentally more acceptable product, and (3) high gasoline yield. The use of ZSM-5 together with reforming catalysts could also improve gasoline octane numbers (Plank *et al.*, 1977a,b, 1979, 1981).

Ethylbenzene is the source material for styrene which in turn is polymerized to polystyrene and styrene copolymers. It is produced by the alkylation of benzene with ethylene. The Badger–Mobil process uses ZSM-5 catalyst. The process works with both dilute and polymer-grade ethylene feeds. The zeolite catalyst is environmentally inert whereas older processes use Friedel–Crafts catalysts such as aluminium chloride with the associated disposal problems. In addition, the Badger–Mobil process is much more economical than the older processes. About one-fourth of the world's ethylbenzene is produced by the Badger–Mobil process (Fagan and Weisz, 1983). The alkylation of toluene to *p*-ethyltoluene is similar to the ethylbenzene process. Mobil scientists also found a way to combine ZSM-5 with FCC catalysts to improve gasoline octane numbers in catalytic cracking (Rosinski and Schwartz, 1982).

A widely acclaimed application of zeolite catalysts is the Mobil process for conversion of methanol into gasoline. A fixed-bed plant has been constructed jointly by the government of New Zealand and Mobil Oil Corporation. The plant went on stream in November, 1985. It produces 14,000 barrels of gasoline per day. This satisfies about one-third of New Zealand's gasoline demand. In February, 1986 the plant was running at 105% design capacity. It will be interesting to watch the economics of this operation; decided at the height of the oil crisis, will it prove worthwhile, even for oilless countries such as New Zeland, when oil prices seem to have retreated durably to below about $18/ barrel of crude? This outstanding application of ZSM-5 has been reviewed by Chang (1983). Other references on this topic include Chang *et al.* (1984b), Chang and Silvestri (1977), Dessau and LaPierre (1982), Ono and Moari (1981), Ceckiewicz (1981), Zatorski *et al.* (1985), Kljueva *et al.* (1985), and Derewinski *et al.* (1985). Ceckiewicz (1981) discusses the reaction over another zeolite, zeolite T.

There are still some disagreements about the first step of the conversion of methanol (and/or dimethyl ether) into hydrocarbons. Is the primary product ethylene or propene? There appears to be a majority opinion in favor of the former (Haag *et al.*, 1982b; Chu and Chang, 1984; Wu and Kaeding, 1984; Espinoza and Mandersloot, 1984). This initial ethylene is further transformed to longer aliphatic alkanes which later cyclize, aromatize, get alkylated by shorter olefinic intermediates (including ethylene and propene), undergo the paring (Sullivan *et al.*, 1961) reaction in reverse, and so on. When ZSM-5 is used as catalyst, the reaction stops at durene because there is no space within the zeolite cavities to form bigger polyalkylaromatics. These higher-order reactions may proceed over other catalysts. Again, the marked advantage of ZSM-5 is that it does not undergo coking, by contrast to other catalysts. Therefore, its lifetime is significantly longer.

Over ZSM-5 zeolites with low acidity the methanol reaction may be stopped before the C_2–C_4 olefinic intermediates are converted to higher molecular weight paraffins and aromatics (Chang and Silvestri, 1977). The modified process could produce light olefins.

XIII. CONCLUSION

The results obtained so far in 25 years of research on zeolite catalysis are extremely impressive. Numerous industrial processes of key economic importance have been radically transformed, mostly by the attendant improvements in selectivity. The basic concept has remained of utmost simplicity: sieve action at the molecular level favors the reactants, products, or transition states most desired.

We venture to predict that zeolites will now see increasing use in the preparation of fine chemicals. Probably, in this more complex and sophisticated field, zeolites will have to be designed with external and internal modifications such as neutralization of superficial active sites, encapsulation of transition metal complexes, inclusion of organophilic cations providing low polarity environments, and association of zeolite crystallites with domains of other minerals, such as pillared clays, providing complementary catalytic action. Future prospects appear to be limited only by the imagination of the chemists. In very many astute ways, the ability of zeolites to deliver into a microreactor (i.e., internal matrix of the zeolite) molecules selected by their shape and size will result in selective and specific formation of target products.

REFERENCES

Anderson, J. R., Foger, K., Mole, T., Rajadhyaksha, R. A., and Sanders. J. V. (1979). *J. Catal.* **58**, 114.

Anonymous (1970). *Hydrocarbon Process.* Sept., 192.

Argauer, R. J., and Landolt, G. R. (1972). U.S. Pat. 3,702,886.

Avnir, D., Farin, D., and Pfeifer, P. (1984). *Nature (London)* **308**, 261.

Barri, S. A. I., Smith, G. W., White, D., and Young, D. (1984). *Nature (London)* **312**, 533.

Barthomeuf, D. (1979). *J. Phys. Chem.* **83**, 249.

Barthomeuf, D. (1980). *Recherche* **11**, 908.

Beagley, B., Dwyer, J., Fitch, F. R., Mann, R., and Walters, J. (1984). *J. Phys. Chem.* **88** 1744.

Bell, W. K., and Chang, C. D. (1982). U.S. Pat. 4, 359,421.

Beltrame, P., Beltrame, P. L., Carniti, P., Forni, L., and Zuretti, G. (1985). *Proc. Int. Sym. Zeolite Catal., Siofok, Hung.*

Bennett, R. N., Elkes, G. J., and Wamless, G. J. (1975). *Oil Gas J.* Jan. 6, 69.

Bonnaci, J. C., and Patterson, R. (1981). U.S. Pat. 4,292,167.

Breck, D. W. (1974). "Zeolite Molecular Sieves", p. 637. Wiley (Interscience), New York.

Butter, S. A. (1976). U.S. Pat. 3,979,472.

Ceckiewicz, S. (1981). *J.C.S. Faraday I* **77**, 269.

Chang, C. D., (1983). Catal. Rev.—Sci. Eng. **25**, 1.

Chang, C. D., and Lang, W. H. (1983). U.S. Pat. 4,380,669.

Chang, C. D., and Lang, W. H. (1984). U.S. Pat. 4,434, 299.

Chang, C. D., and Perkins, P. D. (1983). U.S. Pat. 4,395,554.

Chang, C.D., and Silvestri, A. J. (1977) *J. Catal.* **47**, 249.

Chang, C. D., Chu, C. T. W., Miale, J. N., Bridger, R. F., and Calvert, R. B. (1984a). *J. Am. Chem. Soc.* **106**, 8143.

Chang, C. D., Chu, C. T.-W., and Socha, R. F. (1984b). *J. Catal.* **86**, 289.

Chen, N. Y. (1977). U.S. Pat. 4,002,697.

Chen, N. Y. (1978). U.S. Pat. 4,100,215.

Chen, N. Y., and Garwood, W. E. (1972). U.S. Pat. 3,700,585.

Chen, N. Y., and Garwood, W. E. (1973). *Adv. Chem. Ser.* No. 121, 575.

Chen. N. Y., and Garwood, W. E. (1978a). *J. Catal.* **52**, 453.

Chen, N. Y., and Garwood, W. E. (1978b). *Ind. Eng. Chem., Process Des. Dev.* **17**, 513.

Chen, N. Y., and Reagan, W. J. (1977). U.S. Pat. 4,049,735.

Chen, N. Y., and Reagan, W. J. (1979). U.S. Pat. 4,148,835.

Chen, N. Y., and Reagan, W. J. (1981). U.S. Pat. 4,278,565.

Chen, N. Y. and Weisz, P. B. (1967). *Chem. Eng. Prog.* **63**, 86.

Chen, N. Y., Maziuk, J., Schwartz, A. B., and Weisz, P. B. (1968). *Oil Gas J.* **66** (47), 154.

Chen, N. Y., Lucki, S. J., and Mower, E. B. (1969). *J. Catal.* **13**, 329.

Chen, N. Y., Gorring, R. L., Ireland, H. R., and Stein, T. R. (1977). *Oil Gas J.* **75** (23), 165.

Chen, N. Y., Reagan, W. J., Kokotailo, G. T., and Childs, L. P. (1978a). *In* "Natural Zeolites" (L. B. Sand and F. A. Mumpton, eds.), p. 411. Pergamon, New York.

Chen, N. Y., Miale, J. N., and Reagan, W. J. (1978b). U.S. Pat. 4,112,055.

Chen, N. Y., Kaeding, W. W., and Dwyer, F. G. (1979). *J. Am. Chem. Soc.* **101**, 6783.

Chen, N. Y., Schlenker, J. L., Garwood, W. E., and Kokotailo, G. T. (1984). *J. Catal.* **86**, 24.

Chu, C. T.-W., and Chang, C. D. (1984). *J. Catal.* **86**, 297.

Chu, P. (1973). U.S. Pat. 3,709,979.

Chu, P. (1983). U.S. Pat. 4,397,827.

Cody, I. A. (1983). U.S. Pat. 4,390,414.

Cody, I. A. (1984). U.S. Pat. 4,451,572.

Corbin, D. R., Seidel, W. C., Abrams, L., Herron, N., Stucky, G. D., and Tolman, C. A. (1985). *Inorg. Chem.* **24**, 1800.

Csicsery, S. M. (1967). *J. Chem. Eng. Data* **12**, 118.

Csicsery, S. M. (1969). *J. Org. Chem.* **34**, 3338.

Csicsery, S. M. (1970). *J. Catal.* **19**, 394.

Csicsery, S. M. (1971). *J. Catal.* **23**, 124.

Csicsery, S. M. (1976). *In* "Zeolite Chemistry and Catalysis" (J. A. Rabo, ed.), ACS Monograph No. 171, p. 680. Am. Chem. Soc., Washington, D.C.

Csicsery, S. M. (1979). *Ind. Eng. Chem., Process Des. Dev.* **18**, 191.

Csicsery, S. M. (1984). *Zeolites* **4**, 202.

Csicsery, S. M. (1985). *Chem. Br.* **21**, 473.

Dave, D., Hall, A., and Harold, P. (1982). *Eur. Pat. Appl.*, EP 50,021.

Davies, E. E., and Kolombos, A. J. (1979a). U.S. Pat. 4,175,057.

Davies, E. E., and Kolombos, A. J. (1979b). U.S. Pat. 4,180,689.

Dejaifve, P., Védrine, J. C. Bolis, V., and Derouane, E. G. (1980). *J. Catal.* **63**, 331.

Dejaifve, P., Auroux, A., Gravele, P.C., Védrine, J. C., Gabelica, Z., and Derouane, E. G. (1981). *J. Catal.* **70**, 123.

Derewinski, M., Dzwiggi, S., Haber, J., and Ritter, G. (1985). *Proc. Int. Symp. Zeolite Catal.*, *Siofok, Hung.*

Derouane, E. G. (1980). *In* "Catalysis by Zeolites" (B. Imelik *et al.*, eds.), p. 5. Elsevier, Amsterdam.

Derovane, E. G. (1981). *J. Catal.* **70**, 123.

Derouane, E. G. (1985). *Stud. Surf. Sci. Catal.* **20**, 221.

Derouane, E. G., and Gabelica, Z. (1980). *J. Catal.* **65**, 486.

Derouane, E. G., and Védrine, J. C. (1980). *J. Mol. Catal.* **8**, 479.

Derouane, E. G., Gabelica, Z., and Jacobs, P.A. (1981). *J. Catal.* **70**, 238.

Dessau, R. M. (1982). *J. Catal.* **77**, 304.

Dessau, R. M. (1984). U.S. Pat. 4,443,554.

Dessau, R. M., and LaPierre, R. B. (1982). *J. Catal.* **77**, 249.

Dewing, J., Spencer, M. S., and Whittam, T. V. (1985). *Catal. Rev.—Sci. Eng.* **27**, 461.

Donaldson, K., and Pout, C. R. (1972). *Prepr. Div. Pet. Chem. Am. Chem. Soc.* **17**, G-63.

Dwyer, F. G. (1976). U.S. Pat. 3,941, 871.

Dwyer, F. G. (1981). *Chem. Ind.* (*London*) No. 5, 39.

Dwyer, J., and Dyer, A. (1984). *Chem. Ind.* (*London*) No. 265, 237.

Dwyer, J., Fitch, F. R., Qin, G., and Vickerman, J. C. (1982). *J. Phys. Chem.* **86**, 4574.

Eberly, P. E., Jr. (1976). *In* "Zeolite Chemistry and Catalysis" (J. A. Rabo, ed.), ACS Monograph No. 171, p. 392. Am. Chem. Soc., Washington, D.C.

Eckman, R., and Vega, A. J. (1983). *J. Am. Chem. Soc.* **105**, 4841.

Ernst, S., and Weitkamp, J. (1985). *Proc. Int. Symp. Zeolite Catal.*, *Siofok, Hung.*

Espinoza, R. L., and Mandersloot, W. G. B. (1984). *J. Mol. Catal.* **24**, 127.

Fagan, F. N., and Weisz, P. B. (1983). Jpn. Pet. Inst., Tokyo.

Flanigan, E. M., Bennett, J. M., Grose, R. W., Cohen, J. P., Patton, R. L., and Kirchner, R. M. (1978). *Nature* (*London*) **271**, 512.

Flanigan, E. M., *et al.* (1985). Eur. Pat. Appl. 132,708.

Freeman, J. J., and Unland, M. L. (1978). *J. Catal.* **54**, 183.

Frilette, V. J., Haag, W. O., and Lago, R. M. (1981). *J. Catal.* **67**, 218.

Fyfe, C. A., Gobbi, G. C., Klinowski, J., Thomas, J. M., and Ramdas, S. (1982). *Nature* (*London*) **296**, 530.

Gard, J. A., and Tait, J. M. (1971). *Adv. Chem. Ser.* No. 101, 230.

Garwood, W. E., Rodewald, P. G., and Chen, N. Y. (1983). U.S. Pat. 4,376,036.

Giannetto, G., Perot, G., and Guisnet, M. (1985). *Proc. Int. Symp. Zeolite Catal.*, *Siofok, Hung.*

Gilson, J. P., and Derouane, E. G. (1984). *J. Catal.* **88**, 538.

Gorring, R. L. (1973). *J. Catal.* **31**, 13.

Haag, W. O. (1983). *Proc. Int. Zeolite Conf., 6th, Reno, Nev.* p. 466.

Haag, W. O., and Dessau, R. M. (1984). *Proc. Int. Congr. Catal., 8th, Berlin* **2**, 305.

Haag, W. O., and Lago, R. M. (1982). U.S. Pat. 4,326,994.

Haag, W. O., and Lago, R. M. (1983a). U.S. Pat. 4,418,235.

Haag, W. O., and Lago, R. M. (1983b). U.S. Pat. 4,374,296.

Haag, W. O., and Olson, D. H. (1978). U.S. Pat. 4,117,026.

Haag, W. O., Lago, R. M., and Weisz, P. B. (1984). *Nature* (*London*) **309**, 589.

Haag, W. P., Lago, R. M., and Weisz, P. B. (1982a). *Faraday Discuss. R. Soc. Chem.* No. 72, 317.

Haag, W. P., Lago, R. M., and Rodewald, P. G. (1982b). *J. Mol. Catal.* **17**, 161.

Haggin, J. (1982), *Chem. Eng. News* Dec. 13, p. 9.

Harrison, D., Leach, H. F., and Whan, D. A. (1983). *Proc. Int. Zeolite Conf., 6th, Reno, Nev.*

Heinemann, H. (1981). *Catal. Rev.—Sci. Eng.* **23**, 315.

Herron, N., Tolman, C. A., and Stucky, G. D. (1984). *Chem. Eng. News* Sept. 17, p. 30.

Hilaireau, P., Bearez, C., Chevalier, F., Perot, G., and Guisnet, M. (1982). *Zeolites* **2**, 69.

Ione, K., Echevskii, G. V., and Nosyreva, G. N. (1984). *J. Catal.* **85**, 287.

Jacobs, P. A. (1982). *Catal. Rev.—Sci. Eng.* **24**, 415.

Jacobs, P. A., and von Ballmoos, R. (1982). *J. Phys. Chem.* **86**, 3050.

Jacobs, P. A., and Tielen, M. (1984). *Proc. Int. Congr. Catal., 8th, Berlin* **4**, 357.

Jacobs, P. A., Martens, J. A., Weitkamp, J., and Beyer, H. K. (1982). *Faraday Discuss. R. Soc. Chem.* No. 72, 353.

Johnson, J. A., Weiszman, J. A., Hilder, G. K., and Hall, A. H. P. (1984). *NPRC Annu. Meet., San Antonio, Tex*, No. AM-84-85.

Kaeding, W. W., and Butter, S. A. (1980). *J. Catal.* **61**, 155.

Kaeding, W. W., Chu, C., Young, L. B., Weinstein, B., and Butter, S. A. (1981a). *J. Catal.* **67**, 159.

Kaeding, W. W., Chu, C., Young, L. B., and Butter, S. A. (1981b). *J. Catal.* **69**, 392.

Kaeding, W. W., Young, L. B., and Prapas, A. G. (1982). *Chem.-Tech. (Heidelberg)* **12**, 556.

Kärger, J., and Karo, J. (1977). *J.C.S. Faraday I*, **73**, 1363.

Kärger, J., Pfeifer, H., Rauscher, M., and Walter, A. (1980). *J.C.S. Faraday I* **76**, 717.

Karge, H. G., Sarbak, Z., Hatada, K., Weitkamp, J., and Jacobs, P. A. (1983). *J. Catal.* **82**, 236.

Keown, P. E. (1973). U.S. Pat. 3,751,504.

Kerr, G. T. (1981). *Catal. Rev.—Sci. Eng.* **23**, 281.

Kibby, C. L., Perrotta, A. J., and Massoth, F. E. (1974). *J. Catal.* **35**, 256.

Kikuchi, E., Matsuda, T., Fujiki, H., and Morita, Y. (1984). *Appl. Catal.* **11**, 331.

Kikuchi, E., Nakano, H., *et al.* (1985a). *J. Jpn. Pet. Inst.* **28**, 210.

Kikuchi, E., Nakano, H., *et al.* (1985b). *J. Jpn. Pet. Inst.* **28**, 214.

Klinowski, J., Ramdas, S., and Thomas, J. M. (1982). *J.C.S. Faraday II* **78**, 1025.

Kljueva, N. V., Tien, N. D., and Ione, K. G. (1985). *Proc. Int. Symp. Zeolite Catal., Siofok, Hung.*

Kokotailo, G. T., Lawson, S. L., Olson, D. H., and Maier, W. M. (1978a). *Nature (London)* **272**, 437.

Kokotailo, G. T., Chu, P., Lawton, S. L., and Maier, W. M. (1978b). *Nature (London)* **275**, 119.

Lewis, P. J., and Dwyer, G. (1977). *Oil Gas J.* **75**, 55.

Lok, B. M., Messina, C. A., Patton, R. L., Gajek, R. T., Cannan, T. R., and Flanigan, E. M. (1984). U.S. Pat. 4,440,871.

Martens, J. A., Tielen, M., Jacobs, P. A., and Weitkamp, J. (1984). *Zeolites* **4**, 98.

Martens, J. A., Perez-Pariente, J., and Jacobs, P. A. (1985). *Proc. Int. Symp. Zeolite Catal. Siofok, Hung.*

Meier, W. M., and Olson, D. H. (1978). "Atlas of Zeolite Structure Types," Structure Commission of the International Zeolite Association. Polycryst. Book Serv., Pittsburgh, Pennsylvania.

Melchior, M. T., Vaughan, D. E. W., and Jacobson, A. J. (1982). *J. Am. Chem. Soc.* **104**, 4859.

Miale, J. N., Chen, N. Y., and Weisz, P. B. (1966). *J. Catal.* **6**, 278.

Mikovsky, R. J., and Marshall, J. F. (1977). *J. Catal.* **49**, 120.

Mikovsky, R. J., Marshall, J. F., and Burgess, W. P. (1979). *J. Catal.* **58**, 489.

Nagy, J. B., Gabelica, Z., Derouane, E. G., and Jacobs, P. A. (1982). *Chem. Lett.* p. 2003.

Namba, S., Iwase, O., Takahashi, N., Yashima, T., and Hara, N. (1979). *J. Catal.* **56**, 445.

Namba, S., Inaka, A., and Yashima, T. (1983). *Zeolites* **3**, 106.

Nijs, H. H., Jacobs, P. A., Verdonck, J. J., and Uytterhoeven, J. B. (1980). *Proc. Int. Conf. Zeolites, 5th.* **5**, 632.

Niwa, M., Morimoto, S., Kato, M., Hattori, T., and Murakami, Y. (1984). *Proc. Int. Congr. Catal., 8th, Berlin* **4**, 701.

Olson, D. H., and Haag, W. O. (1984). *ACS Symp. Ser.* No. 248. 275.

Olson, D. H., Haag, W. O., and Lago, R. M. (1980). *J. Catal.* **61**, 390.

Ono, Y., and Mori, T. (1981). *J.C.S. Faraday I* **77**, 2209.

Plank, C. J., Rosinski, E. J., and Rubin, M. K. (1977a). U.S. Pat. 4,046,859.

Plank, C. J., Rosinski, E. J., and Rubin, M. K. (1977b). U.S. Pat. 4,016,245.

Plank, C. J., Rosinski, E. J., and Rubin, M. K. (1978). U.S. Pat. 4,076,842.

Plank, C. J., Rosinski, E. J., and Givens, E. N. (1979). U.S. Pat. 4,141,859.

Plank, C. J., Rosinski, E. J., and Givens, E. N. (1981). U.S. Pat. 4,276,151.

Poole, N., and Whan, D. A. (1984). *Proc. Int, Congr. Catal. 8th, Berlin* **4**, 345.

Pope, C. G. (1981). *J. Catal.* **72**, 174.

Rabo, J. A. (1976). *In* "Zeolite Chemistry and Catalysis" (J. A. Rabo, ed.), ACS Monogaph No. 171, p. 332. Am. Chem. Soc., Washington, D.C.

Ramdas, S., Thomas, J. M., Klinowski, J., Fyfe, C. A., and Hartman, J. S. (1981). *Nature (London)* **292**, 228.

Rodewald, P. G. (1978a). U.S. Pat. 4,100,219.

Rodewald, P. G. (1978b). U.S. Pat. 4,090,981.

Rodewald, P. G. (1979). U.S. Pat. 4,145,315.

Rollmann, L. D. (1978). U.S. Pat. 4,088,605.

Rollmann, L. D. (1979a). *In* "Inorganic Compounds with Unusual Properties II" (R. B. King, ed.), Advances in Chemistry No. 173, p. 387. Am. Chem. Soc., Washington, D.C.

Rollmann, L. D. (1979b). U.S. Pat. 4,148,713.

Rollmann, L. D., and Valyocsik, E. W. (1983). U.S. Pat. 4,423,021.

Rollmann, L. D., and Walsh, D. E. (1982). *In* "Progress in Catalyst Deactivation" (J. L. Figueiredo, ed.), p. 81. Nijhoff, The Hague.

Rosinski, E. J., and Schwartz, A. B. (1982). U.S. Pat. 4,309,280.

Sefcik, M. D. (1979). *J. Am. Chem. Soc.* **101**, 2164.

Sullivan, R. F., Egan, C. J., Langlois, G. E., and Sieg, R. P. (1961). *J. Am. Chem. Soc.* **83**, 1156.

Thomas, J. M., Millward, G. R., Ramdas, S., Bursill, L. A., and Audier, M. (1982). *Faraday Discuss. R. Soc. Chem.* No. 72, 345.

Topsøe, N., Pedersen, K., and Derouane, E. G. (1981). *J. Catal.* **70**, 41.

Unland, M. L. (1978). U.S. Pat. 4,115,424.

Valyocsik, E. W. (1984a). U.S. Pat. 4,481,177.

Valyocsik, E. W. (1984b). U.S. Pat. 4,490,342.

Valyon, J., Mihalyfi, J., Beyer, H. K., and Jacobs, P. A. (1979). *Prepr. Workshop Adsorption Hydrocarbons Berlin, G.D.R.* p. 134.

Védrine, J. C., Auroux, A., Bolix, V., Dejaifve, P., Naccache, C., Wierzchowski, P., Derouane, E. G., Nagy, J. B., Gilson, J. P., Van Haf, J. H. C., Van den Berg, J. P., and Wolthuizen, J. (1979). *J. Catal.* **59**, 248.

Védrine, J. C., Auroux, A., Dejaifve, P., Ducarmen V., Hoser, H., and Zhou, S. (1982). *J. Catal.* **73**, 147.

von Ballmoos, R., and Meier, W. M. (1981). *Nature (London)* **289**, 782.

Walsh, D. E., and Rollmann, L. D. (1979). *J. Catal.* **56**, 195.

Weisz, P. B. (1973). *Chem.-Tech. (Heidelberg)* **3**, 498.

Weisz, P. B. (1980). *Proc. Int. Congr. Catal., 7th, Tokyo* **1**, 3.

Weisz, P. B., and Frilette, V. J. (1960). *J. Phys. Chem.* **64**, 382.

Weisz, P. B., Frilette, V. J., Maatman, R. W., and Mower, E. B. (1962). *J. Catal.* **1**, 307.

Weitkamp, J. (1980). *Prepr., Int. Congr. Catal., 7th,* C7.

Weitkamp, J. (1985). *Proc. Int. Symp. Zeolite Catal., Siofok, Hung.*

Weitkamp, J., Jacobs, P. A., and Martens, J. A. (1983). *Appl. Catal.* **8**, 123.

Whyte, T. E., Jr., and Dalla Betta, R. A. (1982). *Catal. Rev.—Sci. Eng.* **24**, 567.

Wilson, S. T., Lok, B. M., and Flanigen, E. M. (1982). U.S. Pat. 4,310,440.

Wilson, S. T., Lok, B. M., Messina, C. A., Cannan, T. R., and Flanigan, E. M. (1983). *ACS Symp. Ser.* No. 218, 79.

Wu, M. M., and Kaeding, W. W. (1984). *J. Catal.* **88**, 478.

Yasuda, Y. (1982). *J. Phys. Chem.* **86**, 1913.

Yasuda, Y., and Sugasawa, G. (1984). *J. Catal.* **88**, 530.

Yasuda, Y., and Yamamoto, A. (1985). *J. Catal* **93**, 176.

Young, L. B. (1976). U.S. Pat. 3,962,364.

Young, L. B. (1981a). U.S. Pat. 4,283,573.

Young, L. B. (1981b). U.S. Pat 4,301,316.

Young, L. B. (1981c). U.S. Pat. 4,301,317.

Young, L. B., Butter, S. A., and Kaeding, W. W. (1982). *J. Catal.* **76**, 418.

Zatorski, L. W., Wierzchowski, P. T., and Cichowlas, A. A. (1985). *Proc. Int. Symp. Zeolite Catal., Siofok, Hung.*

Part VIII

Clay-Activated Organic Reactions

23 CLAY-ACTIVATED ISOMERIZATION REACTIONS

F. J. A. Kellendonk,
J. J. L. Heinerman,
and R. A. van Santen

Koninklijke/Shell-Laboratorium, Amsterdam
Shell Research B.V.
1003 AA Amsterdam, The Netherlands

I. INTRODUCTION

Natural clays were among the earliest solid acid catalysts used in the oil industry to promote cracking and isomerization reactions. Not long after their introduction in the late 1930s, they were replaced in large part by synthetic amorphous silica–aluminas, which gave more consistent results. During the 1960s the amorphous catalysts were superseded by the more active and shape-selective microporous crystalline aluminosilicates known as zeolites. However, there has been a renewed interest in the use of clays as catalysts because zeolites are not as suitable for the conversion of relatively large molecules. This interest resulted in the synthesis of highly active clays and in the development of the pillared clays, in which the useful surface has been increased and shape selectivity created by separating the layers with the aid of large cationic clusters (pillars).

It is well known that most of the observed catalytic properties of clays are due to their acidic character. Therefore, this chapter starts with a short summary of methods to determine these acidic properties and the main results of these methods. Then the various steps to obtain an active catalyst are discussed, such as synthesis, purification, and activation. This is followed by a discussion of the mechanism of the isomerization reaction and a review of the use of clays as isomerization catalysts.

PREPARATIVE CHEMISTRY
USING SUPPORTED REAGENTS

II. ACIDIC PROPERTIES OF CLAYS

For details of structures of clay minerals and their characterization by x-ray diffraction, we refer to Chapter 11 in this volume. The other characterization methods are fully covered in the monographs of Weaver and Pollard (1973), Theng (1974, 1982), and van Olphen (1977).

The acidity of solid surfaces can be determined by using the well-known Hammett indicators (Tanabe, 1970). This simple technique gives only qualitative information about the number and distribution of acidic sites. More quantitative data can be obtained by monitoring at various temperatures the adsorption of a base such as ammonia, trimethylamine, and pyridine. In combination with IR spectroscopy it is even possible to distinguish between Brønsted and Lewis acidity. For example, the IR spectrum of pyridine adsorbed on Lewis sites differs considerably from that of the pyridinium ion that is formed on Brønsted sites (Aldridge et al., 1973; Forni, 1973).

It has been shown conclusively that the presence of water influences the acidic properties of clays. For cation-exchanged clays, Brønsted acidity is mainly due to the dissociation of adsorbed water. This dissociation is induced by the polarizing power of the cation (e.g., Al^{3+}, Fe^{3+}), which depends on the size and the charge of the ion. When the water content of a clay is reduced to less than ~5% by weight, the acidity increases, because polarization by the cation is less dissipated (Theng, 1982). Removing all of the water from the clay, by calcination above ~200°C, results in a marked reduction in Brønsted acidity. The clay assumes a much more pronounced Lewis acid character, which is due to the transformation of Brønsted acidity into Lewis acidity.

A different kind of acidity is due to H^+ located at exchange positions. The positive charge compensates for the negative charge of the layers induced by isomorphous substitution with lower-valence cations. The effect of calcining H^+ (exchanged) clays depends on the type of layer silicate. In the case of octahedrally substituted clays (montmorillonite type), the protons will migrate into vacancies in the octahedral sheet, where they associate with lattice oxygens (Wright et al., 1972; Davidtz, 1976). In general, these protons are not accessible to catalysis, because the free of diameter of the hexagonal hole in the tetrahedral sheet is only 0.26 nm (Barrer, 1978). After calcination the acidic sites left will be situated at the edges.

However, for tetrahedrally substituted clays, calcination leads to protons attached to surface oxygens of the tetrahedral sheet. This results in acidic sites that are similar to the Brønsted acidic sites on Y zeolite (Uytterhoeven et al., 1965; Barrer, 1978). The work of Plee et al. (1985)

with pillared montmorillonite and beidellite (a tetrahedrally substituted clay) corroborate nicely the previously mentioned features. They observed that the acidity of calcined beidellite compared well with Y zeolite, while the calcined montmorillonite-based catalyst was much less acidic. With IR spectroscopy it was shown that on beidellite mainly Brønsted sites occurred, while montmorillonite exhibited a much lower acidity which was mainly of the Lewis type.

In conclusion, it emerges that the acidic properties are strongly dependent on the type of clay, the exchanged cation, and the H_2O content.

III. METHODS OF CATALYST PREPARATION

A. Purification of Natural Clays

Natural clays often contain quite large amounts of impurities, which should be removed to prevent interference with the catalytic function. As the variation in amount and type of impurities is as large as the variation in clay types, it is not possible to give here a specific purification recipe. Therefore, we limit ourselves to some general remarks and refer to the literature for experimental details (van Olphen, 1977).

Each purification method should involve separation of coarse mineral impurities from the clay. This is best achieved by suspending the sample in water and allowing the large particles to settle. The clay particles can then be siphoned off (van Olphen, 1977). If the sample contains a relatively large amount of organic material, the dry material can be brought in contact with concentrated hydrogen peroxide. Several methods exist for the removal of inorganic impurities. For instance, carbonates and free ferric oxide can be removed by a subsequent treatment with sodium acetate buffer solution and sodium dithionite/sodium citrate (Tzou, 1983).

B. Activation Procedure

The first step in the activation procedure, which is required for both natural and synthetic clays, is to convert the clay into the desired ion form. In the case of metal ions, this can be done with relative ease by ion exchange using a concentrated solution of a suitable salt. The excess salt can be removed by repeated washings employing a centrifuge, followed by dialysis. As an alternative, one can use ion-exchange resins in the suitable ion form. In contrast, it is not as easy to prepare the H^+ form of a clay via ion exchange with an acid or a resin in the H^+ form. The acid attacks the clay structure and part of the aluminum in the structure is transferred to ion-exchange sites. This effect occurs even under relatively mild conditions. For example, van Olphen (1977) found that the conversion of Wyoming bentonite (montmorillonite) with 0.01 N HCl at room temperature led to the occupation of 60% of the exchange sites by Al. The best way to overcome this acid leaching problem is to prepare the NH_4^+ form via ion exchange, followed by a heat treatment at 550°C (Wright *et al.*, 1972). The NH_4^+ ion will decompose, leaving an H^+ ion on the clay. By performing the heat treatment in air all of the residual organic material will be burnt off, which results in a ready-to-use acidic catalyst.

C. Synthetic Clays

The main incentive for using synthetic clays is that several interesting materials are either not available in sufficient quantities in their natural form (beidellite) or do not exist in nature. Most clays can be synthesized via hydrothermal methods (Barrer, 1982). Here we discuss briefly the synthesis of clays that have been used for isomerization reactions.

Koizumi and Roy (1959) have prepared in a pressure vessel at 300°C and 1000 bar, starting with a stoichiometric gel, 100% beidellite with the composition $Na_{0.33}Al_2(Si_{3.67}Al_{0.33}O_{10})(OH)_2$. The very high pressure poses a problem for large-scale preparation of the clay. However, Diddams *et al.* (1984) claim that the reaction also proceeds at 300°C and autogenous pressure.

Granquist (1966) described the synthesis at ~300°C of a novel high-surface-area (140–160 m^2/g) clay with randomly interstratified mica- and beidellite-type layers having the following approximate formula: $Na_{0.69}Al_2(Si_{3.36}Al_{0.64})O_{10}(OH)_2$. This clay has become known as synthetic mica montmorillonite (SMM), although it does not contain any montmorillonite but only beidellite. It was found that partial incorporation of nickel into the octahedral positions of the SMM structure resulted in a dramatic increase in acidity. This became apparent only after reduction of the nickel to the zero-valent state (Swift, 1977; Heinerman *et al.*, 1983; Robschlager *et al.*, 1984). In the original NiSMM synthesis (Granquist, 1974) a relatively large excess of H_2O had been used, which resulted in an inefficient use of the high-pressure autoclave. Heinerman (1985) reported an improved synthesis, which involved the use of (commercial) amorphous silica–alumina as starting material and minimum quantities of water. This resulted in a yield that was three times higher per reaction mixture volume.

Synthesis of trioctahedral clays, where all octahedral positions are occupied with cations, is, in general, somewhat less difficult than the preparation of the dioctahedral materials, where two out of three octahedral positions are filled with cations. For example, chrysotile (trioctahedral) can be prepared by heating a mixture of the appropriate oxides and an excess of H_2O at 250°C for 24 h (Robson, 1973, 1974), whereas preparation of dioctahedral isomorphous kaolinite requires more severe conditions (Barrer, 1982). Furthermore, Sohn and Osaki (1980) have claimed that nickel–talc [$\sim Ni_3Si_4O_{10}(OH)_2$] can be synthesized by precipitation from an acidic solution of a nickel salt–sodium silicate mixture at temperatures as low as 60°C.

Several interesting materials, known as fluorine micas, have been synthesized by employing salt melts as crystallization medium or via a solid-state reaction (Shell and Ivey, 1969; Kitajima et al., 1984). One example is a material with the structural formula $NaMg_{2.5}\square_{0.5}(Si_4O_{10})F_2$, which is swellable in water (Kitajima and Daimon, 1975). Here the charge imbalance is not due to isomorphous substitution but to vacancies in the octahedral layer. Morikawa et al. (1983a) have shown that this material in the Na form is inert, in contrast with other alkali-neutralized clay materials which exhibit some residual acidity. It is, therefore, ideal to study the intrinsic activity of metals and metal ions.

D. Pillared Clays

Pillared clays are materials that are modified by exchanging the interlamellar cations for bulky ions designed to provide increased access to the interior region and shape selectivity at temperatures where water or other solvents are thermally removed. Work on pillaring started by ion exchanging with organic cations such as ethylene–diammonium. This was only partially successful because the ion intercalated with its long axis parallel to clay layers which resulted in a layer separation of only 0.28 nm (Barrer and MacLeod, 1955). A clear advance was the use of the cage-like molecule 1,4-diazobicyclo[2.2.2]octane (DABCO), which resulted in an interlayer distance of 0.53 nm (Mortland and Berkheiser, 1976; Shabtai et al., 1977). However, the thermal stability of these materials appeared to be unsatisfactory. Therefore, attention shifted to inorganic materials as pillars. The best results were obtained with polymeric cationic complexes of aluminum. Several preparation methods have been reported, all of which resulted in samples with a surface area of 200–400 m^2/g and a layer separation of 0.9 nm. All procedures start with the preparation of well-dispersed clay slurries using purified <2-μm materials. The main difference between the methods is the concentration of the clay suspension. For instance, Lahav and Shani (1978) and Lahav et al. (1978) added dropwise aged aluminum hydroxide solutions, prepared with $AlCl_3$ and NaOH (OH/Al = 1.85), to

stirred montmorillonite suspensions of only 50–200 mg/L. After floccula-
tion occurred, the resulting material was allowed to settle for several hours
and was collected by centrifugation. The main disadvantage of this
procedure is the necessity to handle large volumes of suspension. Vaughan
et al. (1979, 1981; Vaughan and Lussier, 1980) brought much higher clay
concentrations (>27 g montmorillonite or beidellite/L) in contact with
solutions containing aluminum hydroxy oligomers at temperatures of the
order of 60°C for about 30 min. The most convenient oligomer source is a
commercial aluminum chlorhydroxide known as Chlorhydrol (supplied by
Reheis Chemical Company). After pillaring the material should be washed
thoroughly in order to obtain an active material. Comparable methods
have been reported by Gaaf and van Santen (1984) and Gaaf *et al.* (1983)
for NiSMM pillaring, and Plee *et al.* (1985) for beidellite pillaring. Similar
clay concentrations have been used by Jacobs *et al.* (1984), who achieved
pillaring by dialyzing clay–aluminum hydroxy slurries at room temperature
for 4 days. After the actual pillaring it is necessary to give the material a
heat treatment at about 550°C, which probably covalently binds the
pillaring agents to the clay layer. The material can now be brought in the
desired (active) ion form. Alternatively, one can try to pillar the active ion
form of the clay. However, this is not always possible because the swelling
depends on the type of interlamellar ion (Barrer, 1978).

Novel pillars with interesting properties have been described in the
literature. Chromium oxide pillars, where chromium oxide has intrinsic
hydrogenation activity, have been claimed to give larger pore size (Tzou,
1984; Pinnavaia *et al.*, 1985). Other novel pillaring materials are silicon
oxide (Lewis *et al.*, 1985), zirconium oxide (Yamanaka and Brindley, 1979;
Vaughan *et al.*, 1979, 1981; Kikuchi *et al.*, 1985), titanium oxide (Vaughan
et al., 1979), and nickel oxide (Tamanaka and Brindley, 1978). Silicon
pillars appear to have better hydrothermal stability and have been used in
the preparation of an isomerization catalyst.

IV. ISOMERIZATION REACTIONS

A. Reaction Mechanism

The isomerization of hydrocarbons over solid acidic catalysts has been
studied extensively. It is now generally accepted that the reaction mechanisms involve carbonium ions. The factors that determine the selectivity of
the reaction are well known as well. Therefore, we give here only a short
review of the relevant principles.

The first step is formation of a surface carbonium ion. This intermediate
can undergo rearrangements by hydrogen-atom and/or carbon shifts. The

former operates in the case of double-bond isomerization. The latter results in skeletal isomerizations of the type given below (Brouwer and Hogeveen, 1972):

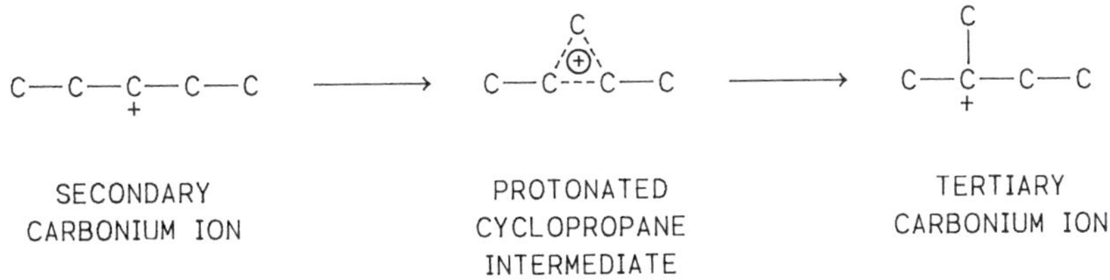

$$C-C-\underset{+}{C}-C-C \longrightarrow C-\underset{C}{\overset{\overset{C}{\triangle}}{C^{\oplus}}{\cdots}C}-C \longrightarrow C-\underset{+}{\overset{\overset{C}{|}}{C}}-C-C$$

| SECONDARY CARBONIUM ION | PROTONATED CYCLOPROPANE INTERMEDIATE | TERTIARY CARBONIUM ION |

The driving force is the relatively high stability of the tertiary carbonium ion compared with the other ions.

Carbonium ions can be formed via several routes:

(1) protonation of unsaturated bonds by Brønsted acidic sites (Weisz, 1970; Kouwenhoven, 1973);

(2) abstraction of a hydride (H^-) ion by a Lewis acidic site (Nace, 1969); and

(3) protonation of saturated compounds at Brønsted sites has been claimed to give five-coordinated carbon atoms (nonclassical carbonium ion) followed by the formation of molecular hydrogen and a classical carbonium ion (Olah, 1973).

The first route, protonation of unsaturated hydrocarbons, proceeds with relative ease compared with the other mechanisms. For that reason lower temperatures can be used, which gives less (thermal) cracking and therefore higher selectivities.

To achieve similarly high selectivities with saturated compounds, the concept of bifunctional catalysis has been developed, in which a (de)-hydrogenation function (for instance, a noble metal) is combined with an acidic support (Sinfelt et al., 1960). The dehydrogenation function produces alkenes resulting in a lowering of the reaction temperature. According to Coonradt and Garwood (1964), the highest selectivities are obtained when there is a balance between the (de)hydrogenation activity and the acidity of the catalyst. Fast dehydrogenation (relative to the isomerization reaction) results in a relatively high steady-state concentration of unsaturated compounds. These compounds will displace the isomerized carbonium ions from the acid sites by competitive adsorption, thus limiting the residence time on the catalyst, which reduces the chances of β-scission of branched carbonium ions. Furthermore, fast hydrogenation of the desorbed rearranged alkenes prevents readsorption and hence removes the possibility of β-scission. The validity of this balance concept has been experimentally demonstrated for several types of acidic supports. For instance: $Pt/SiO_2-Al_2O_3$ (Coonradt and Garwood, 1964), Pt ultrastable Y zeolite (Jacobs et al., 1980), and NiSMM (Robschlager et al., 1984).

In the preceding mechanism, isomerization and β-scission (cracking) are consecutive reactions. It is well known in the field of reactor engineering that the highest yield of the intermediate (in this case isomerized) product is obtained in a reactor configuration where mixing of product is kept to a minimum, i.e., plug-flow operation (Levenspiel, 1972), while mixing in a batch-type operation leads to an increased contribution of consecutive reactions.

B. Alkene Isomerizations

In this section we only deal with gas-phase double bond isomerization of butenes. Other alkene reactions will be discussed by J. Adams in Chapter 27. This rearrangement of butenes is a typical test reaction to detect even the weakest (Brønsted) acidic sites.

The importance of the presence of H_2O in the formation of Brønsted sites in layered silicates without tetrahedral substitution (no intrinsic Brønsted acidity) has been shown by Sohn and Ozaki (1980) for a nickel–talc-like structure $[Ni_3Si_4O_{10}(OH)_2]$. They found that the activity for 1-butene isomerization (at a temperature as low as 20°C) peaks at a pretreatment evacuation temperature of 100°C. The activity ran closely parallel with the Brønsted surface acidity as measured by n-butylamine titration in benzene. At an evacuation temperature of 400°C the catalyst was quite inactive and hardly any Brønsted sites could be detected.

That there are still Brønsted sites present in a tetrahedrally substituted SMM clay (NL Industries, Baroid division) even after activation at 530°C in vacuum has been demonstrated by Hattori et al. (1973). They performed tracer studies with deuterated SMM and nondeuterated butene in a pulse reactor at ambient temperatures under water-free conditions. They inferred from the appearance of deuterium in the reaction products that Brønsted sites are responsible for the isomerization reaction.

The isomerization of cis-2-butene at 150°C over 1,4-diazobicyclo-[2,2,2]octane (DABCO)-exchanged Zn-montmorillonite has been used by Stul et al. (1983) to show that pillaring could make more sites accessible to catalysis. They observed a fourfold increase in initial activity due to the pillaring.

Morikawa et al. (1983b) used a conventional pulse reactor to convert 1-butene to cis- and trans-2-butene (at 380°C/He flow) over fluorotetrasilicic micas (TSM, Topy Industries) exchanged with several metal ions such as Mg^{2+}, Al^{3+}, Pt^{2+}, etc. The Pt^{2+}-exchanged sample was the most active with 70% conversion and 85% selectivity. From the relatively low ratio of *trans*-2- to 1-butene (~1) in the products after *cis*-2-butene isomerization it was concluded that no Brønsted acid site was involved in the reactions. Furthermore, no relation could be found between known Lewis strengths of the ions and the activity. The use of hydrogen instead of helium as a carrier gas resulted in a marked increase in activity, which strongly indicates that the reaction proceeds on the metal ions via the type of half-hydrogenated species well known in the field of metal and metal complex catalyses (Anderson, 1973).

C. Paraffin Isomerizations

The skeletal isomerization of *n*-paraffins is an important reaction especially for the oil industry. For instance, isomerization of paraffins in the light gasoline boiling range results in an increase in octane number. Most work on this reaction has been done with zeolites and γ-Al_2O_3 doped with F and/or Cl (to induce acidity). Clay like materials, i.e., synthetic and pillared clays, have been evaluated as substitution for zeolites to overcome the pore diffusion and steric occlusion problems encountered with micro-porous materials.

In Table I we have summarized the catalysts that have been used for paraffin conversion, together with the relevant data regarding conversion and conditions. The following points deserve further comment:

(1) A relatively high hydrogen pressure has been applied (except in two casses). The rationale is prevention of deactivation due to coke formation on the catalyst surface.

(2) The most active clay catalyst, i.e., NiSMM doped with Pd or Pt, can have an activity comparable to that of the very active 1% Pd–H mordenite (compare nos. 7, 8, and 11).

(3) The high activities are combined with very high selectivities which indicates a bifunctional mechanism.

(4) The very poor performance of natural montmorillonite can be improved markedly by pillaring with an Al oligomer prepared in this case via dialysis (compare nos. 14 and 15). Whether the activity of the pillared clay is comparable with Pt–NiSMM (no. 13) is questionable because Jacobs *et al.* (1984) used a very low H_2 pressure. This normally results in a relatively high initial activity (high alkene concentration) followed by a rapid (coke) deactivation.

The best area for clays as catalysts is the conversion of large molecules as demonstrated by the work of Shabtai *et al.* (1981) on cracking of perhydrophenanthrene over a Ce-exchanged, Al-pillared montmorillonite.

TABLE I

Isomerization of Paraffins with Various Clay Catalysts

No.	Compound	Catalyst	BET surface (m²/g)	Conditions						Reference
				WHSV[a] (g/g h)	H₂/HC[b]	P[c]	Temperature (°C)	Conv.[d] (%)	Sel.[e] (%)	
1	Pentane	0.5% Pd/0.5% Cr-SMM	150	1.2	—	55	342	50	99	Csicsery and Mulaskey (1972)
2		1% Pd–NiSMM	290	2	1.25	30	250	53	99	Heinerman et al. (1983)
3		0.7% Pd–NiSMM	170	2	1.25	30	250	54	99.6	Gaaf and van Santen (1984); Gaaf et al. (1983)
4	Hexane	0.7% Pd–NiSMM Al pillared	230	2	1.25	30	250	65	98.8	Sawyer and Robson (1974)
5		0.5% Pd–20% Al Crysotile	110	1	6	31	250	9.2	—	
6		0.5% Pd-SMM	145	1[f]	2.5	31	260	3.5	100	Swift and Black (1974a,b)
7		0.5% Pd–NiSMM	244	1[f]	2.5	31	260	76	96	Swift and Black (1974a,b)
8		Pt–NiSMM	~170	3.3	22.4	26	250	56	99	Robschlager et al. (1984)
10		0.5% Pt–Ce montmorillonite Si pillared	158	1.2	4.7	59	350	30	—	Lewis et al. (1985)
11		1% Pd–H mordenite[g]	~400	1	2.5	30	250	72	97	Kouwenhoven (1973)
12	n-Decane	0.5% Pd/0.5% Fe-SMM	~150	1.2	—	83	265	31.4	81.8	Csicsery et al. (1972)
13		Pt–NiSMM	~170	2	30	30	250	90	90	Robschlager et al. (1984)
14		Pt-montmorillonite	~30	0.6	71	1	249	1	—	Jacobs et al. (1984)
15		Pt-montmorillonite Al pillared	~300	0.6	71	1	249	65	—	Jacobs et al. (1984)

[a] WHSV = weight hourly space velocity, which is the weight of feed passed over a given weight of catalyst per hour.

[b] H₂/HC = molar ratio of hydrogen and hydrocarbon feed.

[c] P = pressure during reaction.

[d] Conv. = overall conversion.

[e] Sel. = selectivity to uncracked isomerized products.

[f] Here LHSV (ml/ml h) is given (LHSV = liquid hourly space velocity).

[g] Zeolite catalyst included for comparison.

They observed a markedly higher activity of the pillared clay as compared to Y zeolites, while the opposite was found for smaller molecules.

D. α-Pinene Isomerization to Camphene

The effectiveness of various clays for the production of camphene, a raw material for camphor production, has been studied since the 1930s. Tischchenko and Rudakov (1933) used an unspecified clay activated with 10% hydrochloric acid for the α-pinene (**I**) rearrangement to camphene (**II**) and obtained a 62% yield:

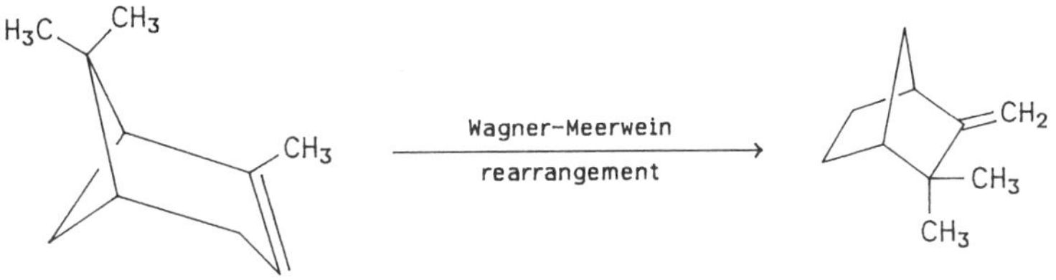

Nazir *et al.* (1976) reported a more detailed investigation with china (kaolin-type) and fire clay samples from Pakistan. The best yield (60%) was obtained with a china clay treated with hot concentrated sulfuric acid. The reaction was carried out at 110°C with 2% by weight catalyst in a stirred round-bottom flask. Other clays catalyzing camphene production are H- and Al-montmorillonite (Dupont and Dolou, 1948; Ovcharenko, 1982), H-vermiculite, H-kaolinite, and H-hydrous mica (Battalova *et al.*, 1977). In all cases yields of around 60% were obtained.

E. Cholestene Rearrangement

During studies concerning the origin of polycyclic compounds present in sediments, shales and crude oils by the group of Albrecht at Strasbourg (Sieskind and Albrecht, 1985) it was discovered that cholest-5-ene (**III**) could be converted quantitatively to backbone-rearranged 20-*R* and 20-*S*-cholest-13(17)-enes (**IV**) at room temperature in the presence of dry K-10 montmorillonite (FLUKA):

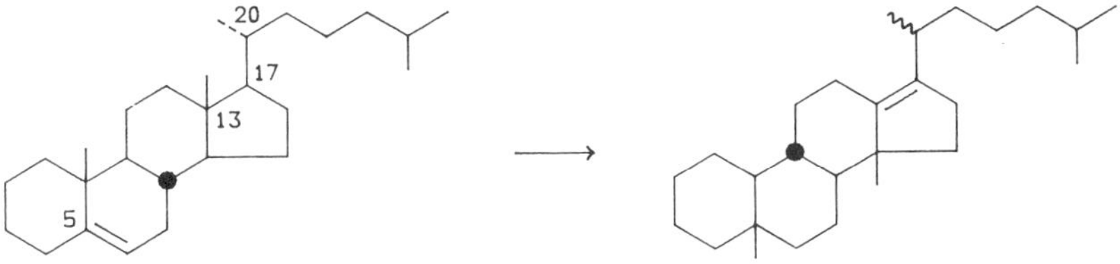

Unfortunately, the "dryness" was not specified. A typical experiment was done in the following way: 750 mg of reactant was added to the same amount of clay suspended in 6 mL of dry cyclohexane, and the mixture was stirred for 1 h. Without the clay catalyst this reaction proceeds only under superacid conditions, e.g., *p*-toluenesulfonic–acetic acid at reflux temperature. Furthermore, the reported yields are only of the order of 50% (Blunt *et al.*, 1969; Kirk and Shaw, 1970). So we have here a clear example of a very effective clay catalyst.

F. Oleic Acid Isomerization

Montmorillonites are industrially used as catalysts for the dimerization of unsaturated fatty acids to dicarboxylic acids (Johnston, 1944). Sometimes considerable amounts of isomerized products can be found in the reaction products. Nakano *et al.* (1985) observed that heating oleic acid at 230°C for 3 h with a montmorillonite clay (Alabama Blue Clay) gave a branched chain isomer yield of 27%. Adding water increased the yield by 10% (higher Brønsted acidity). On $(CH_3)_4N$ pillared montmorillonite mainly cis–trans isomerizations took place, which was probably due to induced shape selectivity (Weiss, 1981).

V. CONCLUSIONS

We did an extensive literature search on clay-catalyzed isomerization reactions. It appears that clays have only scarcely been used in synthetic organic chemistry reactions, while much more effort has been devoted to the industrially important acid-catalyzed paraffin isomerization.

We conclude that Brønsted acidic properties play an important role in clay-catalyzed isomerization reactions. Also, clay selection, pretreatment, and activation procedures are important parameters in obtaining an optimum acidity. This does not seem to be fully recognized in the synthetic organic chemistry literature, where details on clay catalyst preparation, characterization, and activation are often lacking.

In addition, the present review indicates that to obtain H^+-exchanged clay catalysts which combine a high thermal ($\sim 500°C$) stability with a high Brønsted acidity, the negative charge on the clay layers should arise from isomorphous substitution in the tetrahedral sheets, such as is the case for beidellite. This results in active sites with acid strength comparable to those of zeolites. Therefore, clays may be advantageously used either in applications where zeolites are more expensive or in reactions of relatively large molecules that are unable to enter the zeolite pore structure.

REFERENCES

Aldridge, L. P., McLaughlin, J. R., and Pope, C. G. (1973). *J. Catal.* **30**, 409.

Anderson, J. R. (1973). *Adv. Catal.* **23**, 1.

Barrer, R. M. (1978). "Zeolites and Clay Minerals as Sorbents and Molecular Sieves." Academic Press, New York.

Barrer, R. M. (1982). "Hydrothermal Synthesis of Zeolites." Academic Press, New York.

Barrer, R. M., and MacLeod, D. M. (1955). *Trans. Faraday Soc.* **51**, 1290.

Battallova, S. B., Pak, N. D. and Mukitayeva, T. R. (1977). *Izv. Akad. Nauk Kaz. SSR* **2**, 33.

Blunt, J. W., Hartshorn, M. P., and Kirk, D. N. (1969). *Tetrahedron* **25**, 149.

Brouwer, D. M., and Hogeveen, H. (1972). *Prog. Phys. Org. Chem.* **9**, 179.

Coonradt, H. L., and Garwood, W. E. (1964). *Ind. Eng. Chem., Process Des. Dev.* **3**, 38.

Csicsery, S. M., and Mulaskey, B. F. (1972). U.S. Pat. 3,655,798.

Csicsery, S. M., Hickson, D. A., and Jaffe, J. (1972). U.S. Pat. 3,640,904.

Davidtz, J. C. (1976). *J. Catal.* **43**, 260.

Diddams, P. A., Thomas, J. M., Jones, W., Ballantine, J. A., and Purnell, J. H. (1984). *J.C.S. Chem. Commun.* p. 1340.

Dupont, G., and Dolou, R. (1948). *J. Am. Ceram. Soc.* **31**, 143.

Forni, L. (1973). *Catal. Rev.* **8**, 65.

Gaaf, J., and van Santen, R. A. (1984). U.S. Pat. 4,469,813.

Gaaf, J., van Santen, R. A., Knoester, A., and van Wingerden, B. (1983). *J.C.S. Chem. Commun.* p. 655.

Granquist, W. T. (1966). U.S. Pat. 3,252,757.

Granquist, W. T. (1974). U.S. Pat. 3,852,405.

Hattori, H., Milliron, D. L., and Hightower, J. W. (1973). *Prepr., Div. Pet. Chem., Am. Chem. Soc.* **18**, 33.

Heinerman, J. J. L. (1985). U.S. Pat. 4,511,752.

Heinerman, J. J. L., Freriks, I. L. C., Gaaf, J., Pott, G. T., and Coolegen, J. G. F. (1983). *J. Catal.* **80**, 145.

Jacobs, P. A., Uytterhoeven, J. B., Steijns, M., Froment, G., and Weitkamp, J. (1980). *Proc. Int. Conf. Zeolites, 5th, Naples, Italy* p. 607.

Jacobs, P. A., Poncelet, G., and Schutz, A. (1984). U.S. Pat. 4,465,892.

Johnston, W. B. (1944). U.S. Pat. 2,347,562.

Kikuchi, E., Matsuda, T., Ueda, J., and Morita, Y. (1985). *Appl. Catal.* **16**, 401.

Kirk, D. N., and Shaw, M. P. (1970). *J.C.S. Chem. Commun.* p. 806.

Kitajima, K., and Daimon, N. (1975). *J. Chem. Soc. Jpn.* **6**, 991.

Kitajima, K., Shinomiya, Y., and Takusagawa, N. (1984). *Chem. Lett.* p. 1473.

Koizumi, M., and Roy, R. (1959). *Am. Mineral.* **44**, 789.

Kouwenhoven, H. W. (1973). *Adv. Chem. Ser.* No. 121, 529.

Lahav, N., and Shani, U. (1978). *Clays Clay Miner.* **26**, 116.

Lahav, N., Shani, U., and Shabtai, J. (1978). *Clays Clay Miner.* **26**, 107.

Levenspiel, O. (1972). "Chemical Reaction Engineering." Wiley, New York.

Lewis, R. M., Ott, K. C., and van Santen, R. A. (1985). U.S. Pat. 4,510,257.

Morikawa, Y., Takagi, K., Moro-oka, Y., and Ikawa, T. (1983a). *J.C.S. Chem. Commun.* p. 845.

Morikawa, Y., Yasuda, A., Moro-oka, Y., and Ikawa, T. (1983b). *Chem. Lett.* p. 1911.

Mortland, M. M., and Berkheiser, V. (1976). *Clays Clay Miner.* **24**, 60.

Nace, D. M. (1969). *Ind. Eng. Chem., Prod. Res. Dev.* **8**, 31.

Nakano, Y., Foglia, T. A., Kohashi, H., Perlstein, T., and Serota, S. (1985). *J. Am. Oil Chem. Soc.* **62**, 888.

Nazir, M., Ahmad, M., and Chaudhary, F. M. (1976). *Pak. J. Sci. Ind. Res.* **19**, 175.

Olah, G. A. (1973). *Angew. Chem., Int. Ed. Engl.* **12**, 173.

Ovcharenko, F. D. (1982). *Dev. Sedimentol.* **35**, 239.

Pinnavaia, T. J., Tzou, M. S., and Landau, S. D. (1985). *J. Am. Chem. Soc.* **107**, 4783.

Plee, D., Schutz, A., Poncelet, G., and Fripiat, J. J. (1985). *In* "Catalysis by Acids and Bases" (B. Emelik *et al.*, eds.), pp. 343–350. Elsevier, Amsterdam.

Robschlager, K. H. W., Emeis, C. A., and van Santen, R. A. (1984). *J. Catal.* **86**, 1.

Robson, H. E. (1973). U.S. Pat. 3,729,429.

Robson, H. E. (1974). U.S. Pat. 3,804,741; 3,848,018; 3,850,746; 3,852,165; 3,868,316.

Sawyer, W. H., and Robson, H. E. (1974). U.S. Pat. 3,838,041.

Shabtai, J., Frydman, N., and Lazar, R. (1977). *Proc. Int. Congr. Catal., 6th, London, 1976* **2**, 660.

Shabtai, J., Lazar, R., and Oblad, A. G. (1981). *Stud. Surf. Sci. Catal.* **7**, 828.

Shell, H. R., and Ivey, K. H. (1969). *Bull. U.S. Bur. Mines* No. 647.

Sieskind, O., and Albrecht, P. (1985). *Tetrahedron Lett.* **26**, 2135.

Sinfelt, J. H., Hurwitz, H., and Rohrer, J. C. (1960). *J. Phys. Chem.* **64**, 892.

Sohn, J. R., and Ozaki, A. (1980). *J. Catal.* **61**, 29.

Stul, M. S., van Leemput, L., and Uytterhoeven, J. B. (1983). *Clays Clay Miner.* **31**, 158.

Swift, H. E. (1977). *In* "Advanced Materials in Catalysis" (J. J. Burton and R. L. Garten, eds.), pp. 209–233. Academic Press, New York.

Swift, H. E., and Black, E. R. (1974a). *Ind. Eng. Chem., Prod. Res. Dev.* **13**, 106.

Swift, H. E., and Black, E. R. (1974b). *Prepr., Div. Pet. Chem., Am. Chem. Soc.* **19**, 7.

Tanabe, K. (1970). "Solid Acids and Bases." Academic Press, New York.

Theng, B. K. G. (1974). "The Chemistry of Clay-Organic Reactions." Wiley, New York.

Theng, B. K. G. (1982). *Dev. Sedimentol.* **35**, 197.

Tischchenko, V. E., and Rudakov, G. A. (1933). *J. Appl. Chem. USSR (Engl. Transl.)* **6**, 691.

Tzou, M. S. (1983). Ph.D. Thesis, Michigan State Univ., Lansing.

Uytterhoeven, J. B., Christner, L. G., and Hall, W. K. (1965). *J. Phys. Chem.* **69**, 2117.

van Olphen, H. (1977). "An Introduction to Clay Colloid Chemistry." Wiley, New York.

Vaughan, D. E. W., and Lussier, R. J. (1980). *Proc. Int. Conf. Zeolites, 5th, Naples, Italy* p. 94.

Vaughan, D. E. W., Lussier, R. J., and Magee, J. S. (1979). U.S. Pat. 4,176,090.

Vaughan, D. E. W., Lussier, R. J., and Magee, J. S. (1981). U.S. Pat. 4,248,739; 4,271,043.

Weaver, C. E., and Pollard, L. D. (1973). "The Chemistry of Clay Minerals." Elsevier, New York.

Weiss, A. (1981). *Angew. Chem., Int. Ed. Engl.* **20**, 850.

Weisz, P. B. (1970). *Annu. Rev. Phys. Chem.* **21**, 175.

Wright, A. C., Granquist, W. T., and Kennedy, J. V. (1972). *J. Catal.* **25**, 65.

Yamanaka, S., and Brindley, G. W. (1978). *Clays Clay Miner.* **26**, 21.

Yamanaka, S., and Brindley, G. W. (1979). *Clays Clay Miner.* **27**, 119.

24 OXIDATIONS AND CATALYTIC OXIDATIONS

A. McKillop

School of Chemical Sciences
University of East Anglia
Norwich NR4 7TJ, England

D.W. Clissold

Lilly Research Centre Limited
Surrey GU20 6PH, England

I. INTRODUCTION

Adsorption of a number of oxidizing agents on clays not only induces changes in their reactivity and selectivity but also simplifies commonly encountered workup and isolation problems associated with these reagents.

Foremost among the clay minerals used for this purpose has been the hydrous aluminosilicate—montmorillonite. Its adsorptive behavior toward small organic molecules (Theng, 1974) and organic polymers (Theng, 1979) has been described in some detail. The ability of montmorillonite to intercalate a large range of guest molecules (see Chapter 11) is related to its ability to undergo extensive expansion of its lamellar structure (Theng, 1979) as indicated in Fig. 1. The lamellae are composed of aluminosilicate sheets consisting of aluminium ions in an octahedral environment sandwiched between two layers of SiO_4 tetrahedra. In the naturally occurring clay the main interlayer cations are sodium and calcium; however, cation exchange can be effected by treatment with solutions of other ions such as Li^+, Mg^{2+}, Al^{3+}, and NH_4^+ (Theng, 1974). Isomorphous replacement of Al^{3+} ions by Mg^{2+} and Fe^{2+} and of Si^{4+} ions by Al^{3+} results in the layers carrying a permanent negative charge of 0.25–0.6 per formula unit. This charge is balanced by exchangeable counterions which, apart from those associated with the external clay surface, occupy the interlamellar regions

PREPARATIVE CHEMISTRY
USING SUPPORTED REAGENTS

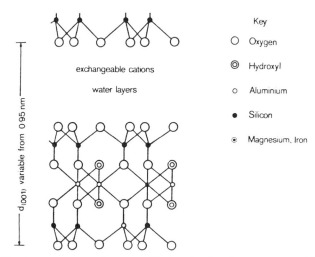

Key

○ Oxygen

◎ Hydroxyl

o Aluminium

● Silicon

◉ Magnesium, Iron

exchangeable cations

water layers

d(001) variable from 0.95 nm

Fig. 1. The layer structure of montmorillonite showing the occurrence of isomorphous replacement and the occupancy of the interlayer space by exchangeable cations and water. [Reprinted with permission from Theng (1979). Copyright 1979 Elsevier.]

of the clay. The cation exchange capacity of montmorillonite is of the order of 1 mol/kg.

Water enters the interlayer spaces of montmorillonite in integral numbers of complete layers. Depending on the type of exchangeable cation present the adsorption of water can result in considerable swelling of the clay. Besides water, many organic compounds can enter the interlayer regions of montmorillonite, and this intercalation is also highly dependent on the nature of the cations in the clay. Consequently, there is scope for the development of novel synthetic reactions via absorption of organic substrates into cation-exchanged montmorillonites (McKillop and Young, 1979).

Montmorillonite clays can be used as acidic catalysts in a number of reactions and there are clear indications that they are more efficient and selective in certain processes than the commonly used Brønsted and Lewis acids (McKillop and Young, 1979).

The introduction of oxidizing agents into or on clays for synthetic purposes has, as yet, received little attention. However, useful advantages, in terms of modification of reactivity and selectivity, may be achieved with concomitant ease of workup and isolation of reaction products.

The clay supported oxidants potassium permanganate (Lee and Noureldin, 1981), thallium(III) trinitrate (McKillop and Young, 1979;

Chiang *et al.*, 1976), and ferric nitrate (Cornélis and Laszlo, 1985) have been studied in most detail and these reagents are discussed here.

II. CLAY-SUPPORTED POTASSIUM PERMANGANATE

Potassium permanganate has been used as a supported reagent on hydrated molecular sieves, alumina, silica gel, copper sulfate, and clays for the oxidation of secondary alcohols to ketones (Koteel and Regen, 1977; Quici and Regen, 1979; Lee and Menger, 1979). Primary alcohols, other than benzyl alcohol, were observed in general to give very low yields of aldehydes (Lee and Noureldin, 1981) and the use of organic polymers as a support for the permanganate tends to lead to degradation of the polymer (Lee and Noureldin, 1981).

A comparative study by Lee and Noureldin (1981) on the use of potassium permanganate adsorbed on bentonite clay and on copper sulfate pentahydrate has revealed that both of these reagents oxidize saturated secondary alcohols to the corresponding ketones in high yields. More notably, these reagents exhibit pronounced selectivity in the oxidation of allylic secondary alcohols to the corresponding ketones, with minimal oxidation of the double bond (Scheme 1) (Lee and Noureldin, 1981).

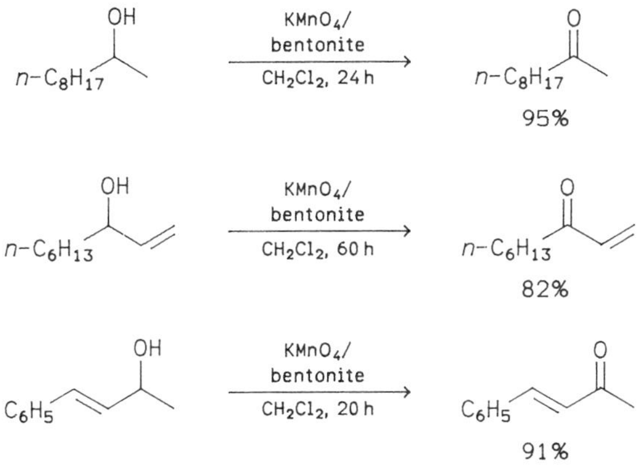

SCHEME 1

These reactions are performed simply by grinding together potassium permanganate (2–10 eq.) with an approximately equal weight of the support and heating, in methylene chloride, under gentle reflux and with efficient stirring, the supported reagent with the substrate for the times indicated. Isolation of the products requires only simple filtration and evaporation of the solvent.

The use of supported potassium permanganate for these oxidations is of obvious synthetic value when it is noted that such reactions are usually achieved by the use of activated manganese dioxide (Fatiadi, 1976). The supported permanganate procedure does not require either any activation procedure or the use of expensive equipment or materials.

The following generalizations were noted (Lee and Noureldin, 1981):

(1) Unsaturated alcohols required more oxidant and longer reflux periods than the corresponding saturated compounds.

(2) The required reaction time was in some cases shorter with the copper sulfate pentahydrate support as compared to the bentonite support.

(3) The presence of a phenyl group in conjugation with the double bond increases the ease of oxidation of the alcohols.

(4) Nonallylic unsaturated alcohols are more difficult to oxidize than the corresponding allylic alcohols.

(5) With a very large excess of oxidant and/or a prolonged reaction time, some cleavage of the carbon–carbon double bond may occur.

III. CLAY-SUPPORTED THALLIUM(III) NITRATE

Thallium(III) trinitrate (TTN) is a useful reagent for the oxidative rearrangement of olefins and enols (McKillop and Taylor, 1976). Chiang et al. (1976) have shown that TTN, when adsorbed on K-10, a readily available and inexpensive acidic montmorillonite clay, is a remarkably effective reagent for the rapid, selective, high-yield, room-temperature oxidation of a variety of unsaturated organic substrates. The TTN/K-10 reagent is readily prepared (McKillop and Young, 1979) by stirring K-10 with a solution of TTN in a mixture of methanol and trimethyl orthoformate followed by evaporation to dryness. The resulting colorless, free-flowing powder can be stored in well-capped bottles for months without any appreciable loss in activity. Oxidations are carried out (Chiang et al., 1976) by stirring a suspension of the TTN/K-10 reagent with a solution of the substrate in an inert solvent (heptane, methylene chloride, carbon tetrachloride, toluene, dioxane) until a starch iodide test for thallium(III) is negative. Products are isolated by removal of the spent reagent system by filtration, washing of the filtrate with aqueous sodium bicarbonate, then water, drying, evaporation of the solvent, and recrystallization or distillation of the crude product.

Oxidative rearrangement of alkyl aryl ketones with TTN/K-10 is a particularly smooth reaction (Chiang et al., 1976). Acetophenones, for instance, are rapidly converted into methyl arylacetates in excellent yields (Scheme 2) (McKillop and Young, 1979; Chiang et al., 1976). Propiophe-

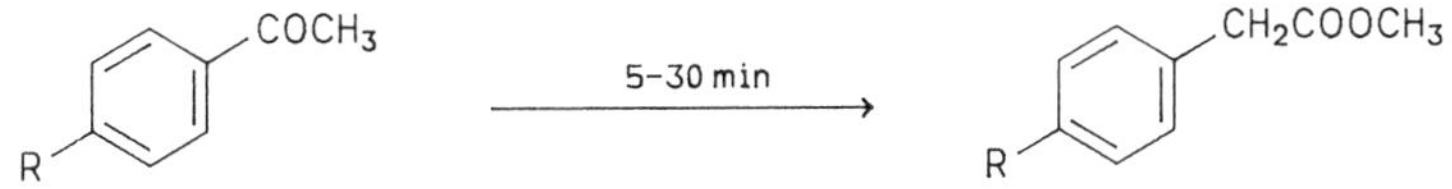

R = H, 86%
R = CH_3, 84%
R = CH_3O, 92%
R = F, 88%
R = Br, 89%

SCHEME 2

none and butyrophenone are cleanly converted into methyl α-methyl- and α-ethylphenylacetate, respectively, under the same reaction conditions (Scheme 3) (McKillop and Young, 1979; Chiang *et al.*, 1976). These results contrast sharply with those obtained using TTN in hot acidic methanol where, for example, propiophenone gives a mixture of methyl α-methylphenylacetate (45%) and α-methoxypropiophenone (32%).

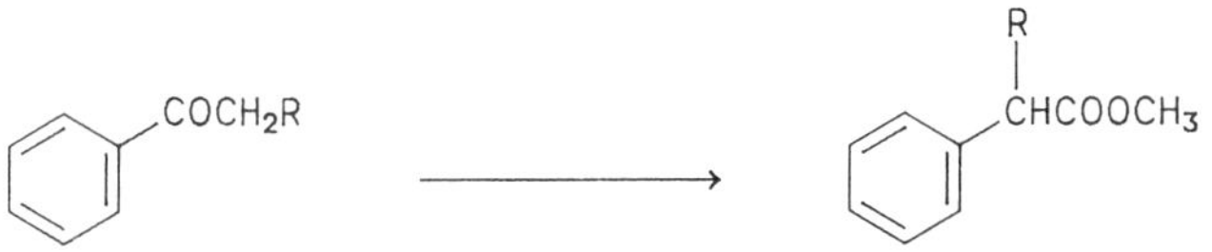

R = CH_3, 98%
R = C_2H_5, 98%

SCHEME 3

Oxidative rearrangement of simple olefins with TTN/K-10 is extremely rapid (Scheme 4) (McKillop and Young, 1979; Chiang *et al.*, 1976), and oxidation of cinnamaldehyde and substituted cinnamaldehydes (Scheme 5) (McKillop and Young, 1979; Chiang *et al.*, 1976) gives clean oxidative

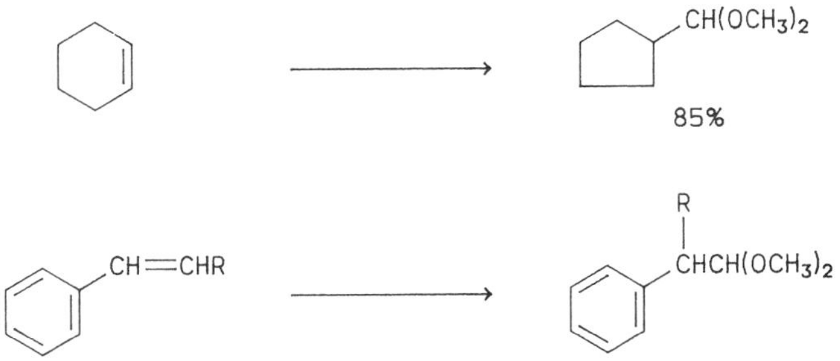

R = H, 92%
R = CH_3, 91%

SCHEME 4

$$R^1 = R^2 = H, \quad 86\%$$
$$R^1 = CH_3O, \quad R^2 = H, \quad 86\%$$
$$R^1 = H, \quad R^2 = CH_3, \quad 83\%$$

SCHEME 5

rearrangements to arylmalondialdehyde tetramethylacetals. This again contrasts with the use of TTN in methanol, which is of no synthetic utility for the oxidation of cinnamaldehydes due to the complex mixtures of products that are produced. Cinnamic esters, which are unaffected by hot solutions of TTN in methanol, are readily oxidized by TTN/K-10 at room temperature (Scheme 6) (McKillop and Young, 1979; Chiang et al., 1976).

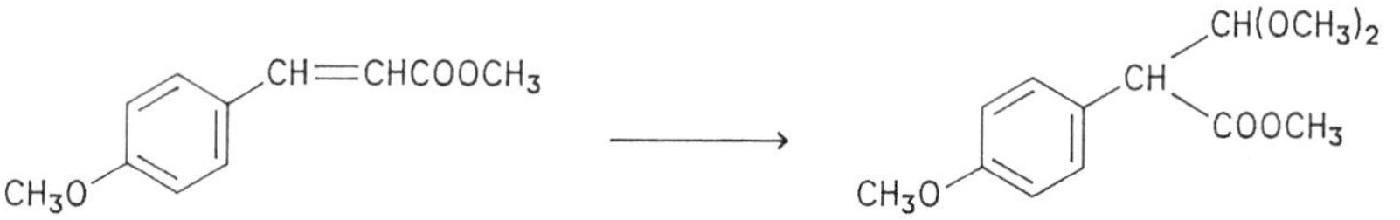

SCHEME 6

These conversions clearly demonstrate the superiority of TTN/K-10 over TTN/methanol for these oxidative rearrangements (Chiang et al., 1976). Moreover, in comparison with more than 20 other supports, K-10 has been found to be the most effective with respect to ease of preparation, rate of reaction, selectivity of oxidation, and general experimental simplicity. Supports with a lamellar structure, and in particular acidic ones such as K-10, are much more effective than nonacidic spongelike supports. Data for the conversion of acetophenone into methyl phenylacetate (Scheme 2), using TTN deposited on a variety of inorganic and organic supports (Table I) (Chiang et al., 1976) clearly show that TTN/K-10 is the reagent of choice for this transformation.

In an interesting extension of this transformation into the heterocyclic field, Adelakun et al. (1984) used TTN to rearrange 3-acetylpyrroles to the corresponding methoxycarbonylpyrroles, which were required for porphyrin synthesis. When a 2-methyl group was also present in the pyrrole facile oxidation to the corresponding 2-formylpyrrole occurred with TTN/K-10 but not with TTN/methanol (Kenner et al., 1973). Furthermore, the presence of the 3-acetyl or methoxycarbonyl functions was not

TABLE I

Oxidation of Acetophenone Using Supported Thallium (III) Trinitrate (TTN)[a]

	Reaction time[b]	Product distribution		
Support		Unreacted $C_6H_5COCH_3$	$C_6H_5CH_2COOCH_3$	(c)
AERO 8020 silica–alumina	1 min	60	20	20
Alcoa F-1 alumina	2 min	50	50	0
K-10 montmorillonite clay	5 min	<1	>99	0
K-306 montmorillonite clay	5 min	10	70	20
Florisil	5 min	40	30	30
Southern bentonite	10 min	20	80	0
Graphite	30 min	25	75	0
Davison 135 neutral silica–alumina	90 min	15	85	0
Amberlyst 15	8 h	50	50	0

[a] Reprinted with permission from Chiang *et al.* (1976). Copyright 1976 American Chemical Society.

[b] Time required for complete utilization of Tl(III) as determined by starch iodide paper.

[c] Indicates other products are also formed.

necessary for this oxidation to occur (Scheme 7) (Kenner *et al.*, 1973; Adelakun *et al.*, 1984). Adelakun *et al.* (1984) state

> The oxidation of the α-methyl groups by the TTN/clay reagent may involve radical intermediates, as has been proposed for halogenation and acetoxylation (by lead tetraacetate). However, exclusive formation of the aldehyde rather than formation of products at the alcohol or carboxylic acid level of oxidation points to a possible two-electron process [in which the thallium is reduced from the (III) to the (I) state]; on the other hand, the reagent oxidizes pyrromethanes to pyrromethenes rather than to pyrroketones. [From Adelakun *et al.* (1984). Reprinted with permission. Copyright 1984 Pergamon Press, Ltd.]

There is as yet no clear understanding of the exact physical nature of the TTN/K-10 system. However, since trimethyl orthoformate is thought to remove some of the water of hydration from TTN (Chiang *et al.*, 1976), a trihydrate, the TTN/K-10 reagent appears to consist of partially hydrated thallium(III) nitrate and methanol adsorbed on the surface of the acidic clay. Thus, the dried reagent corresponds approximately by weight to $Tl(ONO_2)_3/3CH_3OH/K$-10, and the clay can be washed free of both thallium(III) nitrate and methanol by treatment with polar solvents such as acetonitrile.

It is important to note that under standard reaction conditions (nonpolar inert solvents) both the thallium(III) nitrate on the support and the thallium(I) nitrate generated during the reaction are tightly bound to the

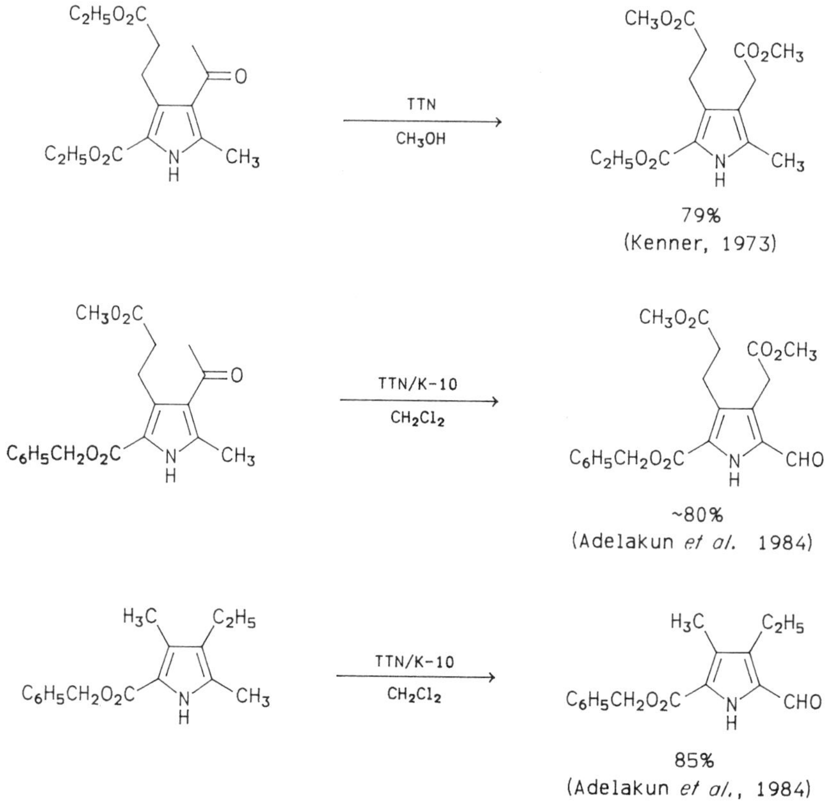

SCHEME 7

support throughout the entire reaction. Consequently, there is no detectable level (atomic absorption spectrocopy) of contamination of either solvent or product by toxic thallium salts (Chiang *et al.*, 1976).

IV. CLAY-SUPPORTED COPPER (II) AND IRON (III) NITRATES

Over the last five years the group led by Laszlo at the University of Liège have been prolific in their development of synthetic uses for these reagents (for which the respective names Claycop and Clayfen have been coined). For a full appreciation of the multifarious synthetic transformations of which these reagents are capable the reader is referred to the recent *Synthesis* review (Cornélis and Laszlo, 1985) as well as to other relevant sections of this volume.

Clayfen is easily prepared (Cornélis and Laszlo, 1980; Cornélis *et al.*, 1984) by deposition of the acetone solvate of iron(III) nitrate, obtained by dissolution of iron(III) nitrate nonahydrate in acetone, on K-10 montmorillonite clay. The free acetone solvate is a highly unstable, deep red oil which decomposes spontaneously in a vigorous exothermic reaction, but deposition on K-10 clay imparts considerable stability (Cornélis and Laszlo, 1985) and facilitates its use as a practical reagent. Claycop is similarly prepared (Balogh *et al.*, 1984a) from copper(II) nitrate and appears to be more stable than Clayfen in that, unlike the latter, no loss of activity is observed on standing and spontaneous decomposition does not occur on heating (full experimental details for the preparation of these reagents and necessary precautionary measures are provided in Cornélis and Laszlo, 1985). K-10 clay was chosen (Cornélis *et al.*, 1983b, 1984) as the support after a comparison with other clays (natural and industrial), sand, silica gel, titanium dioxide, acidic alumina, and zeolites and was preferred both for maintenance of activity of the reagent and for its rheological properties.

In the specific area of oxidative transformations these reagents have been applied to the oxidation of alcohols to aldehydes or ketones (Cornélis and Laszlo, 1980; Cornélis *et al.*, 1982), benzoins to benzils (Besemann *et al.*, 1984), and the oxidative coupling of thiols to disulfides (Cornélis *et al.*, 1983a).

Although there is a large variety of oxidizing agents available for the conversion of alcohols to the corresponding carbonyl compounds there is still a need for simple methods that are mild, inexpensive, and selective, particularly with regard to overoxidation to carboxylic acids. Clayfen is an efficient oxidizing agent for the production of ketones from a variety of secondary aliphatic, alicyclic, and benzylic alcohols (Table II) (Cornélis and Laszlo, 1985), and for the preparation of aromatic aldehydes from the corresponding primary aromatic alcohols (Cornélis and Laszlo, 1980). With primary aliphatic alcohols, complex mixtures of products result (Cornélis and Laszlo, 1980). The overall reaction is best formulated as

$$3 \; \underset{R^2}{\overset{R^1}{\diagdown}}CHOH + 2\,H^+ + 2\,NO_3^- \xrightarrow{\text{Clayfen}} 3 \; \underset{R^2}{\overset{R^1}{\diagdown}}C{=}O + 2\,NO + 4\,H_2O$$

with subsequent air oxidation of NO to NO_2.

The oxidations are readily effected (Cornélis and Laszlo, 1980) by addition of the Clayfen reagent to a solution of the alcohol in pentane or hexane. Evaporation of the solvent and slight heating under vacuum or gentle reflux of the suspension induces the reaction, which is easily detected and monitored by the evolution of reddish nitrous fumes. The product is isolated simply by filtration of the spent reagent and evaporation

TABLE II

Oxidation of Alcohols to Carbonyl Compounds with Clayfen[a]

R^1	R^2	Solvent	Time (h)	Yield (%) Crude	Yield (%) Isolated
C_6H_5	H	$n\text{-}C_5H_{12}$	3	85	76
C_6H_5	CH_3	$n\text{-}C_5H_{12}$	3	89	81
$n\text{-}C_6H_{13}$	CH_3	$n\text{-}C_6H_{14}$	1	98	83
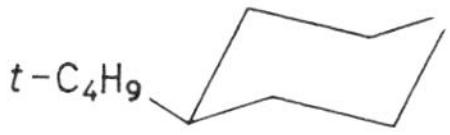 $t\text{-}C_4H_9$ (cyclohexyl)		$n\text{-}C_6H_{14}$	15	74	65
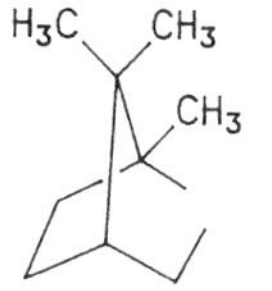 (bicyclic structure)		$n\text{-}C_6H_{14}$	2	92	80
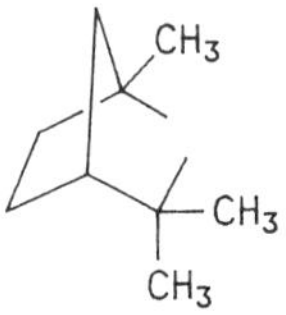 (bicyclic structure)		$n\text{-}C_6H_{14}$	3	89	82
C_6H_5	C_6H_5	$n\text{-}C_6H_{14}$	1.5	100	88

[a] Reprinted with permission from Cornélis and Laszlo (1985).

of the filtrate. Control experiments (Cornélis and Laszlo, 1980) with benzyl alcohol as the substrate using unsupported iron(III) nitrate and the K-10 clay alone have shown that the association of the nitrate with the clay provides a favorable reaction microenvironment with strong reactivity enhancement.

Mechanistic investigations (Cornélis et al., 1982) have shown the intermediacy of nitrous esters which are produced with retention of configuration (Cornélis and Laszlo, 1985) and, in the case of slow reactions (e.g., oxidation of cyclohexanol) which can be easily isolated before complete transformation into the carbonyl compound. By means of preparation (Noyes, 1933) of an authentic sample of cyclohexyl nitrite it has been shown (Cornélis et al., 1982; Cornélis and Laszlo, 1985) that the following occurs:

(1) Under the conditions of the reaction (Clayfen in hexane) the added nitrite gives rise to a mixture of product and starting material.

(2) The addition of the cyclohexyl nitrite to the K-10 support with no iron(III) nitrate present also gives rise to a mixture of cyclohexanone and cyclohexanol.

(3) The whole reagent is necessary to reach the nitrite stage, whereas the acidic K-10 component promotes decomposition of this nitrite intermediate into the carbonyl product, presumably by an acid catalysed pathway (Barton *et al.*, 1967) (Scheme 8).

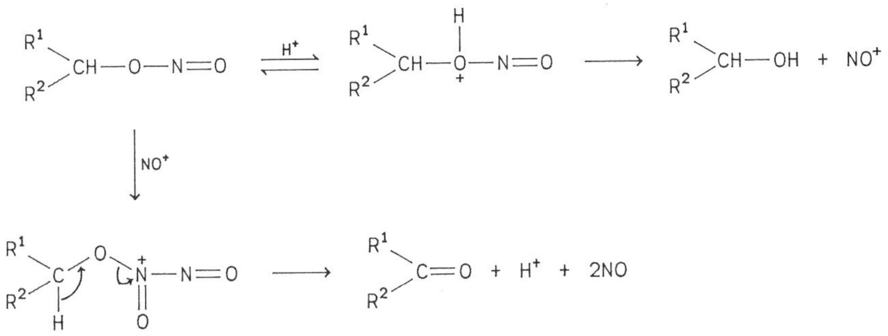

SCHEME 8

These results show, as indicated in the proposed mechanistic pathway, that Clayfen is a source of nitrosonium ions (NO^+) and the reagent has now been used successfully (Cornélis and Laszlo, 1985) as an alternative to reagents such as nitrosonium tetrafluoroborate (Ho and Olah, 1976) in several synthetic applications.

Similar conditions to those used for the oxidation of simple alcohols also give very good results in the oxidation of benzoins to benzils (Cornélis and Laszlo, 1985) (Scheme 9). A broad range of mechanisms has been

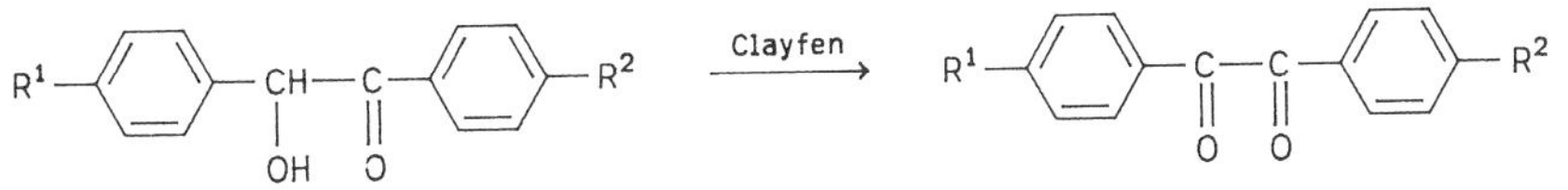

SCHEME 9

proposed (Chaplin *et al.*, 1984) to explain the divergent behavior of benzoins under various oxidizing conditions. However, Clayfen oxidation proceeds as in the oxidation of simple alcohols and yields for a range of substituents compare very favorably with those from Kagan's procedure in which the much more expensive reagent ytterbium(III) nitrate is used as catalyst (Table III) (Cornélis and Laszlo, 1985).

Since the oxidation of alcohols to carbonyl compounds indicated that Clayfen was a potential source of nitrosonium ions (NO^+) Laszlo and co-workers explored by analogy its effect on thiol functions and found that

TABLE III

Comparison of Yields Obtained on Oxidation of
Benzoins $4-R^1C_6H_4CH(OH)COC_6H_4R^2-4$ to
Benzils with Clayfen or Ytterbium (III) Nitrate[a]

		Yield (%)	
R^1	R^2	Clayfen	$Yb(NO_3)_3$
H	H	95	95
H	OCH_3	85	85
CH_3	CH_3	94	93
OCH_3	OCH_3	97	92
H	CH_3	96	92

[a] Reprinted with permission from Cornélis
and Laszlo (1985).

TABLE IV

Formation of Disulfides from Thiols using Clayfen[a]

R	Solvent	Yield (%)
C_6H_5	$n-C_6H_{14}$	97
$4-O_2NC_6H_4$	$C_6H_5CH_3$	58
$2,4,5-Cl_3C_6H_2$	$C_6H_5CH_3$	80
$C_6H_5CH_2$	$C_6H_5CH_3$	85
$(C_6H_5)_3C$	C_6H_6	65
$i-C_3H_7$	$n-C_5H_{12}$	39
$n-C_4H_9$	$n-C_5H_{12}$	88
$t-C_4H_9$	$n-C_5H_{12}$ or $n-C_6H_{14}$	0

[a] Reprinted with permission from Cornélis and Laszlo
(1985).

the reagent efficiently gives coupling to symmetrical disulfides (Table IV:
Scheme 10) (Cornélis, *et al.*, 1983a; Cornélis and Laszlo, 1985). The
reaction appears to proceed via the homolytic cleavage of the sulfur–
nitrogen bond of the unstable, transient thionitrite and subsequent coupling of the thiyl radicals (Cornélis and Laszlo, 1985, and references cited
therein). In certain cases the intermediacy of the thionitrite was proven by
isolation and confirms that, in contrast to the classical transition metal-

$$2 \text{ R—SH} \xrightarrow[\text{Solvent}]{\text{Clayfen}} \text{R—S—S—R}$$

SCHEME 10

catalyzed coupling of thiols to disulfides (Capozzi and Modena, 1974), it is an oxidative process. The reaction is a useful test of the reactivity of Clayfen (Cornélis and Laszlo, 1985) in that the freshly prepared reagent gives complete reaction in less than 1 min whereas deactivated (aged) reagent requires longer times (~10 min or more) and eventually gentle heating is needed to initiate the reaction (Cornélis *et al.*, 1984). Claycop also effects the coupling of thiophenol in high yields (>96%) at room temperature in equally short times (Cornélis and Laszlo, 1985), and it is considered that both reagents offer a new and convenient approach to the activation of thiols via thionitrite formation.

In the short time since the reagents Clayfen and Claycop were discovered by Laszlo and co-workers their application, particularly as cheap readily available sources for the generation of nitrosonium ions, has been demonstrated in a variety of synthetic reactions (Cornélis and Laszlo, 1985) in addition to those surveyed here. These include formation of new carbon–carbon bonds in cycloaddition reactions of the Diels–Alder type (Laszlo and Lucchetti, 1984) and in porphyrin synthesis by condensation of pyrroles with aldehydes (Cornélis *et al.*, 1983b), hydrolytic cleavage of thioacetals (Balogh *et al.*, 1984a), conversion of various imino derivatives to carbonyl compounds (Laszlo and Polla, 1984b, 1985), preparation of azides from hydrazines (Laszlo and Polla, 1984b), aromatization of dihydropyridines (Balogh *et al.*, 1984b), conversion of thiobenzophenones to benzophenones (Chalais *et al.*, 1985) and selective nitration of phenols (Cornélis *et al.*, 1983b,c, 1984). It is interesting to note that some of these reactions are phase-transfer processes, with the clay acting as the catalyst (Cornélis and Laszlo, 1982), and this is perhaps an area where further research will reveal further novel applications of these extraordinarily useful systems.

REFERENCES

Adelakun, E., Jackson, A. H., Ooi, N. S., and Rao, K. R. N. (1984). *Tetrahedron Lett.* **25**, 6049–6050.
Balogh, M., Cornélis, A., and Laszlo, P. (1984a). *Tetrahedron Lett.* **25**, 3313–3316.
Balogh, M., Hermecz, I., Laszlo, P., and Mészáros, Z. (1984b). *Helv. Chim. Acta.* **67**, 2270–2272.
Barton, D. H. R., Ramsay, G. C., and Wege, D. (1967). *J. Chem. Soc. C* 1915–1919.
Basemann, M., Cornélis, A., and Laszlo, P. (1984). *C. R. Acad. Sci. Ser. 2* **299**, 427–428.
Capozzi, G., and Modena, G. (1974). *In* "The Chemistry of the Thiol Group" (S. Patai, ed.), Vol. 2, Chap. 17. Wiley, New York.
Chalais, S., Cornélis, A., Laszlo, P., and Mathy, A. (1985). *Tetrahedron Lett.* **26**, 2327–2328.
Chaplin, R. P., Vorlow, S., Wainwright, M. S., Walpole, A. S., and Zadro, S. (1984). *J. Mol. Catal.* **22**, 269–281.

Chiang, C.-S., McKillop, A., Taylor, E. C., and White, J. F. (1976). *J. Am. Chem. Soc.* **98**, 6750–6752.

Cornélis, A., and Laszlo, P. (1980). *Synthesis* 849–850.

Cornélis, A., and Laszlo, P. (1982). *Synthesis* 162–163.

Cornélis, A., and Laszlo, P. (1985). *Synthesis* 909–918.

Cornélis, A., Herzé, P.-Y., and Laszlo, P. (1982). *Tetrahedron Lett.* **23**, 5035–5038.

Cornélis, A., Depaye, N., Gerstmans, A., and Laszlo, P. (1983a). *Tetrahedron Lett.* **24**, 3103–3106.

Cornélis, A., Laszlo, P., and Pennetreau, P. (1983b). *Clay Miner.* **18**, 437–445 [*C.A.* **100**, 102362v (1984)].

Cornélis, A., Laszlo, P., and Pennetreau, P. (1983c). *J. Org. Chem.* **48**, 4771–4772.

Cornélis, A., Laszlo, P., and Pennetreau, P. (1984). *Bull. Soc. Chim. Belg.* **93**, 961–972.

Fatiadi, A. J. (1976). *Synthesis* 65–104.

Ho, T.-L., and Olah, G. A. (1976). *Synthesis* 610–611.

Kenner, G. W., Smith, K. M., and Unsworth, J. F. (1973). *J.C.S. Chem. Commun.* 43–44.

Koteel, C., and Regen, S. L. (1977). *J. Am. Chem. Soc.* **99**, 3837–3838.

Laszlo, P., and Lucchetti, J. (1984). *Tetrahedron Lett.* **25**, 1567–1570, 2147–2150, 4387–4388.

Laszlo, P., and Polla, E. (1984a). *Tetrahedron Lett.* **25**, 3309–3312.

Laszlo, P., and Polla, E. (1984b). *Tetrahedron Lett.* **25**, 3701–3704.

Laszlo, P., and Polla, E. (1985). *Synthesis* 439–440.

Lee, C., and Menger, F. M. (1979). *J. Org. Chem.* **44**, 3446–3448.

Lee, D. G., and Noureldin, N. A. (1981). *Tetrahedron Lett.* **22**, 4889–4890.

McKillop, A., and Taylor, E. C. (1976). *Endeavour* **35**, 88–93.

McKillop, A., and Young, D. W. (1979). *Synthesis* 401–422, 481–500.

Noyes, W. A. (1933). *J. Am. Chem. Soc.* **55**, 3888–3889.

Quici, S., and Regen, S. L. (1979). *J. Org. Chem.* **44**, 3436–3437.

Theng, B. K. G. (1974). "The Chemistry of Clay-Organic Reactions." Adam Hilger, London.

Theng, B. K. G. (1979). "Formation and Properties of Clay-Polymer Complexes." Elsevier, Amsterdam.

25 CLAY-ACTIVATED CATALYTIC HYDROGENATIONS: CATALYST SYNTHESIS AND FACTORS INFLUENCING SELECTIVITY*

Thomas J. Pinnavaia

Department of Chemistry and
Center for Fundamental Materials Research
Michigan State University
East Lansing, Michigan 48824

I. INTRODUCTION

The study of organic reactions on the surfaces of clay minerals has been directed to the chemistry of soils and petroleum diagenesis (Mortland, 1970; Fripiat and Cruz-Cumplido, 1974; Johns and Shimoyama, 1972), and to test theories of chemical evolution and the origin of life (Bernal, 1951; Fripiat and Poncelet, 1974). More recently, the synthetic chemist has also become interested in reactions on clay surfaces and within intracrystalline clay galleries.

The swelling 2:1 layered silicates (smectites) stand out as particularly important in mediating organic reactions (Mortland, 1970; Pinnavaia, 1983; Thomas, 1982). Typical natural smectites include the abundant montmorillonite, hectorite, and saponite, as well as the less commonly found beidellite. These compounds have layered structures (see Chapter 11) in which two-dimensional polyoxoanions (10-Å-thick layers) are separated by sheets of hydrated cations, typically Na^+ and Ca^{2+}. The hydrated

*The support of this research by the National Science Foundation Division of Materials Research (DMR-8514154) and the Michigan State University Center for Fundamental Materials Research is gratefully acknowledged.

PREPARATIVE CHEMISTRY
USING SUPPORTED REAGENTS

cations, intercalated in the gallery region between more or less rigid oxide layers, are readily ion exchanged with a wide variety of positively charged species. Moreover, the galleries can be swelled by the adsorption of various polar solvents. With multiple layers of solvents the galleries become liquidlike and accessible for chemical reactions.

The gallery region of a smectite offers a novel intracrystalline arena for chemical synthesis. The reactivity patterns of reactions within the galleries can be influenced and in part controlled through choice of clay layer charge, exchange cation, and swelling solvent. The pillared clays (see Chapter 23), which contain robust gallery cations as molecular pillars between the layers, provide another intracrystalline arena for chemical conversions.

The present chapter is based mainly on the author's earlier work. Not only does it give practical examples of chemical synthesis using clays as supported catalysts, but, more important, it illustrates general synthesis principles and catalytic selectivity effects transferable by the synthetic chemist to other reaction systems.

II. SYNTHESIS OF INTERCALATED CLAY CATALYSTS

A. General Considerations

The gallery cations in pristine smectite clays are usually alkali metal and alkaline earth cations (Grim, 1968). For instance, montmorillonites from Wyoming generally contain intercalated Na^+, whereas the same mineral in Arizona normally is found as the Ca^{2+} exchange form. The full unit cell formulas for Wyoming montmorillonite is $Na_{0.60}[Al_{3.23}Fe_{0.42}Mg_{0.47}]$-$(Si_{7.87}Al_{0.13})O_{20}(OH)_4$. The related smectite clay hectorite (California) has a unit cell formula of $Na_{0.66}[Li_{0.66}Mg_{5.34}](Si_{8.00})O_{20}(OH,F)_4$. For these examples, there are 0.6–0.7 exchange equivalents per unit cell. In other smectites, including montmorillonite and hectorite from other localities, the cation exchange equivalents per unit cell may be as low as 0.4 or as high as 1.2. Thus, the layer charge densities and corresponding cation exchange capacities of smectite clays can vary over a considerable range. Despite these variations the charge density remains relatively low, and the average distance between charge centers is large (10–15 Å). Thus, typical transition metal complex cations are sufficiently small to be intercalated into smectite clays without the extent of exchange being limited by the size of the complex.

In preparing an intercalated clay catalyst, we normally replace some or all of the native cations with the desired metal complex. It is important that this metal complex be a cation to benefit from electrostatic binding in the galleries. Neutral complexes can intercalate via physical adsorption, but such interactions usually lead to desorption of the complex. There are two general approaches to the synthesis of intercalated smectite catalysts: (1) *in-situ* gallery assembly wherein a metal precursor is constructed in the clay galleries and (2) direct ion exchange wherein the complex catalyst (or precursor) is first formed in solution and subsequently exchanged into the galleries.

B. *In-Situ* Gallery Assembly

In this method of synthesis the metal complex catalyst is assembled within the clay galleries from a simple, preintercalated precursor. An example is the synthesis of the hydrogenation catalyst $RhH_2(PPh_3)_n^+$-montmorillonite by Pinnavaia *et al.* (1979). Rhodium is introduced into the gallery region as $Rh_2(OAc)_{4-x}^{x+}$ and then converted *in situ* to the desired complex as illustrated in the following scheme:

$$\overline{Na^+} + Rh_2(OAc)_{4-x}^{x+} \xrightarrow{MeOH} \overline{Rh_2(OAc)_{4-x}^{x+} + Na^+} \tag{1}$$

$$\overline{Rh_2(OAc)_{4-x}^{x+} + PPh_3} \xrightarrow{MeOH} \overline{Rh(PPh_3)_n^+ + OAc^-} \tag{2}$$

$$\overline{Rh(PPh_3)_n^+ + H_2(g)} \xrightarrow{MeOH} \overline{RhH_2(PPh_3)n^+} \tag{3}$$

The chemistry observed between $Rh_2(OAc)_{4-x}^{x+}$ and PPh_3 in the clay galleries resembles that observed in homogeneous solution (Wilson and Taube, 1975; Legzdins *et al.*, 1970).

The method does not allow facile control of metal complex stoichiometry. The gallery $Rh_2(OAc)_{4-x}^{x+}$ ions in this example are not uniformly accessible for reaction with stoichiometric quantities of phosphine ligand. To form intercalated $Rh(PPh_3)_n^+$ complexes with $n = 2$ and 3, we must utilize overall PPh_3: Rh ratios of 3 : 1 and 9 : 1, respectively. Thus, even in the presence of an excess of ligand, some of the $Rh_2(OAc)_{4-x}^{x+}$ precursor complexes are shielded from reaction with free ligand by the $Rh(PPh_3)_n^+$ centers in the galleries. Variability in the stoichiometry of gallery complexes can greatly complicate the interpretation of catalytic results. For the $Rh_2(OAc)_{4-x}^{x+}/RH(PPh_3)_n^+$ system, any unreacted $Rh_2(OAc)_{4-x}^{x+}$ in the galleries would be converted under hydrogenation conditions to Rh metal, a potent hydrogenation catalyst with reactivity and selectivity properties dramatically different from $RhH_2(PPh_3)_n^+$.

The *in-situ* gallery synthesis method minimizes the number of manipulations needed to form an intercalated clay catalyst. This could be helpful when dealing with catalyst intermediates that are highly reactive toward trace impurities such as water and oxygen or that are otherwise unstable in solution. However, in view of these disadvantages, the method is not a general route to intercalated clay catalysts.

C. Direct Ion Exchange

It is thus preferable to first generate the metal complex catalyst or catalyst precursor in homogeneous solution and then directly exchange the complex into the gallery region of the host. For example, cationic rhodium–phosphine precursor complexes of specific stoichiometry can be formed according to standard solution methods (Schrock and Osborn, 1971):

$$[Rh(NBD)Cl]_2 + PPh_3 \xrightarrow{\ KPF_6\ } Rh(NBD)(PPh_3)_2^+ \tag{4}$$

where NBD is norbornadiene. The complex precursor can then be exchanged into montmorillonite and activated by reaction with molecular hydrogen (Raythatha and Pinnavaia, 1983):

$$\overline{Na^+} + Rh(NBD)(PPh_3)_2^+ \xrightarrow{\ MeOH\ } \overline{Rh(NBD)(PPh_3)_2^+} + Na^+ \tag{5}$$

$$\overline{Rh(NBD)(PPh_3)_2^+} + H_2(g) \xrightarrow{\ MeOH\ } \overline{RhH_2(PPh_3)_2^+} + norbornane \tag{6}$$

Thus, all of the gallery complexes have identical stoichiometries.

Clay-intercalated rhodium hydrogenation catalysts containing bidentate phosphine ligands (Schrock and Osborn, 1971) also have been prepared by similar methods (Raythatha and Pinnavaia, 1981, 1983):

$$\overline{Na^+} + Rh(NBD)(dppe)^+ \xrightarrow{\ MeOH\ } \overline{Rh(NBD)(dppe)^+} + Na^+ \tag{7}$$

$$\overline{Rh(NBD)(dppe)^+} + H_2(g) \xrightarrow{\ MeOH\ } \overline{Rh(dppe)^+} + norbornane \tag{8}$$

where dppe is 1,2-bis(diphenylphosphino)ethane. In principle, any metal complex cation can be exchanged into smectite galleries provided the complex is sufficiently soluble to drive exchange by mass action.

D. Utilization of Positively Charged Ligands

Immobilization of a metal complex catalyst in smectite galleries demands a net positive charge on the complex. The positive charge ensures electrostatic binding of the complex to the polyanionic layers.

Cationic analogs of neutral metal complex catalysts can be prepared by replacing neutral ligands with positively- charged ligands. Catalyst immobilization in montmorillonite by this means was first demonstrated by Quayle and Pinnavaia (1979) in their preparation of cationic analogs of Wilkinson-type complexes. The positively- charged phosphonium–phosphine ligand $Ph_2PCH_2CH_2\overset{+}{P}Ph_2CH_2Ph$, abbreviated $P—P^+$, was used in place of neutral PPh_3 ligands. Mixed ligand complexes of the type $RhCl(P—P^+)$-$(PPh_3)_x$ $(x = 1, 2)$ were prepared from the cyclooctadiene complex $[RhCl(COD)]_2$ according to Eqs. (9) and (10):

$$[RhCl(COD)]_2 + P—P^+ \longrightarrow RhCl(COD)(P—P^+) \tag{9}$$

$$RhCl(COD)(P—P^+) + PPh_3 \longrightarrow RhCl(P—P^+)(PPh_3)_x \tag{10}$$

The positively charged $RhCl(P—P^+)(PPh_3)_x$ complexes were then incorporated into the montmorillonite by direct ion exchange. Under hydrogen at ambient conditions the mixed ligand complexes were active precursors for the catalytic hydrogenation of 1-hexene.

Rhodium complexes containing the $P—P^+$ ligand have been used as olefin hydroformylation catalysts (Farzaneh and Pinnavaia, 1983). The following reaction scheme was utilized to form the intercalated catalyst:

$$RhCl(COD)P—P^+ + P—P^+ \xrightarrow[\text{1 atm, 25°C}]{CO/H_2} RhCl(CO)(P—P^+)_2 \tag{11}$$

$$\overline{Na^+} + RhCl(CO)(P—P^+)_2 \longrightarrow \overline{RhCl(CO)(P—P^+)_2} + Na^+ \tag{12}$$

$$\overline{RhCl(CO)(P—P^+)_2} \xrightarrow[\text{37 atm, 100°C}]{CO/H_2} \overline{RhH(CO)_x(P—P^+)_2} + HCl \tag{13}$$

There may be a limit to the number of sterically bulky $P—P^+$ ligands that can be used effectively at a metal center intended for catalysis. Quayle and Pinnavaia (1979) found that although it was possible to prepare a $P—P^+$ analog of Wilkinson's complex, namely $RhCl(P—P^+)_3$, this species was inactive as a catalyst precursor for olefin hydrogenation. Mixed-ligand complexes containing $P—P^+$ and PPh_3 [e.g., $RhCl(PPh_3)_x(P—P^+)_{3-x}$] are potentially capable of undergoing ligand distribution reactions to form the parent analogs [e.g., $RhCl(PPh_3)_3$ and $RhCl(P—P^+)_3$]. Such a redistribution would be undesirable because the neutral parent would be desorbed to solution and the positively charged parent, although being retained in the clay galleries, would be inactive for steric reasons. However, most mixed-ligand complexes, including those with distinguishable phosphine ligands, tend to be thermodynamically favored over their parent end members (Crabtree *et al.*, 1977).

Ligand dissociation reactions may also complicate the use of positively charged ligands for immobilizing metal complex catalysts in clay galleries. Consider the intercalated hydroformylation catalyst $RhH(CO)_x(P—P^+)_2$. This species can undergo a ligand dissociation reaction with formation of a neutral metal carbonyl in solution leaving of $P—P^+$ in the galleries of the clay[Eq. (14)]:

$$\overline{RhH(CO)_x(P—P^+)_2} \underset{}{\overset{H_2O/CO}{\rightleftharpoons}} \overline{P—P^+} + \text{Rh-carbonyl} \qquad (14)$$

This equilibrium is very solvent dependent. In acetone and benzene (suitable solvents), the equilibrium lies far to the left, favoring the intercalated catalyst. However, in dimethylformamide (DMF) (and unsuitable solvent) the equilibrium is far to the right.

III. CATALYTIC REACTIONS

A. Alkyne Hydrogenations

1. Choice of Swelling Solvent

In addition to functioning as a medium for the transport of reagents, the solvent chosen should function as a gallery swelling agent to allow access to the intracrystal metal centers. Occasionally, the metal complex itself may be sufficiently robust to serve as a pillar within the galleries (Pinnavaia, 1983) and allow access to the galleries even in the absence of a swelling solvent. This is the case for the complex $Os_3(CO)_{11}(P—P^+)$, which catalyzes the isomerization of 1-hexene (Giannelis and Pinnavaia, 1985).

The importance of gallery swelling is illustrated by the results in Table I for hydrogenation of 2-decyne to cis-2-decene by $RhH_2(PPh_3)_2^+$ intercalated in hectorite (Pinnavaia *et al.*, 1979). The ratio of the reaction rate of the intercalated catalyst relative to that of the homogeneous catalyst (R_I/R_H) parallels the gallery heights determined by each solvent, as indicated by the parameter d_{001}. Methylene chloride is here the best swelling solvent, affording a gallery height of 10.0 Å. Benzene is the poorest solvent, the gallery height of 5.7 Å being essentially that dictated by the intercalated complex itself. The low reactivity observed for the clay catalyst in benzene most likely arises from the small fraction of metal complexes bound at the external surfaces of the clay particles.

TABLE I

Initial Hydrogenation Rates for Reduction of 2-Decyne with $RhH_2(PPh_3)_2{}^+$ as Catalyst under Clay-Intercalated and Homogeneous Reaction Conditions[a]

	Initial rates[b]			
Solvent	Intercalated	Homogeneous	R_1/RH	$\Delta d001$ (Å)
$CH_2Cl_2{}^c$	2800	3300	0.85	10.0
MeOH	1200	2800	0.43	7.7
Et_2/MeOH (3 : 1 v/v)	600	2800	0.24	6.7
$C_6H_6{}^c$	20	1000	0.02	5.7

[a] Reaction conditions: substrate concentration 1.0 M; substrate to Rh ratio 2000 : 1; 25°C; 1 atm pressure.

[b] mL H/min/mmol Rh.

[c] These solvents contained 7% methanol by volume.

2. Transition-State Selectivity

The relationship between reactivity and gallery swelling suggests that spatial factors can be important for the substrate selectivity of an intercalated catalyst (Pinnavaia, 1983). This is illustrated by the data in Table II for the hydrogenation of a series of alkynes by hectorite-intercalated $RhH_2(PPh_3)_2{}^+$ with methanol as the swelling solvent ($\Delta d_{001} = 7.7$ Å). As the spatial requirements on either side of the C≡C bond are increased, the R_1/R_4 values decrease greatly.

The observed substrate selectivity has been attributed to transition-state selectivity due to preferred orientations of the metal complex catalyst in the clay galleries. One possible orientation for the reactive intermediate is shown in Fig. 1, wherein the *trans*-Ph$_3$P—Rh—PPh$_3$ axis is parallel to the

TABLE II

Initial Rates for Hydrogenation of Alkynes in Methanol by Hectorite-Intercalated and Homogeneous $RhH_2(PPh_3)_2{}^{+a}$

Substrate	Intercalated	Homogeneous	R_1/R_H
1-Hexyne	2100	2100	1.0
2-Hexyne	2200	2400	0.92
3-Decyne	1200	2500	0.48
3-Hexyne	360	1800	0.20
PhC≡CPh	<1	100	<0.01

[a] mL H$_2$/min/mmol Rh.

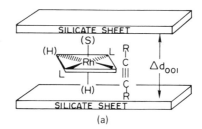

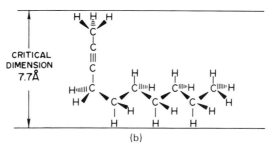

Fig.1. (a) Proposed orientation for the reactive $RhH_2(PPh_3)_2^+$–alkyne intermediate formed in the interlayers of hectorite. (b) The critical dimension, illustrated for 2-decyne, is defined as the minimum distance to be spanned by the substrate to achieve a reactive configuration.

silicate layers and the C≡C is perpendicular to the layers. If the gallery height is greater than the critical dimension needed to achieve the reactive configuration, then reaction will be facile. Gallery heights less then the critical dimension will impede reaction. Thus, by tuning the gallery swelling we can mediate substrate selectivity on the basis of transition-state size. More generally, transition-state selectivity can be designed in clay-catalyzed reactions whenever the complex adopts a preferred orientation within the galleries.

3. Catalyst Longevity

In addition to causing rather dramatic selectivity effects, intercalating metal complexes in smectite clays can result in equally impressive effects on catalyst longevity (Pinnavaia *et al.*, 1979). For the reduction of a terminal alkyne with $RhH_2(PPh_3)_2^+$ in homogeneous solution, the reaction typically ceases after only 1000 catalytic turnovers. However, the complex intercalated in hectorite remains active even after 1500 turnovers. The boost in performance for the intercalated complex is illustrated more clearly by the hydrogen uptake plots shown in Fig. 2 for the hydrogenation

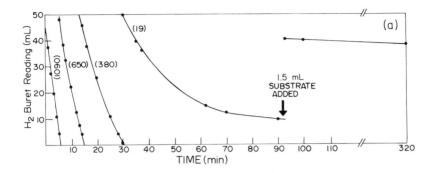

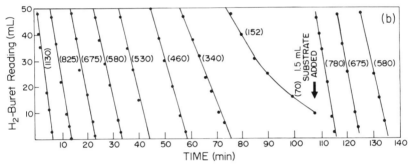

Fig. 2. Hydrogen uptake plots for reduction of 1-hexyne to 1-hexene with (a) homogeneous and (b) intercalated $RhH_2(PPh_3)_2{}^+$. The numbers in parenthesis are uptake rates. The arrows indicate the times of addition of fresh 1-hexyne. The initial 1-hexyne concentration in methanol was $1.0\ M$; the initial 1-hexyne: Rh ratio was 2000:1.

of 1-hexyne. Upon the addition of fresh substrate, the activity of the homogeneous system is not restored, whereas the intercalated system regains about 70% of its original activity. Schrock and Osborn (1976b) have suggested that catalyst deactivation under homogeneous solution conditions may occur by formation of Rh—C≡C—R bonds. The intrinsic Brønsted acidity of clay interlayers (Mortland, 1968) may inhibit dissociation of protons from the relatively acidic terminal alkynes and the subsequent formation of Rh—C≡C—R bonds.

B. Alkene Hydrogenation

1. Surface Equilibria Effects

In addition to being a catalyst for the conversion of alkynes to the corresponding cis-alkenes, $RhH_2(PPh_3)_2{}^+$ also hydrogenates terminal

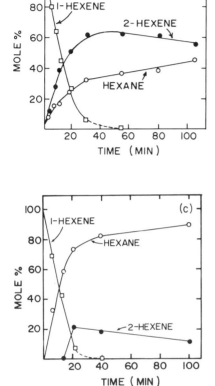

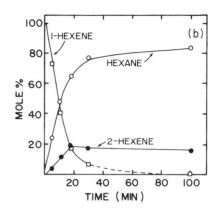

Fig. 3. Hydrogenation–isomerization of 1-hexene in MeOH by $RhH_2(PPh_3)_2^+$. (a) Complex in homogeneous solution; initial concentration of 1-hexene was 1.0 M. (b) Hectorite-intercalated complex; 0.10 M 1-hexene. (c) Hectorite-intercalated complex; 1.0 M 1-hexene. The amount of rhodium complex used was 0.015 mmol in 15 mL of solution. The water content of the methanol was 0.20 wt. % in each case.

olefins to alkanes. Significant isomerization of the terminal olefin to internal olefins accompanies the homogeneous hydrogenation reaction. This is undesirable because the internal olefins are much less reactive toward hydrogenation than are terminal olefins. Figure 3a illustrates the evolution of products obtained from the hydrogenation of 1-hexene under homogeneous conditions in methanol (Raythatha and Pinnavaia, 1981). After 100% conversion of 1-hexene, the amount of 2-hexene formed is almost twice that of hexane. Hydrogenation of 2-hexene to hexane continues after the complete conversion of 1-hexene, but the reaction is very slow.

$RhH_2(PPh_3)_2^+$ in homogeneous solution is involved in the protic equilibrium of Eq. (15) (Schrock and Osborn, 1976a):

$$RhH_2(PPh_3)_2^+ \rightleftharpoons RhH(PPh_3)_2 + H^+ \qquad (15)$$

The dihydride complex is a much better hydrogenation than isomerization catalyst, whereas the reverse is true for the monohydride. Thus, the hexene to 2-hexene product distribution at 100% 1-hexene conversion depends on the relative amounts of monohydride and dihydride complex in solution.

Raythatha and Pinnavaia (1983) have shown that the intercalation of $RhH_2(PPh_3)_2^+$ in hectorite can markedly influence the product distribution obtained in the hydrogenation of terminal olefins. As shown in Fig. 3b and c, the intercalated catalyst dramatically inhibits isomerization. Isomerization is totally inhibited for 0.80 M 1-hexene up to a 1-hexene conversion of 60%. The ability of the clay gallery to inhibit olefin isomerization has been attributed to the Brønsted acidity due to polarization of water molecules by unexchanged gallery Na^+ ions (Mortland *et al.*, 1963):

$$\overline{Na^+ - H_2O} \longrightarrow \overline{[NaOH] + H^+} \tag{16}$$

The equilibrium of Eq. (16) dictates the relative amounts of dihydride and monohydride present in the gallery. Thus, when the galleries are very acidic, the dihydride is favored, and hydrogenation is favored over isomerization.

As might be expected, the position of the equilibrium expressed by Eq. (16) is dependent on the amount of water present in the solvent and on the substrate concentration. Both influence the amounts of water present in the gallery and, hence, the proton activity in the gallery (Mortland and Raman, 1968; Mortland, 1968). The effect of substrate concentration at constant water content (0.20 wt.%) can be seen in the comparison of product distributions shown in Fig. 3b and c. The effect of water on the product distribution is illustrated by the data in Table III. At water levels of 0.10–0.20 wt.% the galleries are highly acidic, because the gallery sodium is incompletely aquated and has high polarizing ability. Under these conditions, the dihydride is favored and hydrogenation predominates over isomerization. At higher water concentrations (0.5–1.0 wt.%) the gallery Na^+ is more heavily solvated, and less acidifying; hence, the extent of hydrogenation via the dihydride route is diminished.

These results show the importance of surface acidity effects in clay galleries in switching between reaction pathways related by protic equilibria. In general, deviations from solution product distributions can be expected whenever the product distributions are controlled by two or more intermediates, because the relative stabilities of the stoichiometrically distinct species almost certainly will be different in the electrically charged clay galleries than in homogeneous solution.

TABLE III

Effect of Water on the $RhH_2(PPh_3)_2{}^+$-Catalyzed Hydrogenation of 0.80 M
1-Hexene in Methanol[a]

wt. % H_2O[b]	Hydrogenation	Isomeri- zation	Hexane: 2-hexene at 50% conversion
	Intercalated catalyst		
0.10	33	0	>95:5
0.20	33	0	>95:5
0.50	4.0	5.3	56:44
1.0	4.0	5.3	50:50
	Homogeneous catalyst		
0.20	11	21	40:60
0.50	10	20	32:68
1.00	11	26	34:66

[a] Rates are expressed as mol/min/mol Rh.

[b] Water content of the methanol solvent.

2. Pathways with Common Intermediates

It is also possible to influence competing reaction pathways in clay galleries when they are not related by equilibria between multiple intermediates. For instance, the 1,2-bis(diphenylphosphino)ethane complex $Rh(dppe)^+$ is a catalyst precursor for olefin hydrogenation and isomerization (Halpern *et al.*, 1977). However, in this case, the metal does not add dihydrogen until olefin is first complexed to the metal.

$$Rh(dppe)^+ \xrightleftharpoons{(ol)} Rh(ol)(dppe)^+ \xrightleftharpoons{H_2} RhH_2(ol)(dppe)^+ \quad (17)$$

$$RhH_2(ol)(dppe)^+ \longrightarrow Rh(dppe)^+ + alkane + olefin\ isomers \quad (18)$$

The hydrogenation and isomerization pathways both proceed through the same $RhH_2(ol)(dppe)^+$ intermediate as depicted in Eq. (18). Isomerization via $Rh(ol)(dppe)^+$ is precluded by the fact that no reaction of olefin is observed in the absence of H_2. Thus, we might expect the distribution of hydrogenated and isomerized products to be similar under homogeneous and intercalated catalyst conditions. However, as seen from the product distributions in Fig. 4a and b, the extent of isomerization is significantly

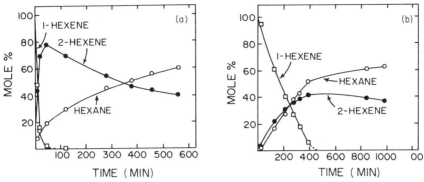

Fig. 4. Hydrogenation of 1.2 M 1-hexene in methanol with Rh(dppe)$^+$ under (a) homogeneous and (b) hectorite-intercalated reaction conditions. The Rh(dppe)$^+$ was derived for Rh(NBD)(dppe)$^+$ as a precursor, where NBD in norbornadiene. The reaction mixture contained 0.015 mmol Rh in 15 mL of solution.

lower for the intercalated catalyst. It has been suggested that electric field effects, spatial factors, or solvation effects which operate in the clay gallery environment alter the relative stabilities of transition states emanating from a common intermediate (Raythatha and Pinnavaia, 1983).

Another relevant reaction system is the hydrogenation of 1,3-butadienes to monoenes with Rh(dppe)$^+$ as catalyst (Schrock and Osborn, 1976b):

$$\underset{R_1 \quad R_2}{\diagup\diagdown} \xrightarrow[\text{Rh(dppe)}^+]{\text{H}_2} \underset{\underbrace{\hspace{2cm}}_{1,2\ \text{Addn.}}}{\diagup\diagdown_{R_1}\diagup\diagdown_{R_2} + \diagup\diagdown_{R_1}\diagup\diagdown_{R_2}} + \underset{\underbrace{\hspace{1cm}}_{1,4\ \text{Addn.}}}{\diagup\diagdown_{R_1}\diagup\diagdown_{R_2}} \qquad (19)$$

The terminal olefins are the result of overall 1,2-hydrogen addition whereas the internal olefins result from overall 1,4-addition.

Mechanistically the conversion of 1,3-butadienes to monoenes is believed to occur via the following reaction sequence:

$$\text{Rh(dppe)}^+ \xrightarrow{\text{diene}} \text{Rh(dppe)(diene)}^+ \xrightarrow{\text{H}_2} \text{HRh(dppe)(allyl)}^+ \qquad (20)$$

$$\text{HRh(dppe)(allyl)}^+ \longrightarrow \text{Rh(dppe)}^+ + \text{monoenes} \qquad (21)$$

The distribution of 1,2- and 1,4-addition products is controlled kinetically by the reductive elimination of olefin from the metalloallyl intermediate of

Eq. (21). Two types of allyl complexes can be evisioned. One is a π-bonded η^3 species of the type wherein the product distribution is

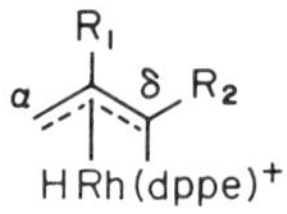

determined by the rates of hydrogen transfer to the δ position (1,2-addition) and the α position (1,4-addition). The other possibility is a mixture of η^1 diastereomers which eliminate to terminal and internal olefins; respectively:

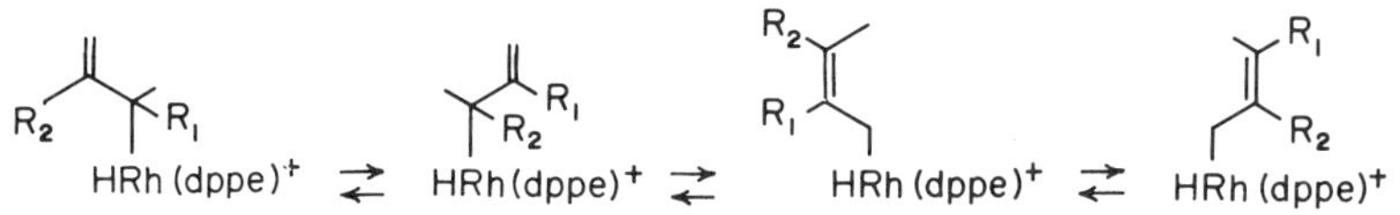

$$(22)$$

Table IV compares the reaction rates and product distributions for hydrogenation of 1,3-butadienes to monoenes with Rh(dppe)$^+$ as catalyst under homogeneous and clay-intercalated reaction conditions (Raythatha and Pinnavaia, 1981). Under homogeneous reaction conditions the rates are solvent dependent, but there is no simple correlation between reaction rate and solvent or substrate properties. However, the relative rates of reaction for the clay-intercalated catalyst correlate with the swelling power of the solvent (acetone > methanol > benzene), again demonstrating the importance of selecting a good swelling solvent in optimizing the reactivity of a clay catalyst. In general, overall 1,4-addition is preferred over 1,2-addition, but the fraction of internal olefin produced is considerably lower than the value expected for reaction under thermodynamic control. Significantly, the yields of terminal olefins produced by the intercalated clay catalyst are 1.5–2.3 times as large as the yields obtained with the homogeneous catalyst. Regardless of whether η^3 or η^1 intermediates are involved in the determination of the product distribution, these results demonstrate that the clay gallery environment can significantly mediate the course of the reductive elimination step [cf. Eq. (21)].

3. Complexes with Positively Charged Ligands

Neutral metal complexes and cationic metal hydrides capable of entering into protic equilibria pose severe limitations for intercalation in clays. To circumvent the limitations, rhodium complexes containing P—P$^+$ have been synthesized [Eqs. (9)–(13)] and investigated as alkene hydrogenation catalysts. The data in Table V for 1-hexene hydrogenation with RhCl(P—P$^+$)(PPh$_3$) illustrate the effectiveness of positively charged

TABLE IV

Reaction Rates and Product Distributions for Hydrogenation of 1,3-Butadienes with Rh(dppe)[+] under Homogeneous and Clay-Intercalated Conditions[a]

Butadiene	Solvent	Homogeneous catalyst			Intercalated catalyst		
		Rate[a]	%1,2-Addition	%1,4-Addition	Rate[a]	%1,2-Addition	%1,4-Addition
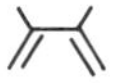 ⋎⧸	Acetone	30	30	70	25	45	55
	MeOH	58	33	67	4.4	60	40
	C_6H_6	1030	33	67	<0.01	—	—
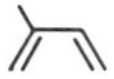 ⋏⧸	Acetone	760	19	81	300	34	66
	MeOH	1070	20	80	126	44	56
	C_6H_6	870	21	79	<0.01	—	—
 ⧸⧹	Acetone	780	17	83	430	32	68
	MeOH	1330	20	80	370	39	61
	C_6H_6	100	19	81	<0.01	—	—

[a] Rates are expressed as mL H_2/min/mmole Rh.

TABLE V

Rates of 1-Hexene Hydrogenation in Methanol with
$RhCl(PPh_3)(P-P^+)$ as Catalyst Precursor[a]

Number of catalyst turnovers	Homogeneous catalyst	Intercalated catalyst
100	100	150
200	90	180
300	100	190
400	90	190
500	80	150
600	80	190
700	70	140
800	60	100
900	60	70

[a] 1-Hexene/Rh = 1500; initial 1-hexene concentration = 1.0 M; H_2 pressure 740 torr. Rates are expressed as mL H_2/min/mmol Rh.

ligands as a method of designing clay-intercalated catalysts (Quayle and Pinnavaia, 1979). Significantly, the intercalated catalyst is somewhat more reactive than the homogeneous catalyst.

C. Asymmetric Hydrogenations

Oftentimes we desire the practical advantages of a supported metal complex catalyst without sacrificing the intrinsic selectivity characteristic of the unsupported catalyst. The catalytic asymmetric hydrogenations of procchiral olefins is a case in point. In these systems the enantiomeric purity of the alkane products is determined by the asymmetry of the ligands in the coordination complex. Rhodium complexes of the type $Rh(diphos^*)^+$, where diphos* is a chiral bidentate phosphine ligand, are asymmetric hydrogenation catalysts suitable for intercalation in clays (Pinnavaia, 1982). For instance, α-acylaminoacrylates are hydrogenated by such catalysts to chiral aminoacid derivatives, some of which have medicinal utility (Kagan and Dang, 1972; Fryzuk and Bosnich, 1978). $Rh((R)$-Prophos)$^+$ and $Rh(4$-Me-(R)-Prophos)$^+$ are particularly effective catalysts for the asymmetric hydrogenation of acrylic acid derivatives (Fryzuk and Bisnich, 1978). The structure of (R)-Prophos is

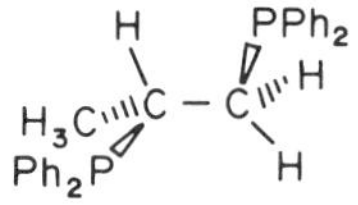

TABLE VI

Optical Yields for the Asymmetric Hydrogenation of Prochiral Olefins with
Rh(4-Me-(R)-Prophos)$^+$ as Catalyst[a]

Substrate	Intercalated catalyst	Homogeneous catalyst
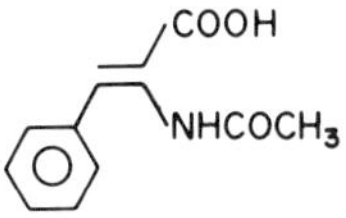	89.6	92.6
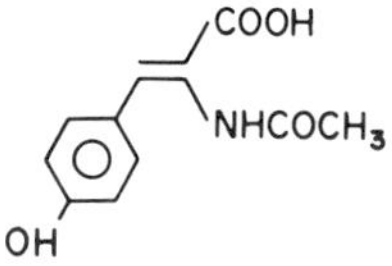	78.5	72.0
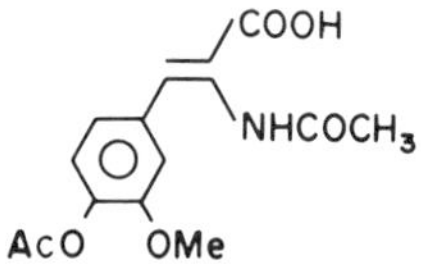	95.1	95.3

[a] Optical yields are expressed as enantiomeric excess. Reaction was carried out at 25°C, 1 atm, in ethanol. Chemical yields were >98%.

The 4-Me-(R)-Prophos ligand is a derivative of (R)-Prophos with all phenyl groups methylated in the para position.

Table VI provides the optical yields obtained for the hydrogenation of related acrylic acid derivatives with [Rh(4-Me-(R)-Prophos)]$^+$ under clay-intercalated and homogeneous solution conditions (Pinnavaia, 1982). The optical yields are very similar for the supported and unsupported catalyst. In a related study, Mozzei et al. (1980) observed a dependence of the optical yields on the nature of the clay support. In general, however, the intercalated clay catalysts exhibit sufficiently high optical yields to make them competitive alternatives to the homogeneous catalyst and other support systems for the large scale synthesis of products of commercial value such as L-DOPA.

D. Hydroformylations

Rh(diene)(PPh$_3$)$_2$$^+$ complexes are rare examples of *cationic* catalyst precursors for homogeneous olefin hydroformylation (Schrock and Osborn, 1971; Oro et al., 1978). These precursors once were thought to generate positively charged catalysts under hydroformylation conditions (Uson et al., 1978), but the active species are actually neutral

$RhH(CO)_x(PPh_3)_2$ complexes with $x = 1, 2, 3$ (Crabtree and Felkin, 1979; Farzaneh and Pinnavaia, 1983). These latter complex catalysts can also be formed from simple rhodium precursors such as $[Rh(diene)Cl]_2$ and $RhCl(diene)(PPh_3)$ (Evans et al., 1968a,b).

Under hydroformylation conditions clays intercalated with Rh(diene)-$(PPh_3)_2^+$ complexes desorb rhodium through the following reaction sequence.

$$\overline{Rh(diene)(PPh_3)} \xrightarrow{CO/H_2} \overline{Rh(CO)_x(PPh_3)_2^+} \tag{23}$$

$$\overline{Rh(CO)_x(PPh_3)_2^+} \underset{H_2}{\rightleftharpoons} \overline{H^+} + HRh(CO)_x(PPh_3)_{2\ (soln)} \tag{24}$$

However, the replacement of PPh_3 by $P—P^+$ affords cationic $HRh(CO)_x$ $(P—P^-)_2$ analogs which remain immobilized in the clay galleries, at least when acetone is the swelling solvent (Farzaneh and Pinnavaia, 1983). With DMF as solvent, ligand exchange with solvent occurs and a neutral carbonyl complex is formed in solution [Eq. (14)].

Table VII provides a comparison of the product distributions obtained from the $[RhCl(COD)]_2 + P—P^+$ precursor system under homogeneous and clay-intercalated reaction conditions (Farzaneh and Pinnavaia, 1983). Somewhat higher normal/branched (n/b) aldehyde ratios are obtained for the clay-intercalated catalyst relative to the homogeneous catalyst in

TABLE VII

Hydroformylation of 1-Hexene with the $[RhCl(COD)]_2 + 2P{\equiv}P^+$ Precursor System[a]

| Solvent | Reaction time (hr) | Conversion (%) | Product distribution (%) | | | | n/b |
			Heptanal	2-Methyl-heptanal	2-Hexene	Other	
			Homogeneous catalyst[b]				
Acetone	3	100	45	26	25	4	1.5
C_6H_6	2	70	29	14	57	—	2.0
			Intercalated catalyst[c]				
Acetone	12	80	53	22	25	—	2.4
C_6H_6	18	0	—	—	—	—	—

[a] Initial 1-hexene concentration, 0.40 M; 100°C; 600 psi; $CO/H_2 = 1$.
[b] 1-Hexene/Rh = 200.
[c] 1-Hexene/Rh = 1300.

acetone. Similar results are obtained when $[Rh(CO)_2Cl]_2$ and $Rh(COD)^+$ are used as precursors, except that the absence of Cl in the $Rh(COD)^+$ precursor system greatly reduces the yields of undesirable 2-hexene isomerization product. Finally, although benzene is a suitable solvent for the homogeneous catalyst, it is an unsuitable swelling medium for the intercalated catalyst.

REFERENCES

Bernal, J. E. (1951). "The Physical Basis of Life." Routledge & Kegan Paul, London.

Crabtree, R. H., and Felkin, H. (1979). *J. Mol. Catal.* **5**, 75.

Crabtree, R. H., Gautier, A., Giordano, G., and Khan, T. (1977). *J. Organomet. Chem.* **141**, 113.

Evans, D., Osborn, J. A., and Wilkinson, G. (1968a). *J. Chem. Soc. AP.* 3133.

Evans, D., Yagupsky, G., and Wilkinson, G. (1968b). *J. Chem. Soc. AP.* 2660.

Farzaneh, F., and Pinnavaia, T. J. (1983). *Inorg. Chem.* **22**, 2216.

Fripiat, J. J., and Cruz-Cumplido, M. I. (1974). *Am. Rev. Earth Planet. Sci.* **2**, 239.

Fripiat, J. J., and Poncelet, G. (1974). *Adv. Org. Geochem., Proc. Int. Meet.*, *6th* p. 875.

Fryzuk, M. D., and Bosnich, B. (1978). *J. Am. Chem. Soc.* **100**, 5491.

Giannelis, E. P., and Pinnavaia, T. J. (1985). *Inorg. Chem.* **24**, 3602.

Grim, R. E. (1968). "Clay Mineralogy." 2nd Ed. McGraw-Hill, New York.

Halpern, J., Riley, D. P. S., Chan, A. S. C., and Pluth, J. J. (1977). *J. Am. Chem. Soc.* **99**, 8055.

Johns, W. D., and Shimoyama, A. (1972). *Am. Assoc. Pet. Geol. Bull.* **56**, 2160.

Kagan, H. B., and Dang, T. P. (1972). *J. Am. Chem. Soc.* **94**, 6429.

Legzdins, P., Mitchell, R. W., Rempel, G. L., Ruddick, J. D., and Wilkinson, G. (1970). *J. Chem. Soc. AP.* 3322.

Mortland, M. M. (1968). *Trans. Annu. Congr. Soil Sci.* **1**, 691.

Mortland, M. M. (1970). *Adv. Agron.* **22**, 75.

Mortland, M. M., and Raman, K. V. (1968). *Clays Clay Miner.* **16**, 393.

Mortland, M. M., Fripiat, J. J., Chaussidon, J., and Uytterhoven, J. (1963). *J. Phys. Chem.* **67**, 248.

Mozzei, M., Marconi, W., and Riocci, M. (1980). *J. Mol. Catal.* **9**, 381.

Oro, L. A., Manrique, A., and Royo, M. (1978). *Trans. Met. Chem.* **3**, 383.

Pinnavaia, T. J. (1982). *ACS Symp. Ser.* No 192, 241.

Pinnavaia. T. J. (1983). *Science* **220**, 365.

Pinnavaia, T.J., Raythatha, R., Lee, J. G.-S., Halloran, L. J., and Hoffman, J. F. (1979). *J. Am. Chem. Soc.* **101**, 6891

Quayle, W. H., and Pinnavaia, T. J. (1979). *Inorg. Chem.* **18**, 2840.

Raythatha, R., and Pinnavaia, T. J. (1981). *J. Organomet. Chem.* **218**, 115.

Raythatha, R., and Pinnavaia, T. J. (1983). *J. Catal.* **80**, 47.

Schrock, R. R., and Osborn, J. A. (1971). *J. Am. Chem. Soc.* **93**, 2397.

Schrock, R. R., and Osborn, J. A. (1976a). *J. Am. Chem. Soc.* **98**, 2134.

Schrock, R. R., and Osborn, J. A. (1976b), *J. Am. Chem. Soc.* **98**, 4450.

Thomas, J. M. (1982). *In* "Intercalation Chemistry" M. S. (Whittingham and A. J. Jacobson, eds.), Chap. 3. Academic Press, New York.

Uson, R., Oro, L. A., Claver, C., and Garralda, M. A. (1978). *J. Mol. Catal.* **4**, 231.

Wilson, C. R., and Taube, H. (1975). *Inorg. Chem.* **14**, 2276.

26 ANIONIC ACTIVATION

André Foucaud

Groupe de Chimie Structurale
Université de Rennes
F- 35042 Rennes, France

I. INTRODUCTION

Clays are composed of layer silicates. There is scope in these structures for isomorphous replacement of Si^{4+} or Al^{3+} by cations of similar size but different and usually lower valence. As a result of such isomorphous substitution, the layers of many clay minerals are negatively charged. This charge is compensated, for the most part, by sorption of cations which may or may not be exchangable. The acid and base properties of clays can be modified by exchanging the cations present in the interlamellar region (Theng, 1974).

The clay surface catalyzes numerous organic reactions. Generally, the reaction is initiated by proton donation from Brønsted acid sites or complexation on the Lewis acid sites of the clay (Swartzen-Allen and Matijevic, 1974). More rarely, the negatively charged lattice has been used to initiate organic reactions. In this chapter, emphasis will be given to the reactions of clays as nucleophiles and as support of reagents for anionic activation.

The surface oxygens of the anionic framework of layer silicates are only weak electron donors. This property is apparent in the high frequencies of the N—H stretching modes of ammonium compounds absorbed on montmorillonite. These frequencies are comparable with those shown by the corresponding perchlorates and are higher than in the halide salts as demonstrated by Greenland and Quirk (1962). However, these negatively

PREPARATIVE CHEMISTRY
USING SUPPORTED REAGENTS

charged oxygen surfaces of montmorillonites can orient intercalated polar molecules and give reactions with electrophilic organic compounds.

II. NUCLEOPHILIC PROPERTIES OF SMECTITES

The reaction of Na-montmorillonite with the sulfonium salt **I**, to give the substitution product **II**, has been described by Deuel *et al.* (1950) [Eq. (1)]:

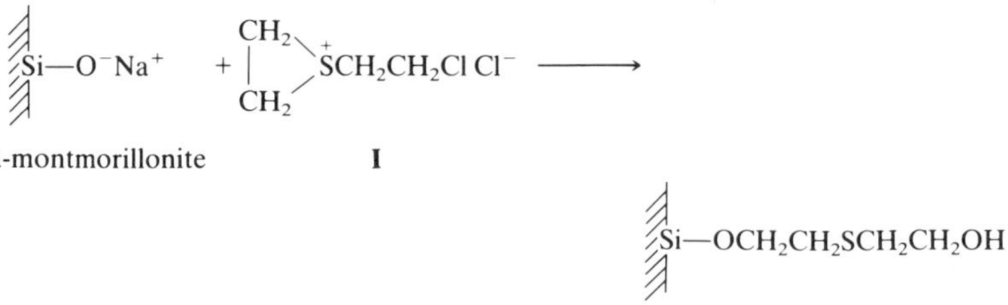

$$\text{Si}\!-\!\text{O}^-\text{Na}^+ \quad + \quad \underset{\text{CH}_2}{\overset{\text{CH}_2}{\diagdown}}\!\!\overset{+}{\text{S}}\text{CH}_2\text{CH}_2\text{Cl Cl}^- \quad \longrightarrow$$

Na-montmorillonite   **I**

$$\text{Si}\!-\!\text{OCH}_2\text{CH}_2\text{SCH}_2\text{CH}_2\text{OH} \quad (1)$$

II

The reaction of Na-montmorillonite with $SOCl_2$ gives the chloroderivative **III** which reacts with various nucleophiles to form stable covalently bonded organic moieties in the interlamellar region [Eq. (2)]:

$$\text{Si}\!-\!\text{O}^-\text{Na}^+ + \text{SOCl}_2 \longrightarrow \text{Si}\!-\!\text{Cl} + \text{SO}_2 + \text{NaCl} \quad (2)$$

III

For example, the reaction of **III** with alcohols gives alkoxymontmorillonites **IV** [Eq. (3)]:

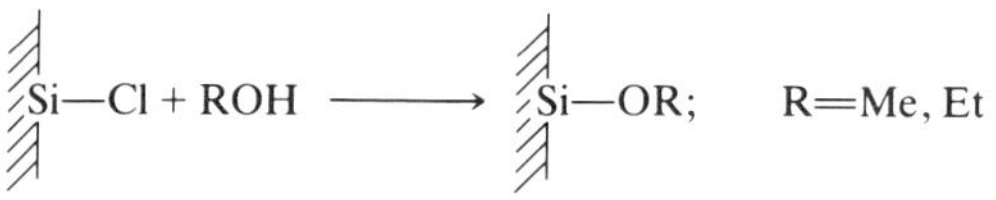

$$\text{Si}\!-\!\text{Cl} + \text{ROH} \longrightarrow \text{Si}\!-\!\text{OR}; \quad \text{R}\!=\!\text{Me, Et} \quad (3)$$

IV

The reaction of **III** with ethylmagnesium iodide gives ethylmontmorillonite **V** (Duel and Huber, 1951). The Friedel–Crafts reaction of **III** with benzene and $AlCl_3$ gives the phenyl montmorillonite **VI**. Substituted montmorillonites **IV–VI** are hydrophobic compounds (Gentili and Deuel, 1957):

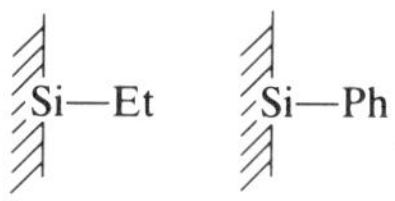

$$\text{Si}\!-\!\text{Et} \qquad \text{Si}\!-\!\text{Ph}$$

V   **VI**

The preparation of rhodium complex catalysts interlayered in smectite clay for hydroformylation of alkenes in the solid state has been reported by Farzaneh and Pinnavaia (1982) (see Chapter 25). Rhodium complexes containing the phosphinophosphonium ligand $Ph_2P(CH_2)_2P^+ Ph_2 CH_2Ph$ should remain electrostatically bonded to the surface of the hectorite; acetone was the swelling solvent. However, some loss of rhodium to solution can be anticipated due to ligand dissociation.

Choudary *et al.* (1985) have reported the preparation of a montmorillonite with a diphenylphosphine unit anchored in the interlamellar region. Sodium montmorillonite is refluxed with chlorodiphenylphosphine in tetrahydrofuran to give the phosphinated montmorillonite **VII** [Eq. (4)]:

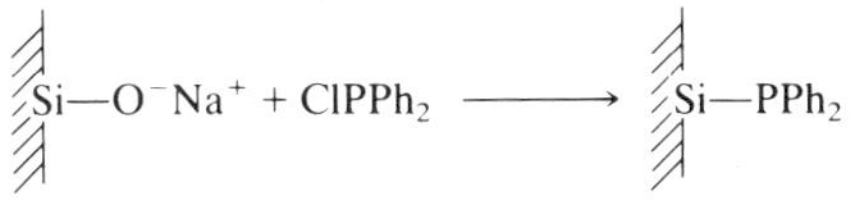

$$\text{Si}-\text{O}^-\text{Na}^+ + \text{ClPPh}_2 \longrightarrow \text{Si}-\text{PPh}_2 \qquad (4)$$

VII

When **VII** is shaken with a solution of $PdCl_2(PhCH_2CN)_2$, a complex **VIII** is formed in the interlamellar region:

$$\text{Si}-\text{O}-\overset{\overset{\text{Ph}}{|}}{\underset{\underset{\text{Ph}}{|}}{\text{P}}}-\text{Pd}^{II}\text{Cl}_2$$

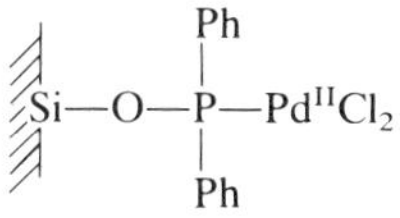

VIII

This complex has been characterized by x-ray diffraction, x-ray photoelectron, and infrared spectroscopies. Complex **VIII** is a catalyst that has been used to hydrogenate terminal alkenes and alkynes selectively.

III. QUATERNARY AMMONIUM CLAYS AS PHASE-TRANSFER CATALYSTS

Phase-transfer catalysis involves an aqueous phase and an organic phase in which the catalyst, usually a quaternary ammonium cation (A^+), will transport a nucleophilic Nu^- from the aqueous layer across the phase boundary to react with a substrate RY in the organic layer [Eqs. (5) and (6)]:

$$\text{aqueous phase: } A^+Y^- + Nu^- \longrightarrow A^+Nu^- + Y^- \qquad (5)$$

$$\text{organic phase: } A^+Nu^- + RY \longrightarrow R-Nu + A^+Y^- \qquad (6)$$

The product R—Nu remains in the organic layer, and the catalyst may then transport Y^- across the phase boundary into the aqueous layer.

The catalyst is introduced as a salt A^+X^-. The counterion X^- must have a low nucleophilicity and a high hydrophilicity.

Quaternary ammonium-exchanged montmorillonite is a suitable phase-transfer catalyst, the counterion X^- being the clay polyanionic sheets. The interlayer water of the clay serves as the aqueous phase.

Thus, benzyltributylammonium-exchanged montmorillonite and sodium hydroxide have been used as catalyst for the reaction of phenol with n-butyl bromide in dichloromethane [Eq. (7)]:

$$PhO^- + Bu—Br \longrightarrow Ph—OBu + Br^- \tag{7}$$

This catalyst has also been used for the formation of dichlorocarbene from chloroform and for the synthesis of an organotellurium compound **IX** from 1,4-dibromobutane and bis(p-ethoxyphenyl)ditelluride.

A good yield of **IX** is obtained in a shorter time than with normal phase transfer catalysis:

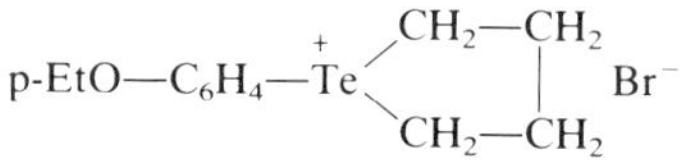

$$p\text{-EtO}—C_6H_4—\overset{+}{Te}\underset{CH_2—CH_2}{\overset{CH_2—CH_2}{\Big\langle}}\Big| \quad Br^-$$

IX

The synthesis of the rhodium complex **X** from chlorotris(triphenyl phosphine)rhodium(I) and sodium acetate and sodium hydroxide in a two-phase system has been catalyzed by benzyltributylammonium-exchanged montmorillonite [Eq. (8)] (Monsef-Mirzai and McWhinnie, 1981):

$$(Ph_3P)_3RhCl + Na^+MeCOO^- \longrightarrow (Ph_3P)_3RhOCOMe + Na^+Cl^- \tag{8}$$

X

Quaternary ammonium-exchanged montmorillonite has been used as α catalyst in the phase-transfer preparation of symmetrical formaldehyde acetals [Eq. (9)] (Cornélis and Laszlo, 1982; Cornélis *et al.*, 1983):

$$CH_2X_2 + 2ROH \xrightarrow[\text{catalyst}]{\text{NaOH}} R—O—CH_2—O—R \tag{9}$$

$(X=Cl, Br)$ 58–95%

IV. CLAY-SUPPORTED REAGENTS

Alkali metal fluorides have been used for promoting hydrogen-bond-assisted alkylations (Clark, 1980; Yakobson and Akhmetova, 1983). Yamawaki and Ando (1979) have examined the effectiveness of various inorganic solids as supports of potassium fluoride for O-alkylation of phenol. Montmorillonite and alumina are more efficient than celite as a support of potassium fluoride. Moderate yield is obtained when KF supported on montmorillonite is used after removal of water, but alumina is a better support than clay (Eq. 10):

$$\text{PhOH} + \text{MeI} \xrightarrow{\text{KF-montmorillonite}} \text{PhOMe} \qquad (10)$$
$$52\%$$

Kaolinite-supported potassium cyanide has also been used for the synthesis of 1-cyanooctane from 1-bromooctane, but alumina appears to be the most effective support for this reaction (Bram *et al.*, 1980).

$$\text{C}_8\text{H}_{17}\text{—Br} + \text{CN}^- \xrightarrow[85°C]{\text{kaolinite}} \text{C}_8\text{H}_{17}\text{—CN} + \text{Br}^-$$
$$19\%$$

In the research of new supports analogous to the aluminosilicates, xonotlite, a hydrothermally synthetized calcium silicate, has been used as support for potassium *t*-butoxide. This system catalyzes the Michael addition of acetylacetone, ethylacetoacetate, or diethylmalonate on methyl vinyl ketone or acrolein (Laszlo and Pennetreau, 1985).

Xonotlite or xonotlite-supported potassium *t*-butoxide catalyzes the Knoevenagel reaction between aromatic aldehydes and malononitrile or alkylcyano acetates (Chalais *et al.*, 1985), although alumina was a better catalyst for this reaction.

[The potential of silicates and aluminosilicates for anionic activation is enormous because of their mineralogical diversity, and because of the numerous possibilities for chemical modification. So far, it has remained virtually untapped.—ED.]

REFERENCES

Bram, G., Fillebeen-Khan, T., and Geraghty, N. (1980). *Synth. Commun.* **10**, 279–289.
Chalais, S., Laszlo, P., and Mathy, A. (1985). *Tetrahedron Lett.* **26**, 4453–4456.
Choudary, B. M., Kumar, K. R., Jamil, Z., and Thygarajan, G. (1985). *J.C.S. Chem. Commun.* 931–932.
Clark, J. H. (1980). *Chem. Rev.* **80**, 429–452.
Cornélis, A., and Laszlo, P. (1982). *Synthesis* 162–163.
Cornélis, A., Laszlo, P., and Pennetreau, P. (1983). *Clay Miner.* **18**, 437–445.
Deuel, H., and Huber, G. (1951). *Helv. Chim. Acta* **34**, 1697–1701.
Deuel, H., Huber, G., and Iberg, R. (1950). *Helv. Chim. Acta* **33**, 1229–1232.
Farzaneh, F., and Pinnavaia, T. J. (1982). *Inorg. Chem.* **22**, 2216–2220.
Gentili, G., and Deuel, H. (1957). *Helv. Chim. Acta* **40**, 106–113.
Greenland, D. J., and Quirk, J. P. (1962). *Clays Clay Miner.* **9**, 484.
Laszlo, P., and Pennetreau, P. (1985). *Tetrahedron Lett.* **26**, 2645–2648.
Monsef-Mirzai, P., and McWhinnie, W. R. (1981). *Inorg. Chim. Acta* **52**, 211–214.
Swartzen-Allen, S. L., and Matijevic, E. (1974). *Chem. Rev.* **74**, 385–400.
Theng, B. K. G. (1984). "The Chemistry of Clay-Organic Reactions." pp. 1–16. Adam Hilger, London.
Yakobson, G. G., and Akhmetova, N. E. (1983). *Synthesis* 169–184.
Yamawaki, J., and Ando, T. (1979). *Chem. Lett.* 755–758.

27 CATIONIC REACTIONS

J. M. Adams*

Edward Davies Chemical Laboratories
University College of Wales
Aberystwyth, SY23 INE
Dyfed, Wales, United Kingdom

I. INTRODUCTION

Most of the cationic reactions catalyzed by montmorillonites make use of the acidic nature of cation-exchanged or acid-treated clays. Both Lewis and Brønsted activities have been noted, the former deriving from aluminum or iron species at crystal edges (Theng, 1974). The Brønsted activity, however, results either from free acid (in some acid-activated clays) or from dissociation of interlayer water molecules coordinated to polarizing interlayer exchangeable cations. The acidity displayed in this latter case increases at low water contents and is enhanced when using interlayer cations of high charge and small radius such as Al^{3+}, Fe^{3+}, or Cr^{3+} (see, e.g., Fripiat and Cruz-Cumplido, 1974; Mortland and Raman, 1968):

$$[M(OH_2)_n]^{m+} \longrightarrow [M(OH_2)_{n-1}OH]^{(m-1)+} + H^+$$

The hydrogen ion concentration between the clay layers may be as high as 10 mol/dm^3 (Weiss, 1981).

For acid-activated clays, a significant amount of the acidity derives from Al^{3+} or Fe^{3+} ions exsolved from the octahedral sites by the acid. These cations relocate in the interlamellar space of the mineral and act as acid

* Present address: Central Laboratories, English China Clays International, St. Austell, Cornwall, England.

PREPARATIVE CHEMISTRY
USING SUPPORTED REAGENTS

centers. Very often about one-half of the exchange capacity of the clay is retained after acid treatment. One other effect of the acid is to increase the surface area of the clay. For example, acid activation of a bentonite clay (the major and active component of which is montmorillonite) from Moosburg, West Germany, can increase the N_2 surface area from ~40 m^2/g to ~250 m^2/g (W. Zschau, 1985 personal communication).

In the past, acid-treated clays were used for the cracking of oil, but in the 1960s they were made obsolete by the introduction of zeolite Y-based materials. Since that time a considerable effort has been made to develop new catalysts having pores rather larger than the 9 Å or so found in zeolite Y. It is ironic that this search led to the discovery of so-called pillared clay-based materials (PILCs), which are claimed to have high cracking activities and good thermal stability (Vaughan et al., 1979). These catalysts consist of smectite layers held apart by pillars of Al_2O_3 (see, e.g., Vaughan et al., 1979; Vaughan and Lussier, 1981) or other oxides, e.g., chromia (Pinnavaia et al., 1985). PILCs are produced by cation exchange of species such as $[Al_{13}O_4(OH)_{24}(H_2O)_{12}]^{7+}$ between the aluminosilicate layers, followed by calcination (Vaughan and Lussier, 1981). At the present state of the art, pores can be produced of comparable sizes to those found in zeolite Y (see, e.g., Pinnavaia et al., 1985).

Although clay catalysts are often used because of their convenience as solid acids, the catalytic organic reactions have often been shown to take place in the interlamellar space. We should expect, therefore, to find instances of unusual factors in these reactions since they occur in a region of high acidity and, possibly more importantly, in a reaction space restricted in one dimension.

II. REACTIVITY

In the reactions studied in Section III the clay catalysts act as Brønsted acids, as Lewis acids or as catalysts for Diels–Alder processes, possibly involving radical cations. Benesi and Winquist (1978) have reviewed methods of acidity measurement in solids and Benesi (1956, 1957) and Frenkel (1974) showed that the strength of the acid sites in montmorillonites (even in their H$^+$ or Al^{3+} form) are not as great as those found in exchanged zeolite Y.

A. Brønsted Acid Catalysis

These reactions involve a nucleophilic attack on a protonated species. If this cationic moiety can rearrange, then a variety of possible products may be formed, whereas if a stable ion [such as $(CH_3)_3C^+$] is formed only one product may result.

After surveying the Brønsted acid reactivity of cation-exchanged clays, Adams *et al.* (1983a) concluded that below 100°C the only reactions to occur are those in which a tertiary or allylic carbocation can be formed from one of the reactants directly on protonation (with the single exception of di-2,2'-alkyl ether formation; see Adams *et al.*, 1978, 1979). At higher temperatures, secondary (and possibly primary) carbocation intermediates are often involved.

Below 100°C 1,4-dioxane is often an efficacious solvent (Adams *et al.*, 1983a) while above that temperature alkane or chloroalkane (McCabe *et al.*, 1985) solvents give better results. In reactions (involving carbocations) above 100°C it is essential that the full polarizing power of the interlayer cations be felt by the remaining water molecules. Consequently, the use of a noncoordinating solvent is indicated.

B. Lewis Acid Catalysis

Potential Lewis acid centers exist in smectite clays. Aluminum(III) and iron(III) ions are normally associated with the octahedral sheets of montmorillonite, but part of the coordination sphere of these ions is likely to be made up of water molecules when they are exposed at crystal edges. Moreover, while it is widely known that certain smectites such as beidellite contain large amounts of tetrahedral aluminum (MacEwan, 1955), it is only as a result of magic angle spinning nuclear magnetic reasonance (MASNMR) studies (see, e.g., Atkins *et al.*, 1983) that it has been discovered that even in montmorillonites 5–10% of the aluminum can be in tetrahedral sites.

When reactions are carried out at low temperatures in aqueous environments or those in which solvents are used which could coordinate to aluminum or iron, then it appears unlikely that Lewis acid catalysis could be occurring. However, in many cases reactions are carried out above 100°C after dehydration of the catalyst by vacuum or heat treatment, or both. In these circumstances, the possibility of Lewis acid rather than Brønsted acid activity exists.

C. Diels–Alder Reactions

Montmorillonites have not been recognized as Diels–Alder catalysts except in the special case of oleic acid dimerization (den Otter, 1970a,b,c; Newton, 1984), and here it can be argued that the function of the clay is to isomerize the acid so as to produce a pair of conjugated double bonds and to transfer hydrogen from one molecule to another. These reactions are usually carried out industrially at ~230°C and at this temperature thermal Diels–Alder activity is expected.

Laszlo and Lucchetti (1984a,b,c) and Adams *et al.* (1985, 1987) have, however, shown that montmorillonites exchanged with transition metal ions or Fe^{3+}-exchanged acid-treated montmorillonite (Fe^{3+}-K10) are effective for a variety of Diels–Alder reactions, while clays exchanged with nontransition metals were not. It appears possible that the clays are not acting as Lewis acid Diels–Alder catalysts, nor as proton donors, but rather a one-electron transfer is involved, leading to a radical cation catalyzed process.

III. REACTIONS AND CONDITIONS

Reaction information is collected together (Tables I–IX) in this section. The reactions have been split into groups for convenience. However, the boundaries of the groupings given are not rigid in any way and often a particular type of reaction could have appeared in more than one section. In certain cases commercial acid-treated clays are mentioned. K-series catalysts and Tonsil are available from Süd-Chemie AG, Munich, West Germany while Filtrol clays are made by Filtrol Corporation, Los Angeles, California. Fulmont, Fulbent, and Fulcat clays are manufactured by Laporte Industries, Widnes, England.

Abbreviations used in the tables include:

IEC—ion-exchanged clays, usually Al^{3+}-, Cr^{3+}-, Fe^{3+}-, or Cu^{2+}-exchanged,

Al^{3+}—EC, etc.—aluminum exchanged clay, etc.,

Al-PILC, etc.—aluminum pillared clay, etc.,

ATC—acid treated clay,

Conv.—conversion, and alkenes(t), alkenes(s)—alkenes capable of forming tertiary or secondary carbocations.

TABLE I

Reactions of Alkenes

	Reactants	Catalyst	Conditions	Selected references
With clay interlayer water	Straight-chain 1-alkenes only (C_5–C_{16})	IEC	60–100°C; 3 h; 100% yield (based on water); hydrocarbon solvent $$H_2O + 2R\text{-}CH\text{=}CH_2 \xrightarrow{\text{Interlayer}} \begin{array}{c} R\text{-}CH\text{-}CH_3 \\ O \\ R\text{-}CH\text{-}CH_3 \end{array}$$	Adams et al. (1978, 1979, 1982a, 1984); Ballantine et al. (1983)
Hydration	Ethene	Al^{3+}-EC or ATC or Al-, B- or Be-PILCs	Gas phase; 250–300°C; 50 bar; 5% conv. (30-min residence time)	Adams et al. (1981a, 1982b); Atkins (1985a,b); Atkins et al. (1983); Gregory and Westlake (1982, 1983); Sommer et al. (1981)
	Propene	Al^{3+}-EC or ATC or Al-1, B-, or Be-PILCs	Gas phase; 100–240°C; 30 bar; 99% selectivity for isopropanol	
	Alkenes(t)	IEC	40–60°C: dioxane solvent; 90% yield in 20 min	
With alcohols	Alkenes(t) + primary alcohols	IEC; ATC; Al-, B-, or Be-PILCs	<100°C; 20–90% yield of the t-ether in 5 h; hydrocarbon or dioxane solvent.	Adams et al. (1982b, 1983b, 1986); Atkins (1985a,b); Ballantine et al. (1983, 1984a); Bylina et al. (1980); Gregory (1983); Gregory and Westlake (1982, 1983); Gulf Oil Canada (1977)
	Alkenes(t) + secondary or tertiary alcohols	IEC; ATC; Al-, B-, or Be-PILCs	Lower yields than above	
	Alkenes(s) + primary or secondary alcohols.	As above plus K10, KSF, K10/SF, and KSF/0	150°C, e.g., 1-hexene + ethanol gives 16% yield of a mixture of 2- and 3-ethoxy hexane	

(continues)

513

TABLE I (*continued*)

	Reactants	Catalyst	Conditions	Selected references
Ester formation	Acid + alkene	IEC, ATC	For alkenes(t) temperatures below 30°C needed. Otherwise up to 200°C, e.g., 90% yield of ethyl acetate (96% selectivity) using Al-EC; 40 h, 200°C	Atkins (1985b); Ballantine et al. (1981a,b,c, 1984c); Gregory (1983); Gregory and Westlake (1982, 1983); Gregory et al. (1982)
Amines formation	Alkene + NH_3	K306	2–4000 psi; 420°C; NH_3 : alkene ≈ 10 : 1. With GHSV ≈ 3000, conv. of propene is ~2%; selectivity ~93%	Peterson and Fales (1983)
Dimerization and oligomerization	Small-chain alkenes (for conjugated dienes see Table VIII)	IEC; ATC; Al- or B-PILCs; Filtrol	Conditions vary widely, depending on the alkene, e.g., isobutene dimerizes at ~60°C, hexenes at ~120°C.	Adams et al. (1981a); Atkins (1985a); Ballantine et al. (1984a); Gregory and Westlake (1984a,b); Idemitsu Petrochemical (1983); Reusser et al. (1982)
	Substituted styrenes	IEC; KSF/0	Refluxing hydrocarbon solvent at 50–150°C. For anethole 90% conv., xylene reflux, 1 h. For α-alkyl styrenes, 80% yield, 98% selectivity (linear dimer); Diphenylethylene gives indane dimer. 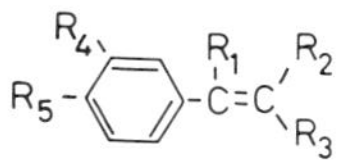	Adams et al. (1977, 1982c); Bittles et al. (1964a,b); Tricker et al. (1975); Wygant (1977)
Lactonization reactions	Ene reaction with diethyloxomalonate → γ-lactones	K10	2-Methyl-2-butene; 80°C, 72 h gives lactones **I** and **II**.	Roudier and Foucaud (1984)

(1) *trans*-stilbene (R₁, R₃, R₄, R₅ = H; R₂ = Ph)
(2) diphenylethylene (R₁ = Ph; R₂, R₃, R₄, R₅ = H)
(3) anethole (R₁, R₃, R₄ = H; R₂ = Me; R₅ = OMe)
(4) isohomognol (R₁, R₃ = H; R₂ = Me; R₄, R₅ = OMe)
(5) α-alkylstyrene (R₁ = various; R₂, R₃, R₄, R₅ = H)

Internal lactonization	IEC	Cyclooctane-5-carboxylic acid gives internal lactones 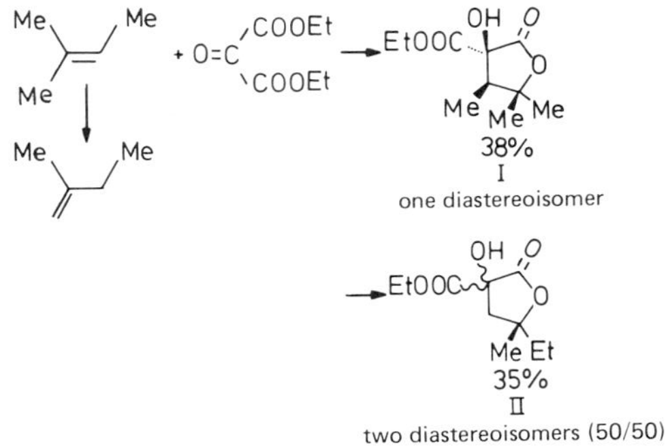	Adams *et al.* (1982c)
Reaction with H_2S	Filtrol 70	Propene + H_2S (1 : 8); 174°C; 4 h; 35% conv. (65% isopropyl thiol; 35% diisopropyl thiol)	Phillips Petroleum (1980)
Reaction with alkyl thiols	IEC	Good yields at 200°C, e.g., 1-hexene + butan-1-thiol gives a mixture of 1-hexyl, 2-hexyl, and 3-hexyl-1-butyl sulfides.	Ballantine *et al.* (1983); Galvin (1983)

Production of thiols + thio ethers

Reaction scheme:

Me₂C=CHMe → Me₂C–CH₂Me (+ O=C(COOEt)(COOEt))

→ EtOOC, OH, Me Me Me — 38% (I) — one diastereoisomer

→ EtOOC, OH, Me Et — 35% (II) — two diastereoisomers (50/50)

TABLE II

Reactions of Alkene Oxides and Tetrahydrofuran

	Reactants	Catalyst	Conditions	Selected References
Dimerization	Ethylene oxide	Al³⁺-EC	200°C; 65% 1,4-dioxane, 16% 2-methyl-1,3-dioxolane	Ballantine et al. (1983)
With alcohols to give ethers	Ethylene or propylene oxide + alcohols	IEC, ATC, Al-, B- or Be-PILCs	Liquid phase, 70°C or gas phase 120°C; 20 bar; 100% conv. (90% selectivity)	Atkins (1985a,b,c); Gregory (1983); Gregory and Westlake (1983); Herold (1975)
	Epichlorhydrin + ROH (R = Me, Et, isoPr, tBu, PhCH₂)	K10	2.5 h reflux in CCl₄; 30–60% yield	Vu Moc Thuy et al. (1982)
Ester formation	Alkene oxide + acid	B- and Be-PILCs		Atkins (1985a,b)
Formation of aldehydes	Alkene oxide + alcohols	Montmorillonite	Epoxystyrene reflux in benzene or CCl₄; 6 h; 8.8% conv.	Ruiz-Hitzky and Casal (1985)
Production of (1,4-diazabicyclo-2.2.2)-octane)	Ethylene oxide + piperazine	ATC	350°C, 1 bar, high yields	Oakes et al. (1973)
Production of dioxolanes	Alkene oxides + aldehydes	K10 and KSF	30–80°C, e.g., butyric aldehyde + styrene oxide (KSF), 4 h, 65% yield of 2-propyl-4-phenyl dioxolane	Steinbrink (1960); Vu Moc Thuy et al. (1980)
Formation of 3-mercapto-propan-1,2-diol	2,3-epoxypropen-1-ol + H₂S	Montmorillonite	30°C, 5-fold excess of H₂S, 90% yield	Kleeman et al. (1981)
Formation of THF/alkene oxide copolymers	THF + alkene oxide	ATC, Tonsil Optimum FF	~75°C, 10–240 min; 5–20% catalyst; ~5% of water or diol (chain terminator)	Del Pesco (1980); Müller (1984)
Formation and depolymerization of polytetra methylene glycol ether.	THF	K10, KSF, KP10, or Tonsil Optimum FF	30°C, 6 h, (carboxylic acid or anhydride terminator) polymers formed have molecular ~1600–2000, reverse reaction at 100°C.	Müller et al. (1979, 1981); Müller and Huchler (1982)

TABLE III

Elimination Reactions

Eliminated	Reactants	Catalyst	Conditions	Selected references
Water	Alcohols	IEC, ATC	200°C; Primary alcohols give di-1, 1-alkyl ethers (30–60%, 4 hr); other alcohols give alkenes (40–90%, 4 h); at 60°C t-butanol gives isobutene	Adams *et al.* (1981b); Atkins (1985a,b); Ballantine *et al.* (1981b,c,d, 1983, 1984a); Gregory (1983); Gregory and Westlake (1982); Gulf Oil Canada (1977)
	2,5-Dimethyl-2,5-hexanediol	ATC	250°C; 2,5-dimethyl-2,4-hexadiene	Petrocine and Harmetz (1985)
	α-hydroxypyrenes + diols	ATC	40–50°C, high yields of the hydroxy ether	Buysch (1973)
	N,N-Dialkylar: no alcohols	ATC	Bis-(N,N-dialkylamino)alkyl ethers formed at 275°C	Kaiser (1984)
NH_3	Primary amines	IEC	Poor yields; high-selectivity intermolecular elimination	Ballantine *et al.* (1981a,b,c, 1983); Gregory (1983).
	Cyclohexylamine	IEC	50% yield; 24 h; 220°C.	
	Tertiary amines	IEC	Poor yield; intramolecular elimination	
	Pyrrolidine	IEC	 Products **III** and **IV** 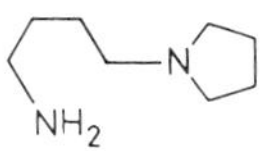	

III

IV

(*continues*)

517

TABLE III (*continued*)

Eliminated	Reactants	Catalyst	Conditions	Selected references
H_2S	Primary and secondary thiols	IEC	Intermolecular elimination; 200°C; 20–80% yields	Ballantine et al. (1981e, 1983)
	Tertiary thiols	IEC	Intramolecular elimination	
	Benzene thiol	IEC	Poly (phenylene methylene) formed	
	Thiophenol	IEC	Diphenyl sulphide + benzene	
Water	α,ω-Dicarboxylicacids	Al^{3+}-EC	Ring anhydrides formed; 5-ring (95% yield. toluene reflux 2.5 h); For larger rings 1,1,2,2-tetrachloroethane solvent useful	McCabe et al. (1985)
Acetic acid	α-Acetoxymethyl pyrroles + α-free pyrroles	Montmorillonite	Dichloromethane solvent; 20°C; 10 min; Without the α-free pyrrole, the symmetrical pyrromethane (**VI**) is formed (**V**) 95% yield (**VI**) 1–5 h	Jackson et al. (1985)

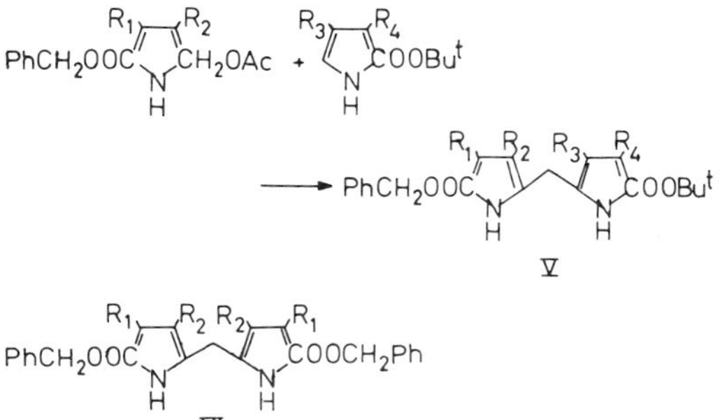

TABLE IV

Coupling Reactions Involving Formaldehyde and Acetone

Reactants	Catalyst	Conditions	Selected References
With aniline or carbanilates			
Formaldehyde (or trioxan, paraformaldehyde, etc.) + aniline	Superfiltrol	Products of type **VII**: 80–300°C; (**VII**) lower temperature forms 4,4'-isomer higher temperature forms 2,4-isomer (**VIII**)	Bentley (1978)
Trioxane + carbanilates	KSF/0	4 hr, 80°C gives product of type **VIII**, R = Et, Y, Z = H; 2,4- and 4,4'-isomers	Wada (1981)
Formaldehyde + phenol, p-cresol or substituted cresols	K10	Dihydroxydiphenyl (90% yield, 4 h, heptane reflux) or dihydroxyditolylmethane (90% yield, 4 h), respectively.	Herdieckerhoff and Sutter (1959)
Phenol + acetone	ATC + small amount of dihydric phenol cocatalyst	"Bis phenol A" produced; 50% yield; 10 h; 75°C	Sun (1977)

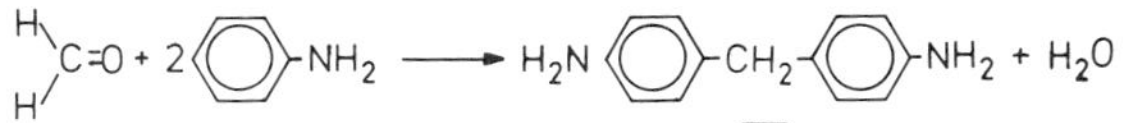

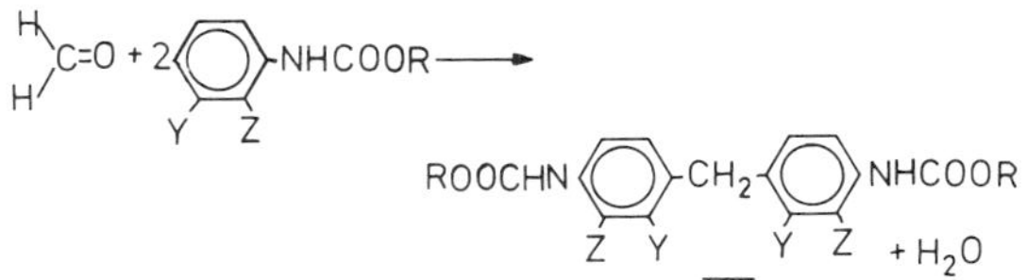

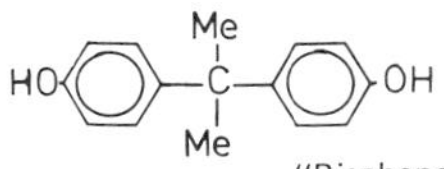

Reaction schemes appearing in the Conditions column:

$$\underset{H}{\overset{H}{C=O}} + 2\ \text{C}_6\text{H}_4\text{-NH}_2 \longrightarrow \text{H}_2\text{N-C}_6\text{H}_4\text{-CH}_2\text{-C}_6\text{H}_4\text{-NH}_2 + \text{H}_2\text{O}$$

VII

$$\underset{H}{\overset{H}{C=O}} + 2\ \text{C}_6\text{H}_3(Y)(Z)\text{-NHCOOR} \longrightarrow \text{ROOCHN-C}_6\text{H}_2(Z)(Y)\text{-CH}_2\text{-C}_6\text{H}_2(Y)(Z)\text{-NHCOOR} + \text{H}_2\text{O}$$

VIII

Structure of "Bisphenol A":

$$\text{HO-C}_6\text{H}_4\text{-C(CH}_3)_2\text{-C}_6\text{H}_4\text{-OH}$$

"Bisphenol A"

TABLE V

Reactions of Carbonyl Compounds

	Reactants	Catalyst	Conditions	Selected references
Ketal and acetal formation	Cyclohexanone + MeOH	Al-EC	MeOH gave 33% yield of dimethyl ketal (30 min)	Ballantine et al. (1983)
	Acreolin + MeOH	KSF, K10	Acetal, once formed, reacts with extra MeOH to form 1,1,3-trimethoxypropane	Süd-Chemie (1985)
	Carbonyl compound + trimethyl or triethyl ortho formate	Al³⁺-EC, KSF, K10	90–100% yield; 2–5 min	Taylor and Chiang (1977); Vu Moc Thuy and Maitte (1975)
Synthesis of enamines	cyclohexanone + variety of bases (e.g., morpholine, di-n-butylamines)	K10	Benzene or toluene reflux; 3–4-h reaction; 35–80% yield	Hunig et al. (1962)
Formation of α,β-unsaturated aldehydes	Aldehydes	IEC, ATC	Heat aldehyde in sealed cell (100°C), e.g., acetaldehyde gives 2-butene-1-al (24%, 8 h)	Ballantine et al. (1981a,b)

TABLE VI

Reactions of Alcohols

	Reactants	Catalyst	Conditions	Selected References
Formation of diol monoesters	Diol + ester	Montmorillonite	Toluene reflux (e.g., $HOCH_2CH_2OH + EtOAc \rightarrow AcOCH_2CH_2OH)$	Mitsui Petrochemical (1984)
Protection of hydroxy group	Alcohol or phenol + ethylal	KSF	$R\text{-}OH + H_2C(OEt)_2 \rightarrow R\text{-}O\text{-}CH_2\text{-}OC_2H_5 + C_2H_5OH$	Schaper (1981)
Formation of methyl-amines	Methanol + NH_3	K306	$420°C$, $CH_3OH : NH_3 = 1 : 2$; the ratio of tri-: di-: monomethylene = $1 : 1 : 5.5$	Ramioulle et al. (1980) Morikawa et al. (1983)
Hydrocarbons from MeOH	MeOH	Ti^{3+}-EC, Ti^{3+}-ATC	$350°C$; 6 h; 90% conv., 95% selectivity to give C_2–C_4 alkenes	Agency for Industrial Sciences and Technology (1982, 1983); Kikuchi et al. (1984); Toyo Soda Mfg. (1983)
Production of trial-kylortho formates	Alcohol + dioxolane	KSF	Compounds of type $HC(OR)(OR_1)(OR_2)$ formed at 60–70°C; typical yield 40%; (95% selectivity) in 7 hr	Eckhardt and Halbritter (1983)

TABLE VII

Alkylation, Acylation, and Nitration of Aromatics

	Reactants	Catalyst	Conditions	Selected References
Alkylation of phenols	Phenol + C_2–C_{30} alkyl halides	Montmorillonite	1–200 psi; 100–170°C; 1–24-h reaction. ~80% conv.	Watts and Schenck (1967)
	Phenol + alkenes (or terpenes)	Fulmont 300C + H_2SO_4, KSF Filtrol No. 1 + H_3PO_4	20 atm; 170°C; $2\frac{1}{2}$ h with butene; 52% of the product was *o*-alkylated, 15% *p*	Lucatello and Smith (1972); Gscheidmeier and Hacker (1977); Boylan and Sturwold (1972); Atkins (1985a); Gregory and Westlake (1983)
	Trans alkylation of substituted benzene	Fulcat 22A + H_3PO_4		Atkinson and Ball (1974)
Alkylation of aromatic amines	niline + alkenes	ATC	Propylene + aniline, 250 atm; 300°C; 7 h (60% mono *o*-product, 20% di *o, p*)	Stroh *et al.* (1959)
	Diphenylamine + styrene	KSF; montmorillonite	4,4'-distyryldiphenylamine produced at 220°C	Rogers (1978); D'Sidocky (1979)
Alkylation of hydrocarbons	Benzene + ethylene	Al-PILC	Mole ratio 2–15 : 1; 300–450°C; 1–25 atm; 60% conv. (residence time < 1 min)	Gregory and Westlake (1983)
	Benzene + propylene	Al-PILC	2–15 : 1; 100–300°C; 20–50 atm; 13% conv. (residence time < 2 h)	
		ATC	2–15 : 1; 70–100°C; 1–10 atm; 34% conv. 4 h	Joris (1960)
	Benzene + dodecene	Al^{3+}-EC	10 : 1; 80°C; 1 atm; 24 h, 88% conv.	NL Industries (1979)

Reaction	Substrate	Catalyst	Conditions	References
	Naphthalene + propylene	ATC *Also* B-PILCs and IEC and KSF	10:1; 110°C; 5 atm; 90% conv.	Elstner et al (1973) *See also* Atkins (1985a,c); Aries (1960); Ballantine et al. (1981a,b); Gregory (1983); Gregory and Westlake (1984a,b); Grosselck and Lotz (1978); Miller (1982); Sachanen and O'Kelley (1941); Westlake et al. (1985); Zuech (1974)
Transalkylation	Porphorins	Montmorillonite	Transmethylation and transethylation of vanadyl octaalkyl prophorins	Bonnett et al. (1972)
Acylation	Substituted aromatic of type **IX** + acylhalide or acidanhydride	K10	Dichloromethene reflux; 36 h; 66% yield of the acylated product 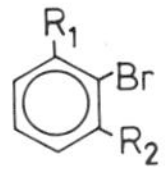	Lee and Jobin (1981)
Nitration and chlorination of aromatics		K306	e.g., toluene:chlorine = 5:1; 150°C; contact time 3 s., 80% yield of chlorotoluene ($o:m:p = 57:4:39$). toluene: $HNO_3 = 3.5:1$; 200°C; contact time = 11 s.; 65% yield	Bakke and Liaskar (1979); Bakke et al. (1982)

TABLE VIII.

Diels–Alder Reactions

Reactants	Catalyst	Conditions	Selected References
With activated and unactivated dienophiles	Cr^{3+} or Fe^{3+}-K10	e.g., cyclopentadiene + methylvinyl ketone, 20°C, dichloromethane solvent, 90% yield; 20 min	Adams *et al.* (1985, 1987); Lazlo and Lucchetti (1984a,b,c)
With conjugated hydrocarbon dienophiles	Cu^+ or Cr^{3+} or Fe^{3+}-EC, Fe^{3+}-KIO + 4-*t*-butylphenol	e.g., butadiene dimerization, 50% conv.; 60°C (90% selectivity)	Downing *et al.* (1978)
Dimerization of unsaturated fatty acids	SuperFiltrol + <5% H_2O	130–160°C; 1 atm; 20% yield	Rummelsburg (1968)
Natural fatty acids (C_{18} mono- or di-unsaturated)	Montmorillonite + LiOH or Li_2CO_3	200–280°C; >1 atm; 70% yield. Head-to-head reaction: cyclohexene ring with substituents $(CH_2)_7COOH$, $(CH_2)_7COOH$, $CH=CH(CH_2)_5CH_3$, $(CH_2)_5CH_3$. Head-to-tail reaction: cyclohexene ring with substituents $(CH_2)_7COOH$, $CH=CH(CH_2)_5CH_3$, $(CH_2)_7COOH$, $(CH_2)_5CH_3$.	Burba and Griebsch (1973); Pirner *et al.* (1972); Schering (1974)

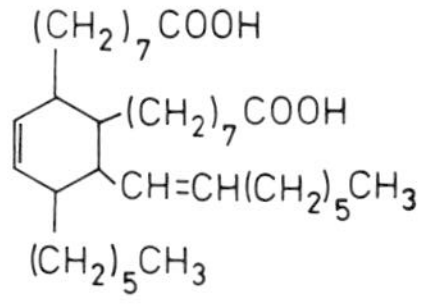

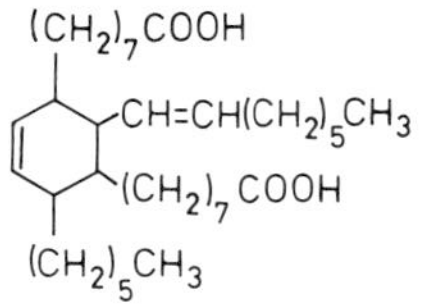

TABLE IX

Miscellaneous Reactions

	Reactants	Catalyst	Conditions	Selected References
Cracking reactions	Crude oil	ATC	Standard fluidized catalytic cracking conditions and regeneration	Gregory and Westlake (1982, 1983); Vaughan and Lussier (1985)
	Esters and fatty acids	IEC	e.g., cyclohexyl esters; xylene reflux; 60–98% yield (7 h)	Adams et al. (1982c)
		Montmorillonite	Much slower than with IEC	Almon and Johns (1975); Sieskind and Ourisson (1972)
Decomposition of per-oxides	Cumene peroxide hydro-peroxides	K10, KSF, etc., B- and Be-PILCs	Phenol and acetone produced	Süd-Chemie (1985)
Synthesis of α,β-unsaturated aldehydes	Acetals and vinyl ethers	K10	$R_1 = $ Ph, Me, Et, nC_6H_{13}; $R_2 = $ H, $R,R_2 = (CH_2)_5$. 50–95% yields in 12 h at 5–40°C 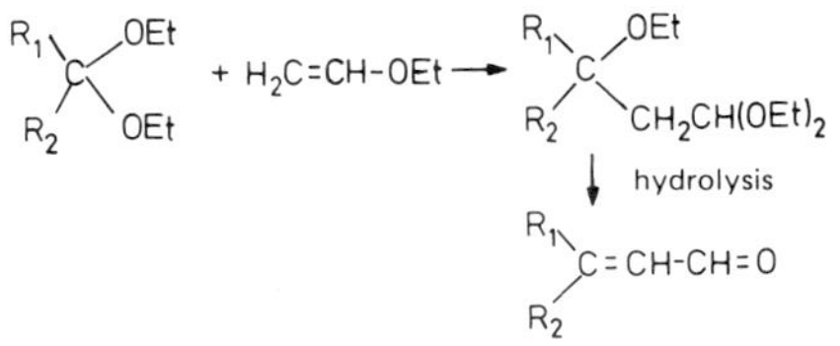	Fishman et al. (1981)
Production of oxydiphenol $(4\text{-}HOC_6H_4)_2O$	Hydroquinones and monohydric phenols	Al^{3+}-EC	Xylene reflux; typically 66% conv., 68% selectivity	Mitsui Petrochemical (1985)
Polyester formation	Lactones + chain termination (polyhydric alcohol)	ATC	$R = $ Me, ET, n-Pr or iso-Pr, H, Cl, or Br; $n = 5$–12. 45°C. 20 h for equilibrium 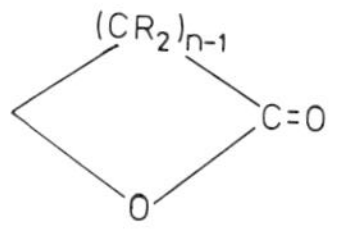	Jenkins and Izzard (1972)
Formation of esters and ethers of glycolic acid	Formaldehyde or precursor + ester of a carboxylic acid + CO	KSF/0	50 atm; mixture of products such as $MeOCH_2COOMe$, $MeOCH_2COOH$, $HOCH_2COOMe$, etc.	Wada et al. (1981)

REFERENCES

Adams, J. M., Graham, S. H., Reid, P. I., and Thomas, J. M. (1977). *J.C.S. Chem. Commun.* 67–68.

Adams, J. M., Ballantine, J. A., Graham, S. H., Laub, R. J., Purnell, J. H., Reid, P. I., Shaman, W. Y. M., and Thomas, J. M. (1978). *Angew. Chem., Int. Ed. Engl.* **17**, 282–283.

Adams, J. M., Ballantine, J. A., Graham, S. H., Laub, R. J., Purnell, J. H., Reid, P. I., Shaman, W. Y. M., and Thomas, J. M. (1979). *J. Catal.* **58**, 238–252.

Adams, J. M., Bylina, A., and Graham, S. H. (1981a). *Clay Miner.* **16**, 325–332.

Adams, J. M., Clement, D. E., and Graham, S. H. (1981b). *J. Chem. Res., Synop.* 254–255.

Adams, J. M., Bylina, A., and Graham, S. H. (1982a). *J. Catal.* **75**, 190–195.

Adams, J. M., Clement, D. E., and Graham, S. H. (1982b). *Clays Clay Miner.* **30**, 129–134.

Adams, J. M., Davies, S. E., Graham, S. H., and Thomas, J. M. (1982c). *J. Catal.* **78**, 197–208.

Adams, J. M., Clapp, T. V., and Clement, D. E. (1983a). *Clay Miner.* **18**, 411–421.

Adams, J. M., Clement, D. E., and Graham, S. H. (1983b). *Clays Clay Miner.* **31**, 129–136.

Adams, J. M., Clapp, T. V., Clement, D. E., and Reid, P. I. (1984). *J. Mol. Catal.* **27**, 179–194.

Adams, J. M., Clapp, T. V., Martin, K., and McCabe, R. W. (1985). *Int. Clay Conf., Denver, Colo. Abstr.*, p. 2.

Adams, J. M., Martin, K., and McCabe, R. W. (1987). *J. Inclusion Phenon.*

Adams, J. M., Martin, K., McCabe, R. W., and Murray, S. (1986). *Clays Clay Miner.* **34**, 597–603.

Agency For Industrial Sciences and Technology (1982). Jpn. Pat. 57,190,650.

Agency For Industrial Sciences and Technology (1983). *Jpn. Kokai Tokkyo Koho* 58, 67, 340.

Almon, W. R., and Johns, W. D. (1975). *Proc. Int. Clay Conf.* pp. 399–409.

Aries, R. S. (1960). U.S. Pat. 2,930,819.

Atkins, M. P. (1985a). Eur. Pat. Appl. 0,130,055.

Atkins, M. P. (1985b). Br. Pat. Appl. 2,151,603.

Atkins, M. P. (1985c). Eur. Pat. Appl. 0,150,897.

Atkins, M. P., Smith, D. J. H., and Westlake, D. J. (1983). *Clay Miner.* **18**, 423–429.

Atkinson, J. H., and Ball, D. (1974). Ger. Pat. 2,423,356.

Bakke, J. M., and Liasker, J. (1979). Ger. Pat. 2,826,433.

Bakke, J. M., Liasker, J., and Lorentzen, G. B. (1922). *J. Prakt. Chem.* **324**, 488–490.

Ballantir ', J. A., Davies, M., Purnell, J. H., Rayanakorn, M., Thomas, J. M., and Williams, K. J. (1981a). *J.C.S. Chem. Commun.* 8–9.

Ballantine, J. A., Purnell, J. H., and Thomas, J. M. (1981b). Eur. Pat. Appl. 0,031,252.

Ballantine, J. A., Purnell, J. H., Westlake, D. J., Gregory, R., and Thomas, J. M. (1981c). Eur. Pat. Appl. 0,031,687.

Ballantine, J. A., Davies, M., Purnell, J. H., Rayanakorn, M., Thomas, J. M., and Williams, K. J. (1981d). *J.C.S. Chem. Commun.* 427–428.

Ballantine, J. A., Galvin, R. P., O'Neil, R. M., Purnell, J. H., Rayanakorn, M., and Thomas, J. M. (1981e). *J.C.S. Chem. Commun.* 695–696.

Ballantine, J. A., Purnell, J. H., and Thomas, J. M. (1983). *Clay Miner.* **18**, 347–356.

Ballantine, J. A., Davies, M., Patel, I., Purnell, J. H., Rayanakorn, M., Williams, K. J., and Thomas, J. M. (1984a). *J. Mol. Catal.* **26**, 37–56.

Ballantine, J. A., Purnell, J. H., and Thomas, J. M. (1984b). *J. Mol. Catal.* **27**, 157–167.

Benesi, H. A. (1956). *J. Am. Chem. Soc.* **78**, 5490–5494.

Benesi, H. A. (1957). *J. Phys. Chem.* **61**, 970–973.

Benesi, H. A., and Winquist, B. H. C. (1978). *Adv. Catal.* **27**, 97–182.

Bentley, F. E. (1978). U.S. Pat. 4,071,558.

Bittles, J. A., Chaudhuri, A. K., and Benson, S. W. (1964a). *J. Polym. Sci.* **2**, 1221–1231.
Bittles, J. A., Chaudhuri, A. K., and Benson, S. W. (1964b). *J. Polym. Sci.* **2**, 1847–1862.
Bonnett, R., Brewer, P., Noro, K., and Noro, T. (1972). *J.C.S. Chem. Commun.* 562–563.
Boylan, J. B., and Sturwold, R. J. (1972). U.S. Pat. 3,679,617.
Burba, C., Griebsch, E. (1973). Ger. Pat. 2,226,397.
Buysch, H. J. (1973). Ger. Pat. 2,154,997.
Bylina, A., Adams, J. M., Graham, S. H., and Thomas, J. M. (1980). *J.C.S. Chem. Commun.* 1003–1004.
Del Pesco, T. W. (1980). U.S. Pat. 4,235,751.
den Otter, M. J. A. M. (1970a). *Fette, Siefen, Anstrichm.* **72**, 667–673.
den Otter, M. J. A. M. (1970b). *Fette, Siefen, Anstrichm.* **72**, 875–883.
den Otter, M. J. A. M. (1970c). *Fette, Siefen, Anstrichm.* **72**, 1056–1066.
Downing, R. S., van Austel, J., and Joustra, A. H. (1978). U.S. Pat. 4,125,483.
D'Sidocky, R. M. (1979). Ger. Pat. 2,842,852.
Eckhardt, H., and Halbritter, K. (1983). Ger Pat. 3,301,711.
Elstner, M., Fremery, M., and Hover, H. (1973). Ger. Pat. 2,208,363.
Fishman, D., Klug, J. T., and Shani, A. (1981). *Synthesis* 137–138.
Frenkel, M. (1974). *Clays Clay Miner.* **22**, 435–441.
Fripiat, J. J., and Cruz-Cumplido, M. I. (1974). *Annu. Rev. Earth Planet. Sci.* **2**, 239–256.
Galvin, R. P. (1983). Ph.D. Thesis, University College, Swansea, Wales.
Gregory, R. (1983). Eur. Pat. Appl. 0,073,141.
Gregory, R., and Westlake, D. J. (1982). Eur. Pat. Appl. 0,045,618.
Gregory, R., and Westlake, D. J. (1983). Eur. Pat. Appl. 0,083,970.
Gregory, R., and Westlake, D. J. (1984a). Eur. Pat. Appl. 0,110,628.
Greogory, R., and Westlake, D. J. (1984b). Eur. Pat. Appl. 0,116,469.
Gregory, R., Smith, D. J. H., and Westlake, D. J. (1983). *Clay Miner.* **18**, 431–435.
Grosselck, J., and Lotz, W. W. (1978). *Z. Naturforsch., B* **33B**, 346–353.
Gscheidmeier, M., and Hacker, K. (1977). Ger. Pat. 2,552,175.
Gulf Oil Canada, Ltd. (1977). Neth. Pat. Appl. 77 06,072.
Herdieckerhoff, E., and Sutton, W. (1959). Ger. Pat. 1,051,864.
Herold, W. (1975). Ger. Pat. 2,406,293.
Hunig, S., Hubner, K., and Benzing, E. (1962). *Chem. Ber.* **95**, 926–936.
Idemitsu Petrochemical Co. (1983). Jpn. Pat. 58,147,496.
Jackson, A. H., Pandey, R. K., Nagaraja Rao, K. R., and Roberts, E. (1985). *Tetrahedron Lett.* **26**, 793–796.
Kenkins, V. F., and Izzard, D. C. (1972). Br. Pat. 1,272,572.
Joris, G. C. (1960). U.S. Pat. 2,945,072.
Kaiser, S. W. (1984). Eur. Pat. Appl. 0,115,071.
Kikuchi, E., Nakano, M., Fukada, I., and Morita, Y. (1984). *Sekiyu Gakkaishi* **27**, 153–158.
Kleeman, S., Nygren, R., and Wagner, R. (1981). Ger. Pat. 3,014,165.
Laszlo, P., and Lucchetti, J. (1984a). *Tetrahedron Lett.* **25**, 2147–2150.
Laszlo, P., and Lucchetti, J. (1984b). *Tetrahedron Lett.* **25**, 4387–4388.
Laszlo, P., and Lucchetti, J. (1984c). *Tetrahedron Lett.* **25**, 1567–1570.
Lee, T. B. K., and Jobin, G. M. (1981). U.S. Pat. 4,304,941.
Lucatello, L. G., and Smith, G. E. (1972). Br. Pat. 1,265,152.
McCabe, R. W., Adams, J. M., and Martin, K. (1985). *J. Chem. Res. S* 356–357.
MacEwan, D. M. C. (1955). *In* "The X-Ray Identification and Crystal Structures of Clay Minerals" (G. Brown, ed.), pp. 143–207. Mineral. Soc. London.
Miller, S. J. (1982). U.S. Pat. 4,361,477.
Mitsui Petrochemical Industries (1984). Jpn. Pat. 59,204,153.
Mitsui Petrochemical Industries (1985). Jpn. Pat. 60,25,946.

Morikawa, Y., Wong, F. L., Moro-Oka, Y., and Ikawa, T. (1983). *Chem. Lett.* 965–968.
Mortland, M. M., and Raman, K. V. (1968). *Clays Clay Miner.* **16**, 393–398.
Müller, H. (1984). Ger. Pat. 3,236,432.
Müller, H., and Huchler, O. H. (1982). Eur. Pat. Appl. 0,052,213.
Müller, H., Huchler, O., and Hoffmann, H. (1979). Ger. Pat. 2,801,578.
Müller, H., Huchler, O., and Hoffmann, H. (1981). U.S. Pat. 4,243,799.
Newton, L. S. (1984). *Spec. Chem.* May, 17–24.
NL Industries (1979). Br. Pat. 1,537,766.
Oakes, M. D., Upson, L. L., and Ziv, M. H. (1973). U.S. Pat. 3,772,293.
Peterson, J. O. H., and Fales, H. S. (1983). U.S. Pat. 4,375,002.
Petrocine, D. V., and Harmetz, R. (1985). U.S. Pat. 4,507,518.
Phillips Petroleum Co. (1981). Belg. Pat. 886,261.
Pinnavaia, T. J., Tzou, M. S., and Landau, S. D. (1985). *J. Am. Chem. Soc.* **107**, 4783–4785.
Pirner, K., Griebsch, E., and Niemann, E. (1972). Ger. Pat. 2,118,702.
Ramioulle, J., Schmitz, G., and Lambert, P. (1980). Ger. Pat. 3,010,791.
Reusser, R. E., Claytor, E. A., and Jones, B. E. (1982). U.S. Pat. 4,351,980.
Rogers, E. A. (1978). Ger. Pat. 2,748,749.
Roudier, J. F., and Foucaud, A. (1984). *Tetrahedron Lett.* **25**, 4375–4378.
Ruiz-Hitzky, E., and Casal, B. (1985). *J. Catal.* **92**, 291–295.
Rummelsburg, A. L. (1969). U.S. Pat. 3,437,650.
Schanen, A. N., and O'Kelly, A. A. (1941). U.S. Pat. 2,242,960.
Schaper, U. A. (1981). *Synthesis* 794–796.
Schering, A. G. (1974). Fr. Pat. 2,199,754.
Sieskind, O., and Ourisson, U. (1972). *C.R. Acad. Sci. Paris* **274**, 2186–2189.
Sommer, A., Heilmann, W., and Brucker, R. (1981). Eur. Pat. Appl. 0,023,071.
Steinbrink, H. (1960). Ger. Pat. 1,086,241.
Stroh, R., Ebersberger, J., Haberland, H., and Hahn, W. (1959). Ger. Pat. 1,051.271.
Süd-Chemie A. G. (1985). "K-Catalysts Handbook," Süd-Chemie A. G., Munich.
Sun, K. K. (1977). U.S. Pat. 4,052,466.
Taylor, E. C., and Chiang, C.-S. (1977). *Synthesis* p. 467.
Theng, B. K. G. (1974). "The Chemistry of Clay-Organic Reactions." Adam Hilger, London.
Toyo Soda Mfg. Co., Ltd. (1983). Jpn. Pat. 58,83,635.
Tricker, M. J., Tennakoon, D. T. B., Thomas, J. M., and Graham, S. H. (1975). *Nature (London)* **253**, 110–111.
Vaughan, D. E. W., and Lussier, R. J. (1981). *Proc. Int. Conf. Zeolites, 5th* pp. 94–101.
Vaughan, D. E. W., and Lussier, R. J. (1985). *Mater. Res. Meet., Boston, Mass. Abstr.,* p. 494.
Vaughan, D. E. W., Lussier, R. J., and Magee, J. B. (1979). U.S. Pat. 4,176,090.
Vu Moc Thuy, and Maitte, P. (1975). *Bull. Soc. Chim. Fr.* **11**, 2558–2560.
Vu Moc Thuy, Petit, H., and Maitte, P. (1980). *Bull. Soc. Chim. Belg.* **89**, 759–761.
Vu Moc Thuy, Petit, H., and Maitte, P. (1982). *Bull. Soc. Chim. Belg.* **91**, 261–262.
Wada, K. (1981). Eur. Pat. Appl. 27,330.
Wada, K., Baka, A., and Wada, N (1981). Ger. Pat. 3,107,518.
Watts, J. M., and Schenck, L. M. (1967). U.S. Pat. 3,360,573.
Weiss, A. (1981). *Angew. Chem., Int. Ed. Engl.* **20**, 850–860.
Westlake, D. J., Atkins, M. P., and Gregory, R. (1985). *Acta Chem. Phys.* **31**, 301–307.
Wygant, J. C. (1977). Ger. Pat. 2,724,491.
Zuech, E. A. (1974). U.S. Pat. 3,849,507.

INDEX

529